NON-INERTIAL FRAMES AND DIRAC OBSERVABLES IN RELATIVITY

Interpreting general relativity relies on a proper description of non-inertial frames and Dirac observables. This book describes global non-inertial frames in special and general relativity. The first part covers special relativity and Minkowski space-time, before covering general relativity, globally hyperbolic Einstein space-time, and the application of the 3+1 splitting method to general relativity. It uses a Hamiltonian description and the Dirac–Bergmann theory of constraints to show the transition between one non-inertial frame and another is a gauge transformation, extra variables describing the frame are gauge variables, and the measureable matter quantities are gauge-invariant Dirac observables. Point particles, fluids, and fields are also discussed, including how to treat the problems of relative times in the description of relativistic bound states, and the problem of relativistic center of mass. Providing a detailed description of mathematical methods, it is perfect for theoretical physicists, researchers, and students working in special and general relativity.

LUCA LUSANNA is a retired research director of the Firenze section of the National Institute for Nuclear Physics (INFN). He is a fellow of the General Relativity and Gravitation Society (SIGRAV), the Italian Physical Society, of which he has been president for five years. Since 2009 he has been the director of the Theoretical Topical Team of the ACES mission of ESA, tasked with putting an atomic clock on the International Space Station.

CAMBRIDGE MONOGRAPHS ON MATHEMATICAL PHYSICS

S. J. Aarseth *Gravitational N-Body Simulations: Tools and Algorithms*†
D. Ahluwalia *Mass Dimension One Fermions*
J. Ambjørn, B. Durhuus and T. Jonsson *Quantum Geometry: A Statistical Field Theory Approach*†
A. M. Anile *Relativistic Fluids and Magneto-fluids: With Applications in Astrophysics and Plasma Physics*
J. A. de Azcárraga and J. M. Izquierdo *Lie Groups, Lie Algebras, Cohomology and Some Applications in Physics*†
O. Babelon, D. Bernard and M. Talon *Introduction to Classical Integrable Systems*†
F. Bastianelli and P. van Nieuwenhuizen *Path Integrals and Anomalies in Curved Space*†
D. Baumann and L. McAllister *Inflation and String Theory*
V. Belinski and M. Henneaux *The Cosmological Singularity*†
V. Belinski and E. Verdaguer *Gravitational Solitons*†
J. Bernstein *Kinetic Theory in the Expanding Universe*†
G. F. Bertsch and R. A. Broglia *Oscillations in Finite Quantum Systems*†
N. D. Birrell and P. C. W. Davies *Quantum Fields in Curved Space*†
K. Bolejko, A. Krasiński, C. Hellaby and M-N. Célérier *Structures in the Universe by Exact Methods: Formation, Evolution, Interactions*
D. M. Brink *Semi-Classical Methods for Nucleus-Nucleus Scattering*†
M. Burgess *Classical Covariant Fields*†
E. A. Calzetta and B.-L. B. Hu *Nonequilibrium Quantum Field Theory*
S. Carlip *Quantum Gravity in 2+1 Dimensions*†
P. Cartier and C. DeWitt-Morette *Functional Integration: Action and Symmetries*†
J. C. Collins *Renormalization: An Introduction to Renormalization, the Renormalization Group and the Operator-Product Expansion*†
P. D. B. Collins *An Introduction to Regge Theory and High Energy Physics*†
M. Creutz *Quarks, Gluons and Lattices*†
P. D. D'Eath *Supersymmetric Quantum Cosmology*†
J. Dereziński and C. Gérard *Mathematics of Quantization and Quantum Fields*
F. de Felice and D. Bini *Classical Measurements in Curved Space-Times*
F. de Felice and C. J. S Clarke *Relativity on Curved Manifolds*†
B. DeWitt *Supermanifolds, 2nd edition*†
P. G. O. Freund *Introduction to Supersymmetry*†
F. G. Friedlander *The Wave Equation on a Curved Space-Time*†
J. L. Friedman and N. Stergioulas *Rotating Relativistic Stars*
Y. Frishman and J. Sonnenschein *Non-Perturbative Field Theory: From Two Dimensional Conformal Field Theory to QCD in Four Dimensions*
J. A. Fuchs *Affine Lie Algebras and Quantum Groups: An Introduction, with Applications in Conformal Field Theory*†
J. Fuchs and C. Schweigert *Symmetries, Lie Algebras and Representations: A Graduate Course for Physicists*†
Y. Fujii and K. Maeda *The Scalar-Tensor Theory of Gravitation*†
J. A. H. Futterman, F. A. Handler, R. A. Matzner *Scattering from Black Holes*†
A. S. Galperin, E. A. Ivanov, V. I. Ogievetsky and E. S. Sokatchev *Harmonic Superspace*†
R. Gambini and J. Pullin *Loops, Knots, Gauge Theories and Quantum Gravity*†
T. Gannon *Moonshine beyond the Monster: The Bridge Connecting Algebra, Modular Forms and Physics*†
A. García-Díaz *Exact Solutions in Three-Dimensional Gravity*
M. Göckeler and T. Schücker *Differential Geometry, Gauge Theories, and Gravity*†
C. Gómez, M. Ruiz-Altaba and G. Sierra *Quantum Groups in Two-Dimensional Physics*†
M. B. Green, J. H. Schwarz and E. Witten *Superstring Theory Volume 1: Introduction*
M. B. Green, J. H. Schwarz and E. Witten *Superstring Theory Volume 2: Loop Amplitudes, Anomalies and Phenomenology*
V. N. Gribov *The Theory of Complex Angular Momenta: Gribov Lectures on Theoretical Physics*†
J. B. Griffiths and J. Podolský *Exact Space-Times in Einstein's General Relativity*†
T. Harko and F. Lobo *Extensions of f(R) Gravity: Curvature-Matter Couplings and Hybrid Metric-Palatini Gravity*
S. W. Hawking and G. F. R. Ellis *The Large Scale Structure of Space-Time*†
F. Iachello and A. Arima *The Interacting Boson Model*†
F. Iachello and P. van Isacker *The Interacting Boson-Fermion Model*†
C. Itzykson and J. M. Drouffe *Statistical Field Theory Volume 1: From Brownian Motion to Renormalization and Lattice Gauge Theory*†
C. Itzykson and J. M. Drouffe *Statistical Field Theory Volume 2: Strong Coupling, Monte Carlo Methods, Conformal Field Theory and Random Systems*†
G. Jaroszkiewicz *Principles of Discrete Time Mechanics*
G. Jaroszkiewicz *Quantized Detector Networks*

C. V. Johnson *D-Branes*†
P. S. Joshi *Gravitational Collapse and Spacetime Singularities*†
J. I. Kapusta and C. Gale *Finite-Temperature Field Theory: Principles and Applications, 2nd edition*†
V. E. Korepin, N. M. Bogoliubov and A. G. Izergin *Quantum Inverse Scattering Method and Correlation Functions*†
J. Kroon *Conformal Methods in General Relativity*
M. Le Bellac *Thermal Field Theory*†
Y. Makeenko *Methods of Contemporary Gauge Theory*†
S. Mallik and S. Sarkar *Hadrons at Finite Temperature*
A. Malyarenko and M. Ostoja-Starzewski *Tensor-Valued Random Fields for Continuum Physics*
N. Manton and P. Sutcliffe *Topological Solitons*†
N. H. March *Liquid Metals: Concepts and Theory*†
I. Montvay and G. Münster *Quantum Fields on a Lattice*†
P. Nath *Supersymmetry, Supergravity, and Unification*
L. O'Raifeartaigh *Group Structure of Gauge Theories*†
T. Ortín *Gravity and Strings, 2nd edition*
A. M. Ozorio de Almeida *Hamiltonian Systems: Chaos and Quantization*†
M. Paranjape *The Theory and Applications of Instanton Calculations*
L. Parker and D. Toms *Quantum Field Theory in Curved Spacetime: Quantized Fields and Gravity*
R. Penrose and W. Rindler *Spinors and Space-Time Volume 1: Two-Spinor Calculus and Relativistic Fields*†
R. Penrose and W. Rindler *Spinors and Space-Time Volume 2: Spinor and Twistor Methods in Space-Time Geometry*†
S. Pokorski *Gauge Field Theories, 2nd edition*†
J. Polchinski *String Theory Volume 1: An Introduction to the Bosonic String*†
J. Polchinski *String Theory Volume 2: Superstring Theory and Beyond*†
J. C. Polkinghorne *Models of High Energy Processes*†
V. N. Popov *Functional Integrals and Collective Excitations*†
L. V. Prokhorov and S. V. Shabanov *Hamiltonian Mechanics of Gauge Systems*
S. Raychaudhuri and K. Sridhar *Particle Physics of Brane Worlds and Extra Dimensions*
A. Recknagel and V. Schiomerus *Boundary Conformal Field Theory and the Worldsheet Approach to D-Branes*
M. Reuter and F. Saueressig *Quantum Gravity and the Functional Renormalization Group*
R. J. Rivers *Path Integral Methods in Quantum Field Theory*†
R. G. Roberts *The Structure of the Proton: Deep Inelastic Scattering*†
P. Romatschke and U. Romatschke *Relativistic Fluid Dynamics In and Out of Equilibrium: And Applications to Relativistic Nuclear Collisions*
C. Rovelli *Quantum Gravity*†
W. C. Saslaw *Gravitational Physics of Stellar and Galactic Systems*†
R. N. Sen *Causality, Measurement Theory and the Differentiable Structure of Space-Time*
M. Shifman and A. Yung *Supersymmetric Solitons*
Y. M. Shnir *Topological and Non-Topological Solitons in Scalar Field Theories*
H. Stephani, D. Kramer, M. MacCallum, C. Hoenselaers and E. Herlt *Exact Solutions of Einstein's Field Equations, 2nd edition*†
J. Stewart *Advanced General Relativity*†
J. C. Taylor *Gauge Theories of Weak Interactions*†
T. Thiemann *Modern Canonical Quantum General Relativity*†
D. J. Toms *The Schwinger Action Principle and Effective Action*†
A. Vilenkin and E. P. S. Shellard *Cosmic Strings and Other Topological Defects*†
R. S. Ward and R. O. Wells, Jr *Twistor Geometry and Field Theory*†
E. J. Weinberg *Classical Solutions in Quantum Field Theory: Solitons and Instantons in High Energy Physics*
J. R. Wilson and G. J. Mathews *Relativistic Numerical Hydrodynamics*†

† Available in paperback

Non-Inertial Frames and Dirac Observables in Relativity

LUCA LUSANNA
National Institute for Nuclear Physics (INFN)

CAMBRIDGE
UNIVERSITY PRESS

CAMBRIDGE
UNIVERSITY PRESS

University Printing House, Cambridge CB2 8BS, United Kingdom

One Liberty Plaza, 20th Floor, New York, NY 10006, USA

477 Williamstown Road, Port Melbourne, VIC 3207, Australia

314-321, 3rd Floor, Plot 3, Splendor Forum, Jasola District Centre, New Delhi - 110025, India

79 Anson Road, #06-04/06, Singapore 079906

Cambridge University Press is part of the University of Cambridge.

It furthers the University's mission by disseminating knowledge in the pursuit of education, learning and research at the highest international levels of excellence.

www.cambridge.org
Information on this title: www.cambridge.org/9781108480826
DOI: 10.1017/9781108691239

First published 2019

A catalogue record for this publication is available from the British Library

Library of Congress Cataloging in Publication data
Names: Lusanna, L., author.
Title: Non-inertial frames and Dirac observables in relativity / Luca Lusanna (National Institute for Nuclear Physics(INFN), Firenze).
Description: Cambridge ; New York, NY : Cambridge University Press, [2019] | Series: Cambridge monographs on mathematical physics | Includes bibliographical references and subject index.
Identifiers: LCCN 2019005996 | ISBN 9781108480826 (hardback : alk. paper)
Subjects: LCSH: Dirac equation. | Wave equation. | Quantum field theory. | General relativity (Physics)
Classification: LCC QC174.45 .L87 2019 | DDC 530.12–dc23
LC record available at https://lccn.loc.gov/2019005996

ISBN 978-1-108-48082-6 Hardback

To Enrica, Haja and Leo: thank you for being always with me!

Contents

Preface *page* xiii

Part I Special Relativity: Minkowski Space-Time

1 Galilei and Minkowski Space-Times **3**
1.1 The Galilei Space-Time of Non-Relativistic Physics and Its Inertial and Non-Inertial Frames 3
1.2 The Minkowski Space-Time: Inertial Frames, Cartesian Coordinates, Matter, Energy–Momentum Tensor, and Poincaré Generators 5
1.3 The 1+3 Approach to Local Non-Inertial Frames and Its Limitations 7

2 Global Non-Inertial Frames in Special Relativity **12**
2.1 The 3+1 Approach to Global Non-Inertial Frames and Radar 4-Coordinates 12
2.2 Parametrized Minkowski Theory for Matter Admitting a Lagrangian Description 19

3 Relativistic Dynamics and the Relativistic Center of Mass **27**
3.1 The Wigner-Covariant Rest-Frame Instant Form of Dynamics for Isolated Systems 31
3.2 The Relativistic Center-of-Mass Problem 36
3.3 The Elimination of Relative Times in Relativistic Systems of Particles and in Relativistic Bound States 39
3.4 Wigner-Covariant Quantum Mechanics of Point Particles 44
3.5 The Non-Inertial Rest-Frames 47

4 Matter in the Rest-Frame Instant Form of Dynamics **54**
4.1 The Klein–Gordon Field 54
4.2 The Electromagnetic Field and Its Dirac Observables 65
4.3 Relativistic Atomic Physics 71
4.4 The Dirac Field 79
4.5 Yang–Mills Fields 96
4.6 Relativistic Fluids, Relativistic Micro-Canonical Ensemble, and Steps toward Relativistic Statistical Mechanics 105

Part II General Relativity: Globally Hyperbolic Einstein Space-Times

5 Hamiltonian Gravity in Einstein Space-Times 123
5.1 Global 3+1 Splittings of Globally Hyperbolic Space-Times without Super-Translations and Asymptotically Minkowskian at Spatial Infinity Admitting a Hamiltonian Formulationof Gravity 123
5.2 The ADM Hamiltonian Formulation of Einstein Gravity and the Asymptotic ADM Poincaré Generators in the Non-Inertial Rest-Frames 127

6 ADM Tetrad Gravity and Its Constraints 138
6.1 ADM Tetrad Gravity, Its Hamiltonian Formulation, and Its First-Class Constraints 138
6.2 The Shanmugadhasan Canonical Transformation to the York Canonical Basis for the Search of the Dirac Observables of the Gravitational Field 145
6.3 The Non-Harmonic 3-Orthogonal Schwinger Time Gauges and the Metrological Interpretation of the Inertial Gauge Variables 149
6.4 Point Particles and the Electromagnetic Field as Matter 155

7 Post-Minkowskian and Post-Newtonian Approximations 167
7.1 The Post-Minkowskian Approximation in the 3-Orthogonal Gauges 167
7.2 The Post-Newtonian Expansion of the Post-Minkowskian Linearization 178
7.3 Dark Matter as a Relativistic Inertial Effect and Relativistic Celestial Metrology 181

Part III Dirac–Bergmann Theory of Constraints

8 Singular Lagrangians and Constraint Theory 189
8.1 Regular Lagrangians and the First Noether Theorem 189
8.2 Singular Lagrangians and Dirac–Bergmann Theory of Hamiltonian First- and Second-Class Constraints 194
8.3 The Gauge Transformations in Field Theory and General Relativity 204

9 Dirac Observables Invariant under the Hamiltonian Gauge Transformations Generated by First-Class Constraints 223
9.1 Shanmugadhasan Canonical Bases Adapted to the Constraints and the Dirac Observables 228

9.2 The Null Eigenvalues of the Hessian Matrix of Singular Lagrangians and Degenerate Cases 232
9.3 The Second Noether Theorem 235
9.4 Constraints in Field Theory 236

10 Concluding Remarks and Open Problems 243

Appendix A Canonical Realizations of Lie Algebras, Poincaré Group, Poincaré Orbits, and Wigner Boosts 248

Appendix B Grassmann Variables and Pseudo-Classical Lagrangians 265

Appendix C Relativistic Perfect Fluids and Covariant Thermodynamics 276

References 293
Index 320

Preface

While in Newtonian physics (NP) both global inertial and rigid non-inertial frames of the Galilei space-time are under control due to the absolute notions of time and Euclidean 3-space, in the Minkowski space-time of special relativity (SR) only global inertial frames centered on inertial observers are defined by using Einstein convention for the synchronization of the clocks present in each point to the one of the inertial observer. Only in this way can instantaneous Euclidean 3-spaces be selected. The Lorentz signature of Minkowski space-time implies that the only intrinsic notion in SR is the conformal structure, namely the light-cone, identifying the allowed paths of the light rays sent by an observer.

However, all physical observers are accelerated and till now there is only the so-called 1+3 point of view for identifying a local instantaneous 3-space in a region around the observer whose radius depends upon the observer's acceleration. With this description it is not possible to define a Cauchy problem in non-inertial frames for classical fields like the Maxwell one.

Therefore, an open theoretical challenge is to find a consistent description of classical particles, fluids, and fields in the non-inertial reference frames of the Minkowski space-time of SR. This accomplishment would allow extension of the existent formulations of these systems in the global inertial frames centered on inertial observers, which are connected in a manifestly covariant way by the transformations of the kinematical Poincaré group (the relativity principle), to a formulation covariant under some family of space-time diffeomorphisms. As a consequence, this description would allow a smooth transition to Einstein space-times of general relativity (GR) with the relativity principle replaced by the equivalence principle forbidding the existence of global inertial frames. In GR only a free-falling observer will recover SR as a local approximation in a neighborhood where tidal effects are negligible.

In SR these steps are preliminary to the transition to the quantum theory of matter, because they allow finding the solution of problems like the elimination of relative times in relativistic bound states, the clarification of the notion of relativistic center of mass (a non-local non-measurable quantity), and a consistent treatment of relativistic atomic physics, relativistic fluids, and relativistic statistical mechanics. The final open challenge will be to find a consistent formulation of quantum gravity (QG). In this book the classical SR and GR space-times are assumed to be nice differentiable 4-manifolds with trivial topology, a notion criticized in many approaches to QG.

The only known mathematical method to define global non-inertial frames with well-defined global instantaneous non-Euclidean 3-spaces is the 3+1 point of view described in the first part of the book, where also the implications for relativistic metrology will be examined. In it one considers an arbitrary observer endowed with a nice foliation of Minkowski space-time whose leaves are the instantaneous (in general non-Euclidean) 3-spaces.

Moreover, the 3+1 point of view allows extension of the Lagrangian description of matter to a formulation (the parametrized Minkowski theory) in which the Lagrangian depends upon the variables identifying the non-inertial frames. The Lagrangian is singular, being invariant under the so-called frame-preserving diffeomorphisms, so that the study of its properties requires the second Noether theorem. As a consequence, one needs Dirac–Bergmann (DB) theory of constraints to study its Hamiltonian formulation and to show that the transition from a non-inertial frame to another (either inertial or non-inertial) frame is a gauge transformation and that the extra variables describing the frame are gauge variables, like those present in Maxwell and Yang–Mills theories, while the measurable matter quantities are the gauge-invariant Dirac observables (DOs).

In the 3+1 point of view it is possible to describe matter in SR separating the unobservable relativistic center of mass and by defining physical relative variables. The ten Poincaré generators can be defined in a consistent way, so that it is possible to define a relativistic quantum mechanics of point particles belonging to irreducible representations of the Poincaré group, as required by every model of particle physics.

While the first part of the book is dedicated to SR, the second part considers GR, where the geometrical view of Einstein implies that the 4-metric of the space-time with Lorentz signature not only determines the chrono-geometrical structure of the space-time by means of the line element, but is also the dynamical field mediating the gravitational interaction. Now the allowed paths of light rays (the conformal structure) are point-dependent and the 4-metric teaches relativistic causality to all the other fields.

Behind Einstein GR there is the principle of general covariance. The Hilbert action is invariant under passive diffeomorphisms (ordinary coordinate transformations), while Einstein equations are form-invariant in every 4-coordinate system (invariance under active diffeomorphisms), so that all the matter fields must have a tensorial character.

The Hamiltonian formulation of GR requires shifting from the Hilbert action to the *ADM* one, in which the passive diffeomorphisms are replaced by local Noether transformations in the framework of the second Noether theorem, so that in phase space one needs to use the DB theory of constraints.

In an arbitrary Einstein space-time one can define only local non-inertial frames around a non-inertial time-like observer. However, the need for a good formulation of the Cauchy problem for Einstein and matter fields requires the notion of instantaneous 3-space, namely the possibility of using the 3+1 point of view like in SR. Moreover, the inclusion of particle physics requires the possibility of introducing a notion of Poincaré algebra. These requirements select a family of Einstein space-times: the globally hyperbolic, asymptotically flat at spatial infinity, and without super-translations. In them there are the asymptotic ADM Poincaré generators: When one switches off the Newton constant they tend to the Poincaré generators of SR.

This class of singularity-free Einstein space-times contains asymptotic inertial observers to be identified with the fixed stars of astronomy. Moreover, the ADM energy generates a real temporal evolution modulo the gauge freedom associated with the first-class constraints. As a consequence there is not a frozen picture like in loop QG, where the 3-spaces are compact manifolds without boundary, implying the absence of a Poincaré algebra.

To be able to include fermions, Einstein 4-metric will be decomposed upon tetrads and the ADM action will be assumed to depend on them. This enlarges the gauge group and the number of first-class constraints due to the extra gauge freedom to orientate three gyroscopes and freedom in how they are transported along time-like trajectories.

The big open problem is to find the DOs of Einstein metric and tetrad gravity, since no one is able to solve the super-Hamiltonian and super-momentum constraints of GR. The best that can be done is to find a Shanmugadhasan canonical transformation adapted to all the first-class constraints except the *four unsolved constraints*.

However, this allows making a preliminary separation before trying to solve the Hamilton equations between physical canonical variables describing tidal effects (the gravitational waves of the linearized theory) and generalized gauge variables describing relativistic inertial effects. The gauge fixing of these inertial gauge variables (including a clock synchronization convention for the choice of the instantaneous 3-spaces) fixes a global non-inertial frame with well-defined relativistic inertial forces, so that the Hamilton equations become deterministic.

As a consequence, one can study post-Minkowskian (PM) and then post-Newtonian (PN) approximations of Einstein equations in the presence of point particles and of the electromagnetic field, with implications for dark matter.

Due to the relevance of DB theory of constraints, in the third part of the book there is a review of what is known on such a theory and on the state of the art in the search for DOs.

It is assumed that readers of this book of relativistic analytical mechanics have a good knowledge of SR and GR and of their mathematical, algebraic, and canonical formalisms, so there is no review of them. See Refs. [1–14] for the status of SR and GR and Refs. [15–17] for the status of the experiments on which they are based. See Refs. [18, 19] for some information on the status of QG. References [20–30] contain the main literature about constraint theory.

This book is the final result of many years of research, mainly together with Dr. David Alba (now a teacher in high schools) and Professor Horace W. Crater, who died last year, meaning I could not ask him to be my co-author. I thank Professor Massimo Pauri for showing me the relevance of trying to solve foundational problems in relativity and for the pleasure of being his collaborator in many works.

Index notation: Greek indices $\mu, \nu, \ldots$ have the values 0, 1, 2, 3 (0 denotes the time axis), while Latin indices $i, j, \ldots$ have the values 1, 2, 3. The flat metric ${}^4\eta_{\mu\nu} = \epsilon\,(1; 0, 0, 0)$ of the inertial frames in SR has $\epsilon = +1$ in the particle physics convention and $\epsilon = -1$ in the general relativity convention. The symbol $\approx$ means Dirac weak equality, while the symbol $\stackrel{\circ}{=}$ means evaluated by using the equations of motion. The symbol $\equiv$ means identically equal. By convention, repeated indices are summed. $\epsilon_{\alpha\beta\gamma\delta}$ and ϵ_{ijk} are completely antisymmetric tensors with $\epsilon_{0123} = \epsilon_{123} = \epsilon\,\epsilon^{123} = 1$.

Dimensions of the quantities appearing in this book: $[\tau] = [x^\mu] = [\vec{\sigma}] = [\vec{\eta}_i] = [l]$, $[\vec{\kappa}_i] = [P^\mu] = [E/c] = [m\,l\,t^{-1}]$, $[{}^4g] = [{}^3g] = [n] = [n_{(a)}] = [{}^3e_{(a)r}] = [\dot{\vec{\eta}}_i] = [\theta_i] = [0]$, $[G = 6.7 \times 10^{-8}\,\mathrm{cm}^3\,\mathrm{s}^{-2}\,\mathrm{g}^{-1}] = [\mathrm{m}^{-1}\,\mathrm{l}^3\,\mathrm{t}^{-2}]$, $[G/c^3] = [\mathrm{m}^{-1}\,\mathrm{t}] \approx 2.5 \times 10^{-39}\,\mathrm{sec/g}$, $[S] = [\hbar] = [J^{AB}] = [\mathrm{m}\,\mathrm{l}^2\,\mathrm{t}^{-1}]$, $[{}^3R] = [{}^3\Omega_{rs(a)}] = [\mathrm{l}^{-2}]$, $[{}^3\omega_{r(a)}] = [{}^3K_{rs}] = [\mathrm{l}^{-1}]$, $[{}^3\pi^r_{(a)}] = [{}^3\tilde{\Pi}^{rs}] = [\mathrm{m}\,\mathrm{l}^{-1}\,\mathrm{t}^{-1}]$, $[T^{AB}] = [\mathcal{M}] = [\mathcal{M}_r] = [\mathcal{H}] = [\mathcal{H}_{(a)}] = [\mathrm{m}\,\mathrm{l}^{-2}\,\mathrm{t}^{-1}]$.

Part I

Special Relativity

Minkowski Space-Time

In this first part, after a review of inertial and non-inertial frames in the non-relativistic Galilei space-time, I will study such frames in the Minkowski space-time of special relativity (SR).

In Newtonian physics, time and space are absolute notions whose metrological units are defined by means of standard clocks and rods, whose structure is not specified. This is satisfactory for the non-relativistic quantum mechanics used in molecular physics and in quantum information, where gravitation effects are described by Newtonian gravity.

However, in atomic physics one needs the description of light, whose quantum nature gives rise to the notion of massless photons whose trajectories do not exist in Galilei space-time. Moreover particle physics has to face high-speed objects. As a consequence, the Minkowski space-time of SR has to be introduced and a new type of metrology has been developed with different standards for length and time [31]. See references [32]–[38] for updated reviews on relativistic metrology on Earth, in the Solar System, and in astronomy.

The fundamental theoretical scale for time is the SI (International System of Units) atomic second, which is the duration of 9 192 631 770 periods of the radiation corresponding to the transition between two hyperfine levels of the ground state of the cesium 133 atom at rest at a temperature of 0K. In practice one uses the International Atomic Time (TAI), defined as a suitable weighted average of the SI kept by (mainly cesium) atomic clocks in about 70 laboratories worldwide.

To introduce a convention for the synchronization of distant clocks one uses the notion of two-way (or round-trip) velocity of light c involving only one clock: The observer emits a ray of light that is reflected somewhere and then reabsorbed by the observer, so that only the clock of the observer is used to measure the time of flight of the ray. It is this velocity that is isotropic and constant in SR (the light postulate) and replaces the standard of length in relativistic metrology. The one-way velocity of light from one observer A to an observer B has a meaning only after a choice of a convention for synchronizing the clock in A with the one in B.

One uses the conventional value $c = 299\,792\,458$ m s^{-1} for the two-way velocity of light. To measure the 3-distance between two objects in an inertial frame, one puts an atomic clock in the first object, then sends a ray of light to the second object, where it is reflected and then reabsorbed by the first object, whose measure of the flight time allows finding the 3-distance. As a consequence, the meter is the length of the path traveled by light in a vacuum in an inertial frame during a time interval of $1/c$ of a second.

1

Galilei and Minkowski Space-Times

In this chapter we review the properties the non-relativistic Galilei space-time and of the relativistic Minkowski one. See Refs. [39–43] for a detailed study of the rotation group and of the kinematical Galilei and Poincaré groups connecting the inertial frames of the respective space-times.

1.1 The Galilei Space-Time of Non-Relativistic Physics and Its Inertial and Non-Inertial Frames

In Newtonian physics the notions of time and space are absolute, so that the chrono-geometrical structure of Galilei space-time is not dynamical. One has at each instant of the absolute time t, registered by an ideal clock, an instantaneous Euclidean 3-space R^3_t with the standard notion of Euclidean distance, measured with ideal rods. The clocks in each point of R^3_t are synchronized at the time t, so that Galilei space-time has the structure $R \times R^3$, where R denotes the time axis and R^3 is an abstract Euclidean 3-space associated to the fixed stars of astronomy. As a consequence, one can parametrize Galilei space-time as the straight trajectory of an inertial observer (the time axis) endowed with a foliation of Euclidean 3-spaces orthogonal to the time axis.

The Galilei relativity principle assumes the existence of preferred rigid inertial frames of reference in uniform translational motion, one with respect to the other with inertial Cartesian coordinates (t, x^i) centered on an inertial observer, whose trajectory is the time axis. In these frames, free bodies move along straight lines (Newton's first law) and Newton's equations take the simplest form. The laws of nature are covariant and there is no preferred observer. The connection between different inertial frames is realized with the kinematical Galilei transformations: $t' = t + \epsilon$, $x'^i = R^{ij}\, x^j + v^i\, t + \epsilon^i$, where ϵ and ϵ^i are the time and space rigid translations, R ($R^{-1} = R^T$, the transposed matrix) is the O(3) matrix describing rigid rotations, and v^i are the parameters of the rigid Galilei boosts.

Due to the absolute nature of Newtonian time, the points on a $t =$ const. section of Galilei space-time are all simultaneous (instantaneous absolute 3-space), whichever inertial system we are using. As a consequence, the causal notions of before and after a certain event are absolute.

A particle of mass m has the trajectory described by inertial Cartesian 3-coordinates $x^i_m(t)$ in Galilei space-time. In the Hamiltonian phase space it has the momentum $p_i = m\, \delta_{ij}\, \frac{d\, x^i(t)}{dt}$. For a free particle the Galilei generators are $H = \vec{p}^2/2\, m$ (energy), p_i (momentum), $K_i = m\, \delta_{ij}\, x^j_m - t\, p_i$ (boost), $J_i = \epsilon_{ijk}\, \delta_{jh}\, x^h_m\, p_k$ (rotation). ϵ^{ijk} is the completely antisymmetric tensor.

For a system of mutually interacting N particles of mass m_k, trajectory $x^i_k(t)$, momenta $p_{k\,i}(t) = m_k\, \delta_{ij}\, \frac{d\, x^j_k(t)}{dt}$, $k = 1, \ldots, N$, the Galilei generators are $H = \sum_{k=1}^N \frac{\vec{p}_k^2}{2\, m_k} + V(\vec{x}_h(t) - \vec{x}_k(t))$, $\vec{P} = \sum_{k=1}^N \vec{p}_k$, $\vec{J} = \sum_{k=1}^N \vec{x}_k(t) \times \vec{p}_k(t)$, $\vec{K} = \sum_{k=1}^N (t\, \vec{p}_k(t) - m_k\, \vec{x}_k(t)) = t\, \vec{P} - m\, \vec{x}$. Here, $\vec{x} = \sum_{k=1}^N \frac{m_k}{m}\, \vec{x}_k(t)$ $(m = \sum_{k=1}^n m_k)$ is the Newton center of mass, whose conjugate variable is $\vec{P}$. Therefore, the conserved Galilei boosts identify the Newtonian center of mass.

Usually the interacting potential depends only on the relative distances of the particles (and not on their velocities) and appears only in the energy (the Hamiltonian) and not in the boosts differently from what happens at the relativistic level with the Poincaré group.

For isolated N-body systems the ten generators of the Galilei group are Noether constants of motion. The Abelian nature of the Noether constants (the 3-momentum) associated to the invariance under translations allow making a global separation of the center of mass from the relative variables (usually the Jacobi coordinates, identified by the centers of mass of subsystems, are preferred): In phase space this can be done with canonical transformations of the point type both in the coordinates and in the momenta.

Instead, the non-Abelian nature of the Noether constants (the angular momentum) associated with the invariance under rotations implies that there is no unique separation [44] of the relative variables in six orientational ones (the body frame in the case of rigid bodies) and in the remaining vibrational (or shape) ones. As a consequence, an isolated deformable body or a system of particles may rotate by changing the shape (the falling cat, the diver).

In Refs. [45, 46] there is a kinematical treatment of non-relativistic N-body systems by means of canonical spin bases and of dynamical body frames, which can be extended to the relativistic case in which the notions of Jacobi coordinates, reduced masses, and tensors of inertia are absent and can be recovered only when extended bodies are simulated with multipolar expansions [47].

Another non-conventional aspect of non-relativistic physics is the many-time formulation of classical particle dynamics [48] with as many first-class constraints as particles. Like in the special relativistic case, a distinction arises between physical positions and canonical configuration variables and a non-relativistic version of the no-interaction theorem (see Chapter 3) emerges.

See Refs. [49, 50] for Newtonian gravity, where the Newton equivalence principle states the equality of inertial and gravitational mass, as a gauge theory of the Galilei group.

To define rigid non-inertial frames, let us consider an arbitrary accelerated observer whose Cartesian trajectory is $y^i(t)$ and let us introduce the rigid non-inertial coordinates (t, σ^i) by imposing $x^i = y^i(t) + R^{ij}(t)\,\sigma^j$, where $R(t)$ is a time-dependent rotation matrix, which can be parametrized with three Euler angles. It describes the rigid rotation of the non-inertial frame. It is convenient to write the 3-velocity of the accelerated observer in the form $v^i(t) = R^{ij}(t)\,\frac{d\,y^j(t)}{dt}$. The angular velocity of the rotating frame is $\Omega^i(t) = \frac{1}{2}\,\epsilon^{ijk}\,\Omega^{jk}$ with $\Omega^{jk}(t) = -\Omega^{kj} = (\frac{d\,R(t)}{dt}\,R^T(t))^{jk}$.

A particle of mass m with trajectory given by the Cartesian 3-coordinates $x^i_m(t)$ is described in the rigid non-inertial frames by 3-coordinates $\eta^r(t)$ such that $x^i_m(t) = y^i(t) + R^{ij}(t)\,\eta^j(t)$.

As shown in every book on Newtonian mechanics, a particle moving in an external potential $V(t, x^k_m(t)) = \tilde{V}(t, \eta^r(t))$ has the equation of motion $m\,\frac{d^2\,x^i_m(t)}{dt^2} = -\frac{\partial\,V(t, x^k_m(t))}{\partial\,x^i_m}$, whose form in the rigid non-inertial frames becomes

$$m\,\frac{d^2\,\vec{\eta}(t)}{dt^2} = -\frac{\partial\,\tilde{V}(t, \eta^k(t))}{\partial\,\vec{\eta}} - m\,\Big[\frac{d\vec{v}(t)}{dt} + \vec{\omega}(t) \times \vec{v}(t) + \frac{d\,\vec{\omega}(t)}{dt} \times \vec{\eta}(t)$$
$$+2\,\vec{\omega}(t) \times \frac{d\,\vec{\eta}(t)}{dt} + \vec{\omega}(t) \times [\vec{\omega}(t) \times \vec{\eta}(t)]\Big]. \tag{1.1}$$

In this equation there are the standard Euler, Jacobi, Coriolis, and centrifugal inertial (or fictitious) forces, proportional to the mass of the body, associated with the acceleration of the non-inertial observer and with the angular velocity of the rotating rigid non-inertial frame.

The extension to non-rigid non-inertial frames with coordinates (t, σ^i) (σ^r are global non-Cartesian 3-coordinates) is done in Ref. [51] by putting the Cartesian 3-coordinates x^i equal to arbitrary functions $\mathcal{A}^i(t, \sigma^r)$, well behaved at spatial infinity: $x^i = \mathcal{A}^i(t, \sigma^r)$. This coordinate transformation must be invertible with inverse $\sigma^r = \mathcal{S}^r(t, x^i)$. The invertibility conditions are $\det J(t, \sigma^r) > 0$, where $J^a_r(t, \sigma^u) = \frac{\partial\,\mathcal{A}^a(t, \sigma^u)}{\partial\,\sigma^r}$ is the three-dimensional Jacobian, whose inverse is $\tilde{J}^r{}_a(t, \sigma^u) = \Big(\frac{\partial\,\mathcal{S}^r(t, x^u)}{\partial\,x^a}\Big)_{x^b = \mathcal{A}^b(t, \sigma^u)}$ $(J^a{}_r(t, \sigma^u)\,\tilde{J}^r{}_b(t, \sigma^u) = \delta^a_b$, $\tilde{J}^s{}_a(t, \sigma^u)\,J^a{}_r(t, \sigma^u) = \delta^s_r)$. The group of Galilei transformations connecting inertial frames is replaced by some subgroup of the 3-diffeomorphisms of the Euclidean 3-space connecting the non-inertial ones. The quantum mechanics of particles in non-rigid non-inertial frames is studied in Ref. [51].

1.2 The Minkowski Space-Time: Inertial Frames, Cartesian Coordinates, Matter, Energy–Momentum Tensor, and Poincaré Generators

The Minkowski space-time of special relativity (SR) is an affine 4-manifold isomorphic to R^4 with Lorentz signature in which neither time nor space are absolute notions. As a consequence there is no unique notion of instantaneous 3-space and one needs some metrological convention about time and space to be able to formulate particle physics in the laboratories on Earth in the approximation of neglecting gravity. The only intrinsic structure of Minkowski space-time is the conformal one connected with the Lorentz signature: It defines the light-cone as the locus of incoming and outgoing radiation.

There is no absolute notion of simultaneity: Given an event, all the points outside the light-cone with vertices in that event are not causally connected with that event (they have space-like separation from it), so that the notions of before and after an event become observer-dependent. Therefore there is no notion of an instantaneous 3-space, of a spatial distance, and of a one-way velocity of light between two observers (the problem of the synchronization of distant clocks). Instead, as already said, there is an absolute chrono-geometrical structure: the light postulate saying that the two-way (or round-trip) velocity of light c (only one clock is needed for its definition) is (1) constant and (2) isotropic. Let us remark that the clocks are assumed to be standard atomic clocks measuring proper time [52–54].

Einstein relativity principle privileges the inertial frames of Minkowski space-time centered on inertial observers endowed with an atomic clock: Their trajectories are the time axis in Cartesian coordinates $x^\mu = (x^o = ct; x^i)$ where the flat metric tensor with Lorentz signature is ${}^4\eta_{\mu\nu} = \epsilon\,(1; -1, -1, -1)$. These inertial frames are in uniform translational motion, one with respect to the other. All special relativistic physical systems, defined in the inertial frames of Minkowski space-time, are assumed to be manifestly covariant under the transformations of the kinematical Poincaré group connecting the inertial frames. The laws of physics are covariant and there is no preferred observer.

The $x^o =$ const. hyper-planes of inertial frames are usually taken as Euclidean instantaneous 3-spaces, on which all the clocks are synchronized. They can be selected with Einstein's convention for the synchronization of distant clocks to the clock of an inertial observer. This inertial observer A sends a ray of light at x^o_i to a second accelerated observer B, who reflects it toward A. The reflected ray is reabsorbed by the inertial observer at x^o_f. The convention states that the clock of B at the reflection point must be synchronized with the clock of A when it signs $\frac{1}{2}\,(x^o_i + x^o_f)$. This convention selects the $x^o =$ const. hyper-planes of inertial frames as simultaneity 3-spaces and implies that only with this synchronization the two-way (A–B–A) and one-way (A–B or B–A) velocities of light coincide and the spatial distance between two simultaneous point is the (3-geodesic) Euclidean distance. However, if observer A is accelerated, the convention can break down due to the possible appearance of coordinate singularities.

Relativistic matter is defined in the relativistic inertial frames of Minkowski space-time centered on inertial observers using Cartesian 4-coordinates. It is in these frames that one defines the matter Lagrangian when it is known. Once one has the Lagrangian $\mathcal{L}(x)$ of a matter system the energy–momentum tensor is defined by replacing the flat 4-metric ${}^4\eta_{\mu\nu}$ appearing in the Lagrangian with a 4-metric ${}^4g_{\mu\nu}(x)$ like the one used in general relativity (GR), so that one gets a new Lagrangian $\mathcal{L}_g(x)$ and by using the definition $T^{\mu\nu}(x) = -\frac{2}{\sqrt{-det^4 g(x)}}\,\frac{\delta\, S_g}{\delta^4 g_{\mu\nu}(x)}$, where $S_g = \int d^4x \, \mathcal{L}_g(x)$. In inertial frames with Cartesian 4-coordinates x^μ, the Poincaré generators, assumed finite due to suitable boundary conditions at spatial infinity, have the following expression: $P^\mu = \int d^3x \, T^{\mu o}(x^o, \vec{x})$, $J^{\mu\nu} = \int d^3x \, [x^\mu \, T^{\nu o}(x^o, \vec{x}) - x^\nu \, T^{\mu o}(x^o, \vec{x})]$.

In Appendix A there are some properties of the Poincaré algebra and group. At the Hamiltonian level the canonical Poincaré generators P^μ, $J^{\mu\nu}$ satisfy the Poisson algebra $\{P^\mu, P^\nu\} = 0$, $\{P^\mu, J^{\alpha\beta}\} = \eta^{\mu\alpha} \, P^\beta - \eta^{\mu\beta} \, P^\alpha$, $\{J^{\alpha\beta}, J^{\mu\nu}\} = C^{\alpha\beta\mu\nu}_{\rho\sigma} \, J^{\rho\sigma}$ ($C^{\alpha\beta\mu\nu}_{\rho\sigma} = \eta^{\alpha\mu} \, \delta^\beta_\rho \, \delta^\nu_\sigma + \eta^{\beta\nu} \, \delta^\alpha_\rho \, \delta^\mu_\sigma - \eta^{\alpha\nu} \, \delta^\beta_\rho \, \delta^\mu_\sigma - \eta^{\beta\mu} \, \delta^\alpha_\rho \, \delta^\nu_\sigma$). If $J^r = -\frac{1}{2} \, \epsilon^{ruv} \, J^{uv}$ is the generator of space rotations and $K^r = J^{ro}$ one of the boosts, the form of the canonical Poincaré algebra becomes $\{J^r, J^s\} = \epsilon^{rst} \, J^t$, $\{K^r, K^s\} = \epsilon^{rsu} \, J^u$, $\{J^r, K^s\} = \{K^r, J^s\} = \epsilon^{rsu} \, K^u$.

To describe point particles with spin, with electric charge and with antiparticles of negative mass in a way that avoids self-reaction divergences at the classical level and gives the correct quantum theory after quantization, one needs a semi-classical approach, named pseudo-classical mechanics, in which these degrees of freedom are described with Grassmann variables. In Appendix B there is a review of this approach and of the needed mathematical tools.

For the detailed mathematical properties of Minkowski space-time, see any book on SR, such as the recent ones of Gourgoulhon Ref. [12, 13].

1.3 The 1+3 Approach to Local Non-Inertial Frames and Its Limitations

Since the actual time-like observers are accelerated, we need some statement correlating the measurements made by them to those made by inertial observers, the only ones with a general framework for the interpretation of their experiments based on Einstein convention for the synchronization of clocks. This statement is usually the hypothesis of locality, which can be expressed in the following terms [55–60]: An accelerated observer at each instant along its world-line is physically equivalent to an otherwise identical momentarily comoving inertial observer, namely a non-inertial observer passes through a continuous infinity of hypothetical momentarily comoving inertial observers.

While this hypothesis is verified in Newtonian mechanics and in those relativistic cases in which a phenomenon can be reduced to point-like coincidences of classical point particles and light rays (geometrical optic approximation), its validity is questionable with moving continuous media (for instance the constitutive equations of the electromagnetic field inside them in non-inertial frames are still unknown) and in the presence of electromagnetic fields when their wavelength is comparable with the acceleration radii of the observer (the observer is not "static" enough to be able to measure the frequency of such a wave). See Refs. [61, 62] for a review of these topics.

The fact that we can describe phenomena only locally near the observer and that the actual observers are accelerated leads to the 1+3 point of view (or threading splitting) [63–70], which tries to solve this problem starting from the local properties of an accelerated observer, whose world-line is assumed to be the time axis of some frame. Given the world-line γ of the accelerated observer, we describe it with Lorentzian coordinates $x^\mu(\tau)$, parametrized with an affine parameter τ, with respect to a given inertial system. Its unit 4-velocity is $u^\mu_\gamma(\tau) = \dot{x}^\mu(\tau)/\sqrt{\epsilon\, \dot{x}^2(\tau)}$ $[\dot{x}^\mu = \frac{dx^\mu}{d\tau}]$. The observer proper time $\tau_\gamma(\tau)$ is defined by $\epsilon\, \dot{\tilde{x}}^2(\tau_\gamma) = 1$ if we use the notations $x^\mu(\tau) = \tilde{x}^\mu(\tau_\gamma(\tau))$ and $u^\mu(\tau) = \tilde{u}^\mu(\tau_\gamma(\tau)) = d\,\tilde{x}^\mu(\tau_\gamma)/d\,\tau_\gamma$, and it is indicated by a comoving standard atomic clock.

By a conventional choice of three spatial axes $E^\mu_{(a)}(\tau) = \tilde{E}^\mu_{(a)}(\tau_\gamma(\tau))$, $a = 1, 2, 3$, orthogonal to $u^\mu(\tau) = E^\mu_{(o)}(\tau) = \tilde{u}^\mu(\tau_\gamma(\tau)) = \tilde{E}^\mu_{(o)}(\tau_\gamma(\tau))$, the non-inertial observer is endowed with an ortho-normal tetrad $E^\mu_{(\alpha)}(\tau) = \tilde{E}^\mu_{(\alpha)}(\tau_\gamma(\tau))$, $\alpha = 0, 1, 2, 3$. This amounts to a choice of three comoving gyroscopes in addition to the comoving standard atomic clock. Usually the spatial axes are chosen to be Fermi–Walker transported as a standard of non-rotation, which takes into account the Thomas precession (see [71]).

Since only the observer 4-velocity is given, this only allows identification of the tangent plane of the vectors orthogonal to this 4-velocity in each point of the world-line. Since there is no notion of a 3-space simultaneous with a point of γ and whose tangent space at that point is $R_{\tilde{u}(\tau_\gamma)}$, this tangent plane is identified with an instantaneous 3-space both in SR and GR (it is the local observer rest-frame at that point). This identification is the basic limitation of this approach because the hyper-planes at different times intersect each other at a distance from the world-line depending on the acceleration of the observer so that the approach works only in a world-tube whose radius is this distance. See Refs. [63–70] for the definition of the (linear and rotational) acceleration radii of the observer. At each point of γ with proper time $\tau_\gamma(\tau)$, the tangent space to Minkowski space-time in that point has the 1+3 splitting of vectors in vectors parallel to $\tilde{u}^\mu(\tau_\gamma)$ and vectors lying in the three-dimensional (so-called local observer rest-frame) subspace $R_{\tilde{u}(\tau_\gamma)}$ orthogonal to $\tilde{u}^\mu(\tau_\gamma)$.

Then Fermi normal coordinates [72–76] are defined on each hyper-plane orthogonal to the observer unit 4-velocity $u^\mu(\tau)$ and are used to define a notion of spatial distance. On each hyper-plane one considers three space-like geodesics as spatial coordinate lines. However, this produces only local coordinates and a notion of simultaneity valid only inside the world-tube. See Refs. [77–79] for variants of this approach, all unable to avoid the coordinate singularity on the world-tube.

To this type of coordinate singularities we have to add the singularities shown by all the rotating coordinate systems (the problem of the rotating disk): In all the proposed uniformly rotating coordinate systems the induced 4-metric expressed in these coordinates has pathologies (the component ${}^4g_{oo}$ vanishes) at the distance R from the rotation axis where $\omega\, R = c$ with ω being the constant angular velocity of rotation. This is the horizon problem: At R the time-like 4-velocity of a disk point becomes light-like, even if there is no real horizon as happens for Schwarzschild black holes. Again, given the unit 4-velocity field of the points of the rotating disk, there is no notion of an instantaneous 3-space orthogonal to the associated congruence of time-like observers, due to the non-zero vorticity of the congruence [71] (see Section 2.1 for the definition of vorticity). Due to the Frobenius theorem, the congruence is (locally) hyper-surface orthogonal, i.e., locally synchronizable [71], if and only if the vorticity vanishes. Moreover, an attempt to use Einstein convention to synchronize the clocks on the rim of the disk fails and one finds a synchronization gap (see Refs. [80–84] and the bibliographies of Refs. [61, 62] for these problems).

One does not know how to define the 3-geometry of the rotating disk, how to measure the length of the circumference, and which time and notion of simultaneity has to be used to evaluate the velocity of (massive or massless) particles in uniform motion along the circumference.

The other important phenomenon connected with the rotating disk is the Sagnac effect (see again Refs. [61, 62, 80–84]), namely the phase difference generated by the difference in the time needed for a round-trip by two light rays, emitted at the same point, one co-rotating and the other counter-rotating with the disk. This effect, which has been tested for light, X-rays, and matter waves (Cooper pairs, neutrons, electrons, and atoms) and must be taken into account for the relativistic corrections to space navigation, has again an enormous number of theoretical interpretations (both in SR and GR). Here the lack of a good notion of simultaneity leads to problems of time discontinuities or desynchronization effects when comparing clocks on the rim of the rotating disk.

In conclusion, in SR inertial frames are a limiting theoretical notion since, also disregarding GR, all the observers on Earth are non-inertial. According to the IAU 2000 Resolutions [32–35], for the physics in the solar system one can consider the Solar System Barycentric Celestial Reference System (BCRS) centered on the barycenter of the Solar System (with the axes identified by fixed stars (quasars) of the Hypparcos catalog) as a quasi-inertial frame. Instead, the Geocentric Celestial Reference System (GCRS), with origin in the center of the geoid, is a non-inertial frame whose axes are non-rotating with respect to the Solar Frame. Instead, every frame centered on an observer fixed on the surface of Earth (using the yet non-relativistic International Terrestrial Reference System [ITRS]) is both non-inertial and rotating. All these frames use notions of time connected to TAI.

Let us also remark that the physical protocols (think of GPS) can establish a clock synchronization convention only inside future light-cone of the physical observer defining the local 3-spaces only inside it, in accord with the 1+3 point of view.

This state of affairs and the need for predictability (a well-posed Cauchy problem for field theory) lead to the necessity of abandoning the 1+3 point of view and shifting to the 3+1 one. In this point of view, besides the world-line of an arbitrary time-like observer, it is given a global 3+1 splitting of Minkowski space-time, namely a foliation of it whose leaves are space-like hyper-surfaces.

Each leaf is both a Cauchy surface for the description of physical systems and an instantaneous Riemannian 3-space, namely a notion of simultaneity implied by a clock synchronization convention different from Einstein's one. Even if it is unphysical (i.e., non-factual) to give initial data (the Cauchy problem) on a non-compact space-like hyper-surface, this is the only way to be able to use the existing existence and uniqueness theorems for the solutions of partial differential equations like the Maxwell ones, needed to test the predictions of the theory.[1] Once we have given the Cauchy data on the initial Cauchy surface (an unphysical process), we can predict the future with every observer receiving the information only from his/her past light-cone (retarded information from inside it; electromagnetic signals on it). As emphasized by Havas [86], the 3+1 approach is based on Møller's formalization [87, 88] of the notion of simultaneity.

For non-relativistic observers the situation is simpler, but the non-factual need to give the Cauchy data on a whole initial absolute Euclidean 3-space is present also in this case for non-relativistic field equations like the Euler equation for fluids.

Moreover, to study relativistic Hamiltonian dynamics one has to follow its formulation given by Dirac [89] with the instant, front (or light), and point forms and the associated canonical realizations of the Poincaré algebra. In the instant form, the simultaneity hyper-surfaces (Cauchy surfaces) defining a parameter for the time evolution are space-like hyper-planes $x^o = \text{const.}$, in the front form hyper-planes $x^- = \frac{1}{2}(x^o - x^3) = \text{const.}$ tangent to future light-cones, while in the point form the future branch of a two-sheeted hyperboloid $x^2 > 0$. In a $6N$-dimensional phase space for N scalar particles the ten generators of the Poincaré algebra are classified into kinematical generators (the generators of the stability group of the simultaneity hyper-surface) and dynamical generators (the only ones to be modified with respect to the free case in the presence of interactions) according to the chosen concept of simultaneity. While in the instant and point forms there are four dynamical generators (in the former energy and boosts, in the latter the 4-momentum), the front form has only three of them.

We will see that the 3+1 approach is the natural framework to implement the instant form of relativistic Hamiltonian dynamics.

Let us add the final remark that both in the 1+3 and in the 3+1 approach we call observer an idealization by means of a time-like world-line whose tangent vector in each point is the 4-velocity of the observer. If the 4-velocity is completed with a spatial triad to form a tetrad in each point of the world-line, we get an idealized observer with both a clock and a gyroscope. While this notion is compatible with the absolute metrology of SR, in GR it corresponds to a test

[1] As far as we know the theorem on the existence and unicity of solutions has not yet been extended starting from data given only on the past light-cone. See Ref. [85] for an attempt to rephrase the instant form of dynamics in a form employing only data from the causal past light-cone of the observer.

observer. To describe dynamical observers we need a model with dynamical matter in both cases. Therefore, an observer, or better a mathematical observer, is an idealization of a measuring apparatus containing an atomic clock and defining, by means of gyroscopes, a set of spatial axes (and then a, maybe orthonormal, tetrad with a convention for its transport) in each point of the world-line. See Ref. [90] for properties of mathematical and dynamical observers.

2

Global Non-Inertial Frames in Special Relativity

In this chapter I will describe the 3+1 approach to global non-inertial frames in Minkowski space-time and the associated parametrized Minkowski theories for the description of matter admitting a Lagrangian description in these frames.

2.1 The 3+1 Approach to Global Non-Inertial Frames and Radar 4-Coordinates

Assume that the world-line $x^\mu(\tau)$ of an arbitrary time-like observer carrying a standard atomic clock is given: τ is an arbitrary monotonically increasing function of the proper time of this clock. Then one gives an admissible 3+1 splitting of Minkowski space-time, namely a nice foliation with space-like instantaneous 3-spaces Σ_τ. It is the mathematical idealization of a protocol for clock synchronization allowing the formulation of the Cauchy problem for the equations of motion of fields: All the clocks in the points of Σ_τ sign the same time of the atomic clock of the observer. The observer and the foliation define a global non-inertial reference frame after a choice of 4-coordinates. On each 3-space Σ_τ one chooses curvilinear 3-coordinates σ^r having the observer as the origin. The quantities $\sigma^A = (\tau; \sigma^r)$ are the Lorentz-scalar and observer-dependent "radar 4-coordinates," first introduced by Bondi [62, 91, 92]. See Ref. [71] for the mathematical aspects of curved spaces like Σ_τ.

If $x^\mu \mapsto \sigma^A(x)$ is the coordinate transformation from the Cartesian 4-coordinates x^μ of an inertial frame centered on a reference inertial observer to radar coordinates, its inverse $\sigma^A \mapsto x^\mu = z^\mu(\tau, \sigma^r)$ defines the embedding functions $z^\mu(\tau, \sigma^r)$ describing the 3-spaces Σ_τ as embedded 3-manifolds into Minkowski space-time. The induced 4-metric on Σ_τ is the following functional of the embedding:

$$ {}^4g_{AB}(\tau, \sigma^r) = z^\mu_A(\tau, \sigma^r)\, {}^4\eta_{\mu\nu}\, z^\nu_B(\tau, \sigma^r), \tag{2.1} $$

where $z^{\mu}_{A}(\tau,\sigma^r) = \partial z^{\mu}(\tau,\sigma^r)/\partial \sigma^A$, $z^A_{\mu}(\tau,\sigma^r) = {}^4g^{AB}(\tau,\sigma^r)\, {}^4\eta_{\mu\nu}\, z^{\nu}_B(\tau,\sigma^r)$, and ${}^4\eta_{\mu\nu} = \epsilon\,(+---)$ is the flat metric.

While the 4-vectors $z^{\mu}_r(\tau,\sigma^u)$ are tangent to Σ_τ, so that the unit normal $l^{\mu}(\tau,\sigma^u)$ is proportional to $\epsilon^{\mu}{}_{\alpha\beta\gamma}\,[z^{\alpha}_1\, z^{\beta}_2\, z^{\gamma}_3](\tau,\sigma^u)$, one has $z^{\mu}_{\tau}(\tau,\sigma^r) = [N\, l^{\mu} + N^r\, z^{\mu}_r](\tau,\sigma^r)$, with $N(\tau,\sigma^r) = \epsilon\,[z^{\mu}_{\tau}\, l_{\mu}](\tau,\sigma^r) = 1 + n(\tau,\sigma^r)$ and $N_r(\tau,\sigma^r) = -\epsilon\, {}^4g_{\tau r}(\tau,\sigma^r)$ being the lapse and shift functions respectively. The unit normal $l^{\mu}(\tau,\sigma^u)$ ($l_{\mu}(\tau,\sigma^u)\, z^{\mu}_r(\tau,\sigma^u) = 0$, $\epsilon\, l^2(\tau,\sigma^u) = 1$), and the space-like 4-vectors $z^{\mu}_r(\tau,\sigma^u)$ identify a (in general non-orthonormal) tetrad in each point of Minkowski space-time. The tetrad in the origin ($l^{\mu}(\tau,0)$ (in general non-parallel to the observer 4-velocity), $z^{\mu}_r(\tau,0)$) is a set of axes carried by the observer; their τ-dependence implies a convention of transport along the world-line. The lapse function measures the proper time interval $N(\tau,\sigma^r)d\tau$ at $z(\tau,\sigma^r) \in \Sigma_\tau$ between Σ_τ and $\Sigma_{\tau+d\tau}$. The shift functions $N^r(\tau,\sigma^r)$ are defined so that $N^r(\tau,\sigma^r)d\tau$ describes the horizontal shift on Σ_τ such that, if $z^{\mu}(\tau+d\tau,\sigma^r+d\sigma^r) \in \Sigma_{\tau+d\tau}$, then $z^{\mu}(\tau+d\tau,\sigma^r+d\sigma^r) \approx z^{\mu}(\tau,\sigma^r) + N(\tau,\sigma^r)\, d\tau l^{\mu}(\tau,\sigma^r) + [d\sigma^s + N^s(\tau,\sigma^r)d\tau] z^{\mu}_s(\tau,\sigma^r)$.

In each point of the 3-spaces Σ_τ, one introduces arbitrary triads ${}^3e^r_{(a)}(\tau,\sigma^u)$ (i.e., three gyroscopes) and the conjugated cotriads ${}^3e_{(a)r}(\tau,\sigma^u)$ $(({}^3e^r_{(a)}\, {}^3e_{(b)r})(\tau,\sigma^u) = \delta_{ab}$, $({}^3e^r_{(a)}\, {}^3e_{(a)s})(\tau,\sigma^u) = \delta^r_s)$.

The components of the 4-metric ${}^4g_{AB}(\tau,\sigma^r)$ and of the inverse 4-metric ${}^4g^{AB}(\tau,\sigma^r)$ $(({}^4g^{AC}\, {}^4g_{cb})(\tau,\sigma^r) = \delta^A_B)$ can be parametrized in the following way ($a = 1,2,3$; $\bar a = 1,2$):

$$
\begin{aligned}
\epsilon\, {}^4g_{\tau\tau}(\tau,\sigma^u) &= \Big(N^2 - N_r\, N^r\Big)(\tau,\sigma^u),\\
-\epsilon\, {}^4g_{\tau r}(\tau,\sigma^u) &= N_r(\tau,\sigma^u) = \Big({}^3g_{rs}\, N^s\Big)(\tau,\sigma^u),\\
{}^3g_{rs}(\tau,\sigma^u) &= -\epsilon\, {}^4g_{rs}(\tau,\sigma^u) = \sum_{a=1}^{3} \Big({}^3e_{(a)r}\, {}^3e_{(a)s}\Big)(\tau,\sigma^u)\\
&= \Big(\tilde\phi^{2/3} \sum_{a=1}^{3} e^{2\sum_{\bar b=1}^{2} \gamma_{\bar b a} R_{\bar b}}\, V_{ra}(\theta^i)\, V_{sa}(\theta^i)\Big)(\tau,\sigma^u),\\
\epsilon\, {}^4g^{\tau\tau}(\tau,\sigma^u) &= N^{-2}(\tau,\sigma^u), \qquad \epsilon\, {}^4g^{\tau r}(\tau,\sigma^u) = -\Big(N^r\, N^{-2}\Big)(\tau,\sigma^u),\\
{}^4g^{rs}(\tau,\sigma^u) &= -\epsilon\Big({}^3g^{rs} - \frac{N^r\, N^s}{N^2}\Big)(\tau,\sigma^u),\\
\text{with}\, {}^3g^{rs}(\tau,\sigma^u) &= \Big({}^3e^r_{(a)}\, {}^3e^s_{(a)}\Big)(\tau,\sigma^u). \qquad (2.2)
\end{aligned}
$$

The quantity $\tilde\phi^2(\tau,\sigma^u) = \det{}^3g_{rs}(\tau,\sigma^u) = \det[-\epsilon\, {}^4g_{rs}(\tau,\sigma^u)] = \gamma(\tau,\sigma^u)$ appearing in the 3-metric is the 3-volume element on Σ_τ. We have $l^{\mu}(\tau,\sigma^u) = (\tilde\phi^{-1}\, \epsilon^{\mu}{}_{\alpha\beta\gamma}\, z^{\alpha}_1\, z^{\beta}_2\, z^{\gamma}_3)(\tau,\sigma^u) = \frac{1}{N(\tau,\sigma^u)}\,(1, N^r(\tau,\sigma^u))$, $l_{\mu}(\tau,\sigma^u) = N(\tau,\sigma^u)\, {}^4\eta_{\mu o} = N(\tau,\sigma^u)\,\epsilon\,(1;000)$, $d^4z = \tilde\phi(\tau,\sigma^u)\, d\tau\, d^3\sigma = \sqrt{\gamma(\tau,\sigma^u)}\, d\tau d^3\sigma$, and $\sqrt{g(\tau,\sigma^u)} = \sqrt{-\det{}^4g_{AB}(\tau,\sigma^u)} = N(\tau,\sigma^u)\,\sqrt{\gamma(\tau,\sigma^u)}$.

The quantities $\lambda_a(\tau,\sigma^r) = [\tilde{\phi}^{1/3}\, e^{\sum_{\bar{b}=1}^{2} \gamma_{\bar{b}a}\, R_{\bar{b}}}](\tau,\sigma^r)$ are the three positive eigenvalues of the 3-metric ${}^3g_{rs}(\tau,\sigma^r)$ ($\gamma_{\bar{a}a}$ are suitable numerical constants satisfying $\sum_a \gamma_{\bar{a}a} = 0$, $\sum_a \gamma_{\bar{a}a}\,\gamma_{\bar{b}a} = \delta_{\bar{a}\bar{b}}$, $\sum_{\bar{a}} \gamma_{\bar{a}a}\,\gamma_{\bar{a}b} = \delta_{ab} - \frac{1}{3}$).

$V_{ra}(\theta^i(\tau,\sigma^r))$ are the components of a diagonalizing rotation matrix depending on three Euler angles, $\theta^i(\tau,\sigma^r)$, $i = 1,2,3$. The gauge Euler angles θ^r give a description of the 3-coordinate systems on Σ_τ from a local point of view, because they give the orientation of the tangents to the three 3-coordinate lines through each point. However, as shown in appendix A of Ref. [94] and in Ref. [95], it is more convenient to replace the three Euler angles with the first kind coordinates $\tilde{\theta}^i(\tau,\sigma^r)$ $(-\infty < \tilde{\theta}^i < +\infty)$ on the O(3) group manifold: With this choice we have $V_{ru}(\tilde{\theta}^i) = (e^{-\sum_i \hat{T}_i\,\tilde{\theta}^i})_{ru}$, where $(\hat{T}_i)_{ru} = \epsilon_{rui}$.

Therefore, starting from the *four* independent embedding functions $z^\mu(\tau,\sigma^r)$, one obtains the ten components ${}^4g_{AB}$ of the 4-metric, which play the role of the inertial potentials generating the relativistic apparent forces in the non-inertial frame. For instance, the shift functions $N_r(\tau,\sigma^u) = -\epsilon\,{}^4g_{\tau r}(\tau,\sigma^u)$ describe inertial forces of the gravito-magnetic type induced by the global non-inertial frame. It can be shown [98–101] that the usual non-relativistic Newtonian inertial potentials are hidden in these functions. The extrinsic curvature tensor ${}^3K_{rs}(\tau,\sigma^u) = [\frac{1}{2N}\,(N_{r|s} + N_{s|r} - \partial_\tau\, \epsilon\, {}^4g_{rs})](\tau,\sigma^u)$, describing the shape of the instantaneous 3-spaces Σ_τ of the non-inertial frame as embedded 3-sub-manifolds of Minkowski space-time, is a secondary inertial potential, functional of the ten inertial potentials ${}^4g_{AB}(\tau,\sigma^r)$.

In this framework a relativistic positive-energy scalar particle with world-line $x^\mu_o(\tau)$ is described by 3-coordinates $\eta^r(\tau)$ defined by $x^\mu_o(\tau) = z^\mu(\tau,\eta^r(\tau))$, satisfying equations of motion containing relativistic inertial forces with the correct non-relativistic limit as shown in Refs. [98–101]. Fields have to be redefined so as to know the clock synchronization convention; for instance, a Klein–Gordon field $\tilde{\phi}(x^\mu)$ has to be replaced with $\phi(\tau,\sigma^r) = \tilde{\phi}(z^\mu(\tau,\sigma^r))$.

The foliation is nice and admissible if it satisfies the following conditions:

1. $N(\tau,\sigma^r) > 0$ in every point of Σ_τ so that the 3-spaces never intersect, avoiding the coordinate singularity of the Fermi coordinates of the 1+3 approach.
2. $\epsilon\,{}^4g_{\tau\tau}(\tau,\sigma^r) = (z^\mu_\tau\,{}^4\eta_{\mu\nu}\, z^\nu_\tau)(\tau,\sigma^r) = (N^2 - N_u\, N^u)(\tau,\sigma^r) > 0$, so as to avoid the coordinate singularity of the rotating disk, and with the positive-definite 3-metric ${}^3g_{rs}(\tau,\sigma^u) = -\epsilon\,{}^4g_{rs}(\tau,\sigma^u)$ having three positive eigenvalues (these are the Møller conditions [87]).
3. All the 3-spaces Σ_τ must tend to the same space-like hyper-plane at spatial infinity with a unit normal ϵ^μ_τ, which is the time-like 4-vector of a set of asymptotic orthonormal tetrads ϵ^μ_A $(\epsilon^\mu_A\,{}^4\eta_{\mu\nu}\,\epsilon^\nu_B = {}^4\eta_{AB} = \epsilon\,(+---))$. These tetrads are carried by asymptotic inertial observers and the spatial axes ϵ^μ_r are identified by the fixed stars of star catalogues. At spatial infinity the lapse function tends to 1, the shift functions vanish, and the 4-metric becomes Euclidean with the same behavior, as will be discussed in relation to general relativity in Part II.

By using the asymptotic tetrads ϵ^μ_A we can give the following parametrization of the embedding functions:

$$\begin{aligned} z^\mu(\tau,\sigma^r) &= x^\mu(\tau) + \epsilon^\mu_A\, F^A(\tau,\sigma^r), \qquad F^A(\tau,0)=0,\\ x^\mu(\tau) &= x^\mu_o + \epsilon^\mu_A\, f^A(\tau), \end{aligned} \tag{2.3}$$

where $x^\mu(\tau)$ is the world-line of the observer. The functions $f^A(\tau)$ determine the 4-velocity $u^\mu(\tau) = \dot x^\mu(\tau)/\sqrt{\epsilon\, \dot x^2(\tau)}$ ($\dot x^\mu(\tau) = \frac{dx^\mu(\tau)}{d\tau}$) and the 4-acceleration $a^\mu(\tau) = \frac{du^\mu(\tau)}{d\tau}$ of the observer. For an inertial frame centered on the inertial observer $x^\mu(\tau) = x^\mu_o + \epsilon^\mu_\tau\, \tau$, the embedding is $z^\mu(\tau,\sigma^r) = x^\mu(\tau) + \epsilon^\mu_r\, \sigma^r$, with ϵ^μ_A being an orthonormal tetrad identifying the Cartesian axes.

It is difficult to construct explicit examples of admissible 3+1 splittings because the Møller conditions are non-linear differential conditions on the functions $f^A(\tau)$ and $F^A(\tau,\sigma^r)$. When these conditions are satisfied, Eq. (2.2) describes a global non-inertial frame in Minkowski space-time.

Till now the solution of Møller conditions is known in the following two cases in which the instantaneous 3-spaces are parallel Euclidean space-like hyper-planes not equally spaced due to a linear acceleration:

1. Rigid non-inertial reference frames with translational acceleration. An example are the following embeddings:

$$\begin{aligned} &z^\mu(\tau,\sigma^u) = x^\mu_o + \epsilon^\mu_\tau\, f(\tau) + \epsilon^\mu_r\, \sigma^r,\\ &{}^4g_{\tau\tau}(\tau,\sigma^u) = \epsilon\left(\frac{df(\tau)}{d\tau}\right)^2, \quad {}^4g_{\tau r}(\tau,\sigma^u) = 0, \quad {}^4g_{rs}(\tau,\sigma^u) = -\epsilon\,\delta_{rs}. \end{aligned} \tag{2.4}$$

   This is a foliation with parallel hyper-planes with normal $l^\mu = \epsilon^\mu_\tau =$ const. and with the time-like observer $x^\mu(\tau) = x^\mu_o + \epsilon^\mu_\tau\, f(\tau)$ as the origin of the 3-coordinates. The hyper-planes have translational acceleration $\ddot x^\mu(\tau) = \epsilon^\mu_\tau\, \ddot f(\tau)$, so that they are not uniformly distributed like in the inertial case $f(\tau) = \tau$.
2. Differentially rotating non-inertial frames without the coordinate singularity of the rotating disk. The embedding defining these frames is as follows:

$$\begin{aligned} &z^\mu(\tau,\sigma^u) = x^\mu(\tau) + \epsilon^\mu_r\, R^r{}_s(\tau,\sigma)\,\sigma^s \to_{\sigma\to\infty} x^\mu(\tau) + \epsilon^\mu_r\,\sigma^r,\\ &R^r{}_s(\tau,\sigma) = R^r{}_s(\alpha_i(\tau,\sigma)) = R^r{}_s(F(\sigma)\,\tilde\alpha_i(\tau)),\\ &\qquad 0 < F(\sigma) < \frac{1}{A\,\sigma}, \qquad \frac{d\,F(\sigma)}{d\sigma} \neq 0 \text{ (Moller conditions)},\\ &z^\mu_\tau(\tau,\sigma^u) = \dot x^\mu(\tau) - \epsilon^\mu_r\, R^r_s(\tau,\sigma)\,\delta^{sw}\,\epsilon_{wuv}\,\sigma^u\,\frac{\Omega^v(\tau,\sigma)}{c},\\ &z^\mu_r(\tau,\sigma^u) = \epsilon^\mu_k\, R^k_v(\tau,\sigma)\left(\delta^v_r + \Omega^v_{(r)u}(\tau,\sigma)\,\sigma^u\right), \end{aligned} \tag{2.5}$$

   where $\sigma = |\vec\sigma|$ and $R^r_s(\alpha_i(\tau,\sigma))$ is a rotation matrix satisfying the asymptotic conditions $R^r_s(\tau,\sigma) \to_{\sigma\to\infty} \delta^r_s$, $\partial_A\, R^r_s(\tau,\sigma) \to_{\sigma\to\infty} 0$, whose Euler angles have the expression $\alpha_i(\tau,\vec{\sigma}) = F(\sigma)\,\tilde{\alpha}_i(\tau)$, $i = 1,2,3$. The unit normal is $l^\mu = \epsilon^\mu_\tau =$ const. and the lapse function is $1+n(\tau,\sigma^u) = \epsilon\,(z^\mu_\tau\, l_\mu)(\tau,\sigma^u) = \epsilon\,\epsilon^\mu_\tau\, \dot{x}_\mu(\tau) > 0$. In Eq. (2.5) one uses the notations $\Omega_{(r)}(\tau,\sigma) = R^{-1}(\tau,\vec{\sigma})\,\partial_r\, R(\tau,\sigma)$ and $\Big(R^{-1}(\tau,\sigma)\,\partial_\tau\, R(\tau,\sigma)\Big)^u_{\;\;v} = \delta^{um}\,\epsilon_{mvr}\,\frac{\Omega^r(\tau,\sigma)}{c}$, with $\Omega^r(\tau,\sigma) = F(\sigma)\,\tilde{\Omega}(\tau,\sigma)\,\hat{n}^r(\tau,\sigma)$ being the angular velocity.[1] The angular velocity vanishes at spatial infinity and has an upper bound proportional to the minimum of the linear velocity $v_l(\tau) = \dot{x}_\mu(\tau)\, l^\mu$ orthogonal to the space-like hyper-planes. When the rotation axis is fixed and $\tilde{\Omega}(\tau,\sigma) = \omega =$ const., a simple choice for the function $F(\sigma)$ is $F(\sigma) = \frac{1}{1+\frac{\omega^2\sigma^2}{c^2}}$.[2]

To evaluate the non-relativistic limit for $c\to\infty$, where $\tau = ct$ with t the absolute Newtonian time, one chooses the gauge function $F(\sigma) = \frac{1}{1+\frac{\omega^2\sigma^2}{c^2}} \to_{c\to\infty} 1 - \frac{\omega^2\sigma^2}{c^2} + O(c^{-4})$. This implies that the corrections to rigidly rotating non-inertial frames coming from Møller conditions are of order $O(c^{-2})$ and become important at the distance from the rotation axis where the horizon problem for rigid rotations appears.

As shown in Ref. [62], global rigid rotations are forbidden in relativistic theories, because, if one uses the embedding $z^\mu(\tau,\sigma^u) = x^\mu(\tau) + \epsilon^\mu_r\, R^r_{\;s}(\tau)\,\sigma^s$ describing a global rigid rotation with angular velocity $\Omega^r = \Omega^r(\tau)$, then the resulting $g_{\tau\tau}(\tau,\sigma^u)$ violates Møller conditions because it vanishes at $\sigma = \sigma_R = \frac{1}{\Omega(\tau)}\Big[\sqrt{\dot{x}^2(\tau) + [\dot{x}_\mu(\tau)\,\epsilon^\mu_r\, R^r_{\;s}(\tau)\,(\hat{\sigma}\times\hat{\Omega}(\tau))^r]^2}\; - \dot{x}_\mu(\tau)\,\epsilon^\mu_r\, R^r_{\;s}(\tau)\,(\hat{\sigma}\times\hat{\Omega}(\tau))^r\Big]$ ($\sigma^u = \sigma\,\hat{\sigma}^u$, $\Omega^r = \Omega\,\hat{\Omega}^r$, $\hat{\sigma}^2 = \hat{\Omega}^2 = 1$). At this distance from the rotation axis the tangential rotational velocity becomes equal to the velocity of light (the horizon problem of the rotating disk). Let us remark that even if in the existing theory of rotating relativistic stars [102] one uses differential rotations, notwithstanding that in the study of the magnetosphere of pulsars often the notion of light cylinder of radius σ_R is still used.

3. The search of admissible 3+1 splittings with non-Euclidean 3-spaces is much more difficult. The simplest case [90] is a parametrization of the embeddings (Eq. 2.1) in terms of Lorentz matrices $\Lambda^A_{\;B}(\tau,\sigma) \to_{\sigma\to\infty} \delta^A_B$ ($\Lambda^A_{\;C}(\tau,\sigma)\,{}^4\eta_{AB}\,\Lambda^B_{\;D}(\tau,\sigma) = {}^4\eta_{CD}$) depending on $\sigma = \sqrt{\sum_r (\sigma^r)^2}$ and with $\Lambda^A_{\;B}(\tau,0)$ finite:[3]

$$
\begin{aligned}
z^\mu(\tau,\sigma^r) &= x^\mu(\tau) + \epsilon^\mu_A\,\Lambda^A_{\;r}(\tau,\sigma)\,\sigma^r \to_{\sigma\to\infty} x^\mu(\tau) + \epsilon^\mu_r\,\sigma^r,\\
z^\mu(\tau,0) &= x^\mu(\tau) = x^\mu_o + \epsilon^\mu_A\, f^A(\tau),
\end{aligned}
$$

[1] $\hat{n}^r(\tau,\sigma)$ defines the instantaneous rotation axis and $0 < \tilde{\Omega}(\tau,\sigma) < 2\max\Big(\dot{\tilde{\alpha}}(\tau), \dot{\tilde{\beta}}(\tau), \dot{\tilde{\gamma}}(\tau)\Big)$.

[2] Nearly rigid rotating systems, like a rotating disk of radius σ_o, can be described by using a function $F(\sigma)$ approximating the step function $\theta(\sigma-\sigma_o)$.

[3] It corresponds to the locality hypothesis of Ref. [55–60], according to which at each instant of time the detectors of an accelerated observer give the same indications as the detectors of the instantaneously comoving inertial observer.

$$\dot{x}^\mu(\tau) = \epsilon^\mu_A\, \dot{f}^A(\tau) = \epsilon^\mu_A\, \alpha(\tau)\, \gamma_x(\tau) \begin{pmatrix} 1 \\ \beta^r_x(\tau) \end{pmatrix}, \qquad \epsilon\, \dot{x}^2(\tau) = \alpha^2(\tau) > 0,$$

$$f^A(\tau) = \int_o^\tau d\tau_1\, \alpha(\tau_1)\, \gamma_x(\tau_1) \begin{pmatrix} 1 \\ \beta^r_x(\tau) \end{pmatrix}, \qquad \gamma_x(\tau) = \frac{1}{\sqrt{1 - \vec{\beta}^2_x(\tau)}}. \tag{2.6}$$

implying $z^\mu_r(\tau, \sigma^u) = \Lambda^\mu_a(\tau, \sigma^u)\, {}^3e_{(a)r}(\tau, \sigma^u)$ and ${}^3g_{rs}(\tau, \sigma^u) = \sum_{a=1}^3 ({}^3e_{(a)r}\, {}^3e_{(a)s})$ (τ, σ^u).

The origin of the 3-coordinates σ^r in the 3-spaces Σ_τ is a time-like accelerated observer with world-line $x^\mu(\tau)$, whose instantaneous 3-velocity, divided by c, is $\vec{\beta}_x(\tau)$ ($|\vec{\beta}_x(\tau)| < 1$) and τ is the proper time of the observer when $\alpha(\tau) = 1$. The 4-velocity of the observer is $u^\mu(\tau) = \epsilon^\mu_A\, \beta^A_x(\tau)/\sqrt{1 - \vec{\beta}^2_x(\tau)}$, $\beta^A_x(\tau) = (1; \beta^r_x(\tau))$.

The Lorentz matrix is written in the form $\Lambda = \mathcal{B}\mathcal{R}$ as the product of a boost $\mathcal{B}(\tau, \sigma)$ and a rotation $\mathcal{R}(\tau, \sigma)$, like the one in Eq. (2.5), ($\mathcal{R}^\tau_\tau = 1$, $\mathcal{R}^\tau_r = 0$, $\mathcal{R}^r_s = R^r_s$). The components of the boost are $\mathcal{B}^\tau_\tau(\tau, \sigma) = \gamma(\tau, \sigma) = 1/\sqrt{1 - \vec{\beta}^2(\tau, \sigma)}$, $\mathcal{B}^\tau_r(\tau, \sigma) = \gamma(\tau, \sigma)\, \beta_r(\tau, \sigma)$, $\mathcal{R}^r_s(\tau, \sigma) = \delta^r_s + \frac{\gamma^2\, \beta^r\, \beta_s}{1+\gamma}(\tau, \sigma)$, with $\beta^r(\tau, \sigma) = G(\sigma)\, \beta^r(\tau)$. The Møller conditions are restrictions on $G(\sigma) \rightarrow_{\sigma \rightarrow \infty} 0$ with $G(0)$ finite, whose explicit form is given in Ref. [90] for the following two cases: (1) boosts with small velocities; and (2) time-independent boosts.

See Ref. [99] for the description of the electromagnetic field and of phenomena like the Sagnac effect and the Faraday rotation in this framework for non-inertial frames. Moreover, the embedding (Eq. 2.5) has been used in Ref. [103] on quantum mechanics in non-inertial frames.

Each admissible 3+1 splitting of space-time allows one to define two associated congruences of time-like observers.

1. The congruence of the Eulerian observers with the unit normal $l^\mu(\tau, \sigma^r) = z^\mu_A(\tau, \sigma^r)\, l^A(\tau, \sigma^r)$ to the 3-spaces embedded in Minkowski space-time as unit 4-velocity. The world-lines of these observers are the integral curves of the unit normal and in general are not geodesics. In adapted radar 4-coordinates the orthonormal tetrads carried by the Eulerian observers are $l^A(\tau, \sigma^r)$, ${}^4\overset{\circ}{\bar{E}}{}^A_{(a)}(\tau, \sigma^r) = (0; {}^3e^u_{(a)}(\tau, \sigma^r))$, where ${}^3e^u_{(a)}(\tau, \sigma^r)$ $(a = 1, 2, 3)$ are triads on the 3-space.

If ${}^4\nabla$ is the covariant derivative associated with the 4-metric ${}^4g_{AB}(\tau, \sigma^r)$ induced by a 3+1 splitting, the equation $(({}^4\nabla_A)^B_C = \delta^B_C\, \partial_A + {}^4\Gamma^B_{AC}$; ${}^3h_{AB}$ $(\tau, \sigma^r) = {}^4g_{AB}(\tau, \sigma^r) - \epsilon\, l_A\, (\tau, \sigma^r) l_B(\tau, \sigma^r))$

$${}^4\nabla_A\, \epsilon\, l_B(\tau, \sigma^r) = \Big(\epsilon\, l_A\, {}^3a_B + \sigma_{AB} + \frac{1}{3}\, \theta\, {}^3h_{AB} - \omega_{AB}\Big)(\tau, \sigma^r) \tag{2.7}$$

defines the acceleration ${}^3a^A(\tau, \sigma^r)$ $({}^3a^A(\tau, \sigma^r)\, l_A(\tau, \sigma^r) = 0)$, the expansion $\theta(\tau, \sigma^r)$, the shear $\sigma_{AB}(\tau, \sigma^r) = \sigma_{BA}(\tau, \sigma^r)$ $(\sigma_{AB}(\tau, \sigma^r)\, l^B(\tau, \sigma^r) = 0)$, and the vorticity or twist $\omega_{AB}(\tau, \sigma^r) = -\omega_{BA}(\tau, \sigma^r)$ $(\omega_{AB}(\tau, \sigma^r)\, l^B(\tau, \sigma^r) = 0)$ of the Eulerian observers with $\omega_{AB}(\tau, \sigma^r) = 0$ since they are surface-forming by construction.

2. The skew congruence with unit 4-velocity $v^\mu(\tau,\sigma^r) = z^\mu_A(\tau,\sigma^r)\, v^A(\tau,\sigma^r)$ (in general it is not surface-forming, i.e., it has a non-vanishing vorticity, like that of a rotating disk). The skew congruence is defined by requiring that its observers have the world-lines (integral curves of the 4-velocity) defined by $\sigma^r = \text{const.}$ for every τ, because the unit 4-velocity tangent to the flux lines $x^\mu_{\vec\sigma_o}(\tau) = z^\mu(\tau,\sigma^r_o)$ is $v^\mu_{\vec\sigma_o}(\tau) = z^\mu_\tau(\tau,\sigma^r_o)/\sqrt{\epsilon\, {}^4g_{\tau\tau}(\tau,\sigma^r_o)}$ (there is no horizon problem because it is everywhere time-like in admissible 3+1 splittings). They carry contro-variant orthonormal tetrads, given in Ref. [104], not adapted to the foliation, connected in each point by a Lorentz transformation to the ones of the Eulerian observer present in this point.

Let us add another motivation for using the 3+1 approach. Let us consider the standard action for a system of N charged scalar particles plus the electromagnetic field:

$$
\begin{aligned}
S &= \sum_{i=1}^N \int d\tau \Big(- m_i c\, \sqrt{\epsilon\, \dot x_i^2(\tau)} - e_i\, \dot x_i^\mu(\tau)\, A_\mu(x_i(\tau))\Big) - \frac{1}{4} \int d^4z\, F^{\mu\nu}(z)\, F_{\mu\nu}(z)\\
&= \int d\tau \Big(L_m + L_I\Big) + \int d^4z\, \mathcal{L}_{em},
\end{aligned} \tag{2.8}
$$

where m_i and e_i are the masses and charges of the particles and $F_{\mu\nu}(z) = \partial_\mu\, A_\nu(z) - \partial_\nu\, A_\mu(z)$. The momenta are

$$
\begin{aligned}
p_{i\mu}(\tau) &= - \frac{\partial (L_m + L_I)(\tau)}{\partial \dot x_i^\mu} = m_i c\, \frac{\dot x^\mu_i(\tau)}{\sqrt{\epsilon\, \dot x_i^2(\tau)}} + e_i\, A_\mu(x_i(\tau)),\\
\pi^\mu(z^o, \vec z) &= - \frac{\partial \mathcal{L}(z^o,\vec z)}{\partial\, \partial_o\, A_\mu(z^o,\vec z)} = F^{o\mu}(z^o,\vec z),
\end{aligned} \tag{2.9}
$$

and we get the following primary constraints:

$$
\begin{aligned}
\bar\phi_i(\tau) &= \epsilon\, \Big(p_i(\tau) - e_i\, A(x_i(\tau))\Big)^2 - m_i^2\, c^2 \approx 0,\\
\pi^o(z^o,\vec z) &\approx 0.
\end{aligned} \tag{2.10}
$$

Since the natural time parameter for the field degrees of freedom is z^o, while the particle world-lines are parametrized by an arbitrary scalar τ, there is no concept of equal time and it is impossible to evaluate the Poisson bracket of these constraints. Also, due to the same reason, the Dirac Hamiltonian, which would be $\bar H_D = \bar H_c + \sum_{i=1}^N \lambda^i(\tau)\, \bar\chi_i(\tau) + \int d^3z \lambda^o(z^o,\vec z)\pi^o(z^o,\vec z)$ with $\bar H_c$ the canonical Hamiltonian and with $\lambda^i(\tau)$, $\lambda^o(z^o,\vec z)$ Dirac's multipliers, does not make sense. This problem is present even at the level of the Euler–Lagrange equations: How does one formulate a Cauchy problem for a system of coupled equations, some of which are ordinary differential equations in the affine parameter τ along the particle world-line, while the others are partial differential equations depending on Minkowski coordinates z^μ?

The standard non-manifestly covariant approach uses z^o as the time parameter by rewriting the action (Eq. 2.8) in the following form:

$$
\begin{aligned}
S &= \int d^4z \left(-\sum_{i=1}^{N} \int d\tau\, \delta^4(x_i(\tau) - z) \left[m_i\, c\, \sqrt{\epsilon\, \dot{x}_i^2(\tau)} + e_i\, \dot{x}_i^\mu(\tau)\, A_\mu(z) \right] \right. \\
&\quad \left. -\frac{1}{4} F^{\mu\nu}(z)\, F_{\mu\nu}(z) \right) \\
&= \int d^4z \left(-\sum_{i=1}^{N} \delta^3(\vec{x}_i(z^o) - \vec{z}) \left[m_i\, c \sqrt{1 - (\frac{d\vec{x}_i(z^o)}{dz^o})^2} + e_i \left[A_o(z) \right.\right.\right. \\
&\quad \left.\left.\left. -\frac{d\vec{x}_i(z^o)}{dz^o} \cdot \vec{A}(z) \right] \right] - \frac{1}{4} F^{\mu\nu}(z)\, F_{\mu\nu}(z) \right),
\end{aligned}
\tag{2.11}
$$

where $\delta(x_i^o(\tau) - z^o) = \delta(\tau - f_i(z^o)) \,/\, |\frac{dx_i^o}{d\tau}|$ has been used as a gauge fixing, eliminating the variables x_i^o. Since the p_i^o are determined by the mass-shell constraints, we remain with six degrees of freedom for particle and no-particle constraint. The net result of this prescription is to break τ-reparametrization invariance.

This problem is due to the lack of a covariant concept of equal time between field and particle variables, so that the 3+1 approach is needed.

2.2 Parametrized Minkowski Theory for Matter Admitting a Lagrangian Description

In the global non-inertial frames of Minkowski space-time it is possible to describe isolated systems (particles, strings, fields, fluids) admitting a Lagrangian formulation by means of parametrized Minkowski theories [98, 99, 105]. The existence of a Lagrangian, which can be coupled to an external gravitational field, makes possible the determination of the matter energy–momentum tensor and of the ten conserved Poincaré generators P^μ and $J^{\mu\nu}$ (assumed finite) of every configuration of the isolated system.

First of all, one must replace the matter variables of the isolated system with new ones knowing the clock synchronization convention defining the 3-spaces Σ_τ. For instance a Klein–Gordon field $\tilde{\phi}(x)$ will be replaced with $\phi(\tau, \sigma^r) = \tilde{\phi}(z(\tau, \sigma^r))$; and the same for every other field. Instead, for a relativistic particle with world-line $x^\mu(\tau)$, one must make a choice of its energy sign, then the positive- (or negative-) energy particle will be described by 3-coordinates $\eta^r(\tau)$ defined by the intersection of its world-line with Σ_τ: $x^\mu(\tau) = z^\mu(\tau, \eta^r(\tau))$. Different from all the previous approaches to relativistic mechanics, the dynamical configuration variables are the 3-coordinates $\eta^r(\tau)$ and not the world-lines $x^\mu(\tau)$ (to rebuild them in an arbitrary frame one needs the embedding defining that frame).

Then, one replaces the external gravitational 4-metric in the coupled Lagrangian with the 4-metric ${}^4g_{AB}(\tau, \sigma^r)$, which is a functional of the embedding defining an admissible 3+1 splitting of Minkowski space-time, and the matter fields with the new ones knowing the instantaneous 3-spaces Σ_τ.

Parametrized Minkowski theories are defined by the resulting Lagrangian depending on the given matter and on the embedding $z^\mu(\tau, \sigma^r)$. The resulting action is invariant under the frame-preserving diffeomorphisms $\tau \mapsto \tau^{'}(\tau, \sigma^u)$, $\sigma^r \mapsto \sigma^{'r}(\sigma^u)$ first introduced in Ref. [106]. As a consequence, there are four first-class constraints with exactly vanishing Poisson brackets (an Abelianized analogue of the super-Hamiltonian and super-momentum constraints of canonical gravity) determining the momenta conjugated to the embeddings in terms of the matter energy–momentum tensor. These constraints ensure the independence of the description from the choice of the foliation and with their help it is possible to rebuild some kind of covariance (Wigner covariance, as we shall see) notwithstanding the 3+1 splitting of flat space-time. This implies that the embeddings $z^\mu(\tau, \sigma^r)$ are gauge variables, so that all the admissible non-inertial or inertial frames are gauge-equivalent, namely physics does not depend on the clock synchronization convention and on the choice of the 3-coordinates σ^r – only the appearances of phenomena change by changing the notion of instantaneous 3-space. Therefore, the embedding configuration variables $z^\mu(\tau, \vec{\sigma})$ will describe all the possible inertial effects compatible with special relativity.

The matter energy–momentum tensor associated with this Lagrangian allows the determination of the ten conserved Poincaré generators P^μ and $J^{\mu\nu}$ of every configuration of the system (in non-inertial frames they are asymptotic generators at spatial infinity, like the ADM ones in general relativity). The behavior of matter at spatial infinity on each 3-space Σ_τ must be such that the Poincaré generators are finite with a 4-momentum not space-like.

As an example, one may consider N free scalar particles with masses m_i. Usually the time-like world-lines of the particles are described by Cartesian 4-coordinates $x_i^\mu(\tau_i)$, depending on the particle proper time in an inertial frame. At the Hamiltonian level one has that the 4-momenta $p_{i\mu}(\tau_i)$ satisfy the mass-shell first-class constraints $\epsilon\, p_i^2(\tau_i) - m_i^2 = 0$. Therefore, there are two solutions with opposite sign for the energies (particles and antiparticles), so that the time components $x_i^o(\tau)$ are gauge variables, sources of the problem of how to eliminate the relative times in relativistic bound states.

In the 3+1 approach the time-like world-lines of the N particles are Cartesian 4-coordinates $x_i^\mu(\tau)$ parametrized with the time τ of Bondi radar 4-coordinates so that we have $x_i^\mu(\tau) = z^\mu(\tau, \eta_i^r(\tau))$, $i = 1, \dots, N$. Therefore, each particle is identified by the three numbers $\sigma^r = \eta_i^r(\tau)$ individuating the intersection of each world-line with the 3-space Σ_τ, and not by four. As a consequence, each particle must have a well-defined sign of the energy, $\eta_i = \mathrm{sign}\, p_i^o = \pm$. Each particle with a definite sign of the energy will be described by the six Lorentz-scalar canonical coordinates $\eta_i^r(\tau)$, $\kappa_{ir}(\tau)$ at the Hamiltonian level, like in Newtonian mechanics.

There are no longer mass-shell constraints, because we cannot describe the two topologically disjoint branches of the mass hyperboloid simultaneously as in the standard manifestly Lorentz-covariant approach. By using Eq. (2.1), the particle 4-velocities can be written in the form $\dot{x}_i^\mu(\tau) = N(\tau, \eta_i^u(\tau))\, l^\mu (\tau, \eta_i^u(\tau)) + [\dot{\eta}_i^r(\tau) + N^r(\tau, \eta_i^u(\tau))]\, z_r^\mu(\tau, \eta_i^u(\tau))$, with $\epsilon\, \dot{x}_i^2(\tau) = \Big[N^2 - {}^3g_{rs} (N^r + \dot{\eta}_i^r(\tau))\, (N^s + \dot{\eta}_i^s)\Big](\tau, \eta_i^u(\tau))$. Therefore the derived usual particle momenta, satisfying $\epsilon\, p_i^2 - m_i^2\, c^2 \approx 0$, are $p_i^\mu(\tau) = m_i\, c\, \dot{x}_i^\mu(\tau)/\sqrt{\epsilon\, \dot{x}_i^2(\tau)} = \eta_i\, \sqrt{m_i^2\, c^2 - {}^3g^{rs}(\tau, \eta_i^u(\tau))\, \kappa_{ir}(\tau)\, \kappa_{is}(\tau)}\, l^\mu(\tau, \eta_i^u(\tau)) + \kappa_{ir}(\tau)\, {}^3g^{rs}(\tau, \eta_i^u(\tau))\, z_s^\mu(\tau, \eta_i^u(\tau))$.

In parametrized Minkowski theories the free particles are described by the following action depending on the configurational variables $\eta_i^r(\tau)$ of the particles with energy sign η_i and on the embedding variables $z^\mu(\tau, \sigma^r)$ of an arbitrary admissible 3+1 splitting of Minkowski space-time:

$$
\begin{aligned}
S &= \int d\tau\, d^3\sigma\, \mathcal{L}(\tau, \sigma^u) = \int d\tau\, L(\tau),\\
\mathcal{L}(\tau, \sigma^u) &= -\sum_{i=1}^N \delta^3(\sigma^u - \eta_i^u(\tau))\\
&\quad m_i c\, \eta_i \sqrt{\epsilon\, [{}^4g_{\tau\tau}(\tau, \sigma^u) + 2\, {}^4g_{\tau r}(\tau, \sigma^u)\, \dot{\eta}_i^r(\tau) + {}^4g_{rs}(\tau, \sigma^u)\, \dot{\eta}_i^r(\tau)\, \dot{\eta}_i^s(\tau)]}\\
&= -\sum_{i=1}^N \delta^3(\sigma^u - \eta_i^u(\tau))\\
&\quad \eta_i\, m_i\, c \sqrt{N^2(\tau, \sigma^u) + \epsilon\, {}^4g_{rs}(\tau, \sigma^u)\, [\dot{\eta}_i^r(\tau) + N^r(\tau, \sigma^u)]\, [\dot{\eta}_i^s(\tau) + N^s(\tau, \sigma^u)]}.
\end{aligned}
\tag{2.12}
$$

This action is invariant under separate τ- and σ^r-reparametrizations.

The resulting canonical momenta and their Poisson brackets are

$$
\begin{aligned}
\rho_\mu(\tau, \sigma^u) &= -\epsilon\, \frac{\partial \mathcal{L}(\tau, \sigma^u)}{\partial z_\tau^\mu(\tau, \sigma^u)} = \sum_{i=1}^N \delta^3(\sigma^u - \eta_i^u(\tau))\, \eta_i\, m_i\, c\\
&\quad \frac{z_{\tau\mu}(\tau, \sigma^u) + z_{r\mu}(\tau, \sigma^u)\, \dot{\eta}_i^r(\tau)}{\sqrt{\epsilon\, [{}^4g_{\tau\tau}(\tau, \sigma^u) + 2\, {}^4g_{\tau r}(\tau, \sigma^u)\, \dot{\eta}_i^r(\tau) + {}^4g_{rs}(\tau, \sigma^u)\, \dot{\eta}_i^r(\tau)\, \dot{\eta}_i^s(\tau)]}}\\
&= [(\rho_\nu\, l^\nu)\, l_\mu + (\rho_\nu\, z_r^\nu)\, {}^3g^{rs}\, z_{s\mu}](\tau, \sigma^u),\\
\kappa_{ir}(\tau) &= -\frac{\partial L(\tau)}{\partial \dot{\eta}_i^r(\tau)} = {}^3g_{rs}(\tau, \eta_i^u(\tau))\, {\kappa_i}^s(\tau)\\
&= \eta_i\, m_i\, c\, \frac{{}^4g_{\tau r}(\tau, \eta_i^u(\tau)) + {}^4g_{rs}(\tau, \eta_i^u(\tau))\, \dot{\eta}_i^s(\tau)}{\sqrt{\epsilon\, [{}^4g_{\tau\tau}(\tau, \eta_i^u(\tau)) + 2\, {}^4g_{\tau r}(\tau, \eta_i^u(\tau))\, \dot{\eta}_i^r(\tau) + {}^4g_{rs}(\tau, \eta_i^u(\tau))\, \dot{\eta}_i^r(\tau)\, \dot{\eta}_i^s(\tau)]}},\\
&\{z^\mu(\tau, \sigma^u), \rho_\nu(\tau, \sigma'^u)\} = -\epsilon\, {}^4\eta^\mu_\nu\, \delta^3(\sigma^u - \sigma'^u),\\
&\{\eta_i^r(\tau), \kappa_{js}(\tau)\} = -\delta_{ij}\, \delta^r_s.
\end{aligned}
\tag{2.13}
$$

The Poincaré generators and the energy–momentum tensor of this system are ($\{z^\mu(\tau,\sigma^u), P^\nu\} = -\epsilon\, {}^4\eta^{\mu\nu}$; in inertial frames we have $T_{\perp\perp} = T^{\tau\tau}$ and $T_{\perp r} = -{}^3g_{rs}\, T^{\tau s}$):

$$
\begin{aligned}
P^\mu &= \int d^3\sigma\, \rho^\mu(\tau,\sigma^u), \qquad J^{\mu\nu} = \int d^3\sigma (z^\mu \rho^\nu - z^\nu \rho^\mu)(\tau,\sigma^u),\\
T^{AB}(\tau,\sigma^u) &= -\frac{2}{\sqrt{-\det\, {}^4g_{CD}(\tau,\sigma^u)}}\, \frac{\delta S}{\delta\, {}^4g_{AB}(\tau,\sigma^u)}, \qquad T^{\mu\nu} = z^\mu_A\, z^\nu_B\, T^{AB},\\
T_{\perp\perp}(\tau,\sigma^u) &= \Big(l_\mu\, l_\nu\, T^{\mu\nu}\Big)(\tau,\sigma^u) = (N\, T^{\tau\tau})(\tau,\sigma^u) = \sum_{i=1}^N \frac{\delta^3(\sigma^u - \eta^u_i(\tau))}{\sqrt{\tilde\phi(\tau,\sigma^u)}}\\
&\qquad \eta_i \sqrt{m_i^2 c^2 + {}^3g^{rs}(\tau,\sigma^u)\, \kappa_{ir}(\tau)\, \kappa_{is}(\tau)},\\
T_{\perp r}(\tau,\sigma^u) &= \Big(l_\mu\, z_{r\nu}\, T^{\mu\nu}\Big)(\tau,\sigma^u)\\
&= -[N\, {}^3g_{rs}\, (T^{\tau\tau}\, N^s + T^{\tau s}](\tau,\sigma^u) = \sum_{i=1}^N \frac{\delta^3(\sigma^u - \eta^u_i(\tau))}{\sqrt{\tilde\phi(\tau,\sigma^u)}}\, \kappa_{ir}(\tau),\\
T_{rs}(\tau,\sigma^u) &= \Big(z_{r\mu}\, z_{s\nu}\, T^{\mu\nu}\Big)(\tau,\sigma^u)\\
&= [N_r\, N_s\, T^{\tau\tau} + (N_r\, {}^3g_{sm} + N_s\, {}^3g_{rm})\, T^{rm} + {}^3g_{rm}\, {}^3g_{sn}\, T^{mn}](\tau,\sigma^u)\\
&= \sum_{i=1}^N \frac{\delta^3(\sigma^u - \eta^u_i(\tau))}{\sqrt{\tilde\phi(\tau,\sigma^u)}}\, \eta_i\, \frac{\kappa_{ir}(\tau)\, \kappa_{is}(\tau)}{\sqrt{m_i^2 c^2 + {}^3g^{vw}(\tau,\sigma^u)\, \kappa_{iv}(\tau)\, \kappa_{iw}(\tau)}}.
\end{aligned} \tag{2.14}
$$

The four first-class constraints implying the gauge nature of the embedding and the gauge equivalence of the description in different non-inertial frames are:[4]

$$
\begin{aligned}
\mathcal{H}_\mu(\tau,\sigma^u) &= \rho_\mu(\tau,\sigma^u) - \sqrt{\tilde\phi(\tau,\sigma^u)}\, \Big[l_\mu\, T_{\perp\perp} - z_{r\mu}\, {}^3g^{rs}\, T_{\perp s}\Big](\tau,\sigma^u) \approx 0,\\
&\{\mathcal{H}_\mu(\tau,\sigma^u), \mathcal{H}_\mu(\tau,\sigma^v)\} = 0,\\
\int d^3\sigma\, \mathcal{H}^\mu(\tau,\sigma^u) &= P^\mu - \sum_{i=1}^N l^\mu(\tau,\eta^u_i(\tau))\, \eta_i \sqrt{m_i^2\, c^2 + {}^3g^{rs}(\tau,\eta^u_i(\tau))\, \kappa_{ir}(\tau)\, \kappa_{is}(\tau)}\\
&\quad - \sum_{i=1}^N z^\mu_r(\tau,\eta^u_i(\tau))\, {}^3g^{rs}(\tau,\eta^u_i(\tau))\, \kappa_{is}(\tau) \approx 0.
\end{aligned} \tag{2.15}
$$

Since the canonical Hamiltonian vanishes, one has the Dirac Hamiltonian ($\lambda^\mu(\tau,\sigma^u)$ are Dirac's multipliers):

[4] Since the constraints (Eq. 2.15) are distributions concentrated at the position of particles, in evaluating their Poisson brackets we must use the embeddings $z^\mu(\tau,\sigma^u)$ at a generic point and not $z^\mu(\tau,\eta^u_i(\tau))$.

$$
\begin{aligned}
H_D &= \int d^3\sigma\, \lambda^\mu(\tau,\sigma^u)\, \mathcal{H}_\mu(\tau,\sigma^u) \\
&= \int d^3\sigma\, \Big(\lambda_\mu(\tau,\sigma^u)\, l^\mu(\tau,\sigma^u)\, \mathcal{H}_\lambda(\tau,\sigma^u) \\
&\quad + \lambda_\mu(\tau,\sigma^u)\, z^\mu_s(\tau,\sigma^u)\, {}^3g^{sr}(\tau,\sigma^u)\, \mathcal{H}_{\lambda\, r}(\tau,\sigma^u)\Big), \\
\mathcal{H}_\lambda(\tau,\sigma^u) &= (l_\mu\, \mathcal{H}^\mu)(\tau,\sigma^u) \approx 0, \\
\mathcal{H}_{\lambda\, r}(\tau,\sigma^u) &= (z_{r\,\mu}\mathcal{H}^\mu)(\tau,\sigma^u) \approx 0, \\
\{\mathcal{H}_{\lambda\, r}(\tau,\sigma^u), \mathcal{H}_{\lambda\, s}(\tau,\sigma^{'u})\} &= \mathcal{H}_{\lambda\, r}(\tau,\sigma^{'u})\, \frac{\partial\delta^3(\sigma^u,\sigma^{'u})}{\partial\sigma^s} + \mathcal{H}_{\lambda\, s}(\tau,\sigma^u)\, \frac{\partial\delta^3(\sigma^u,\sigma^{'u})}{\partial\sigma^r}, \\
\{\mathcal{H}_\lambda(\tau,\sigma^u), \mathcal{H}_{\lambda\, r}(\tau,\sigma^{'u})\} &= \mathcal{H}_\lambda(\tau,\sigma^u)\, \frac{\partial\delta^3(\sigma^u,\sigma^{'u})}{\partial\sigma^r}, \\
\{\mathcal{H}_\lambda(\tau,\sigma^u), \mathcal{H}_\lambda(\tau,\sigma^{'u})\} &= \Big[{}^3g^{rs}(\tau,\sigma^u)\, \mathcal{H}_{\lambda\, s}(\tau,\sigma^u) \\
&\quad + {}^3g^{rs}(\tau,\sigma^{'u})\, \mathcal{H}_{\lambda\, s}(\tau,\sigma^{'u})\Big]\, \frac{\partial\delta^3(\sigma^u,\sigma^{'u})}{\partial\sigma^r},
\end{aligned} \tag{2.16}
$$

and one finds that $\{\mathcal{H}_\mu(\tau,\sigma^u), H_D\} = 0$. Therefore, there are only the four first-class constraints of Eq. (2.16). The constraints $\mathcal{H}_\mu(\tau,\sigma^u) \approx 0$ describe the arbitrariness of the foliation: Physical results do not depend on its choice. In Eq. (2.16) we have also shown the non-holonomic form $\mathcal{H}_\lambda(\tau,\sigma^u) \approx 0$, $\mathcal{H}_{\lambda\, r}(\tau,\sigma^u) \approx 0$ of the constraints, which satisfies the universal Dirac algebra of the super-momentum and super-Hamiltonian constraints of ADM canonical metric gravity (see Chapter 5).

The same description can be given when the matter are not particles but the Klein–Gordon [107] and Dirac [108] fields and for the electromagnetic field [30], starting by their usual description given, for instance, in Ref. [109].

To describe the physics in a given admissible non-inertial frame described by an embedding $z^\mu_F(\tau,\sigma^u)$, one must add the gauge fixings:

$$\zeta^\mu(\tau,\sigma^u) = z^\mu(\tau,\sigma^u) - z^\mu_F(\tau,\sigma^u) \approx 0. \tag{2.17}$$

If we put $z^\mu_F(\tau,\sigma^u) = x^\mu(\tau) + b^\mu_r(\tau)\,\sigma^r$, with $x^\mu(\tau)$ a non-inertial observer (origin of the 3-coordinates σ^r) and with $b^\mu_A(\tau)$ an orthonormal tetrad such that the constant (future-pointing) unit normal to the 3-spaces is $l^\mu = b^\mu_\tau = \epsilon^\mu_{\alpha\beta\gamma}\, b^\alpha_1(\tau)\, b^\beta_2(\tau)\, b^\gamma_3(\tau) = \text{const.}$, we get a foliation of Minkowski space-time with space-like hyper-planes.

If $x^\mu(\tau)$ is the world-line of a time-like inertial observer, the gauge fixings with

$$z^\mu_F(\tau,\sigma^u) = x^\mu(\tau) + b^\mu_r(\tau)\,\sigma^r \tag{2.18}$$

describe inertial frames in Minkowski space-time with $z^\mu_\tau(\tau,\sigma^r) \approx \dot{x}^\mu(\tau) + \dot{b}^\mu_r(\tau)\,\sigma^r$ and $z^\mu_r(\tau) \approx b^\mu_r(\tau)$. Usually the inertial frames have $b^\mu_r(\tau) = \epsilon^\mu_r$ with

ϵ^μ_r constant asymptotic triads. Moreover, we have ${}^3g_{rs}(\tau) = b^\mu_r(\tau)\, {}^4\eta_{\mu\nu}\, b^\nu_s(\tau) = \sum_a {}^3e_{(a)r}(\tau)\, {}^3e_{(a)s}(\tau) = \delta^{rs}$ (in terms of cotriads inside Σ_τ contained in the chosen parametrization of $b^\mu_r(\tau)$) with inverse ${}^3g^{rs}(\tau) = \sum_a {}^3e^r_{(a)}(\tau)\, {}^3e^s_{(a)}(\tau)$ (in terms of the dual triads). The shift and lapse functions have the expression $N_r(\tau, \sigma^u) = -\epsilon\, {}^4g_{\tau r}(\tau, \sigma^u) = (\dot{x}^\mu(\tau) + \dot{b}^\mu_s(\tau)\, \sigma^s)\, {}^4\eta_{\mu\nu}\, b^\nu_r(\tau)$, $N^2(\tau, \sigma^u) = (N^r\, N_r)(\tau, \sigma) + (\dot{x}^\mu(\tau) + \dot{b}^\mu_r(\tau)\, \sigma^r)\, {}^4\eta_{\mu\nu}\, (\dot{x}^\nu(\tau) + \dot{b}^\nu_s(\tau)\, \sigma^s)$.

The second-class constraints implied by the gauge fixings (Eq. 2.17) together with the constraints $\mathcal{H}_\mu(\tau, \sigma^u) \approx 0$ lead to the following Dirac brackets:

$$\{f, g\}^* = \{f, g\} - \int d^3\sigma \Big[\{f, \zeta^\mu(\tau, \sigma^u)\}\, \{\mathcal{H}_\mu(\tau, \sigma^u), g\} - \{f, \mathcal{H}_\mu(\tau, \sigma^u)\}\, \{\zeta^\mu(\tau, \sigma^u), g\}\Big]. \tag{2.19}$$

The preservation in time of the gauge fixings ($\frac{d}{d\tau}\, \zeta^\mu(\tau, \sigma^u) = \{\zeta^\mu(\tau, \sigma^u), H_D\} \approx 0$) implies the following form of the Dirac multipliers: $\lambda^\mu(\tau, \sigma^u) = \tilde{\lambda}^\mu(\tau) + \tilde{\lambda}^\mu_\nu(\tau)\, b^\nu_r(\tau)\, \sigma^r$, with $\tilde{\lambda}^\mu(\tau) = -\dot{x}^\mu(\tau)$, $\tilde{\lambda}^{\mu\nu}(\tau) = \frac{1}{2}\, [\dot{b}^\mu_r(\tau)\, b^\nu_r(\tau) - b^\mu_r(\tau)\, \dot{b}^\nu_r(\tau)]$.

The space-like hyper-planes still depend on ten residual gauge degrees of freedom: (1) the world-line $x^\mu(\tau)$ of the inertial observer chosen as the origin of the 3-coordinates σ^r; and (2) six variables parametrizing an orthonormal tetrad $b^\mu_A(\tau)$. Their ten conjugate variables are contained in the canonical momenta $\rho_\mu(\tau, \sigma^u)$ conjugated to the embedding and are (1) the total 4-momentum P^μ canonically conjugate to $x^\mu(\tau)$, $\{x^\mu, P^\nu\}^* = -\epsilon\, {}^4\eta^{\mu\nu}$; and (2) six momentum variables canonically conjugate to the tetrads $b^\mu_A(\tau)$. They are determined by the constraints (Eq. 2.15). Therefore, after this gauge fixing the Dirac Hamiltonian depends only on the following ten surviving first-class constraints implied by Eq. (2.15):

$$H_D = \tilde{\lambda}^\mu(\tau)\, \tilde{\mathcal{H}}_\mu(\tau) - \frac{1}{2}\, \tilde{\lambda}^{\mu\nu}(\tau)\, \tilde{\mathcal{H}}_{\mu\nu}(\tau),$$

$$\tilde{\mathcal{H}}^\mu(\tau) = \int d^3\sigma\, \mathcal{H}^\mu(\tau, \vec{\sigma}) = P^\mu - l^\mu \sum_{i=1}^N \eta_i\, \sqrt{m_i^2\, c^2 + {}^3g^{rs}(\tau)\, \kappa_{ir}(\tau)\, \kappa_{is}(\tau)} - b^\mu_r(\tau) \sum_{i=1}^N {}^3g^{rs}(\tau)\, \kappa_{is}(\tau) \approx 0,$$

$$\begin{aligned}\tilde{\mathcal{H}}^{\mu\nu}(\tau) &= \int d^3\sigma\, \sigma^r\, \Big[b^\mu_r(\tau)\, \mathcal{H}^\nu(\tau, \vec{\sigma}) - b^\nu_r(\tau)\, \mathcal{H}^\mu(\tau, \vec{\sigma})\Big] \\ &= S^{\mu\nu} - (b^\mu_r(\tau)\, l^\nu - b^\nu(\tau)\, l^\mu) \sum_{i=1}^N \eta^r_i(\tau)) \\ &\quad \eta_i\, \sqrt{m_i^2\, c^2 + {}^3g^{rs}(\tau, \eta^u_i(\tau))\, \kappa_{ir}(\tau)\, \kappa_{is}(\tau)} \\ &\quad - (b^\mu_r(\tau)\, b^\nu_s(\tau) - b^\nu_r(\tau)\, b^\mu_s(\tau)) \sum_{i=1}^N \eta^r_i(\tau))\, \kappa_{is}(\tau) \approx 0,\end{aligned}$$

$$\{\tilde{\mathcal{H}}^\mu(\tau), \tilde{\mathcal{H}}^\nu(\tau)\} = \{\tilde{\mathcal{H}}^\alpha(\tau), \tilde{\mathcal{H}}^{\mu\nu}(\tau)\} = 0,$$
$$\{\tilde{\mathcal{H}}^{\mu\nu}(\tau), \tilde{\mathcal{H}}^{\alpha\beta}(\tau)\} = C^{\mu\nu\alpha\beta}_{\gamma\delta}\, \tilde{\mathcal{H}}^{\gamma\delta}(\tau), \tag{2.20}$$

with $S^{\mu\nu}$ the spin part of the Lorentz generators whose form, implied by Eqs. (2.14) and (2.18), is

$$J^{\mu\nu} = x^\mu(\tau)\, P^\nu - x^\nu(\tau)\, P^\mu + S^{\mu\nu},$$
$$S^{\mu\nu} = \int d^3\sigma\, \sigma^r \Big[b^\mu_r(\tau)\, \rho^\nu(\tau, \vec{\sigma}) - b^\nu_r(\tau)\, \rho^\mu(\tau, \vec{\sigma}) \Big], \tag{2.21}$$

and with $C^{\mu\nu\alpha\beta}_{\gamma\delta} = \delta^\nu_\gamma\, \delta^\alpha_\delta\, {}^4\eta^{\mu\beta} + \delta^\mu_\gamma\, \delta^\beta_\delta\, {}^4\eta^{\nu\alpha} - \delta^\nu_\gamma\, \delta^\beta_\delta\, {}^4\eta^{\mu\alpha} - \delta^\mu_\gamma\, \delta^\alpha_\delta\, {}^4\eta^{\nu\beta}$ being the structure constants of the Lorentz algebra.

Since P^μ is the total conserved 4-momentum of the isolated system, the conjugate variable $x^\mu(\tau)$ describes an inertial observer taking into account its global properties, namely having the role of some kind of relativistic center of mass of the isolated system (see next chapter).

From now on we will use the notation $P^\mu = (P^o = E/c; \vec{P}) = M\, c\, u^\mu(P) = Mc\,(\sqrt{1+\vec{h}^2}; \vec{h}) \stackrel{def}{=} Mc\, h^\mu$, where $\vec{h} = \vec{v}/c$ is an a-dimensional 3-velocity and with $M\, c = \sqrt{\epsilon\, P^2}$.

For an isolated system with total conserved time-like 4-momentum P^μ, the "inertial rest-frame" is the inertial 3+1 splitting with $l^\mu = P^\mu/\sqrt{\epsilon\, P^2}$ and with $b^\mu_r(\tau) = \epsilon^\mu_r$, where $l^\mu = \epsilon^\mu_o$ and ϵ^μ_r are constant orthonormal tetrads.

Instead, the "non-inertial rest-frames" of an isolated system with total conserved time-like 4-momentum P^μ are those admissible non-inertial 3+1 splittings whose 3-spaces Σ_τ tend to space-like hyper-planes perpendicular to P^μ at spatial infinity. In these non-inertial rest-frames the Poincaré generators are asymptotic (constant of the motion) symmetry generators like the asymptotic ADM ones in the asymptotically Minkowskian space-times of general relativity (see Part II).

The non-relativistic limit of the parametrized Minkowski theories gives the parametrized Galilei theories, which are defined and studied in Ref. [51] using also the results of Refs. [49, 50]. Also the inertial and non-inertial frames in Galilei space-time of Section 1.1 are gauge-equivalent in this formulation. In this approach, a non-relativistic particle of 3-coordinates $x^a(t)$ in an inertial frame is described in the non-inertial frames by 3-coordinates $\eta^u(t)$ defined by the equation $x^a(t) = \mathcal{A}^a(t, \eta^u(t))$, where $\mathcal{A}^a$ is the function describing the non-relativistic non-inertial frame (see after Eq. (1.1)). Therefore the standard velocity takes the form

$$\dot{x}^a(t) = \frac{d\mathcal{A}^a(t, \eta^u(t))}{dt} = \frac{\partial\, \mathcal{A}^a(t, \eta^u(t))}{\partial t} + J^a_r(t, \eta^u(t))\, \frac{d\, \eta^u(t)}{dt}. \tag{2.22}$$

In the case of N particles $x^a_i(t), i = 1, \dots, N$, interacting with arbitrary potential $V(t, \vec{x}_1(t), \dots, \vec{x}_N(t))$ the equation of motion $m_i\, \frac{d^2 x^a_i(t)}{dt^2} = -\frac{\partial\, V}{\partial\, x^a_i}(t, \vec{x}_1(t), \dots, \vec{x}_N(t))$ take the form ($\tilde{J}^r_a$ is the inverse of J^a_r)

$$\begin{aligned} m_i \quad & \frac{d}{dt}\Big[\frac{\partial \mathcal{A}^a(t, \eta_i^u(t))}{\partial t} + J^a_r(t, \eta_i^u(t))\, \frac{d\, \eta_i^r(t)}{dt}\Big] \\ & = -\frac{\partial \hat{V}}{\partial x_i^a}|_{x_i^u(t) = \mathcal{A}^u(t, \eta_i^v(t))} = - \tilde{J}^r_a(t, \eta_i^u(t)) \sum_j \frac{\partial \hat{V}}{\partial \eta_j^r}, \end{aligned} \tag{2.23}$$

with $\hat{V} = V_{\vec{x}_i(t) = \vec{\mathcal{A}}(t, \vec{\eta}_i(t))}$.

The Lagrangian of parametrized Galilei theory is

$$L(t) = \int d^3\sigma \sum_{ia} \delta^3(\sigma^u - \eta_i^u(t))\, \frac{1}{2m_i}\, \left(\frac{\partial \mathcal{A}^a(t, \sigma^u)}{\partial t} + J^a_r(t, \sigma^u)\, \frac{d\eta_i^r(t)}{dt}\right)^2 - \hat{V}. \tag{2.24}$$

It depends on the particles and on the non-inertial frame variable $\mathcal{A}^a(t, \sigma^u)$. As shown in Ref. [51] the action $S = \int dt\, L(t)$ is invariant under local Noether transformations $\delta\, \eta_i^k(t) = F^a(t, \eta_i^v(t))\, \tilde{J}^r_a(t, \eta_i^v(t))$, $\delta\, \mathcal{A}^a(t, \sigma^k) = F^a(t, \sigma^k)$ with $F^a(t, \sigma^k)$ arbitrary functions.

Therefore, these are first-class constraints implying that the functions $\mathcal{A}^a(t, \sigma^r)$ are gauge variables. As a consequence, the description of physics in inertial and non-inertial frames is connected by gauge transformations, like in the relativistic case.

3

Relativistic Dynamics and the Relativistic Center of Mass

In this chapter I face the problem of how to get a consistent description of relativistic particle dynamics, eliminating the problem of relative times in relativistic bound states and clarifying the endless problem of which definition of the relativistic center of mass has to be used.

After a review of the attempts to describe interacting relativistic particles taking into account the no-interaction theorem, I will show that the dynamics must be formulated in terms of suitable covariant relative variables after the separation of an external canonical but not covariant center of mass.

The world-tube in Minkowski space-time, where the non-covariance effects are concentrated, has an intrinsic radius, the Møller radius, determined by the value of the Poincaré Casimir invariants associated to the given configuration of the isolated system. It exists due to the Lorentz signature of Minkowski space-time. It is a classical unit of length determined by the system itself, which exists also for classical fields and is a natural candidate for a ultraviolet cutoff in quantization.

To describe the relative motions of an isolated system of interacting particles in a covariant way in special relativity (SR) I use the inertial foliation of Minkowski space-time in which the 3-spaces are space-like hyper-planes orthogonal to the time-like (assumed finite) 4-momentum of the isolated system, namely its inertial rest-frame.

This allows defining canonical bases of Wigner-covariant relative variables inside the 3-spaces (named Wigner hyper-planes) of these rest-frames and defining the Wigner-covariant rest-frame instant form of dynamics. As a consequence, I can develop a new Hamiltonian kinematics for positive-energy scalar and spinning particles.

Before introducing these new developments, let us explore a brief historical review of the theories involving interacting relativistic point particles.

The theory of relativistic bound states and the interpretational problems with the Bethe–Salpeter equation [109] require the understanding of the instantaneous approximations to quantum field theory so as to arrive at an effective relativistic wave equation and to an acceptable scalar product. In turn, the wave equation must also result from the quantization of a relativistic action-at-a-distance two-body problem in relativistic mechanics (but with the particles interpreted as asymptotic states of quantum field theory), since only in this way can we get a solution to the interpretational problems connected to the gauge nature of the relative times.

Usually the particle world-lines $q_i^\mu(\tau_i)$, $i = 1, \ldots, N$, are parametrized with independent affine parameters τ_i and the action principles describing them are invariant under separate reparametrizations of each world-line, since this is geometrically possible even in the presence of interactions with a finite time delay to avoid instantaneous action at a distance. Since the dynamical correlation among the points on the particle's world-lines is not in general one-to-one in these approaches, it is impossible to develop a Hamiltonian formulation starting from the Euler–Lagrange integro-differential equations of motion implied by the delay. The natural development of these approaches was field theory – for instance the study of the coupled system of relativistic charged particles plus the electromagnetic field.

As an alternative to field theory, there was the development of relativistic mechanics with action-at-a-distance interactions described by suitable potentials implying a one-to-one correlation among the world-lines [110–113]. Geometrically each particle has its world-line described by a 4-vector (the 4-position) $q_i^\mu(\tau_i)$, $i = 1, \ldots, N$, parametrized with an independent arbitrary affine scalar parameter τ_i.[1] By inverting $q_i^o(\tau_i)$ to get $\tau_i = \tau_i(q_i^o)$, we can identify the world-line in a non-manifestly covariant way with $\vec{q}_i = \vec{q}_i(q_i^o)$: In this form they are named predictive coordinates. The instant form amounts to putting $q_1^o = \ldots = q_N^o = x^o$ and to describing the world-lines with the functions $\vec{q}_i(x^o)$. Each one of these configuration variables has a different associated notion of velocity: $\frac{dq_i^\mu(\tau_i)}{d\tau_i}$ (or $\frac{dq_i^\mu(\tau)}{d\tau}$), $\frac{d\vec{q}_i(q_i^o)}{dq_i^o}$ (predictive velocities), $\frac{d\vec{q}_i(x^o)}{dx^o}$; and of acceleration: $\frac{d^2 q_i^\mu(\tau_i)}{(d\tau_i)^2}$ (or $\frac{d^2 q_i^\mu(\tau)}{d\tau^2}$), $\frac{d^2\vec{q}_i(q_i^o)}{(dq_i^o)^2}$ (predictive accelerations), $\frac{d^2\vec{q}_i(x^o)}{(dx^o)^2}$.

Bel's non-manifestly covariant predictive mechanics [114–117] is the attempt to describe relativistic mechanics with N-time predictive equations of motion for the predictive coordinates $\vec{q}_i(q_i^o)$ in Newtonian form: $m_i \frac{d^2\vec{q}_i(q_i^o)}{(dq_i^o)^2} \overset{\circ}{=} \vec{\mathcal{F}}_i(q_k^o, \vec{q}_k(q_k^o), \frac{d\vec{q}_k(q_k^o)}{dq_k^o})$. Since the left-hand side of these equations depends only on q_i^o, the predictive forces must satisfy the predictive conditions $\frac{d\vec{\mathcal{F}}_i}{dq_k^o} = 0$ for $k \neq i$. Moreover they must be invariant under space translations and behave like space three-vectors under spatial rotations. Finally, they must satisfy

[1] The standard choice in the manifestly covariant approach with a $4N$-dimensional configuration space is $\tau_1 = \ldots = \tau_N = \tau$. Another possibility is the choice of proper times $\tau_i = \tau_{iPT}$.

the Currie–Hill equations [118, 119] (or *Currie–Hill world-line conditions*), whose satisfaction implies that the predictive positions $\vec{q}_i(x^o)$ behave under Lorentz boosts like the spatial components of 4-vectors. Bel [114–117] proved that these equations constitute the necessary and sufficient conditions that guarantee that the dynamics is Lorentz invariant with respect to finite Lorentz transformations. However, the Currie–Hill equations are so non-linear that it is practically impossible to find consistent predictive forces and develop this point of view.

The first well-posed Hamiltonian formulation of relativistic mechanics was given by Dirac [89] with the instant, front (or light), and point forms of relativistic Hamiltonian dynamics and the associated canonical realizations of the Poincaré algebra. His non-manifestly covariant Hamiltonian instant form has a well-defined non-relativistic limit and $1/c$ expansions containing the deviations (potentials) from the free case.

However, the development of Hamiltonian models was blocked by the no-interaction theorem of Currie, Jordan, and Sudarshan [120–122] (see Ref. [123] for a review). Its original form was formulated in the Hamiltonian Dirac instant form in the $6N$-dimensional phase space $\left(\vec{q}_i(x^o), \vec{p}_i(x^o)\right)$ of N particles. The no-interaction theorem states that in the hypotheses: (1) the configuration variables $\vec{q}_i(x^o)$ are canonical, i.e., $\{\vec{q}_i(x^o), \vec{q}_j(x^o)\} = 0$; (2) the Lorentz boosts can be implemented as canonical transformations (existence of a canonical realization of the Poincaré group) and the $\vec{q}_i(x^o)$ are the space components of 4-vectors; and (3) the system is non-singular (the transformation from positions and velocity to canonical coordinates is non-singular; the existence of a Lagrangian it is not assumed). This implies only free motion.

As a consequence of the theorem, if we denote $x_i^\mu(\tau)$, $p_{i\mu}(\tau)$ the canonical coordinates of the manifestly covariant approach and $\vec{x}_i(x^o)$, $\vec{p}_i(x^o)$ their equal time restriction in the instant form, we have $\vec{x}_i(x^o) \neq \vec{q}_i(x^o)$ except for free motion. Let us remark that, since the manifestly covariant approach gives the classical basis for the theory of covariant wave equations, the 4-coordinates $x_i^\mu(\tau)$ (and not the geometrical 4-positions $q_i^\mu(\tau)$) are the coordinates locally minimally coupled to external fields.

Many attempts were made to avoid this theorem by relaxing one of its hypotheses or by renouncing the concept of the world-line till when Droz Vincent's many-time Hamiltonian formalism [124–128] (a refinement of the manifestly covariant non-manifestly predictive approach; it is the origin of the multi-temporal equations of Ref. [48]), Todorov's quasi-potential approach to bound states [130–133], and Komar's study of toy models for general relativity (GR) [134–137] converged toward manifestly covariant models based on singular Lagrangians and/or the Dirac–Bergmann theory of constraints [138–140].

Since the Lagrangian formulation is usually not known, a system of N relativistic scalar particles is usually described in a manifestly covariant

$8N$-dimensional phase space with coordinates $\left(x_i^\mu(\tau), p_{i\mu}(\tau)\right)$ $[\{x_i^\mu(\tau), p_{j\nu}(\tau)\} = -\delta_{ij}\,\delta_\nu^\mu$, $\{x_i^\mu(\tau), x_j^\nu(\tau)\} = \{p_{i\mu}(\tau), p_{j\nu}(\tau)\} = 0]$, where τ is a scalar evolution parameter. The description is independent of the choice of τ: The Lagrangian (even if usually not explicitly known) is assumed τ-reparametrization invariant, so that at the Hamiltonian level the canonical Hamiltonian vanishes identically, $H_c \equiv 0$. Since the physical degrees of freedom for N scalar particles are $6N$, there are constraints, which, in the case of N free scalar particles of mass m_i are just the mass-shell conditions $\phi_i(q,p) = \epsilon\, p_i^2 - m_i^2 \approx 0$, $i = 1, \ldots, N$. These constraints say that the time variables $x_i^o(\tau)$ are the gauge variables of a τ-reparametrization invariant theory with canonical Hamiltonian $H_c \equiv 0$. The Dirac Hamiltonian is $H_D = \sum_{i=1}^N \lambda^i(\tau)\,\phi_i$ if all the first-class constraints are primary. The final constraint sub-manifold is the union of 2^N (for generic masses m_i) disjoint sub-manifolds corresponding to the choice of either the positive- or negative-energy branch of each two-sheeted mass-shell hyperboloid. Each branch is a non-compact sub-manifold of phase space on which each particle has a well-defined sign of the energy and 2^N is a topological number (the zeroth homotopy class of the constraint sub-manifold).[2]

However, only in the case of two-body systems it is known how to introduce interactions (due to the Droz Vincent–Todorov–Komar model) with an arbitrary action-at-a-distance interaction instantaneous in the rest-frame described by the two first-class constraints $\phi_i = \epsilon\, p_i^2 - m_i^2\, c^2 + V(r_\perp^2) \approx 0$, $i = 1, 2$, with $r_\perp^\mu = (\eta^{\mu\nu} - p^\mu p^\nu/\epsilon\, p^2) r_\nu$, $r^\mu = x_1^\mu - x_2^\mu$, $p_\mu = p_{1\mu} + p_{2\mu}$. For $N > 2$ a closed form of the N first-class constraints is not known explicitly (there is only an existence proof): Only versions of the model with explicit gauge fixings, so that all the constraints except one are second class, are known.

This model has been completely understood both at the classical and quantum level [139]. The no-interaction theorem is initially avoided due to the singular nature of the Lagrangian: There is a canonical realization of the Poincaré group and the canonical coordinates x_i^μ are 4-vectors. However, when we restrict ourselves to the constraint sub-manifold and look for canonical coordinates adapted to it and to the Poincaré group, it turns out that among the final canonical coordinates will always appear the canonical non-covariant center of mass of the particle system. Therefore, all these models have the following properties: (1) the canonical and predictive 4-positions do not coincide (except in the free case); and (2) the decoupled canonical center of mass is not covariant.

Let us see how to overcome these problems.

[2] Let us remark that when the particles are coupled to weak external fields the 2^N sub-manifolds are deformed but remain disjoint. But when the strength of the external fields increases, the various sub-manifolds may intersect each other and this topological discontinuity is the signal that we are entering a non-classical regime where quantum pair production becomes relevant due to the disappearance of mass gaps.

3.1 The Wigner-Covariant Rest-Frame Instant Form of Dynamics for Isolated Systems

Given an inertial observer described by a 4-coordinate $x^\mu(\tau)$ assumed canonically conjugated to the finite time-like 4-momentum P^μ, $\epsilon\, P^2 > 0$ ($\{x^\mu, P^\nu\} = -\epsilon\, {}^4\eta^{\mu\nu}$) of an isolated system to get the inertial frame of the rest-frame instant form, we must add the following gauge fixings on the orthonormal tetrads $b^\mu_A(\tau)$ appearing in Eq. (2.18) and use the following new Dirac brackets:

$$\begin{aligned}
&\delta^\mu_A(\tau) = b^\mu_A(\tau) - \epsilon^\mu_A(u(P)) \approx 0, \quad \Rightarrow z^\mu_F(\tau, \sigma^u) = x^\mu(\tau) + \epsilon^\mu_r(u(P))\,\sigma^r,\\
&\{f, g\}^{**} = \{f, g\}^* - \frac{1}{4}\Big[\{f, \tilde{\mathcal{H}}^{\mu\nu}\}^*\Big({}^4\eta_{\mu\sigma}\,\epsilon^A_\nu(u(P)) - {}^4\eta_{\nu\sigma}\,\epsilon^D_\mu(u(P))\Big)\{\delta^\sigma_D, g\}^*\\
&\qquad + \{f, \delta^\sigma_D\}^*\Big({}^4\eta_{\sigma\nu}\,\epsilon^D_\mu(u(P)) - {}^4\eta_{\sigma\mu}\,\epsilon^D_\nu(u(P))\Big)\{\tilde{\mathcal{H}}^{\mu\nu}, g\}^*\Big],\\
&\{x^\mu, P^\nu\}^{**} = -\epsilon\,{}^4\eta^{\mu\nu}, \quad \{\eta^r_i, \kappa_{js}\}^{**} = \delta_{ij}\,\delta^r_s,
\end{aligned} \tag{3.1}$$

where $\epsilon^\mu_A(u(P)) = \epsilon^\mu_A(\vec{h}) = L^\mu_A(P, \overset{\circ}{P})$ ($\epsilon^A_\mu(u(P))\,\epsilon^\nu_A(u(P)) = \delta^\nu_\mu$) are the columns of the standard Wigner boost sending the time-like 4-momentum P^μ to its rest-frame form $P^\mu = L^\mu_\nu(P, \overset{\circ}{P})\, \overset{\circ}{P}{}^\nu$, $\overset{\circ}{P}{}^\mu = M\, c\,(1; \vec{0})$, which are given in Eqs. (A.7)–(A.10) of Appendix A. In the Euclidean 3-spaces of the rest-frame, the Lorentz scalar 3-coordinates σ^r can be chosen as Cartesian 3-vectors $\vec{\sigma}$.

As a consequence of the P^μ-dependence of the gauge fixings (Eq. 3.1), as shown in Ref. [105], the Lorentz-scalar 3-vectors $\eta^r_i(\tau)$, $\kappa_{ir}(\tau)$ giving the Hamiltonian description of particles in the rest-frame ($x^\mu_i(\tau) = x^\mu(\tau) + \epsilon^\mu_r(u(P))\,\eta^r_i(\tau)$; $\kappa_{ir}(\tau)$ is given in Eq. (2.13), restricted to the rest-frame) become Wigner spin-1 3-vectors. They transform under the Wigner rotations of Eqs. (A.11) and (A.12) of Appendix A. If a Lorentz transformation $x'^\mu_i(\tau) = \Lambda^\mu_\nu\, x^\nu_i(\tau)$ is done, we get $\eta^r_i(\tau) \rightarrow \eta'^r_i(\tau) = R^r_s(\Lambda, P)\,\eta^s_i(\tau)$, $\epsilon^\mu_r(u(P))\,\eta^r_i(\tau) \rightarrow \Lambda^\mu_\nu\,\epsilon^\nu_r(u(\Lambda\, P))\,\eta'^r_i(\tau)$. Therefore the scalar product of two such vectors is a Lorentz scalar. The same happens for all the 3-vectors living inside these instantaneous 3-spaces. Therefore the instantaneous 3-spaces Σ_τ are orthogonal to P^μ: They are named Wigner hyper-planes.

From now on, $i, j \dots$ will denote Euclidean indices, while $r, s \dots$ will denote Wigner spin-1 indices. With the notation introduced after Eq. (2.20) with $\vec{h} = \vec{v}/c = \vec{P}/Mc$ we have $\epsilon^\mu_\tau(u(P)) = u^\mu(P) = P^\mu/Mc = \epsilon^\mu_\tau(\vec{h}) = h^\mu$, $\epsilon^\mu_r(u(P)) = \Big(-h_r; \delta^i_r - \frac{h^i\, h_r}{1+\sqrt{1+\vec{h}^2}}\Big) = \epsilon^\mu_r(\vec{h}) = (\sqrt{1+\vec{h}^2}; \vec{h})$, ${}^4\eta^{\mu\nu} = \epsilon^\mu_A(\vec{h})\,{}^4\eta^{AB}\,\epsilon^\nu_B(\vec{h}) = \epsilon\,(h^\mu\, h^\nu - \epsilon^\mu_r(\vec{h})\,\epsilon^\nu_r(\vec{h}))$.

The P^μ-dependent gauge fixing (Eq. 3.1) changes the interpretation of the residual gauge variables, because the constraints (Eq. 2.15) are reduced to the following remaining four first-class constraints (being in an inertial frame we have $T^{\tau\tau} = T_{\perp\perp}$ and $T^{\tau r} = -{}^3g^{rs}\, T_{\perp s}$):

$$\begin{aligned}
\tilde{\mathcal{H}}^\mu(\tau) &= \int d^3\sigma\, \mathcal{H}^\mu(\tau,\vec{\sigma})\\
&= P^\mu - u^\mu(P) \int d^3\sigma\, T^{\tau\tau}(\tau,\vec{\sigma}) - \epsilon^\mu_r(u(P)) \int d^3\sigma\, T^{\tau r}(\tau,\vec{\sigma}) \approx 0,\\
&\Downarrow\\
&\quad Mc = \sqrt{\epsilon\, P^2} \approx \int d^3\sigma\, T^{\tau\tau}(\tau,\vec{\sigma}),\\
\mathcal{P}^r_{(int)} &= \int d^3\sigma\, T^{\tau r}(\tau,\vec{\sigma}) \approx 0,
\end{aligned} \tag{3.2}$$

which implies that M is the mass of the isolated system and that its total 3-momentum vanishes in the global rest-frame.

At this stage, only the four degrees of freedom $x^\mu(\tau)$ of the original embedding $z^\mu(\tau,\sigma^r)$ are still free parameters. However, due to the dependence of the gauge fixings (Eq. 3.1) upon P^μ, the final canonical gauge variable conjugated to P^μ is not the 4-vector x^μ, but the following non-covariant position variable $\tilde{x}^\mu(\tau)$ [105] (see Section A.5 of Appendix A; T_s is the Lorentz-scalar rest time):

$$\begin{aligned}
\tilde{x}^\mu &= x^\mu - \frac{1}{M\,c\,(P^o + M\,c)} \Big[P_\nu\, S^{\nu\mu} + M\,c\,\Big(S^{o\mu} - S^{o\nu}\, \frac{P_\nu\, P^\mu}{M^2\,c^2}\Big)\Big],\\
u(P)\cdot\tilde{x} &= u(P)\cdot x = c\,T_s, \qquad \{\tilde{x}^\mu(\tau), P^\nu\} = -{}^4\eta^{\mu\nu},\\
\{T_s, M\} &= \{u(P)\cdot\tilde{x}, \sqrt{\epsilon\, P^2}\} = -\epsilon.
\end{aligned} \tag{3.3}$$

Let us remark that, since the Poincaré generators P^μ, $J^{\mu\nu}$ are global quantities (they know the whole instantaneous 3-space), collective variables like $\tilde{x}^\mu(\tau)$, being defined in terms of them, are also global non-local quantities: As a consequence, they are non-measurable with local means [98–101, 141–143] (see also the review papers in Refs. [144–147]). This is a fundamental difference from the non-relativistic 3-center of mass.

As shown in Ref. [105] after the gauge fixing (Eq. 3.1), the final form of the external Poincaré generators (Eq. 2.14) of an arbitrary isolated system in the rest-frame instant form is:

$$\begin{aligned}
P^\mu,\ J^{\mu\nu} &= \tilde{x}^\mu\, P^\nu - \tilde{x}^\nu\, P^\mu + \tilde{S}^{\mu\nu},\\
P^o &= \sqrt{M^2\,c^2 + \vec{P}^2} = Mc\,\sqrt{1+\vec{h}^2}, \qquad \vec{P} = Mc\,\vec{h},\\
\tilde{S}^r &= \bar{S}^r = \frac{1}{2}\,\epsilon^{ruv}\,\bar{S}^{uv}, \qquad \tilde{S}^{0r} = \frac{\epsilon^{rsu}\, P^s\, \bar{S}^u}{Mc + P^o},\\
J^{ij} &= \tilde{x}^i\, P^j - \tilde{x}^j\, P^i + \epsilon^{iju}\,\bar{S}^u = z^i\, h^j - z^j\, h^i + \epsilon^{iju}\,\bar{S}^u,\\
K^i &= J^{oi} = \tilde{x}^o\, P^i - \tilde{x}^i\,\sqrt{M^2\,c^2 + \vec{P}^2} - \frac{\epsilon^{isu}\, P^s\, \bar{S}^u}{M\,c + \sqrt{M^2\,c^2 + \vec{P}^2}}\\
&= -\sqrt{1+\vec{h}^2}\, z^i + \frac{(\vec{\bar{S}}\times\vec{h})^i}{1+\sqrt{1+\vec{h}^2}},
\end{aligned} \tag{3.4}$$

with $\tilde{S}^{\mu\nu}$ function only of the 3-spin $\bar{S}^r$ of the rest spin tensor $\bar{S}_{AB} = \epsilon^\mu_A(u(P))\,\epsilon^\nu_B(u(P))\,S_{\mu\nu}$ (both the spin tensors $\tilde{S}^{\mu\nu}$ and $\bar{S}^{AB}$ satisfy the Lorentz algebra). It is this external realization that implements the Wigner rotations on the Wigner hyper-planes through the last term in the Lorentz boosts.

Note that both $\tilde{L}^{\mu\nu} = \tilde{x}^\mu\,P^\nu - \tilde{x}^\nu\,P^\mu$ and $\tilde{S}^{\mu\nu} = J^{\mu\nu} - \tilde{L}^{\mu\nu}$ are conserved. Since we assume $\epsilon\,P^2 > 0$, the Pauli–Lubanski invariant is $W^2 = -\epsilon\,P^2\,\vec{\bar{S}}^2$.

The Lorentz boosts are differently interaction-dependent from the Galilei ones.

Let us remark that this realization is universal in the sense that it depends on the nature of the isolated system only through a $U(2)$ algebra [142], whose generators are the invariant mass M (which in turn depends on the relative variables and on the type of interaction) and the internal spin $\vec{\bar{S}}$, which is interaction-independent, being in an instant form of dynamics.

The Dirac Hamiltonian is now $H_D = \lambda(\tau)\,\Big(Mc - \int d^3\sigma\,T^{\tau\tau}(\tau,\vec{\sigma})\Big) + \vec{\lambda}(\tau)\cdot\vec{\mathcal{P}}_{(int)}$ and the embedding of Wigner hyper-planes is

$$z^\mu_W(\tau,\vec{\sigma}) = x^\mu(\tau) + \epsilon^\mu_r(u(P))\,\sigma^r, \tag{3.5}$$

with x^μ a function of $\tilde{x}^\mu$, P^μ, and $S^{\mu\nu}$ (or $\bar{S}^{AB}$) according to Eq. (3.3). The gauge freedom in the choice of x^μ is connected with the arbitrariness of the spin boosts $\bar{S}^{\tau r}$ (or S^{oi}).

Eq. (3.2) implies that the eight variables $\tilde{x}^\mu(\tau)$ and P^μ are restricted only by a first-class constraint identifying $Mc = \sqrt{\epsilon\,P^2}$ with the invariant mass of the isolated system, evaluated by using its energy–momentum tensor: Therefore, these eight variables are to be reduced to six physical variables describing the external decoupled relativistic center of mass of the isolated system ($\sqrt{\epsilon\,P^2}$ is determined by the constraint and its conjugate variable, the rest time $T_s = u(P)\cdot\tilde{x}(\tau)$ is a gauge variable). The choice of a gauge fixing for the rest time T_s is equivalent to the identification of a clock carried by the inertial observer. The natural inertial observer for this description is the external Fokker–Pryce center of inertia (the only covariant collective variable), which is also a function of τ, $\vec{z}$, and $\vec{h}$ (see Eq. (3.8)). Therefore, there are three collective position degrees of freedom hidden in the embedding field $z^\mu(\tau,\vec{\sigma})$, which become non-local non-measurable physical variables.

As a consequence, every isolated system (i.e., a closed universe) can be visualized as a decoupled non-covariant collective (non-local) pseudo-particle (the external center of mass), described by canonical variables identifying the frozen Jacobi data $\vec{z}$, $\vec{h}$ (see Eq. (3.8)), carrying a "pole–dipole structure," namely the invariant mass $M\,c$ (the Lorentz-scalar monopole) and the rest spin $\vec{\bar{S}}$ (the dipole, a Wigner spin-1 3-vector) of the system, and with an associated *external* realization of the Poincaré group:[3]

[3] The last term in the Lorentz boosts induces the Wigner rotation of the 3-vectors inside the Wigner 3-spaces.

$$P^\mu = M\,c\,h^\mu = M\,c\Big(\sqrt{1+\vec{h}^2};\vec{h}\Big),$$

$$J^{ij} = z^i\,h^j - z^j\,h^i + \epsilon^{ijk}\,S^k, \qquad K^i = J^{oi} = -\sqrt{1+\vec{h}^2}\,z^i + \frac{(\vec{S}\times\vec{h})^i}{1+\sqrt{1+\vec{h}^2}}. \tag{3.6}$$

The universal breaking of Lorentz covariance is connected to this decoupled non-local collective variable and is irrelevant because all the dynamics of the isolated system lives inside the Wigner 3-spaces and is Wigner-covariant. The invariant mass and the rest spin are built in terms of the Wigner-covariant variables of the given isolated system (the $6N$ variables $\vec{\eta}_i(\tau)$ and $\vec{\kappa}_i(\tau)$ for a system of N particles) living inside the Wigner 3-spaces [98–101, 140].

The presence of the three first-class constraints $\vec{\mathcal{P}}_{(int)} \approx 0$ (the rest-frame conditions) on these variables implies that each Wigner 3-space is a rest-frame of the isolated system whose Wigner spin-1 3-vector describing the internal 3-center of mass, is a gauge variable. In this way we avoid a double counting of the center of mass and the dynamics inside the Wigner hyper-planes is described only by Wigner spin-1 internal relative variables.

Let us now add the τ-dependent gauge fixing,

$$c\,T_s - \tau \approx 0, \tag{3.7}$$

where $c\,T_s = u(P)\cdot\tilde{x} = u(P)\cdot x$ is the Lorentz-scalar rest time, to the first-class constraint $Mc - \int d^3\sigma\,T^{\tau\tau}(\tau,\vec{\sigma}) \approx 0$ of Eq. (3.2). After this imposition (from now on τ/c is the rest time T_s, which satisfies $\{T_s, M\} = \{u(P)\cdot\tilde{x}, \sqrt{\epsilon\,P^2}\} = -\epsilon$), we have the following properties [100, 101]:

1. Due to the τ-dependence of the gauge fixing $c\,T_s - \tau \approx 0$, the Dirac Hamiltonian given before Eq. (3.5) is replaced by the Hamiltonian $H_D = \vec{\lambda}(\tau)\cdot\vec{\mathcal{P}}_{(int)} \approx 0$: There is a frozen description of dynamics in the external space of the frozen external center of mass modulo the three first-class constraints $\vec{\mathcal{P}}_{(int)} \approx 0$, as in the standard Hamilton–Jacobi description.
2. Inside the Wigner 3-space the evolution in $\tau = cT_s$ of the relative variables after the elimination of the internal center of mass is governed by the total mass M of the isolated system, i.e., by the energy generator of the internal Poincaré group, as the natural physical Hamiltonian. The 4-momentum P^μ becomes the 4-momentum of the isolated system: $P^\mu = Mc(\sqrt{1+\vec{h}^2};\vec{h}) = Mc\,h^\mu$ with $\vec{h}$ arbitrary a-dimensional 3-velocity and $Mc \equiv \int d^3\sigma\,T^{\tau\tau}(\tau,\vec{\sigma})$.
3. The six physical degrees of freedom describing the external decoupled relativistic center of mass are the non-evolving Jacobi data:

$$\vec{z} = Mc\left(\vec{\tilde{x}} - \frac{\vec{P}}{P^o}\,\tilde{x}^o\right), \qquad \vec{h} = \frac{\vec{P}}{Mc}, \qquad \{z^i, h^j\} = \delta^{ij}. \tag{3.8}$$

The 3-vector $\vec{x}_{NW} = \vec{z}/M\,c$ is the *external non-covariant canonical 3-center of mass* (the classical counterpart of the ordinary Newton–Wigner position operator) and is canonically conjugate to the 3-momentum $M\,c\,\vec{h}$. We have $\tilde{x}^\mu(\tau) = (\tilde{x}^o(\tau); \vec{x}_{NW} + \frac{\vec{P}}{P^o}\,\tilde{x}^o(\tau))$.

From appendix B of Ref. [139] we have the following transformation properties under Poincaré transformations (a, Λ):

$$
\begin{aligned}
h^\mu &= u^\mu(P) = (\sqrt{1+\vec{h}^2}; \vec{h}) \mapsto h^{'\,\mu} = \Lambda^\mu_\nu\, h^\nu,\\
z^i &\mapsto z^{'\,i} = \Big(\Lambda^i_j - \frac{\Lambda^i_\mu\, h^\mu}{\Lambda^o_\nu\, h^\nu}\,\Lambda^o_j\Big)\, z^j + \Big(\Lambda^i_\mu - \frac{\Lambda^i_\nu\, h^\nu}{\Lambda^o_\rho\, h^\rho}\,\Lambda^o_\mu\Big)\,(\Lambda^{-1}\, a)^\mu,\\
\tau &\mapsto \tau^{'} + h_\mu\,(\Lambda^{-1}\, a)^\mu.
\end{aligned} \tag{3.9}
$$

As a consequence, under Lorentz transformations we have $\vec{h}' \cdot \vec{z}' = \vec{h}\cdot\vec{z} + \frac{\Lambda^o_j\, z^j}{\Lambda^o_\mu\, h^\mu}$.

In each Wigner 3-space Σ_τ there is an unfaithful internal realization of the Poincaré algebra, whose generators are built by using the energy–momentum tensor (Eq. 2.14) of the isolated system.

$$
\begin{aligned}
\mathcal{P}^\tau_{(int)} &= M\, c,\\
\vec{\mathcal{P}}_{(int)} &\approx 0,\\
J^r_{(int)} &= \bar{S}^r,\\
\mathcal{K}^r_{(int)} &= \bar{S}^{or} = -\bar{S}^{ro},
\end{aligned} \tag{3.10}
$$

with $\bar{S}^r$ and $\bar{S}^{or}$ given in Eq. (3.4). While the internal energy and angular momentum are Mc and $\vec{\bar{S}}$ respectively, the internal 3-momentum vanishes: It is the rest-frame condition. M and the internal boosts are interaction-dependent.

An important remark is that the internal space of relative variables is independent from the orientation of the total 4-momentum of the isolated system. As shown in Ref. [140], the formalism is built in such a way that the Wigner rotation induced on the relative variables by a Lorentz transformation connecting $\overset{\circ}{P}{}^\mu = Mc\,(1;\vec{0})$ to $P^\mu = M c\, h^\mu = L^\mu_\nu(P, \overset{\circ}{P})\, \overset{\circ}{P}{}^\nu$ is the identity, because from Eq. (A.11) of Appendix A we get that the Wigner rotation $R(\Lambda, P) = L(\overset{\circ}{P}, P)\,\Lambda^{-1}\, L(\Lambda\, P, \overset{\circ}{P})$ satisfies $R(L(P, \overset{\circ}{P}), P) = 1$. Therefore the space of the relative variables is an abstract internal space insensitive to Lorentz transformations carried by the external center of mass

After the gauge fixing (Eq. 3.7) the final gauge freedom in the embedding (Eq. 3.5) are the three degrees of freedom that identify which point of Wigner 3-space is the origin of the σ^r coordinates, namely which inertial observer is chosen.

They are conjugated to the last three first-class constraints $\mathcal{P}^s_{(int)} \approx 0$ of Eq. (3.2), which describe the freedom in the choice of $x^\mu(\tau)$, i.e., of the inertial observer origin of the 3-coordinates on the Wigner hyper-planes of the family of Wigner-covariant inertial rest-frames. The natural choice would be to identify the point $\sigma^r = q_+^r(\tau)$, which is the location of the internal canonical center of mass of the isolated system built in terms of Wigner-covariant variables.

As shown in equation 3.3 of Ref. [142] (using results from Ref. [41–43]) the internal canonical center of mass of the isolated system inside the Wigner 3-spaces[4] is built in terms of the generators of the internal Poincaré group of Eq. (3.10):

$$\vec{q}_+ = -\frac{\vec{\mathcal{K}}_{(int)}}{\sqrt{M^2 c^2 - \vec{\mathcal{P}}^2_{(int)}}} + \frac{\vec{J}_{(int)} \times \vec{\mathcal{P}}_{(int)}}{\sqrt{M^2 c^2 - \vec{\mathcal{P}}^2_{(int)}}\,(Mc + \sqrt{M^2 c^2 - \vec{\mathcal{P}}^2_{(int)}})} + \frac{}{Mc\sqrt{M^2 c^2 - \vec{\mathcal{P}}^2_{(int)}}\,(Mc + \sqrt{M^2 c^2 - \vec{\mathcal{P}}^2_{(int)}})}. \tag{3.11}$$

The constraints $\vec{\mathcal{P}}_{(int)} \approx 0$ imply $\vec{q}_+ \approx -\vec{\mathcal{K}}_{(int)}/\mathcal{P}^\tau_{(int)}$. Therefore, if we add the gauge fixings $\vec{\mathcal{K}}_{(int)} \approx 0$, we get $\vec{q}_+ \approx 0$.

In equation 4.4 of of Ref. [142] it is shown that the 3-coordinate σ^r_Y in each Wigner 3-space of the external Fokker–Pryce center of inertia (defined in Section 3.2) is proportional to q^r_+ when $\tau = cT_s$. Therefore, the gauge fixing $\mathcal{K}^r_{(int)} = \bar{S}^{\tau r} = -\bar{S}^{r\tau} \approx 0$ (whose time preservation using the H_D defined after Eq. (3.4) gives $\vec{\lambda}(\tau) = 0$) implies that the inertial observer $x^\mu(\tau)$ origin of the σ^r coordinates inside each Wigner 3-space coincides with the only global notion of covariant (but not canonical) relativistic center of mass, i.e., the Fokker–Pryce center-of-inertia 4-vector, and that the embedding (Eq. 3.5) of the Wigner hyperplanes becomes

$$z^\mu_W(\tau, \vec{\sigma}) = Y^\mu(\tau) + \epsilon^\mu_r(\vec{h})\,\sigma^r. \tag{3.12}$$

A check of the consistency of this identification can be easily done by putting Eq. (3.12) into Eq. (2.13): Eq. (3.4) for the external Poincaré generators will be recovered if the rest-frame conditions $\vec{\mathcal{P}}_{(int)} \approx 0$ and $\vec{\mathcal{K}}_{(int)} \approx 0$ hold.

In Section 3.2 we will describe the only three notions of relativistic center of mass, which can be built by using only the external Poincaré generators of the isolated system. If one relaxes this requirement, it is possible to define an endless set of notions of relativistic center of mass.

3.2 The Relativistic Center-of-Mass Problem

As shown in Refs. [105, 141–146], for every isolated relativistic system there are only three collective variables (replacing the unique non-relativistic 3-center of mass), which can be constructed by using only the generators P^μ, $J^{\mu\nu}$ of the associated realization of the Poincaré group. They are the external covariant non-canonical Fokker–Pryce 4-center of inertia $Y^\mu(\tau)$, the external non-covariant

[4] One can show [100, 101, 137, 138] that one has $\vec{\mathcal{K}}_{(int)} = -M\vec{R}_+$, where $\vec{R}_+$ is the internal Møller 3-center of energy inside the Wigner 3-spaces. The rest-frame condition $\vec{\mathcal{P}}_{(int)} \approx 0$ implies $\vec{R}_+ \approx \vec{q}_+ \approx \vec{y}_+$, where $\vec{q}_+$ is the internal 3-center of mass and $\vec{y}_+$ the internal Fokker–Pryce 3-center of inertia. The gauge fixing $\vec{\mathcal{K}}_{(int)} \approx 0$ implies $\vec{R}_+ \approx \vec{q}_+ \approx \vec{y}_+ \approx 0$.

canonical 4-center of mass (also called center of spin) $\tilde{x}^\mu(\tau)$, and the external non-covariant non-canonical Møller 4-center of energy $R^\mu(\tau)$.[5] All of them have unit 4-velocity $h^\mu = u^\mu(P)$, but only $Y^\mu(\tau) = Y^\mu(0) + u^\mu(P)\,\tau = Y^\mu(0) + \left(\sqrt{1+\vec{h}^2};\vec{h}\right)\tau$ is a 4-vector, whose world-line can be used as an inertial observer.

In Ref. [105] it is shown that the three relativistic collective variables, originally defined in terms of the Poincaré generators in Refs. [41–43], can be expressed in terms of the variables τ, $\vec{z}$, $\vec{h}$, M, and $\vec{\bar{S}}$ (the spin or total baricenter angular momentum of isolated system) in the following way (one uses [105] the value $\vec{\tilde{\sigma}} = \frac{-\vec{\bar{S}}\times\vec{h}}{Mc\,(1+\sqrt{1+\vec{h}^2})}$):

1. The pseudo-world-line of the canonical non-covariant 4-center of mass (or center of spin) is

$$\begin{aligned}\tilde{x}^\mu(\tau) &= \left(\tilde{x}^o(\tau);\vec{\tilde{x}}(\tau)\right) = \left(\sqrt{1+\vec{h}^2}\left(\tau+\frac{\vec{h}\cdot\vec{z}}{Mc}\right);\frac{\vec{z}}{Mc}+\left(\tau+\frac{\vec{h}\cdot\vec{z}}{Mc}\right)\vec{h}\right)\\ &= z^\mu_W(\tau,\vec{\tilde{\sigma}}) = Y^\mu(\tau) + \left(0;\frac{-\vec{\bar{S}}\times\vec{h}}{Mc\,(1+\sqrt{1+\vec{h}^2})}\right),\end{aligned}\tag{3.13}$$

so that we get $Y^\mu(0) = \left(\sqrt{1+\vec{h}^2}\,\frac{\vec{h}\cdot\vec{z}}{Mc};\frac{\vec{z}}{Mc}+\frac{\vec{h}\cdot\vec{z}}{Mc}\vec{h}+\frac{\vec{\bar{S}}\times\vec{h}}{Mc\,(1+\sqrt{1+\vec{h}^2})}\right)$.

2. The world-line of the non-canonical covariant Fokker–Pryce 4-center of inertia is

$$\begin{aligned}Y^\mu(\tau) &= \left(\tilde{x}^o(\tau);\vec{Y}(\tau)\right)\\ &= \left(\sqrt{1+\vec{h}^2}\left(\tau+\frac{\vec{h}\cdot\vec{z}}{Mc}\right);\frac{\vec{z}}{Mc}+\left(\tau+\frac{\vec{h}\cdot\vec{z}}{Mc}\right)\vec{h}\right.\\ &\quad\left.+\frac{\vec{\bar{S}}\times\vec{h}}{Mc\,(1+\sqrt{1+\vec{h}^2})}\right) = z^\mu_W(\tau,\vec{0}).\end{aligned}\tag{3.14}$$

3. The pseudo-world-line of the non-canonical non-covariant Møller 4-center of energy is

$$\begin{aligned}R^\mu(\tau) &= \left(\tilde{x}^o(\tau);\vec{R}(\tau)\right)\\ &= \left(\sqrt{1+\vec{h}^2}\left(\tau+\frac{\vec{h}\cdot\vec{z}}{Mc}\right);\frac{\vec{z}}{Mc}+\left(\tau+\frac{\vec{h}\cdot\vec{z}}{Mc}\right)\vec{h}\right.\\ &\quad\left.-\frac{\vec{\bar{S}}\times\vec{h}}{Mc\sqrt{1+\vec{h}^2}\left(1+\sqrt{1+\vec{h}^2}\right)}\right) = z^\mu_W(\tau,\vec{\sigma}_R)\end{aligned}$$

[5] In non-relativistic physics, the center-of-mass variable takes on two essential roles. First, it is a locally observable 3-vector with the necessary transformation properties under the Galilei group. Second, it, together with the total momentum, forms a canonical pair. In relativity, the first two properties are split respectively between the Fokker–Pryce 4-center of inertia $Y^\mu(\tau)$ and $\tilde{x}^\mu(\tau)$. Both as well as $R^\mu(\tau)$ move with constant velocity h^μ for an isolated system in analogy to what happens in non-relativistic physics.

$$= Y^\mu(\tau) + \left(0; \frac{-\vec{\bar{S}} \times \vec{h}}{Mc\sqrt{1+\vec{h}^2}}\right). \tag{3.15}$$

If one centers the inertial rest frame on the world–line of the Fokker-Planck center of inertia thought as an inertial observer, then the corresponding embedding has the expression of Eq. (3.12) [105, 143–146].

In each Lorentz frame one has different pseudo-world-lines describing R^μ and $\tilde{x}^\mu$: The canonical 4-center of mass $\tilde{x}^\mu$ lies in between Y^μ and R^μ in every (non-rest)-frame. As discussed in subsection IIF of Ref. [100], this leads to the existence of the Møller non-covariance world-tube, around the world-line Y^μ of the covariant non-canonical Fokker–Pryce 4-center of inertia Y^μ. The invariant radius of the tube is $\rho = \sqrt{-\epsilon\, W^2}/P^2 = |\vec{S}|/\sqrt{\epsilon\, P^2}$ where $W^2 = -P^2\,\vec{S}^2$ is the Pauli–Lubanski invariant for time-like P^μ. This classical intrinsic radius is a non-local effect of Lorentz signature absent in Euclidean spaces and delimits the non-covariance effects (the pseudo-world-lines) of the canonical 4-center of mass $\tilde{x}^\mu$.[6] They are not detectable because the Møller radius is of the order of the Compton wavelength: An attempt to test its interior would mean entering the quantum regime of pair production. The Møller radius ρ is also a remnant of the energy conditions of GR in flat Minkowski space-time [105].

The existence of the Møller world-tube for rotating systems is a consequence of the Lorentz signature of Minkowski space-time. It identifies a region that cannot be explored without breaking manifest Lorentz covariance: This implies a limitation on the localization of the canonical center of mass due to its frame-dependence. Moreover, it leads to the identification of a fundamental length, the Møller radius, associated with every configuration of an isolated system and built with its global Poincaré Casimirs. This unit of length is really remarkable for the following two reasons:

1. At the quantum level ρ becomes the Compton wavelength of the isolated system times its spin eigenvalue $\sqrt{s(s+1)}$, $\rho \mapsto \hat{\rho} = \sqrt{s(s+1)}\hbar/M = \sqrt{s(s+1)}\lambda_M$, with $M = \sqrt{\epsilon\, P^2}$ the invariant mass and $\lambda_M = \hbar/M$ its associated Compton wavelength. Therefore the region of frame-dependent localization of the canonical center of mass is also the region where classical relativistic physics is no longer valid, because any attempt to make a localization more precise than the Compton wavelength at the quantum level leads to pair production. The interior of the classical Møller world-tube must be described by using quantum mechanics and this suggests that after quantization the Newton Wigner operator could be non-self-adjoint!

[6] In the rest-frame the world-tube is a cylinder: In each instantaneous 3-space there is a disk of possible positions of the canonical 3-center of mass orthogonal to the spin. In the non-relativistic limit the radius ρ of the disk tends to zero and one recovers the non-relativistic center of mass.

In string theory, extended Heisenberg relations $\triangle x = \frac{\hbar}{\triangle p} + \frac{\triangle p}{\alpha'} = \frac{\hbar}{\triangle p} + \frac{l_s^2 \triangle p}{\hbar}$ ($l_s = \sqrt{\hbar \alpha'}$ is the fundamental string length) have been proposed [147] to get the lower bound $\triangle x > l_s$ (due to the $y + 1/y$ form), forbidding the exploration of distances below l_s. By replacing l_s with the Møller radius of an isolated system and x with its Newton Wigner 3-center of mass $\vec{x}_{NW} = \vec{z}/Mc$ in the modified Heisenberg relations, $\triangle x^r_{NW} = \frac{\hbar}{\triangle p^r} + \frac{\rho^2 \triangle p^r}{\hbar}$, we could obtain the impossibility ($\triangle x^r > \rho$) of exploring the interior of the Møller world-tube of the isolated system. This would be compatible with a non-self-adjoint Newton–Wigner position operator after quantization.

Moreover, the Møller radius of a field configuration (consider the radiation field discussed in the next chapter) could be a candidate for a physical (configuration-dependent) ultraviolet cutoff in quantum field theory (QFT) [143–146].

2. As shown in Refs. [87, 88], where the Møller world-tube was introduced in connection with the Møller center of energy $R^\mu(\tau)$, an extended rotating relativistic isolated system with the material radius smaller than its intrinsic radius ρ has: (1) a peripheral rotation velocity that can exceed the velocity of light; and (2) its classical energy density that cannot be positive definite everywhere in every frame.[7] Therefore, the Møller radius ρ is also a remnant of the energy conditions of GR in flat Minkowski space-time [143–146].

3.3 The Elimination of Relative Times in Relativistic Systems of Particles and in Relativistic Bound States

The world-lines of the positive-energy particles are parametrized by the Wigner spin-1 3-vectors $\vec{\eta}_i(\tau)$, $i = 1, 2, \ldots, N$, and are given by

$$x^\mu_i(\tau) = z^\mu_W(\tau, \vec{\eta}_i(\tau)) = Y^\mu(\tau) + \epsilon^\mu_r(\tau)\, \eta^r_i(\tau). \qquad (3.16)$$

The world-lines $x^\mu_i(\tau)$ are derived (interaction-dependent) quantities. Also the standard particle 4-momenta are derived quantities, whose expression is $p^\mu_i(\tau) = \epsilon^\mu_A(\vec{h})\, \kappa^A_i(\tau) = h^\mu \sqrt{m_i^2 c^2 + \vec{\kappa}_i^2(\tau)} - \epsilon^\mu_r(\vec{h})\, \kappa_{ir}(\tau)$, with $\epsilon\, p_i^2 = m_i^2 c^2$ in the free case.

In general the world-lines $x^\mu_i(\tau)$ do not satisfy vanishing Poisson brackets (they are relativistic predictive non-canonical coordinates, see Ref. [142]): Already at the classical level a non-commutative structure emerges due to the Lorentz signature of the space-time [140].

[7] Classically, energy density is always positive and the stress–energy tensor for all classical fields satisfies the weak energy condition $T_{\mu\nu}\, u^\mu\, u^\nu \geq 0$, where u^μ is any time-like or null vector. In a sense the Møller world-tube is a classical version of the Epstein, Glaser, Jaffe theorem [148] in QFT: If a field $Q(x)$ satisfies $< \Psi|Q(x)|\Psi > \geq 0$ for all states and if $< \Omega|Q(x)|\Omega > = 0$ for the vacuum state, then $Q(x) = 0$. Therefore, in QFT the weak energy condition does not hold for the renormalized stress–energy tensor. Since it has by definition a null vacuum expectation value, there are states $|Y>$ such that $< Y|T_{\mu\nu}\, u^\mu\, u^\nu|Y > < 0$. This holds both for the scalar field and for the squeezed state of the electromagnetic field.

The three pairs of second-class (interaction-dependent) constraints $\vec{\mathcal{P}}_{(int)} \approx 0$, $\vec{\mathcal{K}}_{(int)} \approx 0$, eliminate the internal 3-center of mass and its conjugate momentum inside the Wigner 3-spaces, as already said: This avoids a double counting of the collective variables (external and internal center of mass). As a consequence the dynamics inside the Wigner 3-spaces is described in terms of internal Wigner-covariant relative variables.

For N free particles the external Poincaré generators are given in Eq. (3.6) and the internal Poincaré generators have the following expression:

$$
\begin{aligned}
M\,c &= \frac{1}{c}\,\mathcal{E}_{(int)} = \sum_{i=1}^{N} \sqrt{m_i^2\,c^2 + \vec{\kappa}_i^2},\\
\vec{\mathcal{P}}_{(int)} &= \sum_{i=1}^{N} \vec{\kappa}_i \approx 0,\\
\vec{S} = \vec{\mathcal{J}}_{(int)} &= \sum_{i=1}^{N} \vec{\eta}_i \times \vec{\kappa}_i,\\
\vec{\mathcal{K}}_{(int)} &= -\sum_{i=1}^{N} \vec{\eta}_i\,\sqrt{m_i^2\,c^2 + \vec{\kappa}_i^2} \approx 0.
\end{aligned}
\tag{3.17}
$$

Since one is in an instant form of the dynamics, in the interacting case only Mc and $\vec{\mathcal{K}}_{(int)}$ become interaction-dependent.

In the case of N relativistic particles, one defines the following canonical transformation [105] (see Refs. [137, 141, 142] for other variants) ($m = \sum_{i=1}^{N} m_i$):

$$
\begin{aligned}
&\vec{\eta}_+ = \sum_{i=1}^{N} \frac{m_i}{m}\,\vec{\eta}_i, \qquad \vec{\kappa}_+ = \vec{\mathcal{P}}_{(int)} = \sum_{i=1}^{N} \vec{\kappa}_i \approx 0,\\
&\vec{\rho}_a = \sqrt{N}\sum_{i=1}^{N} \gamma_{ai}\,\vec{\eta}_i, \qquad \vec{\pi}_a = \frac{1}{\sqrt{N}} \sum_{i=1}^{N} \Gamma_{ai}\,\vec{\kappa}_i, \qquad a = 1, \dots, N-1,\\
&\vec{\eta}_i = \vec{\eta}_+ + \frac{1}{\sqrt{N}} \sum_{a-1}^{N-1} \Gamma_{ai}\,\vec{\rho}_a, \qquad \vec{\kappa}_i = \frac{m_i}{m}\,\vec{\kappa}_+ + \sqrt{N}\sum_{a=1}^{N-1} \gamma_{ai}\,\vec{\pi}_a,
\end{aligned}
\tag{3.18}
$$

with the following canonicity conditions[8] on the numerical parameters γ_{ai}, Γ_{ai}:

$$
\begin{aligned}
&\sum_{i=1}^{N} \gamma_{ai} = 0, \qquad \sum_{i=1}^{N} \gamma_{ai}\,\gamma_{bi} = \delta_{ab}, \qquad \sum_{a=1}^{N-1} \gamma_{ai}\,\gamma_{aj} = \delta_{ij} - \frac{1}{N},\\
&\Gamma_{ai} = \gamma_{ai} - \sum_{k=1}^{N} \frac{m_k}{m}\,\gamma_{ak}, \qquad \gamma_{ai} = \Gamma_{ai} - \frac{1}{N} \sum_{k=1}^{N} \Gamma_{ak},\\
&\sum_{i=1}^{N} \frac{m_i}{m}\,\Gamma_{ai} = 0, \qquad \sum_{i=1}^{N} \gamma_{ai}\,\Gamma_{bi} = \delta_{ab}, \qquad \sum_{a=1}^{N-1} \gamma_{ai}\,\Gamma_{aj} = \delta_{ij} - \frac{m_i}{m}.
\end{aligned}
\tag{3.19}
$$

[8] Eq. (3.18) describes a family of canonical transformations, because the γ_{ai} depend on $\frac{1}{2}(N-1)(N-2)$ free independent parameters.

Since the embedding (Eq. 3.16) is determined by the gauge fixings $\vec{\mathcal{K}}_{(int)} \approx 0$, i.e., by the vanishing of the internal canonical center of mass (Eq. 3.11), $\vec{q}_+ \approx 0$, the variable $\vec{\eta}_+$ has to be expressed in terms of $\vec{q}_+$. In Ref. [142] there is the definition both of the canonical transformation (Eq. 3.18) and of a canonical transformation different from Eq. (3.18), in which $\vec{\eta}_i$, $\vec{\kappa}_i$ go in the canonical variables $\vec{q}_+$, $\vec{\kappa}_+ \approx 0$, $\vec{\rho}_{qa}$, $\vec{\pi}_{qa}$. In the case of two free particles this canonical transformation and its inverse are

$$
\begin{aligned}
Mc &= \sqrt{m_1^2c^2 + \vec{\kappa}_1^2} + \sqrt{m_2^2c^2 + \vec{\kappa}_2^2}, \qquad \vec{S}_q = \vec{\rho}_q \times \vec{\pi}_q,\\
\vec{q}_+ &= \frac{\sqrt{m_1^2c^2 + \vec{\kappa}_1^2}\,\vec{\eta}_1 + \sqrt{m_2^2c^2 + \vec{\kappa}_2^2}\,\vec{\eta}_2}{\sqrt{M^2c^2 - \vec{p}^2}} + \frac{(\vec{\eta}_1 \times \vec{\kappa}_1 + \vec{\eta}_2 \times \vec{\kappa}_2) \times \vec{p}}{\sqrt{M^2c^2 - \vec{p}^2}\,(Mc + \sqrt{M^2c^2 - \vec{p}^2})}\\
&\quad - \frac{(\sqrt{m_1^2c^2 + \vec{\kappa}_1^2}\,\vec{\eta}_1 + \sqrt{m_2^2c^2 + \vec{\kappa}_2^2}\,\vec{\eta}_2) \cdot \vec{p}\,\vec{p}}{Mc\sqrt{M^2c^2 - \vec{p}^2}\,(Mc + \sqrt{M^2c^2 - \vec{p}^2})},\\
\vec{p} &= \vec{\kappa}_1 + \vec{\kappa}_2 \approx 0,\\
\vec{\pi}_q &= \vec{\pi} - \frac{\vec{p}}{\sqrt{M^2c^2 - \vec{p}^2}}\Big[\frac{1}{2}(\sqrt{m_1^2c^2 + \vec{\kappa}_1^2} - \sqrt{m_2^2c^2 + \vec{\kappa}_2^2})\\
&\quad - \frac{\vec{p}\cdot\vec{\pi}}{\vec{p}^2}\,(Mc - \sqrt{M^2c^2 - \vec{p}^2})\Big] \approx \vec{\pi} = \frac{1}{2}(\vec{\kappa}_1 - \vec{\kappa}_2),\\
\vec{\rho}_q &= \vec{\rho} + \left(\frac{\sqrt{m_1^2c^2 + \vec{\kappa}_1^2}}{\sqrt{m_2^2c^2 + \vec{\pi}_q^2}} + \frac{\sqrt{m_2^2c^2 + \vec{\kappa}_2^2}}{\sqrt{m_1^2c^2 + \vec{\pi}_q^2}}\right)\frac{\vec{p}\cdot\vec{\rho}\,\vec{\pi}_q}{Mc\sqrt{M^2c^2 - \vec{p}^2}} \approx \vec{\rho} = \vec{\eta}_1 - \vec{\eta}_2,\\
&\Rightarrow Mc = \sqrt{\mathcal{M}^2c^2 + \vec{p}^2} \approx \mathcal{M}c = \sqrt{m_1^2c^2 + \vec{\pi}_q^2} + \sqrt{m_2^2c^2 + \vec{\pi}_q^2},\\
\vec{q}_+ &\approx \frac{\vec{\eta}_1\sqrt{m_1^2c^2 + \vec{\pi}^2} + \vec{\eta}_2\sqrt{m_2^2c^2 + \vec{\pi}^2}}{\mathcal{M}c},
\end{aligned}
\tag{3.20}
$$

$$
\begin{aligned}
\vec{\eta}_i &= \vec{q}_+ - \frac{\vec{S}_q \times \vec{p}}{\sqrt{\mathcal{M}^2c^2 + \vec{p}^2}\,(\mathcal{M}c + \sqrt{\mathcal{M}^2c^2 + \vec{p}^2}} + \frac{1}{2}\Big[(-1)^{i+1}\\
&\quad - \frac{2\,\mathcal{M}c\,\vec{\pi}_q\cdot\vec{p} + (m_1^2c^2 - m_2^2c^2)\sqrt{\mathcal{M}^2c^2 + \vec{p}^2}}{\mathcal{M}^2c^2\sqrt{\mathcal{M}^2c^2 + \vec{p}^2}}\Big]\\
&\quad \left[\vec{\rho}_q - \frac{\vec{\rho}_q\cdot\vec{p}\,\vec{\pi}_q}{\mathcal{M}c\sqrt{\mathcal{M}^2c^2 + \vec{p}^2}\left(\frac{\sqrt{m_1^2c^2+\vec{\kappa}_1^2}}{\sqrt{m_2^2c^2+\vec{\pi}_q^2}} + \frac{\sqrt{m_2^2c^2+\vec{\kappa}_2^2}}{\sqrt{m_1^2c^2+\vec{\pi}_q^2}}\right)^{-1}} + \vec{\pi}_q\cdot\vec{p}\right]\\
&\approx \vec{q}_+ + \frac{1}{2}\left[(-1)^{i+1} - \frac{m_1^2c^2 - m_2^2c^2}{\mathcal{M}^2c^2}\right]\vec{\rho}_q \approx \frac{1}{2}\left[(-1)^{i+1} - \frac{m_1^2c^2 - m_2^2c^2}{\mathcal{M}^2c^2}\right]\vec{\rho},\\
\vec{\kappa}_i &= \Big[\frac{1}{2} + \frac{(-1)^{i+1}}{\mathcal{M}c\sqrt{\mathcal{M}^2c^2 + \vec{p}^2}}\Big(\vec{\pi}_q\cdot\vec{p}\Big[1 - \frac{\mathcal{M}c}{\vec{p}^2}(\sqrt{\mathcal{M}^2c^2 + \vec{p}^2} - \mathcal{M}c)\Big]\\
&\quad + (m_1^2c^2 - m_2^2c^2)\sqrt{\mathcal{M}^2c^2 + \vec{p}^2}\Big)\Big]\,\vec{p} + (-1)^{i+1}\,\vec{\pi}_q \approx (-1)^{i+1}\,\vec{\pi}_q \approx (-)^{i+1}\,\vec{\pi},\\
&\Rightarrow \quad \vec{\kappa}_i^2 \approx \vec{\pi}^2.
\end{aligned}
\tag{3.21}
$$

Due to the constraints $\vec{\kappa}_+ \approx 0$ from the comparison of these two canonical bases, one gets the following result (see equation 5.27 of Ref. [142]):

$$\vec{q}_+ = \vec{\eta}_+ - \frac{1}{\sqrt{N}}\, \frac{\sum_{a=1}^{N-1} \vec{\rho}_a}{\sqrt{m_i^2\, c^2 + \sum_{ab}^{1...N-1} \gamma_{ai}\, \gamma_{bi}\, \vec{\pi}_a \cdot \vec{\pi}_b}}. \tag{3.22}$$

Therefore,

$$\begin{aligned}
\vec{\eta}_+ &\approx \frac{1}{\sqrt{N}}\, \frac{\sum_{a=1}^{N-1} \vec{\rho}_a}{\sqrt{m_i^2\, c^2 + \sum_{ab}^{1...N-1} \gamma_{ai}\, \gamma_{bi}\, \vec{\pi}_a \cdot \vec{\pi}_b}},\\
\vec{\rho}_{qa} &\approx \vec{\rho}_a, \qquad \vec{\pi}_{qa} \approx \vec{\pi}_a, \qquad \vec{\bar{S}} \approx \sum_{a=1}^{N-1} \vec{\rho}_a \times \vec{\pi}_a,\\
M\, c &= \sqrt{m_i^2\, c^2 + \sum_{ab}^{1...N-1} \gamma_{ai}\, \gamma_{bi}\, \vec{\pi}_a \cdot \vec{\pi}_b}.
\end{aligned} \tag{3.23}$$

Therefore, the invariant mass Mc and the rest spin $\vec{\bar{S}}$ become functions only of the $N-1$ pairs of relative canonical variables $\vec{\rho}_a$, $\vec{\pi}_a$.

As a consequence, Eqs. (3.16)–(3.18) imply that the world-lines $x_i^\mu(\tau)$ can be expressed in terms of the Jacobi data $\vec{z}$, $\vec{h}$, and of the relative variables $\vec{\rho}_a(\tau)$, $\vec{\pi}_a(\tau)$, $a = 1, \ldots, N-1$.

Therefore the 3-positions $\vec{\eta}_i(\tau)$ (and also their conjugate momenta $\vec{\kappa}_i(\tau)$) become functionals only of a set of relative coordinates and momenta satisfying Hamilton equations governed by the invariant mass Mc. Once these equations are solved and the orbits $\vec{\eta}_i(\tau)$ are reconstructed, Eq. (3.16) leads lead to world-lines $x_i^\mu(\tau)$ functions of τ, of the non-evolving Jacobi data $\vec{z}$, $\vec{h}$, and of the solutions for the relative variables (Eq. (3.12) has to be used for $Y^\mu(\tau)$).

Since $\vec{q}_+$ tends to the non-relativistic center of mass $\vec{q}_{NR} = \sum_i m_i\, \vec{\eta}_i / \sum_i m_i$, the non-relativistic limit of this description [100, 101, 141] is Newton mechanics with the Newton center of mass decoupled from the relative variables and moreover after a canonical transformation to the frozen Hamilton–Jacobi description of the center of mass.

The non-covariant $\vec{z}, \vec{h}$ of the external center of mass and the Wigner-covariant relative variables $\vec{\rho}_a(\tau)$, $\vec{\pi}_a(\tau)$, $a = 1, \ldots, N-1$, are a canonical basis of Dirac observables (together with those for the electromagnetic field if present) for the rest-frame instant form of parametrized Minkowski theories. The Dirac observables in other either inertial or non-inertial frames are obtained from them, with canonical transformations depending on the chosen frame.

In the case of interacting particles, either with action-at-a-distance (AAAD) mutual interaction or coupled to dynamical fields, the reconstruction of the world-lines requires a complex interaction-dependent procedure delineated in Refs. [141–143], where there is also a comparison of the present approach with the other formulations of relativistic mechanics developed for the study of the problem of relativistic bound states. With the Wigner-covariant rest-frame instant form, the problem of the elimination of relative times both in relativistic bound states and in the scattering of the particles, which would produce non-physical states after quantization, is completely solved at the classical level.

As shown in Ref. [105] AAAD interactions inside the Wigner hyper-plane may be introduced either under (scalar and vector potentials) or outside (scalar potential like the Coulomb one) the square roots appearing in the free Hamiltonian. Since a Lagrangian density in the presence of AAAD mutual interactions is not known and since we are working in an instant form of dynamics, the potentials in the constraints restricted to hyper-planes must be introduced by hand (see, however, Refs. [98–101, 149–151] for their evaluation starting from the Lagrangian density for the electromagnetic interaction). The only restriction is that the Poisson brackets of the modified constraints generators must generate the same algebra as the free ones.

In the rest-frame instant form the most general two-body Hamiltonian with AAAD interactions (allowing both bound states and scattering) is

$$M_{(AAAD)}c = \sqrt{m_1^2c^2 + U_1 + [\vec{\kappa}_1 - \vec{V}_1]^2} + \sqrt{m_2^2c^2 + U_2 + [\vec{\kappa}_2 - \vec{V}_2]^2} + V, \quad (3.24)$$

where $U_i = U_i(\vec{\kappa}_1, \vec{\kappa}_2, \vec{\eta}_1 - \vec{\eta}_2)$, $\vec{V}_i = \vec{V}_i(\vec{\kappa}_{j\neq i}, \vec{\eta}_1 - \vec{\eta}_2)$, $V = V_o(|\vec{\eta}_1 - \vec{\eta}_2|) + V'(\vec{\kappa}_1, \vec{\kappa}_2, \vec{\eta}_1 - \vec{\eta}_2)$.

If we use the canonical transformation (Eq. 3.20) defining the relativistic canonical internal 3-center of mass (now $\vec{q}_+^{(int)}$ is interaction-dependent) and relative variables on the Wigner hyper-plane, together with the rest-frame conditions $\vec{\mathcal{P}} \approx 0$, the rest-frame Hamiltonian for the relative motion becomes $M_{(AAAD)}c \approx \sqrt{m_1^2c^2 + \tilde{U}_1 + [\vec{\pi}_q - \tilde{\vec{V}}_1]^2} + \sqrt{m_2^2c^2 + \tilde{U}_2 + [-\vec{\pi}_q - \tilde{\vec{V}}_2]^2} + \tilde{V}$, where $\tilde{U}_i = U_i(\vec{\pi}_q, \vec{\rho}_q)$, $\tilde{V} = V_o(|\vec{\rho}_q|) + V'(\vec{\pi}_q, \vec{\rho}_q)$, $\tilde{\vec{V}}_1 = \vec{V}_1(-\vec{\pi}_q, \vec{\rho}_q)$, $\tilde{\vec{V}}_2 = \vec{V}_2(\vec{\pi}_q, \vec{\rho}_q)$. However, since the 3-center $\vec{q}_+$ becomes interaction dependent, the final canonical basis $\vec{q}_+$, $\vec{p}$, $\vec{\rho}_q$, $\vec{\pi}_q$ is not explicitly known in the interacting case. As a consequence, we cannot use the gauge fixing $\vec{q}_+ \approx 0$, but we must find $\vec{q}_+ \approx \vec{f}(\vec{\rho}_q, \vec{\pi}_q)$ from the gauge fixings on the internal boosts $\vec{\mathcal{K}}_{(int)} \approx 0$. The relativistic orbits are rebuilt in terms of the Wigner-covariant variables $\vec{\eta}_i(\tau) \approx \vec{q}_+(\vec{\rho}_q, \vec{\pi}_q) + \frac{1}{2}\left[(-)^{i+1} - \frac{m_1^2 - m_2^2}{\mathcal{M}^2}\right]\vec{\rho}_q$, $\vec{\kappa}_i(\tau) \approx (-)^{i+1}\,\vec{\pi}_q(\tau)$.

In Ref. [141] there is the explicit construction of a two-particle model with AAAD interactions (including also Coulomb-like potentials) with internal Poincaré generators $Mc = M_{(AAAD)}c = \sqrt{m_1^2c^2 + \vec{\kappa}_1^2 + \Phi(\vec{\rho}^2)} + \sqrt{m_2^2c^2 + \vec{\kappa}_2^2 + \Phi(\vec{\rho}^2)}$, $\vec{\mathcal{P}}_{(int)} = \vec{\kappa}_1 + \vec{\kappa}_2$, $\vec{J}_{(int)} = \vec{\eta}_1 \times \vec{\kappa}_1 + \vec{\eta}_2 \times \vec{\kappa}_2$, $\vec{\mathcal{K}}_{(int)} = -\vec{\eta}_1\sqrt{m_1^2c^2 + \vec{\kappa}_1^2 + \Phi(\vec{\rho}^2)} - \vec{\eta}_2\sqrt{m_2^2c^2 + \vec{\kappa}_2^2 + \Phi(\vec{\rho}^2)}$, with $\Phi = \Phi(\vec{\rho}^2)$ ($\vec{\rho} = \vec{\eta}_1 - \vec{\eta}_2$), and it is shown how to rebuild the orbits of relative motions and the world-lines of the particles.

See Section 3.5 and Refs. [98, 99] for the extension to non-inertial frames.

Finally, (Eq. 2.14) for the external Poincaré generators and the energy–momentum tensor can be used to extend the multipolar expansions of Refs. [152–154] to this framework for relativistic isolated systems as is shown in Ref. [47] (see also Ref. [155]), where Dixon's multipoles for a system of N relativistic positive-energy scalar particles are evaluated in the rest-frame instant form of dynamics. This multipolar expansion is mainly used to make pole–dipole approximations of extended objects, used for instance to build templates for gravitational waves from binary systems.

See Ref. [107] for the collective and relative variables of the Klein–Gordon field and Ref. [101] for such variables for the electromagnetic field in the radiation gauge. For these systems one can give for the first time the explicit closed form of the interaction-dependent Lorentz boosts.

3.4 Wigner-Covariant Quantum Mechanics of Point Particles

A new formulation of relativistic quantum mechanics in the Wigner 3-spaces of the inertial rest frame is developed in Ref. [140] in the absence of the electromagnetic field. It incorporates all the known results about relativistic bound states (absence of relative times) and avoids the causality problems of the Hegerfeldt theorem [156, 157] (the instantaneous spreading of wave packets) and the problem of how to connect the quantizations on two space-like hyper-planes connected by a Lorentz transformation [158–162].

In it one quantizes the frozen Jacobi data $\vec{z}$ and $\vec{h}$ of the canonical non-covariant decoupled external center of mass and the Wigner-covariant relative variables in the Wigner 3-spaces. Since the external center of mass is decoupled, its non-covariance is irrelevant: As for the wave function of the universe, who will observe it? Maybe only relative motions have to be quantized and the notion of mathematical observer has to be abandoned, as in Rovelli's approach [18].

The resulting Hilbert space has the following tensor product structure: $H = H_{com,HJ} \otimes H_{rel}$, where $H_{com,HJ}$ is the Hilbert space of the external center of mass (in the Hamilton–Jacobi formulation due to the use of frozen Jacobi data) while H_{rel} is the Hilbert space of the relative variables in the abstract internal space living in the Wigner 3-spaces. While at the non-relativistic level this presentation of the Hilbert space is unitarily equivalent to the tensor product of the Hilbert spaces H_i of the individual particles[9] $H = H_1 \otimes H_2 \otimes \ldots$, this is not true at the relativistic level.

If one considers two scalar quantum particles with Klein–Gordon wave functions belonging to Hilbert spaces $\mathcal{H}_{x_i^o}$ in the tensor product Hilbert space $(\mathcal{H}_1)_{x_1^o} \otimes (\mathcal{H}_2)_{x_2^o} \otimes \ldots$ there is no correlation among the times of the particles

[9] This is the zeroth postulate of quantum mechanics [163], which relies on a non-relativistic notion of separability [164, 165].

(their clocks are not synchronized) so that in most of the states there are some particles in the absolute future of the others. As a consequence, the two types of Hilbert spaces ($H_{com,HJ} \otimes H_{rel}$ and $H_1 \otimes H_2 \otimes \dots$) lead to unitarily non-equivalent descriptions and have different scalar products (compare Refs. [140] and [136, 137]).

Therefore, at the relativistic level, the zeroth postulate of non-relativistic quantum mechanics [163] does not hold: The Hilbert space of composite systems is not the tensor product of the Hilbert spaces of the subsystems. Contrary to Einstein's notion of separability (separate objects have their independent real states), one gets a kinematical spatial non-separability induced by the need for clock synchronization for eliminating the relative times and to be able to formulate a well-posed relativistic Cauchy problem.

Moreover, one has the non-locality of the non-covariant external center of mass, which implies its non-measurability with local instruments.[10] While its conjugate momentum operator must be well defined and self-adjoint, because its eigenvalues describe the possible values for the total momentum of the isolated system (the momentum basis is therefore a preferred basis in the Hilbert space), it is not clear whether it is meaningful to define center-of-mass wave packets.

The non-locality and kinematical spatial non-separability (not existing in non-relativistic entanglement) are due to the Lorentz signature of Minkowski space-time, and this fact reduces the relevance of quantum non-locality in the study of the foundational problems of quantum mechanics, which have to be rephrased in terms of relative variables.

An open problem is that already in non-relativistic quantum mechanics it is not clear what is the meaning of the localization of a quantum particle. There is no consensus on whether this wave function describes the given quantum system or is only information on a statistical ensemble of such systems. Moreover, it could be that only the density matrix, the statistical operator determining the probabilities, makes sense and not the wave function. There is no accepted interpretation for the theory of measurements. In experiments we have macroscopic semi-classical objects as sources and detectors of quantities named quantum particles (or atoms) and the results shown by the pointers of the detectors are the end-point of a macroscopic (many-body) amplification of the interaction of the quantum object with some microscopic constituent of the detector (for instance, an α particle interacting with a water molecule followed by the formation of a droplet as the amplification allowing detection of the particle trajectory in bubble chambers). Usually one invokes the theory of decoherence with its uncontrollable coupled environment for the emergence of robust classical aspects explaining the well-defined position of the pointer in a measurement. This state of affairs is in accord with Bohr's point of view, according to which

[10] In Ref. [140] it was shown that the quantum Newton–Wigner position should not be a self-adjoint operator, but only a symmetric one, with an implication of bad localization.

we need a classical description of the experimental apparatus. It seems that all the realizable experiments must admit a quasi-classical description not only of the apparatus but also of the quantum particles: They are present in the experimental area as classical massive particles with a mean trajectory and a mean value of 4-momentum (measured with time-of-flight methods).

As shown in Ref. [166], by using Refs. [47, 155], if one assumes that the wave function describes the given quantum system (no ensemble interpretation), the statement of Bohr can be justified by noting that the wave functions used in the preparation of particle beams (semi-classical objects with a mean classical trajectory and a classical mean momentum determined with time-of-flight methods) are a special subset of the wave functions solutions of the Schroedinger equation for the given particles. Their associated density matrix, pervading the whole 3-space, admits a multipolar expansion around a classical trajectory having zero dipole. This implies that in this case the equations of the Ehrenfest theorem give rise to the Newton equations for the Newton trajectory (the monopole) with a classical force augmented by forces of quantum nature coming from the quadrupole and the higher multipoles (they are proportional to powers of the Planck constant). As shown in Ref. [166], the mean trajectories of the prepared beams of particles and of the particles revealed by the detectors are just these classically emerging Newton trajectories implied by the Ehrenfest theorem for wave functions with zero dipole. Also all the intuitive descriptions of experiments in atomic physics are compatible with this emergence of classicality. In these descriptions an atom is represented as a classical particle delocalized in a small sphere, whose origin can be traced to the effect of the higher-multipole forces in the emerging Newton equations for the atom trajectory. The wave functions without zero dipole do not seem to be implementable in feasible experiments.

Then in Ref. [166] there is a discussion on how to extend these results to relativistic particles and to relativistic quantum mechanics.

See Ref. [167] for a discussion on relativistic entanglement and on the localization of particles. It is shown that at the non-relativistic level in real experiments only relative variables are measured, with their directions determined by the effective mean classical trajectories of particle beams present in the experiment. The existing results about the non-relativistic and relativistic localization of particles and atoms support the view that detectors only identify effective particles following this type of trajectory: These objects are the phenomenological emergent aspect of the notion of particle defined by means of the Fock spaces of QFT.

The quantization defined in Ref. [140] leads to a first formulation of a theory for relativistic entanglement, which is deeply different from the non-relativistic entanglement due to the kinematical non-locality and spatial non-separability. To have control of the Poincaré group, one needs an isolated system containing all the relevant entities (for instance, both Alice and Bob) of the experiment under investigation and also the environment when needed. One has to learn to reason in terms of relative variables adapted to the experiment like molecular physicists do when they look to the best system of Jacobi coordinates adapted to the main chemical bonds in the given molecule. This theory has still to be developed, together with its extension to non-inertial rest frames.

In Ref. [168] in the framework of non-inertial relativistic quantum mechanics it is also proposed to replace the usual postulated mathematical observers with dynamical ones, whose world-lines belong to interacting particles of the isolated system.

Till now non-inertial relativistic quantum mechanics could be defined only in the class of global non-inertial frames with space-like hyper-planes as 3-spaces of the 3+1 approach by using the multi-temporal quantization approach developed in Ref. [103]. As shown in Ref. [169, 170], in this type of quantization one quantizes only the 3-coordinates $\eta_i^r(\tau)$ of the particles and not the gauge variables (i.e., the inertial effects like the Coriolis and centrifugal ones): They remain c-numbers describing the appearances of phenomena (they can be considered as gauge time variables to be added to the standard time). The known results in atomic and nuclear physics are reproduced.

3.5 The Non-Inertial Rest-Frames

As already said, the family of non-inertial rest-frames for an isolated system consists of all the admissible 3+1 splittings of Minkowski space-time whose instantaneous 3-spaces Σ_τ tend to space-like hyper-planes orthogonal to the conserved 4-momentum of the isolated system at spatial infinity. Therefore they tend to the Wigner 3-spaces of the inertial rest frame asymptotically.

As shown in Refs. [98, 99], these non-inertial frames can be centered on the external Fokker–Pryce center of inertia like the inertial ones and are described by the following embeddings:

$$\begin{aligned}
&z^\mu(\tau,\sigma^u) \approx z^\mu_F(\tau,\sigma^u) = Y^\mu(\tau) + u^\mu(\vec{h})\, g(\tau,\sigma^u) + \epsilon^\mu_r(\vec{h})\, [\sigma^r + g^r(\tau,\sigma^u)],\\
&\qquad \to_{|\vec{\sigma}|\to\infty} z^\mu_W(\tau,\sigma^u) = Y^\mu(\tau) + \epsilon^\mu_r(\vec{h})\, \sigma^r, \qquad x^\mu(\tau) = z^\mu_F(\tau, 0^u),\\
&g(\tau,0^u) = g^r(\tau,0^u) = 0, \qquad g(\tau,\sigma^u) \to_{|\vec{\sigma}|\to\infty} 0, \qquad g^r(\tau,\sigma^u) \to_{|\vec{\sigma}|\to\infty} 0.
\end{aligned} \tag{3.25}$$

These embeddings are a special case of (Eq. 2.17), with $x^\mu(\tau) = Y^\mu(\tau)$ and $z^\mu_F(\tau,\sigma^u) - Y^\mu(\tau) = \epsilon^\mu_\tau(\vec{h})\, g(\tau,\sigma^u) + \epsilon^\mu_r(\vec{h})\, [\sigma^r + g^r(\tau,\sigma^u)]$, $\epsilon^\mu_\tau(\vec{h}) = h^\mu = \dot{Y}^\mu(\tau)$.

For the induced metric, we have

$$\begin{aligned}
z^\mu_\tau(\tau,\sigma^u) &\approx z^\mu_{F\,\tau}(\tau,\sigma^u) = h^\mu\, [1 + \partial_\tau\, g(\tau,\sigma^u)] + \epsilon^\mu_r(\vec{h})\, \partial_\tau\, g^r(\tau,\sigma^u),\\
z^\mu_r(\tau,\sigma^u) &\approx z^\mu_{F\,r}(\tau,\sigma^u) = h^\mu\, \partial_r\, g(\tau,\sigma^u) + \epsilon^\mu_s(\vec{h})\, [\delta^s_r + \partial_r\, g^s(\tau,\sigma^u)],\\
\epsilon\, g_{F\,\tau\tau}(\tau,\sigma^u) &= [1 + \partial_\tau\, g(\tau,\sigma^u)]^2 - \sum_r [\partial_\tau\, g^r(\tau,\sigma^u)]^2\\
&= \Big[(1+n_F)^2 - h^{rs}_F\, n_{F\,r}\, n_{F\,s}\Big](\tau,\sigma^u),
\end{aligned}$$

$$\begin{aligned}
\epsilon\, g_{F\,\tau u}(\tau,\sigma^u) &= [1+\partial_\tau\, g(\tau,\sigma^u)]\, \partial_u\, g(\tau,\sigma^u) - \sum_r \partial_\tau\, g^r(\tau,\sigma^u)\, [\delta^r_u + \partial_u\, g^r(\tau,\sigma^u)]\\
&= \Big([1+\partial_\tau\, g]\, \partial_u\, g - \partial_\tau\, g^u - \sum_r \partial_\tau\, g^r\, \partial_u\, g^r\Big)(\tau,\sigma^u) = -n_{F\,u}(\tau,\sigma^u),\\
\epsilon\, g_{F\,uv}(\tau,\sigma^u) &= -h_{F\,uv}(\tau,\sigma^u)\\
&= \partial_u\, g(\tau,\sigma^u)\, \partial_v\, g(\tau,\sigma^u) - \sum_r [\delta^r_u + \partial_u\, g^r(\tau,\sigma^u)]\, [\delta^r_v + \partial_v\, g^r(\tau,\sigma^u)]\\
&= -\delta_{uv} + \Big(\partial_u\, g\, \partial_v\, g - (\partial_u\, g^v + \partial_v\, g^u) - \sum_r \partial_u\, g^r\, \partial_v\, g^r\Big)(\tau,\sigma^u).
\end{aligned} \tag{3.26}$$

The admissibility conditions for the foliation, plus the requirement $1 + n_F(\tau,\sigma^u) > 0$, can be written as restrictions on the functions $g(\tau,\sigma^u)$ and $g^r(\tau,\sigma^u)$.

The unit normal $l^\mu_F(\tau,\sigma^u)$ and the tangent 4-vectors $z^\mu_{F\,r}(\tau,\sigma^u)$ to the instantaneous 3-spaces Σ_τ can be projected on the asymptotic tetrad $h^\mu = \epsilon^\mu_\tau(\vec{h})$, $\epsilon^\mu_r(\vec{h})$:

$$\begin{aligned}
z^\mu_{F\,r}(\tau,\sigma^u) &= \Big[\partial_r\, g\, h^\mu + \partial_r\, g^s\, \epsilon^\mu_s(\vec{h})\Big](\tau,\sigma^u)\\
l^\mu_F(\tau,\sigma^u) &= \Big[\frac{1}{\sqrt{\gamma}}\, \epsilon^\mu_{\alpha\beta\gamma}\, z^\alpha_{F1}\, z^\beta_{F2}\, z^\gamma_{F3}\Big](\tau,\sigma^u)\\
&= \frac{1}{\sqrt{\gamma(\tau,\sigma^u)}}\, \Big[\det\,(\delta^s_r + \partial_r\, g^s)\, h^\mu\\
&\quad -\delta^{ra}\, \epsilon_{asu}\, \epsilon_{vwt}\, \partial_v\, g\, \partial_w\, g^s\, \partial_t\, g^u\, \epsilon^\mu_r(\vec{h})\Big](\tau,\sigma^u),\\
1+n_F(\tau,\sigma^u) &= \epsilon\, z^\mu_\tau(\tau,\sigma^u)\, l_{F\mu}(\tau,\sigma^u)\\
&= \frac{1}{\sqrt{\gamma(\tau,\sigma^u)}}\, \Big[(1+\partial_\tau\, g\, \det\,(\delta^s_r + \partial_r\, g^s)\\
&\quad -\partial_\tau\, g^r\, \epsilon_{rsu}\, \epsilon_{vwt}\, \partial_v\, g\, \partial_w\, g^s\, \partial_t\, g^u\Big](\tau,\sigma^u),\\
l^2_F(\tau,\sigma^u) = \epsilon, \Rightarrow\ \gamma_F(\tau,\sigma^u) &= \Big[\Big(\det\,(\delta^s_r + \partial_r\, g^s)\Big)^2\\
&\quad -2\, \epsilon_{vwt}\, \partial_v\, g\, \partial_w\, g^s\, \partial_t\, g^u\, \epsilon_{hmn}\, \partial_h\, g\, \partial_m\, g^s\, \partial_n\, g^u\Big](\tau,\sigma^u).
\end{aligned} \tag{3.27}$$

To define the non-inertial rest-frame instant form we must find the form of the internal Poincaré generators replacing the ones of the inertial rest-frame one, given in Eq. (2.21).

Eqs. (2.14) and (2.20) imply

$$\begin{aligned}
P^\mu &= Mc\, h^\mu = \int d^3\sigma\, \rho^\mu(\tau,\sigma^u)\\
&\approx h^\mu \int d^3\sigma\, \sqrt{\gamma(\tau,\sigma^u)}\, \Big(\frac{\det\,(\delta^s_r + \partial_r\, g^s)}{\sqrt{\gamma_F}}\, T_{F\,\perp\perp}\\
&\quad -\partial_r\, g\, h^{rs}_F\, T_{F\,\perp s}\Big)(\tau,\sigma^u)
\end{aligned}$$

$$
\begin{aligned}
&+\epsilon^\mu_u(\vec{h}) \int d^3\sigma \Big(- \frac{\delta^{ua}\, \epsilon_{asr}\, \epsilon_{vwt}\, \partial_v\, g\, \partial_w\, g^s\, \partial_t\, g^r}{\sqrt{\gamma_F}}\, T_{F\perp\perp} \\
&-(\delta^u_r + \partial_r\, g^u)\, h^{rs}_F\, T_{F\perp s}\Big)(\tau, \sigma^u) \\
&\stackrel{\text{def}}{=} \int d^3\sigma\, \mathcal{T}^\mu_F(\tau, \sigma^u),
\end{aligned}
\tag{3.28}
$$

so that the internal mass and the rest-frame conditions become (Eq. (3.10) is recovered for the inertial rest-frame)

$$
\begin{aligned}
Mc &= \int d^3\sigma \Big(\frac{\det(\delta^s_r + \partial_r\, g^s)}{\sqrt{\gamma}}\, T_{F\perp\perp} - \partial_r\, g\, h^{rs}_F\, T_{F\perp s}\Big)(\tau, \sigma^u), \\
\hat{\mathcal{P}}^u &= \int d^3\sigma \Big(- \frac{\delta^{ua}\, \epsilon_{asr}\, \epsilon_{vwt}\, \partial_v\, g\, \partial_w\, g^s\, \partial_t\, g^r}{\sqrt{\gamma_F}}\, T_{F\perp\perp} \\
&\quad -(\delta^u_r + \partial_r\, g^u)\, h^{rs}_F\, T_{F\perp s}\Big)(\tau, \sigma^u) \approx 0.
\end{aligned}
\tag{3.29}
$$

By using Eq. (2.21) for the angular momentum, we get $J^{\mu\nu} \approx \int d^3\sigma \Big(z^\mu_F\, \rho^\nu_F - z^\nu_F\, \rho^\mu_F\Big)(\tau, \sigma^u)$, with $\rho^\mu_F(\tau, \sigma^u) = \Big[\sqrt{\gamma_F}\, \Big(T_{\perp\perp}\, l^\mu_F - T_{\perp s}\, h^{sr}_F\, z^\mu_{Fr}\Big)\Big](\tau, \sigma^u)$, where z^μ_F, z^μ_{Fr}, and l^μ_F are given in Eqs. (3.25), (3.26), and (3.27) respectively. The description of the isolated system as a pole–dipole carried by the external center of mass $\vec{z}$ requires that we must identify the previous J^{ij} and J^{oi} with the expressions like the ones given in Eq. (2.21), now functions of $\vec{z}$, $\vec{h}$, Mc of Eq. (3.25) and of an effective spin $\vec{\tilde{S}}$. This identification will allow finding the effective spin $\vec{\tilde{S}}$ and three constraints $\tilde{\mathcal{K}}^r \approx 0$ eliminating the internal 3-center of mass: In the limit of the inertial rest-frame they must reproduce the quantities in Eq. (3.10).

By using Eq. (3.29), this procedure implies ($\hat{\mathcal{K}}^r$ and $\hat{\mathcal{S}}^r$ are the analogue of the quantities defined in Eq. (3.17) for the embedding (Eq. 3.25)):

$$
\begin{aligned}
J^{\mu\nu} &\approx \int d^3\sigma \Big(z^\mu_F\, \rho^\nu_F - z^\nu_F\, \rho^\mu_F\Big)(\tau, \sigma^u) \\
&= Mc \Big(Y^\mu(0)\, h^\nu - Y^\nu(0)\, h^\mu\Big) + \hat{\mathcal{P}}^u \Big(Y^\mu(0)\, \epsilon^\nu_u(\vec{h}) - Y^\nu(0)\, \epsilon^\mu_u(\vec{h})\Big) \\
&\quad + \Big(\tau\, \hat{\mathcal{P}}^u + \hat{\mathcal{K}}^u\Big) \Big(h^\mu\, \epsilon^\nu_u(\vec{h}) - h^\nu\, \epsilon^\mu_u(\vec{h})\Big) + \delta^{un}\, \epsilon_{nvr}\, \hat{\mathcal{S}}^r\, \epsilon^\mu_u(\vec{h})\, \epsilon^\nu_v(\vec{h}) \\
&\approx Mc \Big(Y^\mu(0)\, h^\nu - Y^\nu(0)\, h^\mu\Big) + \hat{\mathcal{K}}^u \Big(h^\mu\, \epsilon^\nu_u(\vec{h}) - h^\nu\, \epsilon^\mu_u(\vec{h})\Big) \\
&\quad + \delta^{un}\, \epsilon_{nvr}\, \hat{\mathcal{S}}^r\, \epsilon^\mu_u(\vec{h})\, \epsilon^\nu_v(\vec{h}),
\end{aligned}
$$

so that we get

$$
\begin{aligned}
J^{ij} &= z^i\, h^j - z^j\, h^i + \delta^{iu}\, \delta^{jv}\, \epsilon_{uvk}\, \tilde{S}^k \\
&\approx Mc \Big(Y^i(0)\, h^j - Y^j(0)\, h^i\Big) + \hat{\mathcal{K}}^u \Big(h^i\, \epsilon^j_u(\vec{h}) - h^j\, \epsilon^i_u(\vec{h})\Big)
\end{aligned}
$$

$$
\begin{aligned}
&+ \delta^{un}\, \epsilon_{nvr}\, \hat{S}^r\, \epsilon^i_u(\vec{h})\, \epsilon^j_v(\vec{h}),\\
J^{oi} &= -\sqrt{1+\vec{h}^2}\, z^i + \frac{\delta^{in}\, \epsilon_{njk}\, \tilde{S}^j\, h^k}{1+\sqrt{1+\vec{h}^2}}\\
&\approx Mc\left(Y^o(0)\, h^i - Y^i(0)\, h^o\right) + \hat{\mathcal{K}}^u\left(h^o\, \epsilon^i_u(\vec{h}) - h^i\, \epsilon^o_u(\vec{h})\right)\\
&+ \delta^{un}\, \delta^{vm}\, \epsilon_{nmr}\, \hat{S}^r\, \epsilon^o_u(\vec{h})\, \epsilon^i_v(\vec{h}).
\end{aligned} \tag{3.30}
$$

As a consequence, by using the expression of the Fokker–Pryce 4-center of inertia given in Eq. (3.14) for $Y^\mu(0)$, the constraints eliminating the 3-center of mass and the effective spin are

$$
\begin{aligned}
\hat{\mathcal{K}}^u &= \int d^3\sigma\, \Big(g\left[\delta^{ur}\, \partial_r\, g\, T_{F\perp\perp} - (\delta^u_r + \partial_r\, g^u)\, h^{rs}_F\, T_{F\perp s}\right]\\
&-(\sigma^u + g^u)\left[\frac{\det(\delta^s_r + \partial_r\, g^s)}{\sqrt{\gamma}}\, T_{F\perp\perp} - \partial_r\, g\, h^{rs}_F\, T_{F\perp s}\right]\Big)(\tau, \sigma^u) \approx 0,\\
\tilde{S}^r \approx \hat{S}^r &= \frac{1}{2}\, \delta^{rn}\, \epsilon_{nuv} \int d^3\sigma\, \Big((\sigma^u + g^u)\left[\delta^{vm}\, \partial_m\, g\, T_{F\perp\perp} - (\delta^v_r + \partial_r\, g^v)\, h^{rs}_F\, T_{F\perp s}\right]\\
&-(\sigma^v + g^v)\left[\delta^{um}\, \partial_m\, g\, T_{F\perp\perp} - (\delta^u_r + \partial_r\, g^u)\, h^{rs}_F\, T_{F\perp s}\right]\Big)(\tau, \sigma^u),
\end{aligned} \tag{3.31}
$$

and these formulas allow recovery of Eq. (3.4) of the inertial rest-frame.

Therefore, the non-inertial rest-frame instant form of dynamics is well defined.

We now have to find which is the effective Hamiltonian of the non-inertial rest-frame instant form replacing Mc of the inertial rest-frame one. The gauge fixing (Eq. 3.31) is a special case of Eq. (2.17). To be able to impose this gauge fixing, let us put $F^\mu(\tau, \sigma^u) = z^\mu_F(\tau, \sigma^u) - x^\mu(\tau) = h^\mu\, g(\tau, \sigma^u) + \epsilon^\mu_r(\vec{h})\, [\sigma^r + g^r(\tau, \sigma^u)]$ in Eq. (2.17), but let us leave $x^\mu(\tau)$ as an arbitrary time-like observer to be restricted to $Y^\mu(\tau)$ at the end. We will only assume that $x^\mu(\tau)$ is canonically conjugate with $P^\mu = \int d^3\sigma\, \rho^\mu(\tau, \sigma^u)$, $\{x^\mu(\tau), P^\nu\} = -\epsilon\, \eta^{\mu\nu}$.

Due to the dependence of $F^\mu(\tau, \sigma^u)$ and of $Y^\mu(\tau)$ on $\vec{h} = \vec{P}/\sqrt{\epsilon\, P^2}$, we must develop a different procedure for the identification of the Dirac Hamiltonian.

In this case, the constraints (Eq. 2.15) can be rewritten in the following form ($\mathcal{T}^\mu_F(\tau, \sigma^u)$ is defined in Eq. (3.28)):

$$
\begin{aligned}
&\mathcal{H}^\mu(\tau, \sigma^u) = \tilde{\mathcal{H}}^\mu(\tau, \sigma^u) + \delta^3(\sigma^u) \int d^3\sigma_1\, \mathcal{H}^\mu(\tau, \sigma^u_1) \approx 0,\\
&\text{with} \qquad \int d^3\sigma\, \tilde{\mathcal{H}}^\mu(\tau, \sigma^u) \equiv 0,\\
&\Downarrow\\
&\rho^\mu(\tau, \sigma^u) \approx P^\mu\, \delta^3(\sigma^u) + \left[\mathcal{T}^\mu_F(\tau, \sigma^u) - \delta^3(\sigma^u)\, \mathcal{R}^\mu_F(\tau)\right]\\
&\qquad = \delta^3(\sigma^u)\, H^\mu(\tau) + \mathcal{T}^\mu_F(\tau, \sigma^u),\\
&H^\mu(\tau) = P^\mu - \mathcal{R}^\mu_F(\tau) \approx 0, \qquad \mathcal{R}_F(\tau) \overset{def}{=} \int d^3\sigma\, \mathcal{T}^\mu_F(\tau, \sigma^u).
\end{aligned} \tag{3.32}
$$

In this way, the original canonical variables $z^\mu(\tau,\vec\sigma)$, $\rho^\mu(\tau,\vec\sigma)$ are replaced by the observer $x^\mu(\tau)$, P^μ and by relative variables with respect to it.

From Eq. (3.25) we get the following:

1. The gauge fixing to the constraints $\tilde{\mathcal{H}}^\mu(\tau,\sigma^u)\approx 0$ is

$$\psi^\mu_r(\tau,\sigma^u) = \frac{\partial\,\chi^\mu(\tau,\sigma^u)}{\partial\,\sigma^r} = \Big(z^\mu_r - \epsilon^\mu_s(\vec h)\,\Big[\delta^s_r + \frac{\partial\,g^s}{\partial\,\sigma^r} - u^\mu(\vec h)\,\frac{\partial\,g}{\partial\,\sigma^r}\Big]\Big)(\tau,\sigma^u)\approx 0. \tag{3.33}$$

2. The gauge fixing to the constraints $H^\mu(\tau) = P^\mu - \mathcal{R}^\mu_F\approx 0$ is $\chi^\mu(\tau,0) = z^\mu(\tau,0) - Y^\mu(\tau) = x^\mu(\tau) - Y^\mu(\tau)\approx 0$.

The gauge fixing (Eq. 3.33) has the following Poisson brackets with the collective variables $x^\mu(\tau)$, P^μ:

$$\begin{aligned}
&\{P^\mu,\psi^\nu_r(\tau,\sigma^u)\} = 0,\\
&\{x^\mu(\tau),\psi^\nu_r(\tau,\sigma^u)\} = -\frac{\partial\,\epsilon^\nu_s(\vec h)}{\partial\,P_\mu}\Big(\delta^s_r + \frac{\partial\,g^s(\tau,\sigma^u)}{\partial\,\sigma^r}\Big) - \frac{\partial\,\epsilon^\nu_\tau(\vec h)}{\partial\,P_\mu}\,\frac{\partial\,g(\tau,\sigma^u)}{\partial\,\sigma^r}\neq 0.
\end{aligned} \tag{3.34}$$

Therefore, $x^\mu(\tau)$ is no more a canonical variable after the gauge fixing $\psi^\mu_r(\tau,\sigma^u)\approx 0$.

By introducing the notation ($\epsilon^A_\mu = \eta^{AB}\,\epsilon_{B\mu}\ \Rightarrow\ \epsilon^\tau_\mu(\vec h) = \epsilon\,h_\mu$),

$$\mathcal{T}^\mu_F(\tau,\sigma^u)h^\mu\,\mathcal{T}^\tau_F(\tau,\sigma^u) + \epsilon^\mu_r(\vec h)\,\mathcal{T}^r_F(\tau,\sigma^u), \qquad \Rightarrow \mathcal{T}^A_F(\tau,\sigma^u) = \epsilon^A_\mu(\vec h)\,\mathcal{T}^\mu_F(\tau,\sigma^u), \tag{3.35}$$

the angular momentum generator of Eq. (2.14) takes the form

$$\begin{aligned}
J^{\mu\nu} &= x^\mu(\tau)\,P^\nu - x^\nu(\tau)\,P^\mu + S^{\mu\nu},\\
S^{\mu\nu} &\approx \epsilon^\mu_r(\vec h)\,\epsilon^\nu_s(\vec h)\,\int d^3\sigma\,\Big[(\sigma^r + g^r)\,\mathcal{T}^s - (\sigma^s + g^s)\,\mathcal{T}^r\Big](\tau,\sigma^u)\\
&\quad + \Big(\epsilon^\mu_r(\vec h)\,\epsilon^\nu_\tau(\vec h) - \epsilon^\nu_r(\vec h)\,\epsilon^\mu_\tau(\vec h)\Big)\,\int d^3\sigma\,\Big[(\sigma^r + g^r)\,\mathcal{T}^\tau + g\,\mathcal{T}^r\Big](\tau,\sigma^u)\\
&= \epsilon^\mu_A(\vec h)\,\epsilon^\nu_B(\vec h)\,S^{AB},\\
S^{rs} &= \int d^3\sigma\,\Big[(\sigma^r + g^r)\,\mathcal{T}^s - (\sigma^s + g^s)\,\mathcal{T}^r\Big](\tau,\sigma^u)\delta^{rn}\,\epsilon_{nsu}\,\mathcal{J}^u,\\
S^{\tau r} &= -S^{r\tau} = -\int d^3\sigma\,\Big[(\sigma^r + g^r)\,\mathcal{T}^\tau + g\,\mathcal{T}^r\Big](\tau,\sigma^u)\mathcal{K}^r,
\end{aligned} \tag{3.36}$$

where only the constraints $\tilde{\mathcal{H}}^\mu(\tau,\sigma^u)\approx 0$ have been used.

Since we have

$$\begin{aligned}
&\{x^\mu(\tau), S^{\alpha\beta}\} = 0,\\
&\Big\{\frac{\partial\,z^\mu(\tau,\sigma^u)}{\partial\,\sigma^r}, S^{\alpha\beta}\Big\} = \Big(\frac{\partial\,z^\beta}{\partial\,\sigma^r}\,\eta^{\mu\alpha} - \frac{\partial\,z^\alpha}{\partial\,\sigma^r}\,\eta^{\mu\beta}\Big)(\tau,\sigma^u)\\
&\qquad\approx \Big(\Big[\epsilon^\beta_s(\vec h)\,(\delta^s_r + \frac{\partial\,g^s}{\partial^r}) + h^\beta\,\frac{\partial\,g}{\partial\,\sigma^r}\Big]\,\eta^{\mu\alpha}\Big)(\tau,\sigma^u),
\end{aligned} \tag{3.37}$$

after the gauge fixing the new canonical variable for the observer becomes

$$\tilde{x}^\mu(\tau) = x^\mu(\tau) - \frac{1}{2}\,\epsilon_{\sigma\,A}(\vec{h})\,\frac{\partial\,\epsilon^A_\rho(\vec{h})}{\partial\,P_\mu}\,S^{\sigma\rho}, \qquad \{\tilde{x}^\mu(\tau), \psi^\nu_r(\tau,\vec{\sigma})\} = 0. \tag{3.38}$$

If we eliminate the relative variables by going to Dirac brackets with respect to the second-class constraints $\tilde{\mathcal{H}}^\mu(\tau,\sigma^u) \approx 0$, $\psi^\mu_r(\tau,\sigma^u) \approx 0$, the canonical variables $z^\mu(\tau,\sigma^u)$, $\rho_\mu(\tau,\sigma^u)$ are reduced to the canonical variables $\tilde{x}^\mu(\tau)$, P^μ.

By defining $\mathcal{R}_F(\tau) = \epsilon\,h_\mu\,\mathcal{R}^\mu_F(\tau) \approx Mc = \sqrt{\epsilon\,P^2}$, the remaining constraints are

$$\begin{aligned} H^\mu(\tau) &= h^\mu\left(\sqrt{\epsilon\,P^2} - \mathcal{R}_f(\tau)\right) + \epsilon^\mu_r(\vec{h})\,\hat{\mathcal{P}}^r, \\ \text{or}\quad \epsilon\,h^\mu\,H_\mu(\tau) &= \sqrt{\epsilon\,P^2} - \mathcal{R}_F(\tau) \approx 0, \qquad \epsilon^r_\mu(\vec{h})H^\mu(\tau) = \hat{\mathcal{P}}^r \approx 0. \end{aligned} \tag{3.39}$$

Like after Eq. (2.19), this reduction implies that the following form of the Dirac multiplier $\lambda^\mu(\tau,\sigma^u)$ in the Dirac Hamiltonian becomes

$$\begin{aligned} \lambda_\mu(\tau,\sigma^u) &= \epsilon\,h_\mu\left(\lambda_\tau(\tau) - \frac{\partial g(\tau,\sigma^u)}{\partial\tau}\right) + \epsilon\,\epsilon_{\mu\,r}(\vec{h})\left(\lambda^r(\tau) - \frac{\partial g^r(\tau,\sigma^u)}{\partial\tau}\right) \\ &\overset{\circ}{=} -\epsilon\,\frac{\partial z^\mu_F(\tau,\sigma^u)}{\partial\tau}. \end{aligned} \tag{3.40}$$

At this stage, the Dirac Hamiltonian depends only on the residual Dirac multipliers $\lambda_\tau(\tau)$ and $\vec{\lambda}(\tau)$:

$$H_D = \lambda_\tau(\tau)\,(\sqrt{\epsilon\,P^2} - \mathcal{R}_F) - \vec{\lambda}(\tau)\cdot\hat{\vec{\mathcal{P}}} + \int d^3\sigma\,\left(\frac{\partial\,g^r}{\partial\,\tau}\,\mathcal{T}_{F\,r} + \frac{\partial g}{\partial\tau}\,\mathcal{T}_{F\,\tau}\right)(\tau,\sigma^u), \tag{3.41}$$

where we introduced the notation $\mathcal{T}_{F\,A}(\tau,\sigma^u)\epsilon_{\mu A}(\vec{h})\,\mathcal{T}^\mu_F(\tau,\sigma^u)$ so that $\mathcal{T}^\tau_F = \mathcal{T}_{F\,\tau}$, $\mathcal{T}^r_F = -\epsilon\,\mathcal{T}_{F\,r}$.

To implement the gauge fixing $x^\mu(\tau) - Y^\mu(\tau) \approx 0$ requires two other steps:

1. We impose the gauge fixing $\tilde{x}^\mu(\tau)\,h_\mu = \epsilon\,\tau$. It implies $\lambda_\tau(\tau) = -1$ and $\sqrt{\epsilon\,P^2} = Mc \equiv \mathcal{R}_F$. The Dirac Hamiltonian becomes

$$\begin{aligned} H_{F\,D} &= \mathcal{M}\,c - \vec{\lambda}(\tau)\cdot\hat{\vec{\mathcal{P}}} + \int d^3\sigma\,\left[\mu\,\pi^\tau - A_\tau\,\Gamma\right](\tau,\sigma^u), \\ \mathcal{M}\,c &= Mc + \int d^3\sigma\,\left(\frac{\partial\,g^r}{\partial\,\tau}\,\mathcal{T}_{F\,r} + \frac{\partial g}{\partial\tau}\,\mathcal{T}_{F\,\tau}\right)(\tau,\sigma^u). \end{aligned} \tag{3.42}$$

2. We add the gauge fixing $\hat{\mathcal{K}}^r \approx 0$ to the rest-frame conditions $\hat{\mathcal{P}}^r \approx 0$: This implies $\vec{\lambda}(\tau) = 0$. In this way we get $x^\mu(\tau) \approx Y^\mu(\tau)$ and we also eliminate the internal 3-center of mass. Having chosen the Fokker–Pryce external 4-center of inertia $Y^\mu(\tau)$ as the origin of the 3-coordinates, the constraints $\hat{\mathcal{K}}^r \approx 0$ correspond to the requirement $S^{\tau r} \approx 0$.

In conclusion, the effective Hamiltonian $\mathcal{M}\,c$ of the non-inertial rest-frame instant form is not the internal mass Mc, since Mc describes the evolution from the point of view of the asymptotic inertial observers. There is an additional term interpretable as an inertial potential producing relativistic inertial effects (see Eq. (3.27) for $1 + n_F(\tau, \sigma^u)$ and Eq. (3.26) for $n_{F\,r}(\tau, \sigma^u)$):

$$\begin{aligned}
\mathcal{M}c &= Mc + \int d^3\sigma \left(\frac{\partial g^r}{\partial \tau} \mathcal{T}_{F\,r} + \frac{\partial g}{\partial \tau} \mathcal{T}_{F\,\tau} \right)(\tau, \sigma^u) \\
&= \int d^3\sigma\, \epsilon \left(\left[h_\mu \left(1 + \frac{\partial g}{\partial \tau} \right) + \epsilon_{\mu\,r} \frac{\partial g^r}{\partial \tau} \right] \mathcal{T}^\mu_F \right)(\tau, \sigma^u) \\
&= \int d^3\sigma\, \sqrt{\gamma(\tau, \sigma^u)} \left((1 + n_F)\, T_{F\perp\perp} + n^r_F\, T_{F\perp r} \right)(\tau, \sigma^u), \qquad (3.43)
\end{aligned}$$

where the energy–momentum tensor is given in Eq. (2.14) in the case of scalar particles.

4

Matter in the Rest-Frame Instant Form of Dynamics

Till now I have used scalar massive point particles to describe matter in the Wigner-covariant rest-frame instant form of dynamics. In Appendix B there is a description in this framework of:

1. massive particles with either positive or negative mass (the two branches of the mass shell describing particles and antiparticles) [95–97];
2. massive particles with either positive or negative electric charge [98–101, 105, 149–151]; and
3. massive spinning particles whose quantization reproduces the Dirac equation [171–173].

This is done by means of the use of Grassmann variables, as suggested by Berezin [174] in the framework of pseudo-classical mechanics. Also, the photon can be described in this framework [175].

See Ref. [176] for the description of massless particles, Ref. [177] for two-level atoms and Refs. [178–182] for the open and closed Nambu string in the rest-frame instant form.

In this chapter I shall describe the Klein–Gordon field [107], the electromagnetic field and its coupling to scalar point particles and its use in atomic physics [98–101, 149–151], the Yang–Mills field [93, 94], and the Dirac field [108]. Beside their rest-frame instant form description I shall look for the Dirac observables (DOs) of these fields. With these methods one can treat the Higgs field [183, 184] and the standard $SU(3) \otimes SU(2) \otimes U(1)$ model [185, 186].

4.1 The Klein–Gordon Field

Let us reformulate the real Klein–Gordon field on arbitrary space-like hypersurfaces Σ_τ, leaves of a 3+1 splitting of Minkowski space-time [107]. Let $\phi(\tau, \sigma^u) = \tilde{\phi}(z(\tau, \sigma^u))$ be the new field variable.

The action of a scalar Klein–Gordon field reads ($\dot{\phi}(\tau, \sigma^u) = \partial\phi(\tau, \sigma^u)/\partial\tau$):

$$
\begin{aligned}
S &= \int d\tau d^3\sigma\, N(\tau, \sigma^u)\, \sqrt{\gamma(\tau, \sigma^u)}\, \mathcal{L}(\tau, \sigma^u) = \int d\tau d^3\sigma\, N(\tau, \sigma^u)\, \sqrt{\gamma(\tau, \sigma^u)} \\
&\quad \frac{1}{2}\Big[{}^4g^{\tau\tau}\, \dot{\phi}^2 + 2\, {}^4g^{\tau r}\, \dot{\phi}\, \partial_r\, \phi + {}^4g^{rs}\, \partial_r\, \phi\, \partial_s\, \phi - m^2c^2\, \phi^2\Big](\tau, \sigma^u) \\
&= \int d\tau d^3\sigma\, \sqrt{\gamma(\tau, \sigma^u)}\, \frac{1}{2}\Big[\frac{1}{N}\, [\partial_\tau - N^r\, \partial_r]\, \phi\, [\partial_\tau - N^s\, \partial_s]\, \phi \\
&\quad + N\, [{}^3g^{rs}\, \partial_r\, \phi\, \partial_s\, \phi - m^2c^2\, \phi^2]\Big](\tau, \sigma^u),
\end{aligned}
\tag{4.1}
$$

where the configuration variables are $z^\mu(\tau, \sigma^u)$ and $\phi(\tau, \sigma^u)$. A self-interaction would correspond to the addition of a term $-2\, V(\phi(\tau, \sigma^u))$. As said, the new fields $\phi = \tilde{\phi} \circ z$ contain the non-local information about the 3+1 splitting of Minkowski space-time M^4, that is they have a built-in definition of equal time associated with the Lorentz-scalar time parameter τ which labels the leaves of the foliation.

The canonical momenta are ($\gamma = \det {}^3g$):

$$
\begin{aligned}
\pi(\tau, \sigma^u) &= \frac{\partial \mathcal{L}(\tau, \sigma^u)}{\partial \partial_\tau \phi(\tau, \sigma^u)} = \frac{\sqrt{\gamma(\tau, \sigma^u)}}{N(\tau, \sigma^u)}\, \Big[\dot{\phi} - N^r\, \partial_r\, \phi\Big](\tau, \sigma^u), \\
&\Rightarrow \dot{\phi}(\tau, \sigma^u) = \Big[\frac{N}{\sqrt{\gamma}}\, \pi + N^r\, \partial_r\, \phi\Big](\tau, \sigma^u), \\
\rho_\mu(\tau, \sigma^u) &= -\frac{\partial \mathcal{L}(\tau, \sigma^u)}{\partial \partial_\tau z^\mu(\tau, \sigma^u)} \\
&= l_\mu(\tau, \sigma^u)\, \Big[\frac{\sqrt{\gamma}}{2}\Big(\frac{1}{N^2}\, (\dot{\phi} - N^r\, \partial_r\, \phi)^2 - {}^3g^{rs}\, \partial_r\, \phi\, \partial_s\, \phi + m^2\, c^2\, \phi^2\Big)\Big]\, (\tau, \sigma^u) \\
&\quad + z_{s\mu}(\tau, \sigma^u)\, {}^3g^{sr}(\tau, \sigma^u)\, \Big[\frac{\sqrt{\gamma}}{N}\, \partial_r\, \phi\, (\dot{\phi} - N^u\, \partial_u\, \phi)\Big](\tau, \sigma^u).
\end{aligned}
\tag{4.2}
$$

We have the following primary constraints and Dirac Hamiltonian ($\lambda^\mu(\tau, \sigma^u)$ are the Dirac multiplier):

$$
\begin{aligned}
\bar{\mathcal{H}}_\mu(\tau, \sigma^u) &= \rho_\mu(\tau, \sigma^u) - l_\mu(\tau, \sigma^u)\, \Big[\frac{\pi^2}{2\, \sqrt{\gamma}} - \frac{\sqrt{\gamma}}{2}\, ({}^3g^{rs}\, \partial_r\, \phi\, \partial_s\, \phi - m^2\, c^2\, \phi^2)\Big](\tau, \sigma^u) \\
&\quad - z_{s\mu}(\tau, \sigma^u)\, \gamma^{rs}(\tau, \sigma^u)\, [\pi\, \partial_r\, \phi](\tau, \sigma^u) \approx 0, \\
\bar{H}_D &= \int d^3\sigma\, \lambda^\mu(\tau, \sigma^u)\, \bar{\mathcal{H}}_\mu(\tau, \sigma^u).
\end{aligned}
\tag{4.3}
$$

By using the Poisson brackets

$$
\{z^\mu(\tau, \sigma^u), \rho_\nu(\tau, \sigma'^u)\} = \epsilon\, \eta^\mu_\nu\, \delta^3(\sigma^u - \sigma'^u), \quad \{\phi(\tau, \sigma^u), \pi(\tau, \sigma'^u)\} = \delta^3(\sigma^u - \sigma'^u),
\tag{4.4}
$$

we find that the time constancy of the primary constraints does not imply the existence of new secondary constraints and that the constraints are first class and have vanishing Poisson brackets. The conserved Poincaré generators have the standard form (Eq. 2.14):

$$
\begin{aligned}
P^\mu_\phi &= \frac{1}{2}\int d^3\sigma\,\Big[\pi\,\partial^\mu\,\phi - \frac{1}{2}\,{}^4\eta^{o\mu}\,(\pi^2 - (\vec{\partial}\,\phi)^2 - m^2\,c^2\,\phi^2)\Big](\tau,\sigma^u),\\
J^{ij}_\phi &= \int d^3\sigma\,\Big[\pi\,(\sigma^i\,\partial^j - \sigma^j\,\partial^i)\,\phi\Big](\tau,\sigma^u),\\
J^{oi}_\phi &= \tau\int d^3\sigma\,(\pi\,\partial^i\,\phi)(\tau,\vec{\sigma}) - \frac{1}{2}\int d^3\sigma\,\sigma^i\,[\pi^2 + (\vec{\partial}\,\phi)^2 + m^2\,c^2\,\phi^2](\tau,\sigma^u).
\end{aligned} \tag{4.5}
$$

After the restriction to space-like hyper-planes $z^\mu(\tau,\sigma^u) = x^\mu(\tau) + b^\mu_r(\tau)\,\sigma^r$ and to the associated Dirac brackets (Eq. 2.19), the constraints are reduced to the following ten:

$$
\begin{aligned}
\tilde{\bar{\mathcal{H}}}^\mu(\tau) &= \int d^3\sigma\,\bar{\mathcal{H}}^\mu(\tau,\sigma^u) = P^\mu - b^\mu_A(\tau)\,\tilde{P}^A_\phi = P^\mu - l^\mu\,\tilde{P}^\tau_\phi - b^\mu_r(\tau)\,P^r_\phi\\
&= P^\mu - l^\mu\,\frac{1}{2}\int d^3\sigma\,\Big[\pi^2 + (\vec{\partial}\,\phi)^2 + m^2\,\phi^2\Big](\tau,\sigma^u)\\
&\quad - b^\mu_r(\tau)\int d^3\sigma\,\Big(\pi\,\partial_r\,\phi\Big)(\tau,\sigma^u) \approx 0,\\
\tilde{\bar{\mathcal{H}}}^{\mu\nu}(\tau) &= b^\mu_r(\tau)\int d^3\sigma\,\sigma^r\,\bar{\mathcal{H}}^\nu(\tau,\sigma^u) - b^\nu_r(\tau)\int d^3\sigma\,\sigma^r\,\bar{\mathcal{H}}^\mu(\tau,\sigma^u)\\
&= S^{\mu\nu}_s - \Big(b^\mu_r(\tau)\,l^\nu - b^\nu_r(\tau)\,l^\mu\Big)\,\frac{1}{2}\int d^3\sigma\,\sigma^r\,\Big[\pi^2 + (\vec{\partial}\,\phi)^2 + m^2\,\phi^2\Big](\tau,\sigma^u)\\
&\quad - \Big(b^\mu_r(\tau)\,b^\nu_s(\tau) - b^\nu_r(\tau)\,b^\mu_s(\tau)\Big)\int d^3\sigma\,\sigma^r\,\Big(\pi\,\partial^s\,\phi\Big)(\tau,\sigma^u)\\
&= S^{\mu\nu} - b^\mu_r(\tau)\,b^\nu_s(\tau)\,S^{rs}_\phi - \Big(b^\mu_r(\tau)\,l^\nu - b^\nu_r(\tau)\,l^\mu\Big)\,S^{\tau r}_\phi \approx 0,
\end{aligned} \tag{4.6}
$$

where $\tilde{P}^A_\phi = (\tilde{P}^\tau_\phi; P^r_\phi)$ is the 4-momentum of the field configuration and $S^{rs}_\phi = J^{rs}_\phi|_{\vec{P}_\phi=0}$, $S^{\tau r}_\phi = J^{or}_\phi|_{\vec{P}_\phi=0}$ its spin tensor.

If we select all the configurations of the system with time-like total momentum ($\epsilon\,P^2 > 0$), we can restrict ourselves to the Wigner hyper-planes $\Sigma_{W\tau}$ of the rest-frame instant form orthogonal to P^μ and arrive at the Dirac brackets (Eq. 3.1). Due to Eq. (3.3) the final phase space is spanned only by the variables $\tilde{x}^\mu_s(\tau)$, P^μ, $\phi(\tau,\vec{\sigma})$, and $\pi(\tau,\vec{\sigma})$, with standard Dirac brackets ($\{\phi(\tau,\vec{\sigma}),\pi(\tau,\vec{\sigma}^{'})\}^{**} = \delta^3(\vec{\sigma},\vec{\sigma}^{'})$; $\vec{\sigma}$ are Cartesian 3-coordinates).

Instead of $\tilde{x}^\mu_s$, P^μ, we can use the canonical variables Mc, $T_s = u(P)\cdot\tilde{x}/c$, $\vec{h}$, $\vec{z}$ of Eq. (3.8). The four surviving constraints are (now $\vec{P}_\phi = \{P^r_\phi\}$ is a spin-1 Wigner 3-vector)

$$
\begin{aligned}
\bar{\mathcal{H}}(\tau) &= Mc - P^\tau_\phi = Mc - \frac{1}{2}\int d^3\sigma\,\Big[\pi^2 + (\vec{\partial}\,\phi)^2 + m^2c^2\,\phi^2\Big](\tau,\vec{\sigma}) \approx 0,\\
\vec{\bar{\mathcal{H}}}_p(\tau) &= \vec{P}_\phi = \int d^3\sigma\,\Big(\pi\,\vec{\partial}\,\phi\Big)(\tau,\vec{\sigma}) \approx 0,
\end{aligned} \tag{4.7}
$$

where $\tilde{P}^A_\phi = (\tilde{P}^\tau_\phi; \vec{P}_\phi)$ is the 4-momentum of the field configuration. In the presence of a self-interaction P^τ_ϕ is replaced by $\tilde{P}^\tau_\phi = P^\tau_\phi + \int d^3\sigma\, V(\phi)(\tau,\vec{\sigma})$. The Dirac Hamiltonian is now $\bar{H}_D = \lambda(\tau)\,\bar{\mathcal{H}} + \vec{\lambda}(\tau)\cdot\vec{\bar{\mathcal{H}}}_p$.

By defining $\bar{S}^{AB}_s = \epsilon^A_\mu(u(p_s))\,\epsilon^B_\nu(u(p_s))\,S^{\mu\nu}_s$, we can show that on the Wigner hyper-planes we have $\bar{S}^{AB}_s \equiv S^{AB}_\phi = J^{AB}_\phi|_{P^D_\phi=0}$ (J^{AB}_ϕ is the angular momentum of the field configuration) so that the generators of the external realization of the Poincaré group in the rest-frame Wigner-covariant instant form of dynamics are given by Eq. (3.4) with

$$\bar{S}^{rs}_s \equiv S^{rs}_\phi = J^{rs}_\phi|_{\vec{P}_\phi=0} = \int d^3\sigma \left[\pi\left(\sigma^r\,\partial^s - \sigma^s\,\partial^r\right)\phi\right]|_{\vec{P}_\phi=0}(\tau,\vec{\sigma}). \tag{4.8}$$

The Hamiltonian for the τ-evolution after the gauge fixing (Eq. 3.7) $c\,T_s - \tau \approx 0$ is (P^τ_ϕ is replaced by $\tilde{P}^\tau_\phi$ in the presence of self-interaction)

$$\begin{aligned} \bar{H} &= \bar{M}_\phi c - \vec{\lambda}(\tau)\cdot\vec{\bar{\mathcal{H}}}_p(\tau),\\ M\,c = \bar{M}_\phi\,c = P^\tau_\phi &= \frac{1}{2}\int d^3\sigma\left[\pi^2 + (\vec{\partial}\,\phi)^2 + m^2\,c^2\,\phi^2\right](\tau,\vec{\sigma}), \end{aligned} \tag{4.9}$$

where $\bar{M}_\phi$ is the invariant mass of the field configuration. It, the spin vector (Eq. 4.8), and the boost generators $\mathcal{K}^r_{(int)} = \bar{S}^{\tau r}_s \equiv \frac{1}{2}\int d^3\sigma\,\sigma^r\left[\pi^2 + (\vec{\partial}\,\phi)^2 + m^2c^2\,\phi^2\right](\tau,\vec{\sigma})$ are the relevant generators of the unfaithful internal realization of the Poincaré group.

In the gauge $\vec{\lambda}(\tau) = 0$ and in the presence of a self-interaction, the Hamilton equations are ($\triangle = -\vec{\partial}^2$)

$$\begin{aligned} \partial_\tau\,\phi(\tau,\vec{\sigma}) &\mathrel{\circeq} \pi(\tau,\vec{\sigma}),\\ \partial_\tau\,\pi(\tau,\vec{\sigma}) &\mathrel{\circeq} -\left(\triangle + m^2c^2\right)\phi(\tau,\vec{\sigma}) - \frac{\partial V(\phi)}{\partial\phi}(\tau,\vec{\sigma}),\\ \Rightarrow\quad \left(\partial^2_\tau + \triangle + m^2c^2\right)\phi(\tau,\vec{\sigma}) &\mathrel{\circeq} -\frac{\partial V(\phi)}{\partial\phi}(\tau,\vec{\sigma}). \end{aligned} \tag{4.10}$$

We get a description in which the non-covariant canonical external center-of-mass 3-variables $\vec{z}$, $\vec{h}$, move freely and are decoupled from the Klein–Gordon field variables $\phi(\tau,\vec{\sigma})$, $\pi(\tau,\vec{\sigma})$, living on the Wigner hyper-plane and restricted by $\vec{P}_\phi \approx 0$.

We can use the external and internal realizations of the Poincaré group to build:

1. the three external 4-positions $\tilde{x}^\mu_s$ (the canonical non-covariant 4-center of mass), R^μ_s (the non-covariant non-canonical center of energy), and Y^μ_s (the non-canonical covariant center of inertia); and
2. the three internal analogous 3-positions $\vec{q}_+ \approx \vec{R}_+ \approx \vec{y}_+$ with the internal boost becoming $\vec{\mathcal{K}}_{(int)} = -\bar{M}_\phi\,c\,\vec{R}_+$.

To reduce the field variables only to relative degrees of freedom we must add the gauge fixings $\vec{q}_+ \approx 0$ (or $\vec{\mathcal{K}}_{(int)} \approx 0$).

To find the internal relative variables with respect to the internal 3-center of mass (see Eqs. (3.21)–(3.24) in the particle case) we must first find an analogue of the canonical transformation (Eq. 3.20) of the particle case. Here, the role of the naive internal particle center of mass $\vec{\eta}_+$ will be played by the space part $\vec{X}_\phi$ of the Longhi–Materassi center of phase defined in Ref. [187] (using mathematical results from Ref. [188]), after the reformulation on the Wigner hyper-planes.

To find this center we define the Fourier transform of the fields $\phi(\tau, \vec{\sigma})$, $\pi(\tau, \vec{\sigma})$ restricted to the solutions of the Klein–Gordon (Eq. 4.10),

$$\begin{aligned}
\phi(\tau, \vec{\sigma}) &\stackrel{\circ}{=} \int d\tilde{k}\, [a(\tau, \vec{k})\, e^{-ik_A\, \sigma^A} + \tilde{a}^*(\tau, \vec{k})\, e^{ik_A\, \sigma^A}],\\
\pi(\tau, \vec{\sigma}) &\stackrel{\circ}{=} -i \int d\tilde{k}\, \omega(\vec{k})\, [\tilde{a}(\tau, \vec{k})\, e^{-i\, k_A\, \sigma^A} - \tilde{a}^*(\tau, \vec{k})\, e^{i\, k_A\, \sigma^A}].
\end{aligned} \tag{4.11}$$

In the exponents we have $k_A\, \sigma^A = \omega(\vec{k})\, \tau - \vec{k}\cdot\vec{\sigma}$, which is a Lorentz scalar defined by using $k^A = (k^\tau = \omega(\vec{k}) = \sqrt{m^2\, c^2 + \vec{k}^2};\, k^r)$ with k^τ a Lorentz scalar and $\vec{k}$ a spin-1 Wigner 3-vector like $\vec{\sigma}$.

The Fourier coefficients $a(\tau, \vec{k})$, $a^*(\tau, \vec{k})$ and the modulus-phase variables $I(\tau, \vec{k})$, $\varphi(\tau, \vec{k})$ (in the free case they are τ-independent) have the expression ($d\tilde{k} = d^3k/\Omega(k)$, $\Omega(k) = (2\pi)^3\, \omega(k)$):

$$\begin{aligned}
a(\tau, \vec{k}) &= \int d^3\sigma\, \Big[\omega(\vec{k})\, \phi(\tau, \vec{\sigma}) + i\, \pi(\tau, \vec{\sigma})\Big]\, e^{i\,[\omega(\vec{k})\, \tau - \vec{k}\cdot\vec{\sigma}]} = \sqrt{I(\tau, \vec{k})}\, e^{i\, \varphi(\tau, \vec{k})},\\
a^*(\tau, \vec{k}) &= \int d^3\sigma\, \Big[\omega(\vec{k})\, \phi(\tau, \vec{\sigma}) - i\, \pi(\tau, \vec{\sigma})\Big]\, e^{-i\,[\omega(\vec{k})\, \tau - \vec{k}\cdot\vec{\sigma}]} = \sqrt{I(\tau, \vec{k})}\, e^{-i\, \varphi(\tau, \vec{k})},\\
I(\tau, \vec{k}) &= a^*(\tau, \vec{x})\, a(\tau, \vec{k})\\
&= \int d^3\sigma \int d^3\sigma'\, e^{i\, \vec{k}\cdot(\vec{\sigma} - \vec{\sigma}')}\, \Big[\omega(\vec{k})\phi(\tau, \vec{\sigma}) - i\, \pi(\tau, \vec{\sigma})\Big]\\
&\quad \Big[\omega(\vec{k})\phi(\tau, \vec{\sigma}') + i\, \pi(\tau, \vec{\sigma}')\Big],\\
\varphi(\tau, \vec{k}) &= \frac{1}{2i}\, \ln \left[\frac{a(\tau, \vec{k})}{a^*(\tau, \vec{k})}\right]\\
&= \frac{1}{2i}\, \ln \left[\frac{\int d^3\sigma\, \Big[\omega(\vec{k})\, \phi(\tau, \vec{\sigma}) + i\, \pi(\tau, \vec{\sigma})\Big]\, e^{i\,(\omega(\vec{k})\, \tau - \vec{k}\cdot\vec{\sigma})}}{\int d^3\sigma'\, \Big[\omega(\vec{k})\, \phi(\tau, \vec{\sigma}') - i\, \pi(\tau, \vec{\sigma}')\Big]\, e^{-i\,(\omega(\vec{k})\, \tau - \vec{k}\cdot\vec{\sigma}')}}\right]\\
&= \omega(\vec{k})\tau + \frac{1}{2i}\, \ln \left[\frac{\int d^3\sigma\, \Big[\omega(\vec{k})\phi(\tau, \vec{\sigma}) + i\, \pi(\tau, \vec{\sigma})\Big]\, e^{-i\, \vec{k}\cdot\vec{\sigma}}}{\int d^3\sigma'\, \Big[\omega(\vec{k})\phi(\tau, \vec{\sigma}') - i\, \pi(\tau, \vec{\sigma}')\Big]\, e^{i\, \vec{k}\cdot\vec{\sigma}'}}\right],
\end{aligned} \tag{4.12}$$

$$\phi(\tau,\vec{\sigma}) = \int d\tilde{k}\,\sqrt{I(\tau,\vec{k})}\,\Big[e^{i\,\varphi\tau,\vec{k})-i\,[\omega(\vec{k})\,\tau-\vec{k}\cdot\vec{\sigma}]} + e^{-i\,\varphi(\tau,\vec{k})+i\,[\omega(\vec{k})\,\tau-\vec{k}\cdot\vec{\sigma}]}\Big],$$
$$\pi(\tau,\vec{\sigma}) = -i\int d\tilde{k}\,\omega(\vec{k})\,\sqrt{I(\tau,\vec{k})}\,\Big[e^{i\,\varphi(\tau,\vec{k})-i\,[\omega(\vec{k})\,\tau-\vec{k}\cdot\vec{\sigma}]} - e^{-i\,\varphi(\tau,\vec{k})+i\,[\omega(\vec{k})\,\tau-\vec{k}\cdot\vec{\sigma}]}\Big]. \tag{4.13}$$

The 4-momentum and angular momentum of the field configuration on the Wigner hyper-planes are

$$\begin{aligned}
P^\tau_\phi &= \frac{1}{2}\int d^3\sigma\,\Big[\pi^2 + (\vec{\partial}\,\phi)^2 + m^2\phi^2\Big](\tau,\vec{\sigma})\\
&= \int d\tilde{k}\,\omega(\vec{k})\,a^*(\tau,\vec{k})\,a(\tau,\vec{k}) = \int d\tilde{k}\,\omega(\vec{k})\,I(\tau,\vec{k}),\\
\vec{P}_\phi &= \int d^3\sigma\,\Big(\pi\,\vec{\partial}\,\phi\Big)(\tau,\vec{\sigma}) = \int d\tilde{k}\,\vec{k}\,a^*(\tau,\vec{k})\,a(\tau,\vec{k}) = \int d\tilde{k}\,\vec{k}\,I(\vec{k}),\\
J^{rs}_\phi &= \int d^3\sigma\,\Big[\pi\,(\sigma^r\,\partial^s - \sigma^s\,\partial^r)\,\phi\Big](\tau,\vec{\sigma})\\
&= -i\int d\tilde{k}\,a^*(\tau,\vec{k})\,\Big(k^i\,\frac{\partial}{\partial k^j} - k^j\,\frac{\partial}{\partial k^i}\Big)\,a(\tau,\vec{k})\\
&= \int d\tilde{k}\,I(\vec{k})\,\Big(k^i\,\frac{\partial}{\partial k^j} - k^j\,\frac{\partial}{\partial k^i}\Big)\,\varphi(\vec{k}),\\
J^{\tau r}_\phi &= -\tau\int d^3\sigma\,(\pi\,\partial^r\,\phi)(\tau,\vec{\sigma}) + \frac{1}{2}\int d^3\sigma\,\sigma^r\,\Big[\pi^2 + (\vec{\partial}\,\phi)^2 + m^2c^2\,\phi^2\Big](\tau,\vec{\sigma})\\
&= -\tau\,P^r_\phi + i\int d\tilde{k}\,a^*(\tau,\vec{k})\,\omega(\vec{k})\,\frac{\partial}{\partial k^i}\,a(\tau,\vec{k})\\
&= -\tau\,P^r_\phi - \int d\tilde{k}\,I(\vec{k})\,\omega(\vec{k})\,\frac{\partial}{\partial k^i}\,\varphi(\vec{k}).
\end{aligned} \tag{4.14}$$

The classical analogue of the occupation number is

$$\begin{aligned}
N_\phi &= \int d\tilde{k}\,a^*(\tau,\vec{k})\,a(\tau,\vec{k}) = \int d\tilde{k}\,I(\tau,\vec{k})\\
&= \frac{1}{2}\int d^3\sigma\,\Big(\pi\,\frac{1}{\sqrt{m^2c^2+\triangle}}\pi + \phi\,\sqrt{m^2c^2+\triangle}\phi\Big)(\tau,\vec{\sigma}).
\end{aligned} \tag{4.15}$$

We must require the following behaviors also on the Wigner hyper-plane ($|q| = \sqrt{\vec{q}^{\,2}}$):

$$\begin{aligned}
&1.\ q\to\infty,\quad \sigma>0,\quad |a(\tau,\vec{q})|\to q^{-\frac{3}{2}-\sigma},\quad |I(\tau,\vec{q})|\to q^{-3-\sigma},\quad |\varphi(\tau,\vec{q})|\to q,\\
&2.\ q\to 0,\quad \epsilon>0,\ \eta>-\epsilon,\quad |a(\tau,\vec{q})|\to q^{-\frac{3}{2}+\epsilon},\quad |I(\tau,\vec{q})|\to q^{-3+\epsilon},\\
&\qquad |\varphi(\tau,\vec{q})|\to q^\eta
\end{aligned} \tag{4.16}$$

in order to have the Poincaré generators and the occupation number finite.

In Ref. [107] the existence of the following sequence of canonical transformations is demonstrated:

$$\boxed{\begin{array}{c}\phi(\tau,\vec{\sigma})\\ \pi(\tau,\vec{\sigma})\end{array}} \longrightarrow \boxed{\begin{array}{c}a(\tau,\vec{k})\\ a^*(\tau,\vec{k})\end{array}} \longrightarrow \boxed{\begin{array}{c}\varphi(\tau,\vec{k})\\ I(\tau,\vec{k})\end{array}},$$

$$\{a(\tau,\vec{k}), a^*(\tau,\vec{q})\} = -i\,\Omega(\vec{k})\,\delta^3(\vec{k}-\vec{q}), \qquad \{I(\tau,\vec{k}),\varphi(\tau,\vec{q})\} = \Omega(\vec{k})\,\delta^3(\vec{q}-\vec{k}). \tag{4.17}$$

On the Wigner hyper-planes the center of phase of Ref. [187] can be expressed in terms of four variables $X^A_\phi[\phi,\pi] = (X^\tau_\phi; \vec{X}_\phi)$ canonically conjugated to $P^A_\phi[\phi,\pi] = (P^\tau_\phi; \vec{P}_\phi)$, which are functionals of the phases

$$\begin{aligned} X^\tau_\phi &= \int d\tilde{q}\,\omega(\vec{q})\,F^\tau(q)\varphi(\tau,\vec{q}), \qquad \vec{X}_\phi = \int d\tilde{q}\,\vec{q}\,F(q)\varphi(\tau,\vec{q}),\\ &\Rightarrow \{X^r_\phi, X^s_\phi\} = 0, \qquad \{X^\tau_\phi, X^r_\phi\} = 0, \end{aligned} \tag{4.18}$$

depending on the two Lorentz-scalar functions $F^\tau(q)$, $F(q)$, whose form will be restricted by the following requirements implying that X^A_ϕ and P^A_ϕ are canonical variables:

$$\{P^\tau_\phi, X^\tau_\phi\} = 1,\quad \{P^r_\phi, X^s_\phi\} = -\delta^{rs},\quad \{P^r_\phi, X^\tau_\phi\} = 0,\quad \{P^\tau_\phi, X^r_\phi\} = 0. \tag{4.19}$$

Since $\{P^\tau_\phi, X^\tau_\phi\} = \int d\tilde{q}\,\omega^2(\vec{q})\,F^\tau(q)$ and $\{P^r_\phi, X^s_\phi\} = \int d\tilde{q}\,q^r\,q^s\,F(q)$, we must require the following normalizations for $F^\tau(q)$, $F(q)$:

$$\int d\tilde{q}\,\omega^2(\vec{q})\,F^\tau(q) = 1, \qquad \int d\tilde{q}\,q^r\,q^s\,F(q) = -\delta^{rs}. \tag{4.20}$$

Moreover, $\{P^r_\phi, X^\tau_\phi\} = \int d\tilde{q}\,\omega(\vec{q})\,q^r\,F^\tau(q)$ and $\{P^\tau_\phi, X^r_\phi\} = \int d\tilde{q}\,\omega(\vec{q})\,q^r\,F(q)$, implying the following conditions:

$$\int d\tilde{q}\,\omega(\vec{q})\,q^r\,F^\tau(q) = 0, \qquad \int d\tilde{q}\,\omega(\vec{q})\,q^r\,F(q) = 0, \tag{4.21}$$

which are automatically satisfied because $F^\tau(q)$, $F(q)$, $q = |\vec{q}|$ are even under $q^r \to -q^r$. A solution to these equations is

$$F^\tau(q) = \frac{16\,\pi^2}{mc\,q^2\,\sqrt{m^2c^2+q^2}}\,e^{-\frac{4\pi}{m^2c^2}q^2}, \qquad F(q) = -\frac{48\,\pi^2}{mc\,q^4}\,\sqrt{m^2c^2+q^2}\,e^{-\frac{4\pi}{m^2c^2}q^2}. \tag{4.22}$$

The singularity in $\vec{q} = 0$ requires $\varphi(\tau,\vec{q}) \to_{q\to 0} q^\eta$, $\eta > 0$, (and not $\eta > -\epsilon$ as in Eq. (4.16)) for the existence of X^τ_ϕ, $\vec{X}_\phi$.

Let us define an auxiliary relative action variable and an auxiliary relative phase variable:

$$\begin{aligned} \hat{I}(\tau,\vec{q}) &= I(\tau,\vec{q}) - F^\tau(q)\,P^\tau_\phi\,\omega(\vec{q}) + F(q)\,\vec{q}\cdot\vec{P}_\phi,\\ \hat{\varphi}(\tau,\vec{q}) &= \varphi(\tau,\vec{q}) - \omega(\vec{q})\,X^\tau_\phi + \vec{q}\cdot\vec{X}_\phi. \end{aligned} \tag{4.23}$$

The previous conditions on $F^\tau(q)$, $F(q)$, imply

$$\int d\tilde{q}\, \omega(\vec{q})\, \hat{I}(\tau,\vec{q}) = 0, \qquad \int d\tilde{q}\, q^r\, \hat{I}(\tau,\vec{q}) = 0,$$
$$\int d\tilde{q}\, F^\tau(q)\, \omega(\vec{q})\, \hat{\varphi}(\tau,\vec{q}) = 0, \qquad \int d\tilde{q}\, F(q)\, q^r\, \hat{\varphi}(\tau,\vec{q}) = 0. \tag{4.24}$$

These auxiliary variables have the following non-zero Poisson bracket:

$$\{\hat{I}(\tau,\vec{k}), \hat{\varphi}(\tau,\vec{q})\} = \Delta(\vec{k},\vec{q}) = \Omega(\vec{k})\, \delta^3(\vec{k}-\vec{q}) - F^\tau(k)\, \omega(\vec{k})\, \omega(q) + F(k)\vec{k}\cdot\vec{q}. \tag{4.25}$$

The distribution $\Delta(\vec{k},\vec{q})$ has the semigroup property $\int d\tilde{q}\, \Delta(\vec{k},\vec{q})\, \Delta(\vec{q},\vec{k}') = \Delta(\vec{k},\vec{k}')$ and satisfies the constraints:

$$\int d\tilde{q}\, \omega(\vec{q})\, \Delta(\vec{q},\vec{k}) = 0, \qquad \int d\tilde{q}\, q^r\, \Delta(\vec{q},\vec{k}) = 0,$$
$$\int d\tilde{q}\, F^\tau(q)\, \omega(\vec{q})\, \Delta(\vec{k},\vec{q}) = 0, \qquad \int d\tilde{q}\, q^r\, F(q)\, \Delta(\vec{k},\vec{q}) = 0. \tag{4.26}$$

At this stage the canonical variables $I(\tau,\vec{q})$, $\varphi(\tau,\vec{q})$ of Eq. (4.12) are replaced by the non-canonical set X^τ_ϕ, P^τ_ϕ, $\vec{X}_\phi$, $\vec{P}_\phi$, $\hat{I}(\tau,\vec{q})$, $\hat{\varphi}(\tau,\vec{q})$, whose non-null Poisson brackets are $\{P^\tau_\phi, X^\tau_\phi\} = 1$, $\{P^r_\phi, X^s_\phi\} = -\delta^{rs}$, and $\{\hat{I}(\tau,\vec{k}), \hat{\varphi}(\tau,\vec{q})\} = \Omega(\vec{k})\, \delta^3 (\vec{k}-\vec{q}) - F^\tau(k)\, \omega(\vec{k})\, \omega(\vec{q}) + F(k)\, \vec{k}\cdot\vec{q}$.

The generators of the Lorentz group are already decomposed into two parts, the collective and the relative ones, each satisfying the Lorentz algebra and having vanishing mutual Poisson brackets:

$$\begin{aligned}
J^{rs}_\phi &= L^{rs}_\phi + \hat{S}^{rs}_\phi,\\
&\qquad L^{rs}_\phi = X^r_\phi\, P^s_\phi - X^s_\phi\, P^r_\phi, \qquad \hat{S}^{rs}_\phi = \int d\tilde{q}\, \hat{I}(\tau,\vec{q}) \left(q^r \frac{\partial}{\partial q^s} - q^s \frac{\partial}{\partial q^r}\right) \hat{\varphi}(\tau,\vec{q}),\\
J^{\tau r}_\phi &= L^{\tau r}_\phi + \hat{S}^{\tau r}_\phi,\\
&\qquad L^{\tau r}_\phi = [X^\tau_\phi - \tau]\, P^r_\phi - X^r_\phi\, P^\tau_\phi, \qquad \hat{S}^{\tau r}_\phi = -\int d\tilde{q}\, \omega(\vec{q}) \hat{I}(\tau,\vec{q})\, \frac{\partial}{\partial q^r}\, \hat{\varphi}(\tau,\vec{q}).
\end{aligned} \tag{4.27}$$

We must now find the canonical relative variables hidden inside the auxiliary ones, which are not free but satisfy Eq. (4.24).

Let us introduce the following differential operator:

$$\mathcal{D}_{\vec{q}} = 3 - m^2c^2 \left[\sum_{i=1}^{3} \left(\frac{\partial}{\partial q^i}\right)^2 + \frac{2}{m^2c^2} \sum_{i=1}^{3} q^i \frac{\partial}{\partial q^i} + \frac{1}{m^2c^2} \left(\sum_{i=1}^{3} q^i \frac{\partial}{\partial q^i}\right)^2\right], \tag{4.28}$$

which is a scalar on the Wigner hyper-plane, has the null modes $\omega(\vec{q})$ and $\vec{q}$ and has a Green's function $\mathcal{G}(\vec{q},\vec{k})$ ($\mathcal{D}_{\vec{q}}\, \mathcal{G}(\vec{q},\vec{k}) = \Omega(\vec{k})\, \delta^3(\vec{k}-\vec{q})$) given explicitly in Ref. [107].

The operators $\mathcal{D}_{\vec{q}}$ and $\Delta(\vec{k},\vec{q})$ allow replacing the auxiliary functions $\hat{I}(\tau,\vec{q})$ and $\hat{\varphi}(\tau,\vec{q})$ with the following new functions $\mathbf{H}(\tau,\vec{q})$, $\mathbf{K}(\tau,\vec{q})$:

$$
\begin{aligned}
\hat{I}(\tau,\vec{q}) &= \mathcal{D}_{\vec{q}}\,\mathbf{H}(\tau,\vec{q}), \qquad \mathbf{H}(\tau,\vec{q}) = \int d\tilde{k}\, \mathcal{G}(\vec{q},\vec{k})\, \hat{I}(\tau,\vec{k}),\\
&\mathbf{H}(\tau,\vec{q}) \rightarrow_{q\rightarrow 0} q^{-1+\epsilon}, \qquad \epsilon > 0,\\
&\mathbf{H}(\tau,\vec{q}) \rightarrow_{q\rightarrow \infty} q^{-3-\sigma}, \qquad \sigma > 0,\\
\hat{\varphi}(\tau,\vec{q}) &= \int d\tilde{k} \int d\tilde{k}'\, \mathbf{K}(\tau,\vec{k})\, \mathcal{G}(\vec{k},\vec{k}')\, \Delta(\vec{k}',\vec{q}),\\
\mathbf{K}(\tau,\vec{q}) &= \mathcal{D}_{\vec{q}}\, \hat{\varphi}(\tau,\vec{q}) = \mathcal{D}_{\vec{q}}\, \varphi(\tau,\vec{q}),\\
&\mathbf{K}(\tau,\vec{q}) \rightarrow_{q\rightarrow 0} q^{-2+\epsilon}, \qquad \epsilon > 0,\\
&\mathbf{K}(\tau,\vec{q}) \rightarrow_{q\rightarrow \infty} q^{1-\sigma}, \qquad \sigma > 0,
\end{aligned}
$$

$$
\begin{aligned}
&\{\mathbf{H}(\tau,\vec{q}), X^\tau_\phi\} = 0, \quad \{\mathbf{H}(\tau,\vec{q}), P^\tau_\phi\} = 0, \quad \{\mathbf{H}(\tau,\vec{q}), X^r_\phi\} = 0, \quad \{\mathbf{H}(\tau,\vec{q}), P^r_\phi\} = 0,\\
&\{\mathbf{K}(\tau,\vec{q}), X^\tau_\phi\} = 0, \quad \{\mathbf{K}(\tau,\vec{q}), P^\tau_\phi\} = 0, \quad \{\mathbf{K}(\tau,\vec{q}), X^r_\phi\} = 0, \quad \{\mathbf{K}(\tau,\vec{q}), P^r_\phi\} = 0,\\
&\{\mathbf{H}(\tau,\vec{q}), \mathbf{K}(\tau,\vec{q}')\} = \Omega(\vec{q})\, \delta^3(\vec{q}-\vec{q}').
\end{aligned} \tag{4.29}
$$

The final decomposition of the Lorentz generators is

$$
\begin{aligned}
J^{rs}_\phi &= L^{rs}_\phi + S^{rs}_\phi,\\
&L^{rs}_\phi = X^r_\phi\, P^s_\phi - X^s_\phi\, P^r_\phi, \quad S^{rs}_\phi = \int d\tilde{k}\, \mathbf{H}(\tau,\vec{k}) \left(k^r \frac{\partial}{\partial k^s} - k^s \frac{\partial}{\partial k^r}\right) \mathbf{K}(\tau,\vec{k}),\\
J^{\tau r}_\phi &= L^{\tau r}_\phi + S^{\tau r}_\phi,\\
&L^{\tau r}_\phi = (X^\tau_\phi - \tau)\, P^r_\phi - X^r_\phi\, P^\tau_\phi, \quad S^{\tau r}_\phi = -\int d\tilde{q}\, \omega(\vec{q})\, \mathbf{H}(\tau,\vec{q})\, \frac{\partial}{\partial q^r}\, \mathbf{K}(\tau,\vec{q}).
\end{aligned} \tag{4.30}
$$

The final canonical transformation found in Ref. [107] is

$$
\begin{aligned}
I(\tau,\vec{q}) &= F^\tau(q)\, \omega(\vec{q})\, P^\tau_\phi - F(q)\, \vec{q}\cdot \vec{P}_\phi + \mathcal{D}_{\vec{q}}\, \mathbf{H}(\tau,\vec{q}),\\
\varphi(\tau,\vec{q}) &= \int d\tilde{k} \int d\tilde{k}'\, \mathbf{K}(\tau,\vec{k})\, \mathcal{G}(\vec{k},\vec{k}')\, \Delta(\vec{k}',\vec{q}) + \omega(\vec{q})\, X^\tau_\phi - \vec{q}\cdot \vec{X}_\phi,\\
N_\phi &= P^\tau_\phi \int d\tilde{q}\, \omega(\vec{q})\, F^\tau(q) - \vec{P}_\phi \cdot \int d\tilde{q}\, \vec{q}\, F(q) + \int d\tilde{q}\, \mathcal{D}_{\vec{q}}\, \mathbf{H}(\tau,\vec{q})\\
&= \tilde{c}\, \frac{P^\tau_\phi}{m} + \int d\tilde{q}\, \mathcal{D}_{\vec{q}}\, \mathbf{H}(\tau,\vec{q}), \quad \tilde{c} = m \int d\tilde{q}\, \omega(\vec{q})\, F^\tau(q)\\
&= 2 \int_0^\infty \frac{dq}{\sqrt{m^2c^2+q^2}}\, e^{-\frac{4\pi}{m^2} q^2},
\end{aligned} \tag{4.31}
$$

with the two functions $F^\tau(q)$, $F(q)$ given in Eq. (4.22). In Ref. [107] the conditions are given on the solutions of the Klein–Gordon equation for the existence of this canonical transformation. Its inverse is

$$
P^\tau_\phi = \int d\tilde{q}\, \omega(\vec{q})\, I(\tau,\vec{q}) = \frac{1}{2} \int d^3\sigma \left[\pi^2 + (\vec{\partial}\phi)^2 + m^2\phi^2\right](\tau,\vec{\sigma}),
$$

$$
\begin{aligned}
\vec{P}_\phi &= \int d\tilde{q}\, \vec{q}\, I(\tau,\vec{q}) = \int d^3\sigma \left[\pi\, \vec{\partial}\, \phi\right](\tau,\vec{\sigma}),\\
X^\tau_\phi &= \int d\tilde{q}\, \omega(\vec{q})\, F^\tau(q)\, \varphi(\tau,\vec{q}) = \tau + \frac{1}{2\pi\, i\, m} \int d^3q\, \frac{e^{-\frac{4\pi}{m^2c^2}q^2}}{q^2\,\omega(\vec{q})}\\
&\quad \ln\left[\frac{\omega(\vec{q}) \int d^3\sigma\, e^{i\,\vec{q}\cdot\vec{\sigma}}\, \phi(\tau,\vec{\sigma}) + i \int d^3\sigma\, e^{i\,\vec{q}\cdot\vec{\sigma}}\, \pi(\tau,\vec{\sigma})}{\omega(\vec{q}) \int d^3\sigma\, e^{-i\,\vec{q}\cdot\vec{\sigma}}\, \phi(\tau,\vec{\sigma}) - i \int d^3\sigma\, e^{-i\,\vec{q}\cdot\vec{\sigma}}\, \pi(\tau,\vec{\sigma})}\right]\\
&\overset{\text{def}}{=} \tau + \tilde{X}^\tau_\phi, \qquad \Rightarrow \qquad L^{\tau r}_\phi = \tilde{X}^\tau_\phi\, P^r_\phi - X^r_\phi\, P^\tau_\phi,\\
\vec{X}_\phi &= \int d\tilde{q}\, \vec{q}\, F(q)\, \varphi(\tau,\vec{q}) = \frac{2\,i}{\pi\, m} \int d^3q\, \frac{q^i}{q^4}\, e^{-\frac{4\pi}{m^2c^2}q^2}\\
&\quad \ln\left[\frac{\sqrt{m^2c^2+q^2} \int d^3\sigma\, e^{i\,\vec{q}\cdot\vec{\sigma}}\, \phi(\tau,\vec{\sigma}) + i \int d^3\sigma\, e^{i\,\vec{q}\cdot\vec{\sigma}}\, \pi(\tau,\vec{\sigma})}{\sqrt{m^2c^2+q^2} \int d^3\sigma\, e^{-i\,\vec{q}\cdot\vec{\sigma}}\, \phi(\tau,\vec{\sigma}) - i \int d^3\sigma\, e^{-i\,\vec{q}\cdot\vec{\sigma}}\, \pi(\tau,\vec{\sigma})}\right],\\
\mathbf{H}(\tau,\vec{q}) &= \int d\tilde{k}\, \mathcal{G}(\vec{q},\vec{k})\, [I(\tau,\vec{k}) - F^\tau(k)\, \omega(\vec{k}) \int d\tilde{q}_1\, \omega(\vec{q}_1)\, I(\tau,\vec{q}_1)\\
&\quad + F(k)\, \vec{k}\cdot \int d\tilde{q}_1\, \vec{q}_1\, I(\tau,\vec{q}_1)]\\
&= \int d^3\sigma_1 d^3\sigma_2 \Big[\pi(\tau,\vec{\sigma}_1)\, \pi(\tau,\vec{\sigma}_2) \int d\tilde{k}\, \mathcal{G}(\vec{q},\vec{k}) \int d\tilde{k}_1\, \triangle(\vec{k},\vec{k}_1)\, e^{i\,\vec{k}_1\cdot(\vec{\sigma}_1-\vec{\sigma}_2)}\\
&\quad + \phi(\tau,\vec{\sigma}_1)\, \phi(\tau,\vec{\sigma}_2) \int d\tilde{k}\, \mathcal{G}(\vec{q},\vec{k}) \int d\tilde{k}_1\, \omega^2(\vec{k}_1)\, \triangle(\vec{k},\vec{k}_1)\, e^{i\,\vec{k}_1\cdot(\vec{\sigma}_1-\vec{\sigma}_2)}\\
&\quad - i\Big(\pi(\tau,\vec{\sigma}_1)\, \phi(\tau,\vec{\sigma}_2) + \pi(\tau,\vec{\sigma}_2)\, \phi(\tau,\vec{\sigma}_1)\Big)\\
&\quad \int d\tilde{k}\, \mathcal{G}(\vec{q},\vec{k}) \int d\tilde{k}_1\, \omega(\vec{k}_1)\, \triangle(\vec{k}.\vec{k}_1)\, e^{i\,\vec{k}_1\cdot(\vec{\sigma}_1-\vec{\sigma}_2)}\Big],\\
\mathbf{K}(\tau,\vec{q}) &= \mathcal{D}_{\vec{q}}\, \hat{\varphi}(\tau,\vec{q}) = \mathcal{D}_{\vec{q}}\, \varphi(\tau,\vec{q})\\
&= \frac{1}{2\,i}\, \mathcal{D}_{\vec{q}} \ln \frac{\int d^3\sigma\, [\omega(\vec{q})\, \phi(\tau,\vec{\sigma}) + i\, \pi(\tau,\vec{\sigma})]\; e^{-i\,\vec{q}\cdot\vec{\sigma}}}{\int d^3\sigma'\, [\omega(\vec{q})\, \phi(\tau,\vec{\sigma}') - i\, \pi(\tau,\vec{\sigma}')]\; e^{i\,\vec{q}\cdot\vec{\sigma}'}}.
\end{aligned} \tag{4.32}
$$

We get the following expression of the other canonical variables $a(\tau,\vec{q})$, $\phi(\tau,\vec{\sigma})$, $\pi(\tau,\vec{\sigma})$, in terms of the final ones:

$$
\begin{aligned}
a(\tau,\vec{q}) &= \sqrt{F^\tau(q)\, \omega(\vec{q})\, P^\tau_\phi - F(q)\, \vec{q}\cdot \vec{P}_\phi + \mathcal{D}_{\vec{q}}\, \mathbf{H}(\tau,\vec{q})}\\
&\quad e^{i\,[\omega(\vec{q})\, X^\tau_\phi - \vec{q}\cdot\vec{X}_\phi] + i \int d\tilde{k} \int d\tilde{k}'\, \mathbf{K}(\tau,\vec{k})\, \mathcal{G}(\vec{k},\vec{k}')\, \triangle(\vec{k}',\vec{q})},\\
N_\phi &= \tilde{c}\, \frac{P^\tau_\phi}{m\, c} + \int d\tilde{k}\, \mathcal{D}_{\vec{k}}\, \mathbf{H}(\tau,\vec{k}),
\end{aligned} \tag{4.33}
$$

$$
\begin{aligned}
\phi(\tau,\vec{\sigma}) &= \int d\tilde{q} \sqrt{F^\tau(q)\, \omega(\vec{q})\, P^\tau_\phi - F(q)\, \vec{q}\cdot \vec{P}_\phi + \mathcal{D}_{\vec{q}}\, \mathbf{H}(\tau,\vec{q})}\\
&\quad \Big[e^{-i\,[\omega(\vec{q})\,(\tau - X^\tau_\phi) - \vec{q}\cdot(\vec{\sigma}-\vec{X}_\phi)] + i \int d\tilde{k} \int d\tilde{k}'\, \mathbf{K}(\tau,\vec{k})\, \mathcal{G}(\vec{k},\vec{k}')\, \triangle(\vec{k}',\vec{q})}\\
&\quad + e^{i\,[\omega(\vec{q})\,(\tau - X^\tau_\phi) - \vec{q}\cdot(\vec{\sigma}-\vec{X}_\phi)] - i \int d\tilde{k} \int d\tilde{k}'\, \mathbf{K}(\tau,\vec{k})\, \mathcal{G}(\vec{k},\vec{k}')\, \triangle(\vec{k}',\vec{q})}\Big]
\end{aligned}
$$

$$
\begin{aligned}
&= 2 \int d\tilde{q}\, \mathbf{A}_{\vec{q}}(\tau; P^A_\phi, \mathbf{H}]\, \cos\Big[\vec{q}\cdot\vec{\sigma} + \mathbf{B}_{\vec{q}}(\tau; X^A_\phi, \mathbf{K}]\Big],\\
\pi(\tau,\vec{\sigma}) &= -i \int d\tilde{q}\, \omega(\vec{q})\, \sqrt{F^\tau(q)\,\omega(\vec{q})\, P^\tau_\phi - F(q)\,\vec{q}\cdot\vec{P}_\phi + \mathcal{D}_{\vec{q}}\mathbf{H}(\tau,\vec{q})}\\
&\quad \Big[e^{-i\,[\omega(\vec{q})\,(\tau - X^\tau_\phi) - \vec{q}\cdot(\vec{\sigma} - \vec{X}_\phi)] + i\int d\tilde{k}\int d\tilde{k}'\, \mathbf{K}(\tau,\vec{k})\, \mathcal{G}(\vec{k},\vec{k}')\, \triangle(\vec{k}',\vec{q})}\\
&\quad - e^{+i\,[\omega(\vec{q})\,(\tau - X^\tau_\phi) - \vec{q}\cdot(\vec{\sigma} - \vec{X}_\phi)] - i\int d\tilde{k}\int d\tilde{k}'\, \mathbf{K}(\tau,\vec{k})\, \mathcal{G}(\vec{k},\vec{k}')\, \triangle(\vec{k}',\vec{q})}\Big]\\
&= -2 \int d\tilde{q}\omega(q)\, \mathbf{A}_{\vec{q}}(\tau; P^A_\phi, \mathbf{H}]\, \sin\Big[\vec{q}\cdot\vec{\sigma} + \mathbf{B}_{\vec{q}}(\tau; X^A_\phi, \mathbf{K}]\Big],\\
\mathbf{A}_{\vec{q}}(\tau; P^A_\phi, \mathbf{H}] &= \sqrt{F^\tau(q)\,\omega(\vec{q})\, P^\tau_\phi - F(q)\,\vec{q}\cdot\vec{P}_\phi + \mathcal{D}_{\vec{q}}\mathbf{H}(\tau,\vec{q})} = \sqrt{I(\tau,\vec{q})},\\
\mathbf{B}_{\vec{q}}(\tau; X^A_\phi, \mathbf{K}] &= -\vec{q}\cdot\vec{X}_\phi - \omega(\vec{q})\,(\tau - X^\tau_\phi) + \int d\tilde{k} d\tilde{k}'\, \mathbf{K}(\tau,\vec{k})\, \mathcal{G}(\vec{k},\vec{k}')\, \triangle(\vec{k}',\vec{q})\\
&= \varphi(\tau,\vec{q}) - \omega(\vec{q})\,\tau. \qquad (4.34)
\end{aligned}
$$

The Klein–Gordon field configuration is described by:

1. its energy P^τ_ϕ and the conjugate field time-variable X^τ_ϕ, which is equal to τ plus some kind of internal time $\tilde{X}^\tau_\phi$, which does not exist for relativistic particles;
2. the conjugate reduced canonical variables of the internal gauge 3-position $\vec{X}_\phi$, $\vec{P}_\phi \approx 0$; and
3. an infinite set of canonically conjugate relative variables $\mathbf{H}(\tau,\vec{q})$, $\mathbf{K}(\tau,\vec{q})$.

While sets 1 and 2 describe some kind of monopole field configuration (it is the Dixon monopole on the Wigner hyper-planes, see Ref. [107]), which depends only on eight degrees of freedom, like a scalar particle at rest ($\vec{P}_\phi \approx 0$) and with mass $\epsilon_s \approx \sqrt{(P^\tau_\phi)^2 - \vec{P}^2_\phi} \approx P^\tau_\phi$, corresponding to the decoupled collective variables of the field configuration, set 3 describes an infinite set of canonical relative variables with respect to the relativistic collective variables of sets 1 and 2.

If we add the gauge-fixings $\vec{X}_\phi \approx 0$ to $\vec{P}_\phi \approx 0$ (this implies $\vec{\lambda}(\tau) = 0$ in the Dirac Hamiltonian like $\vec{q}_+ \approx 0$ in relativistic mechanics) and go to Dirac brackets, the rest-frame instant-form Klein–Gordon canonical variables in the gauge $\tau \equiv c\,T_s$ are (in the following formulas one has $c\,T_s - X^\tau_\phi = -\tilde{X}^\tau_\phi$):

$$
\begin{aligned}
a(T_s,\vec{q}) &= \sqrt{F^\tau(q)\,\omega(\vec{q})\, P^\tau_\phi + \mathcal{D}_{\vec{q}}\mathbf{H}(T_s,\vec{q})}\\
&\quad e^{i\,[\omega(\vec{q})\,\tilde{X}^\tau_\phi + \vec{q}\cdot\vec{\sigma}] + i\int d\tilde{k}\int d\tilde{k}'\, \mathbf{K}(T_s,\vec{k})\, \mathcal{G}(\vec{k},\vec{k}')\, \triangle(\vec{k}',\vec{q})},\\
N_\phi &= \tilde{c}\,\frac{P^\tau_\phi}{m} + \int d\tilde{q}\, \mathcal{D}_{\vec{q}}\mathbf{H}(T_s,\vec{q}),\\
\phi(T_s,\vec{\sigma}) &= \int d\tilde{q}\, \sqrt{F^\tau(q)\,\omega(\vec{q})\, P^\tau_\phi + \mathcal{D}_{\vec{q}}\mathbf{H}(T_s,\vec{q})}\\
&\quad \Big[e^{i\,[\omega(\vec{q})\,\tilde{X}^\tau_\phi + \vec{q}\cdot\vec{\sigma}] + i\int d\tilde{k}\int d\tilde{k}'\, \mathbf{K}(T_s,\vec{k})\, \mathcal{G}(\vec{k},\vec{k}')\, \triangle(\vec{k}',\vec{q})}
\end{aligned}
$$

$$
\begin{aligned}
&\quad + e^{-i\,[\omega(\vec q)\,\tilde X^\tau_\phi + \vec q\cdot\vec\sigma] - i\int d\tilde k \int d\tilde k'\, \mathbf{K}(T_s,\vec k)\,\mathcal{G}(\vec k,\vec k')\,\Delta(\vec k',\vec q)}\Big]\\
&= 2\int d\tilde q\, \mathbf{A}_{\vec q}(T_s; P^\tau_\phi, \mathbf{H}]\, \cos\Big[\vec q\cdot\vec\sigma + \mathbf{B}_{\vec q}(T_s; \tilde X^\tau_\phi, \mathbf{K}]\Big],\\
\pi(T_s,\vec\sigma) &= -i\int d\tilde q\, \omega(\vec q)\,\sqrt{F^\tau(q)\,\omega(\vec q)\,P^\tau_\phi + \mathcal{D}_{\vec q}\mathbf{H}(T_s,\vec q)}\\
&\quad \Big[e^{i\,[\omega(\vec q)\,\tilde X^\tau_\phi + \vec q\cdot\vec\sigma] + i\int d\tilde k \int d\tilde k'\, \mathbf{K}(T_s,\vec k)\,\mathcal{G}(\vec k,\vec k')\,\Delta(\vec k',\vec q)}\\
&\quad - e^{-i\,[\omega(\vec q)\,\tilde X^\tau_\phi + \vec q\cdot\vec\sigma] - i\int d\tilde k \int d\tilde k'\, \mathbf{K}(T_s,\vec k)\,\mathcal{G}(\vec k,\vec k')\,\Delta(\vec k',\vec q)}\Big]\\
&= -2\int d\tilde q\, \omega(\vec q)\, \mathbf{A}_{\vec q}(T_s; P^\tau_\phi, \mathbf{H}]\, \sin\Big[\vec q\cdot\vec\sigma + \mathbf{B}_{\vec q}(T_s; \tilde X^\tau_\phi, \mathbf{K}]\Big],\\
\mathbf{A}_{\vec q}(T_s; P^\tau_\phi, \mathbf{H}] &= \sqrt{F^\tau(q)\,\omega(\vec q)\,P^\tau_\phi + \mathcal{D}_{\vec q}\mathbf{H}(T_s,\vec q)} = \sqrt{I(T_s,\vec q)},\\
\mathbf{B}_{\vec q}(T_s; X^\tau_\phi, \mathbf{K}] &= \int d\tilde k d\tilde k'\, \mathbf{K}(T_s,\vec k)\mathcal{G}(\vec k,\vec k')\,\Delta(\vec k',\vec q) + \omega(\vec q)\,\tilde X^\tau_\phi\\
&= \varphi(T_s,\vec q) - \omega(\vec q)\,T_s.
\end{aligned}
\tag{4.35}
$$

As in the particle case, one can reintroduce an evolution in T_s inside the Wigner hyper-plane with the Hamiltonian $\tilde M_\phi = P^\tau_\phi$. In the free case $\mathbf{H}(T_s,\vec q)$, $\mathbf{K}(T_s,\vec q)$ are constants of the motion.

By adding the two second-class constraints $X^\tau_\phi - T_s \approx 0$, $P^\tau_\phi - \text{const.} \approx 0$, and by going to Dirac brackets, we get the rest-frame Hamilton–Jacobi formulation corresponding to the given constant value of the total energy and by putting the gauge fixing $\vec X_\phi \approx 0$, one chooses the Fokker–Pryce center of inertia as the inertial observer origin of the $\vec\sigma$ 3-coordinates on the Wigner hyper-planes.

Let us remark that the collective and relative canonical variables can be defined in closed form only in the absence of self-interactions of the field.

4.2 The Electromagnetic Field and Its Dirac Observables

On a space-like hyper-surface Σ_τ with embedding $x^\mu = z^\mu(\tau,\sigma^u)$, we describe [105] the electromagnetic potential $\tilde A_\mu(x^o,\vec x)$ and field strength $\tilde F_{\mu\nu}(x^0,\vec x) = \partial_\mu\,\tilde A_\nu(x^o,\vec x) - \partial_\nu\,\tilde A_\mu(x^o,\vec x)$ with Lorentz-scalar variables $A_A(\tau,\sigma^u)$ and $F_{AB}(\tau,\sigma^u)$ respectively, defined by

$$
\begin{aligned}
A_A(\tau,\sigma^u) &= z^\mu_A(\tau,\sigma^u)\,\tilde A_\mu(z(\tau,\sigma^u)),\\
F_{AB}(\tau,\sigma^u) &= \partial_A\,\tilde A_B(\tau,\sigma^u) - \partial_B\,\tilde A_A(\tau,\sigma^u)\\
&= z^\mu_A(\tau,\sigma^u)\,z^\nu_B(\tau,\sigma^u)\,\tilde F_{\mu\nu}(z(\tau,\sigma^u)).
\end{aligned}
\tag{4.36}
$$

The electric and magnetic fields are $E_r(\tau,\sigma^u) = F_{r\tau}(\tau,\sigma^u)$ and $B_r(\tau,\sigma^u) = \frac{1}{2}\,\epsilon_{rst}\,F_{st}(\tau,\sigma^u)$, $F_{rs} = \epsilon_{rsu}\,B_u$, $F^{AB} = {}^4g^{AC}\,{}^4g^{BD}\,F_{CD}$.

In inertial frames the fields are usually assumed to obey the boundary conditions at spatial infinity $\tilde A_\mu(x^o,\vec x) \to_{r\to\infty} \frac{a_\mu}{r^{1+\epsilon}} + O(r^{-2})$ (ϵ small, $r = |\vec x|$), implying a finite action. However, as discussed in Section 4.4, to have a good behavior of the Hamiltonian gauge transformations (and to avoid the Gribov ambiguity in Yang–Mills theory) it is convenient to use the boundary conditions $\tilde{A}_o(x^o, \vec{x}) \rightarrow_{r\rightarrow\infty} \frac{a_o}{r^{1+\epsilon}} + O(r^{-2})$ $\tilde{A}_i(x^o, \vec{x}) \rightarrow_{r\rightarrow\infty} \frac{a_i}{r^{2+\epsilon}} + O(r^{-3})$.

In parametrized Minkowski theories these boundary conditions become covariant in a natural way. In the Wigner 3-spaces of the rest-frame we have $r = |\vec{\sigma}|$ and $A_\tau(\tau, \vec{\sigma}) \rightarrow_{r\rightarrow\infty} \frac{a_\mu}{r^{1+\epsilon}} + O(r^{-2})$, $A_r(\tau, \vec{\sigma}) \rightarrow_{r\rightarrow\infty} \frac{a_\mu}{r^{2+\epsilon}} + O(r^{-3})$.

The system is described by the action

$$S = \int d\tau d^3\sigma\, \mathcal{L}(\tau, \sigma^u),$$
$$\mathcal{L}(\tau, \sigma^u) = -\frac{1}{4} \sqrt{g(\tau, \sigma^u)}\, {}^4g^{AC}(\tau, \sigma^u)\, {}^4g^{BD}(\tau, \sigma^u)\, F_{AB}(\tau, \sigma^u)\, F_{CD}(\tau, \sigma^u), \tag{4.37}$$

where the configuration variables are $z^\mu(\tau, \sigma^u)$ and $A_A(\tau, \sigma^u)$. The explicit expression of the Lagrangian density is

$$\begin{aligned}
\mathcal{L}(\tau, \sigma^h) &= -\Big(\frac{1}{4} \sqrt{g}\, {}^4g^{AC}\, {}^4g^{BD}\, F_{AB}\, F_{CD}\Big)(\tau, \sigma^h) \\
&= -\frac{1}{4} \Big(\sqrt{g} \Big[2\, ({}^4g^{\tau\tau}\, {}^4g^{rs} - {}^4g^{\tau r}\, {}^4g^{\tau s})\, F_{\tau r}\, F_{\tau s} \\
&\quad + 4\, {}^4g^{rs}\, {}^4g^{\tau u}\, F_{\tau r}\, F_{su} + {}^4g^{ru}\, {}^4g^{sv}\, F_{rs}\, F_{uv}\Big]\Big)(\tau, \sigma^h) \\
&= -\Big(\sqrt{\gamma} \Big[\frac{1}{2} \sqrt{\frac{\gamma}{g}}\, F_{\tau r}\, {}^3g^{rs}\, F_{\tau s} - \sqrt{\frac{\gamma}{g}}\, {}^4g_{\tau v}\, {}^3g^{vr}\, F_{rs}\, {}^3g^{su}\, F_{\tau u} \\
&\quad + \frac{1}{4} \sqrt{\frac{g}{\gamma}}\, {}^3g^{rs}\, F_{ru}\, F_{sv}\, ({}^3g^{uv} + 2 \frac{\gamma}{g}\, {}^4g_{\tau m}\, {}^3g^{mu}\, {}^4g_{\tau n}\, {}^3g^{nv})\Big]\Big)(\tau, \sigma^h).
\end{aligned} \tag{4.38}$$

The action is invariant under separate τ- and $\vec{\sigma}$-reparametrizations, since $A_\tau(\tau, \sigma^u) = z^\mu_\tau(\tau, \sigma^u)\, \tilde{A}_\mu(z(\tau, \sigma^u))$ transforms as a τ-derivative.

The canonical momenta are (for ${}^4g_{AB} \rightarrow {}^4\eta_{AB}$ one gets $\pi^r = -E_r = E^r$):

$$\begin{aligned}
\pi^\tau(\tau, \sigma^u) &= \frac{\partial \mathcal{L}(\tau, \sigma^u)}{\partial \partial_\tau A_\tau(\tau, \sigma^u)} = 0, \\
\pi^r(\tau, \sigma^u) &= \frac{\partial \mathcal{L}(\tau, \sigma^u)}{\partial \partial_\tau A_r(\tau, \sigma^u)} = -\frac{\gamma(\tau, \sigma^u)}{\sqrt{g(\tau, \sigma^u)}}\, \gamma^{rs}(\tau, \sigma^u) \Big(F_{\tau s} - {}^4g_{\tau v}\, {}^3g^{vu}\, F_{us}\Big)(\tau, \sigma^u) \\
&= \frac{\gamma(\tau, \sigma^u)}{\sqrt{g(\tau, \sigma^u)}}\, {}^3g^{rs}(\tau, \sigma^u) \Big(E_s(\tau, \sigma^u) \\
&\quad + {}^4g_{\tau m}(\tau, \sigma^u)\, {}^3g^{mn}(\tau, \sigma^u)\, \epsilon_{nst}\, B_t(\tau, \sigma^u)\Big),
\end{aligned} \tag{4.39}$$

and the following Poisson brackets are assumed:

$$\begin{aligned}
\{z^\mu(\tau, \sigma^u), \rho_\nu(\tau, \sigma^{u'}\} &= -\epsilon\, {}^4\eta^\mu_\nu\, \delta^3(\sigma^u - \sigma'^u), \\
\{A_A(\tau, \sigma^u), \pi^B(\tau, \sigma'^u)\} &= \epsilon\, {}^4\eta^B_A\, \delta^3(\sigma - \sigma'^u).
\end{aligned} \tag{4.40}$$

The five primary constraints are (see Eq. (2.14) for $T_{AB}(\tau,\sigma^u)$):

$$\bar{\mathcal{H}}_\mu(\tau,\sigma^u) = \rho_\mu(\tau,\sigma^u) - l_\mu(\tau,\sigma^u)\, T_{\tau\tau}(\tau,\sigma^u) - z_{r\mu}(\tau,\sigma^u)\,\gamma^{rs}(\tau,\sigma^u)\, T_{\tau s}(\tau,\sigma^u) \approx 0,$$
$$\pi^\tau(\tau,\sigma^u) \approx 0, \tag{4.41}$$

where

$$\begin{aligned} T_{\tau\tau}(\tau,\sigma^u) &= -\frac{1}{2}\Big(\frac{1}{\sqrt{\gamma}}\pi^r\, {}^3g_{rs}\,\pi^s - \frac{\sqrt{\gamma}}{2}\, {}^3g^{rs}\, {}^3g^{uv}\, F_{ru}\, F_{sv}\Big)(\tau,\sigma^u),\\ T_{\tau s}(\tau,\sigma^u) &= -F_{st}(\tau,\sigma^u)\,\pi^t(\tau,\sigma^u) = -\epsilon_{stu}\,\pi^t(\tau,\sigma^u)\, B_u(\tau,\sigma^u)\\ &= [\vec{\pi}(\tau,\sigma^u)\times\vec{B}(\tau,\sigma^u)]_s, \end{aligned} \tag{4.42}$$

are the energy density and the Poynting vector, respectively. We use the notation $(\vec{\pi}\times\vec{B})_{\check{s}} = (\vec{E}\times\vec{B})_{\check{s}}$ because it is consistent with $\epsilon_{stu}\,\pi^t\, B_u$ in the flat metric limit ${}^4g_{AB}\to{}^4\eta_{AB}$; in this limit $T_{\tau\tau}\to\frac{1}{2}(\vec{E}^2+\vec{B}^2)$.

Since the canonical Hamiltonian is (we assume boundary conditions for the electromagnetic potential such that all the surface terms can be neglected):

$$\begin{aligned} \bar{H}_c &= \int d^3\sigma\,\Big(\pi^A(\tau,\sigma^u)\,\partial_\tau\, A_A(\tau,\sigma^u) - \rho_\mu(\tau,\sigma^u)\, z^\mu_\tau(\tau,\sigma^u) - \mathcal{L}(\tau,\sigma^u)\Big)\\ &= \int d^3\sigma\,\Big(\partial_r\,(\pi^r(\tau,\sigma^u)\, A_\tau(\tau,\sigma^u)) - A_\tau(\tau,\sigma^u)\,\bar{\Gamma}(\tau,\sigma^u)\Big)\\ &= -\int d^3\sigma\, A_\tau(\tau,\sigma^u)\,\bar{\Gamma}(\tau,\sigma^u), \end{aligned} \tag{4.43}$$

with

$$\bar{\Gamma}(\tau,\sigma^u) = \partial_r\,\pi^r(\tau,\sigma^u), \tag{4.44}$$

we have the Dirac Hamiltonian

$$\bar{H}_D = \int d^3\sigma\,\Big(\lambda^\mu(\tau,\sigma^u)\,\bar{\mathcal{H}}_\mu(\tau,\sigma^u) + \lambda_\tau(\tau,\sigma^u)\,\pi^\tau(\tau,\sigma^u) - A_\tau(\tau,\sigma^u)\,\bar{\Gamma}(\tau,\sigma^u)\Big). \tag{4.45}$$

The Lorentz-scalar constraint $\pi^\tau(\tau,\sigma^u)\approx 0$ is generated by the gauge invariance of S under the electromagnetic gauge transformations $\delta\, A_A(\tau,\sigma^u) = \partial_A\,\alpha(\tau,\sigma^u)$ $(\alpha(\tau,\sigma^u \to_{\sigma\to\infty} \frac{a}{\sigma^{3+\epsilon}} + O(r^{-4}),\ \sigma = |\vec{\sigma}|,\ \epsilon > 0)$; its time constancy will produce the only secondary constraint (Gauss's law):

$$\bar{\Gamma}(\tau,\sigma^u)\approx 0. \tag{4.46}$$

The six constraints $\bar{\mathcal{H}}_\mu(\tau,\sigma^u)\approx 0$, $\pi^\tau(\tau,\sigma^u)\approx 0$, $\bar{\Gamma}(\tau,\sigma^u)\approx 0$ are first class with the only non-vanishing Poisson brackets:

$$\begin{aligned} \{\bar{\mathcal{H}}_\mu(\tau,\sigma^u), \bar{\mathcal{H}}_\nu(\tau,\sigma'^u)\} &= \Big([l_\mu(\tau,\sigma^u)\, z_{r\nu}(\tau,\sigma^u) - l_\nu(\tau,\sigma^u)\, z_{r\mu}(\tau,\sigma^u)]\,\frac{\pi^r(\tau,\sigma^u)}{\sqrt{\gamma(\tau,\sigma^u)}}\\ &\quad - z_{u\mu}(\tau,\sigma^u)\,\gamma^{ur}(\tau,\sigma^u)\, F_{rs}(\tau,\sigma^u)\, {}^3g^{sv}(\tau,\sigma^u)\, z_{v\nu}(\tau,\sigma^u)\Big)\\ &\quad \bar{\Gamma}(\tau,\sigma^u)\,\delta^3(\sigma^u - \sigma'^u)\approx 0. \end{aligned} \tag{4.47}$$

The ten conserved Poincaré generators have the standard form (Eq. 2.14).

On arbitrary space-like hyper-planes determined by the gauge fixings (Eq. 2.18), i.e., $z^\mu(\tau,\sigma^u) \approx x^\mu(\tau) + b^\mu_r(\tau)\,\sigma^r$, we remain with the variables $x^\mu, P^\mu, b^\mu_A, S^{\mu\nu}$ (see Eq. (2.21)), A_A, π^A, and the 12 constraints (we choose Cartesian 3-coordinates $\vec{\sigma}$):

$$\begin{aligned}
\tilde{\bar{\mathcal{H}}}^\mu(\tau) &= P^\mu - l_\mu \frac{1}{2}\int d^3\sigma\,[\vec{\pi}^2(\tau,\vec{\sigma}) + \vec{B}^2(\tau,\vec{\sigma})] \\
&\quad - b_{r\mu}(\tau)\int d^3\sigma\,[\vec{\pi}(\tau,\vec{\sigma})\times\vec{B}(\tau,\vec{\sigma})]_r \approx 0,\\
\tilde{\bar{\mathcal{H}}}^{\mu\nu}(\tau) &= S^{\mu\nu}(\tau) - [b^\mu_r(\tau)\,b^\nu_\tau - b^\nu_r(\tau)\,b^\mu_\tau]\left[\frac{1}{2}\int d^3\sigma\,\sigma^r\,[\vec{\pi}^2(\tau,\vec{\sigma}) + \vec{B}^2(\tau,\vec{\sigma})\right]\\
&\quad + [b^\mu_r(\tau)\,b^\nu_s(\tau) - b^\nu_r(\tau)\,b^\mu_s(\tau)]\int d^3\sigma\,\sigma^r\,[\vec{\pi}(\tau,\vec{\sigma})\times\vec{B}(\tau,\vec{\sigma})]_s \approx 0,\\
\pi^\tau(\tau,\vec{\sigma}) &\approx 0,\\
\bar{\Gamma}(\tau,\vec{\sigma}) &\approx 0,
\end{aligned} \tag{4.48}$$

with Poisson algebra ($C^{\mu\nu\alpha\beta}_{\gamma\delta}$ are the structure constants of the Lorentz algebra):

$$\begin{aligned}
\{\tilde{\bar{\mathcal{H}}}^\mu(\tau), \tilde{\bar{\mathcal{H}}}^\nu(\tau)\}^* &= \int d^3\sigma\,\Big([b^\mu_\tau\,b^\nu_r(\tau) - b^\nu_\tau\,b^\mu_r(\tau)]\,\pi_r(\tau,\vec{\sigma})\\
&\quad - b^\mu_r(\tau)\,F_{rs}(\tau,\vec{\sigma})\,b^\nu_s(\tau)\Big)\bar{\Gamma}(\tau,\vec{\sigma}),\\
\{\tilde{\bar{\mathcal{H}}}^\mu(\tau), \tilde{\bar{\mathcal{H}}}^{\alpha\beta}(\tau)\}^* &= -\int d^3\sigma\,\sigma^t\,\Big([(b^\alpha_r(\tau)\,b^\beta_t(\tau) - b^\alpha_t(\tau)\,b^\beta_r(\tau))\,b^\mu_\tau\\
&\quad - (b^\alpha_\tau\,b^\beta_t(\tau) - b^\beta_\tau\,b^\alpha_t(\tau))\,b^\mu_r(\tau)]\,\pi^r(\tau,\vec{\sigma})\\
&\quad + b^\mu_r(\tau)\,F_{rs}(\tau,\vec{\sigma})\,[b^\alpha_t(\tau)\,b^\beta_s(\tau) - b^\beta_t(\tau)\,b^\alpha_s(\tau)]\Big)\bar{\Gamma}(\tau,\vec{\sigma}),\\
\{\tilde{\bar{\mathcal{H}}}^{\mu\nu}(\tau), \tilde{\bar{\mathcal{H}}}^{\alpha\beta}(\tau)\}^* &= C^{\mu\nu\alpha\beta}_{\gamma\delta}\,\tilde{\bar{\mathcal{H}}}^{\gamma\delta}(\tau) + \int d^3\sigma\,\sigma^{\check{u}}\,\sigma^v\\
&\quad \Big(b^\mu_u(\tau)\,b^\alpha_v(\tau)\,([b^\nu_\tau\,b^\beta_r(\tau) - b^\beta_\tau\,b^\nu_r(\tau)]\,\pi_r(\tau,\vec{\sigma})\\
&\quad - b^\nu_r(\tau)\,F_{rs}(\tau,\vec{\sigma})\,b^\beta_s(\tau)) - b^\mu_u(\tau)\,b^\beta_v(\tau)\,([b^\nu_\tau\,b^\alpha_r(\tau)\\
&\quad - b^\alpha_\tau\,b^\nu_r(\tau)]\,\pi_r(\tau,\vec{\sigma}) - b^\nu_r(\tau)\,F_{rs}(\tau,\vec{\sigma})\,b^\alpha_s(\tau))\\
&\quad - b^\nu_u(\tau)\,b^\alpha_v(\tau)\,([b^\mu_\tau\,b^\beta_r(\tau) - b^\beta_\tau\,b^\mu_r(\tau)]\,\pi_r(\tau,\vec{\sigma})\\
&\quad - b^\mu_r(\tau)\,F_{rs}(\tau,\vec{\sigma})\,b^\beta_s(\tau)) + b^\nu_u(\tau)\,b^\beta_v(\tau)\,([b^\mu_\tau\,b^\alpha_r(\tau)\\
&\quad - b^\alpha_\tau\,b^\mu_r(\tau)]\,\pi_r(\tau,\vec{\sigma}) - b^\mu_r(\tau)\,F_{rs}(\tau,\vec{\sigma})\,b^\alpha_s(\tau))\Big)\bar{\Gamma}(\tau,\vec{\sigma}).
\end{aligned} \tag{4.49}$$

The gauge fixing (Eq. 3.1), i.e., $z^\mu(\tau,\vec{\sigma}) \approx x^\mu(\tau) + \epsilon^\mu_r(u(P))\,\sigma^r$, gives the Wigner-covariant rest-frame instant form. In the Wigner hyper-planes the new variables are $\tilde{x}^\mu$, P^μ, A_τ, π^τ, A_r, π^r, with "r" being a Wigner spin-1 index, and we get the spin tensor

$$\begin{aligned}\bar{S}^{AB} &= \epsilon^A_\mu(u(P))\,\epsilon^B_\nu(u(P))\,S^{\mu\nu} \approx [b^A_r(\tau)\,b^B_\tau \\ &\quad - b^B_r(\tau)\,b^A_\tau]\,\frac{1}{2}\int d^3\sigma\,\sigma^r\,[\vec{\pi}^2(\tau,\vec{\sigma}) + \vec{B}^2(\tau,\vec{\sigma})] \\ &\quad - [b^A_r(\tau)\,b^B_s(\tau) - b^B_r(\tau) b^A_s(\tau)]\int d^3\sigma\,\sigma^r\,[\vec{\pi}(\tau,\vec{\sigma}) \times \vec{B}(\tau,\vec{\sigma})]_s. \end{aligned} \tag{4.50}$$

The generators of the external Poincaré group now have the standard form of Eq. (3.4), while for the spin tensors we get

$$\begin{aligned}\bar{S}^{AB} &\approx (\eta^A_r\,\eta^B_\tau - \eta^B_r\,\eta^A_\tau)\,\frac{1}{2}\int d^3\sigma\,\sigma^r\,[\vec{\pi}^2(\tau,\vec{\sigma}) + \vec{B}^2(\tau,\vec{\sigma})] \\ &\quad -(\eta^A_r\,\eta^B_s - \eta^B_r\,\eta^A_s)\int d^3\sigma\,\sigma^r\,[\vec{\pi}(\tau,\vec{\sigma}) \times \vec{B}(\tau,\vec{\sigma})]_s, \\ \bar{S}^{rs} &\approx \int d^3\sigma\,(\sigma^r\,[\vec{\pi}(\tau,\vec{\sigma}) \times \vec{B}(\tau,\vec{\sigma})]^s - \sigma^s\,[\vec{\pi}(\tau,\vec{\sigma}) \times \vec{B}(\tau,\vec{\sigma})]^r), \\ \bar{S}^{\tau r} &\approx -\bar{S}^{r\tau} = -\frac{1}{2}\int d^3\sigma\,\sigma^r\,[\vec{\pi}^2(\tau,\vec{\sigma}) + \vec{B}^2(\tau,\vec{\sigma})]. \end{aligned} \tag{4.51}$$

Only the following six first-class constraints are left:

$$\begin{aligned} \tilde{\bar{\mathcal{H}}}^\mu(\tau) &= P^\mu - u^\mu(u(P))\,\frac{1}{2}\int d^3\sigma\,[\vec{\pi}^2(\tau,\vec{\sigma}) + \vec{B}^2(\tau,\vec{\sigma})] \\ &\quad - \epsilon^\mu_r(u(P))\int d^3\sigma\,[\vec{\pi}(\tau,\vec{\sigma}) \times \vec{B}(\tau,\vec{\sigma})]^r \approx 0, \\ \pi^\tau(\tau,\vec{\sigma}) &\approx 0, \\ \bar{\Gamma}(\tau,\vec{\sigma}) &\approx 0, \\ \{\tilde{\bar{\mathcal{H}}}^\mu, \tilde{\bar{\mathcal{H}}}^\nu\}^{**} &= \int d^3\sigma\,\Big([\epsilon^\mu_\tau(u(P))\,\epsilon^\nu_r(u(P)) \\ &\quad - \epsilon^\nu_\tau(u(P))\,\epsilon^\mu_r(u(P))]\,\pi_r(\tau,\vec{\sigma}) \\ &\quad + \epsilon^\mu_r(u(P))\,F_{rs}(\tau,\vec{\sigma})\,\epsilon^\nu_s(u(P))\Big)\bar{\Gamma}(\tau,\vec{\sigma}), \end{aligned} \tag{4.52}$$

or

$$\begin{aligned} \bar{\mathcal{H}}(\tau) &= \mathcal{P}^\tau_{(int)} - \frac{1}{2}\int d^3\sigma\,[\vec{\pi}^2(\tau,\vec{\sigma}) + \vec{B}^2(\tau,\vec{\sigma})] \approx 0, \\ \tilde{\bar{\mathcal{H}}}_p(\tau) &= \int d^3\sigma\,\vec{\pi}(\tau,\vec{\sigma}) \times \vec{B}(\tau,\vec{\sigma}) \approx 0, \\ \pi^\tau(\tau,\vec{\sigma}) &\approx 0, \\ \bar{\Gamma}(\tau,\vec{\sigma}) &\approx 0, \end{aligned} \tag{4.53}$$

where $\frac{1}{2}\int d^3\sigma\,[\vec{\pi}^2(\tau,\vec{\sigma}) + \vec{B}^2(\tau,\vec{\sigma})]$ and $\int d^3\sigma\,\vec{\pi}(\tau,\vec{\sigma}) \times \vec{B}(\tau,\vec{\sigma})$ are the rest-frame field energy and three-momentum respectively (now we have $\vec{\pi}(\tau,\vec{\sigma}) = \vec{E}(\tau,\vec{\sigma})$).

The Dirac Hamiltonian is

$$\bar{H}_D = \lambda(\tau)\,\bar{\mathcal{H}} - \vec{\lambda}(\tau)\,\bar{\vec{\mathcal{H}}}_p + \int d^3\sigma\,\Big(\lambda_\tau(\tau,\vec{\sigma})\,\pi^\tau(\tau,\vec{\sigma}) - A_\tau(\tau,\vec{\sigma})\,\bar{\Gamma}(\tau,\vec{\sigma})\Big). \tag{4.54}$$

The Shanmugadhasan canonical transformation (see Chapter 9) adapted to the electromagnetic first-class constraints is ($\triangle_\sigma = -\partial^2_\sigma$ is the Laplace–Beltrami operator; $\Box = \partial^2_\tau$ is the D'Alambertian; $\partial_A = \partial/\partial\,\sigma^A$):

$$A^r(\tau,\vec{\sigma}) = \frac{\partial}{\partial\sigma^r}\,\eta_{em}(\tau,\vec{\sigma}) + A^r_\perp(\tau,\vec{\sigma}),\quad A^r_\perp(\tau,\vec{\sigma}) = \left(\delta^{rs} + \frac{\partial^r_\sigma\,\partial^s_\sigma}{\triangle_\sigma}\right)A_s(\tau,\vec{\sigma}),$$

$$\pi^r(\tau,\vec{\sigma}) = \pi^r_\perp(\tau,\vec{\sigma}) + \frac{1}{\triangle_\sigma}\,\frac{\partial}{\partial\sigma^r}\,\bar{\Gamma}(\tau,\vec{\sigma}),\quad \pi^r_\perp(\tau,\vec{\sigma}),\quad = \left(\delta^{rs} + \frac{\partial^r_\sigma\,\partial^s_\sigma}{\triangle_\sigma}\right)\pi_s(\tau,\vec{\sigma}),$$

$$\eta_{em}(\tau,\vec{\sigma}) = -\frac{1}{\triangle_\sigma}\,\frac{\partial}{\partial\vec{\sigma}}\cdot\vec{A}(\tau,\vec{\sigma}),$$

$$\begin{aligned}
\{\eta_{em}(\tau,\vec{\sigma}),\bar{\Gamma}(\tau,\vec{\sigma}^{'})\}^{**} &= -\delta^3(\vec{\sigma}-\vec{\sigma}^{'}),\\
\{A^r_\perp(\tau,\vec{\sigma}),\pi^s_\perp(\tau,\vec{\sigma}^{'})\}^{**} &= -\left(\delta^{rs} + \frac{\partial^r_\sigma\,\partial^s_\sigma}{\triangle_\sigma}\right)\delta^3(\vec{\sigma}-\vec{\sigma}^{'}),
\end{aligned}$$

$$\boxed{\begin{matrix} A_A(\tau,\vec{\sigma}) \\ \pi^A(\tau,\vec{\sigma}) \end{matrix}} \longrightarrow \boxed{\begin{matrix} A_\tau(\tau,\vec{\sigma}) & , & \eta_{em}(\tau,\vec{\sigma}) \\ \pi^\tau(\tau,\vec{\sigma})\,(\approx 0) & & \bar{\Gamma}(\tau,\vec{\sigma}) \end{matrix} \,\Bigg|\, \begin{matrix} \vec{A}_\perp(\tau,\vec{\sigma}) \\ \vec{\pi}_\perp(\tau,\vec{\sigma}) \end{matrix}} \tag{4.55}$$

In the rest-frame instant form the pairs of conjugate variables $A_\tau(\tau,\vec{\sigma})$, $\pi^\tau(\tau,\vec{\sigma}) \approx 0$, $\eta_{em}(\tau,\vec{\sigma})$, $\bar{\Gamma}(\tau,\vec{\sigma}) \approx 0$ span a Lorentz-scalar gauge subspace of the phase space, while the Wigner spin-1 3-vectors $\vec{A}_\perp(\tau,\vec{\sigma})$, $\vec{\pi}_\perp(\tau,\vec{\sigma}) = \vec{E}(\tau,\vec{\sigma})$ ($\vec{B}(\tau,\vec{\sigma}) = \vec{\partial}\times\vec{A}_\perp(\tau,\vec{\sigma})$) give a canonical basis of electromagnetic DOs. All these results are the final end-point of the original suggestions of Dirac [189].

Having decoupled the electromagnetic gauge variables in a Lorentz-scalar way, the canonical basis $\tilde{x}^\mu(\tau), P^\mu, \vec{A}_\perp(\tau,\vec{\sigma}), \vec{\pi}_\perp(\tau,\vec{\sigma})$ spans the phase space, where, due to the analogue of Eq. (3.10) for the internal Poincaré generators, the remaining gauge-invariant four first-class constraints in Eq. (4.53) have the form

$$\begin{aligned}
\hat{\bar{\mathcal{H}}}(\tau) &= \mathcal{P}^\tau_{(int)} - \frac{1}{2}\int d^3\sigma\,[\vec{\pi}^2_\perp(\tau,\vec{\sigma}) + \vec{B}^2[\vec{A}_\perp(\tau,\vec{\sigma})]] = \epsilon_s - \bar{M}_{em}\,c \approx 0,\\
\hat{\bar{\vec{\mathcal{H}}}}_p(\tau) &= \vec{\mathcal{P}}_{(int)} = \vec{P}_{em} = \int d^3\sigma\,\vec{\pi}_\perp(\tau,\vec{\sigma})\times\vec{B}[\vec{A}_\perp(\tau,\vec{\sigma})] \approx 0,\\
&\{\hat{\bar{\mathcal{H}}},\hat{\bar{\mathcal{H}}}^r_p\}^{**} = \{\hat{\bar{\mathcal{H}}}^r_p,\hat{\bar{\mathcal{H}}}^s_p\}^{**} = 0,
\end{aligned} \tag{4.56}$$

where $\bar{M}_{em}$ is the invariant mass of the configuration of the electromagnetic field. Now the Dirac Hamiltonian is $\bar{H}_D = \lambda(\tau)\,\hat{\bar{\mathcal{H}}}(\tau) - \vec{\lambda}(\tau)\cdot\hat{\bar{\vec{\mathcal{H}}}}_p(\tau)$.

The rest-frame spin tensor becomes

$$\begin{aligned}
\bar{S}^{rs}_s = \epsilon^{rsu}\,\bar{S}_{em} = \int d^3\sigma\,\Big(&\sigma^r\,[\vec{\pi}_\perp(\tau,\vec{\sigma})\times\vec{B}[\vec{A}_\perp(\tau,\vec{\sigma})]]^s\\
&-\sigma^s\,[\vec{\pi}_\perp(\tau,\vec{\sigma})\times\vec{B}[\vec{A}_\perp(\tau,\vec{\sigma})]]^r\Big),
\end{aligned}$$

$$\bar{S}_s^{\tau r} = -\bar{S}_s^{r\tau} = K_{em}^r = -\frac{1}{2}\int d^3\sigma\,\sigma^r\left(\vec{\pi}_\perp^2(\tau,\vec{\sigma}) + \vec{B}^2[\vec{A}_\perp(\tau,\vec{\sigma})]\right). \quad (4.57)$$

By reintroducing the evolution inside the Wigner hyper-planes with the Dirac Hamiltonian $H_D = \bar{M}_{em}\,c + \int d^3\sigma\,(\lambda_\tau\,\pi^\tau - A_\tau\,\bar{\Gamma})(\tau,\vec{\sigma})$, where $\lambda_\tau(\tau,\sigma)$ is the Dirac multiplier associated with the electromagnetic gauge freedom, one gets the following Hamilton equations:

$$\partial_\tau\,A_\tau(\tau,\vec{\sigma}) \circeq \lambda_\tau(\tau,\vec{\sigma}), \quad \partial_\tau\,\eta(\tau,\vec{\sigma}) \circeq A_\tau(\tau,\vec{\sigma}), \quad \partial_\tau\,A_{\perp\,r}(\tau,\vec{\sigma}) \circeq -\,\pi_{\perp\,r}(\tau,\vec{\sigma}),$$
$$\partial_\tau\,\pi_\perp^r(\tau,\vec{\sigma}) \circeq \triangle\,A_\perp^r(\tau,\vec{\sigma}), \quad \Rightarrow \quad \Box\,A_{\perp\,r}(\tau,\vec{\sigma}) \circeq 0. \quad (4.58)$$

To fix the electromagnetic gauge we must only add a gauge fixing $\eta_{em}(\tau,\vec{\sigma}) \approx 0$ to Gauss's law, for the fixation of the gauge variable η_{em}. Its time constancy will generate the gauge fixing for the gauge variable $A_\tau(\tau,\vec{\sigma})$: $\partial_\tau\,\eta_{em}(\tau,\vec{\sigma}) + \{\eta_{em}(\tau,\vec{\sigma}), \bar{H}_D\} = A_\tau(\tau,\vec{\sigma}) \approx 0$. Finally, the time constancy $\partial_\tau\,A_\tau(\tau,\vec{\sigma}) + \{A_\tau(\tau,\vec{\sigma}), \bar{H}_D\} \approx 0$ will determine the Dirac multiplier $\lambda_\tau(\tau,\vec{\sigma})$. By adding these two gauge fixing constraints to the first-class constraints $\pi^\tau(\tau,\vec{\sigma}) \approx 0$, $\Gamma(\tau,\vec{\sigma}) \approx 0$, one gets two pairs of second-class constraints allowing the elimination of the gauge degrees of freedom so that only the DOs survive. This is the Wigner-covariant radiation gauge in the rest-frame instant form. Instead, in the standard formulation one usually uses the Lorentz gauge $\partial_\mu\,A^\mu(x^o,\vec{x}) \approx 0$ to preserve the manifest covariance of the formulation, but it does not allow finding covariant DOs.

If we add the gauge fixing $T_s - \tau \approx 0$, we can eliminate ϵ_s, T_s and we get the rest-frame canonical variables $\vec{A}_\perp(\tau,\vec{\sigma})$, $\vec{\pi}_\perp(\tau,\vec{\sigma})$ restricted by $\hat{\vec{\mathcal{H}}}_p(\tau) \approx 0$ and with the dynamics ruled by the Hamiltonian $\bar{H} = \frac{1}{2}\int d^3\sigma\,[\vec{\pi}_\perp^2(\tau,\vec{\sigma}) + \vec{B}^2[\vec{A}_\perp(\tau,\vec{\sigma})]] + \vec{\lambda}(\tau)\cdot\int d^3\sigma\,\vec{\pi}_\perp(\tau,\vec{\sigma}) \times \vec{B}[\vec{A}_\perp(\tau,\vec{\sigma})]$; in the gauge $\vec{\lambda}(\tau) = 0$, the Hamilton equations are $\frac{\partial}{\partial T}\,\vec{A}_\perp(T,\vec{\sigma}) \circeq \vec{\pi}_\perp(T,\vec{\sigma})$, $\frac{\partial}{\partial T}\,\vec{\pi}_\perp(T,\vec{\sigma}) \circeq -\triangle\,\vec{A}_\perp(T,\vec{\sigma})$, so that we recover $\Box\,\vec{A}_\perp(T,\vec{\sigma}) \circeq 0$.

The external Poincaré generators have the standard form (Eq. 3.6) with $\bar{M}_{em}$ and $\vec{\bar{S}}_{em}$ given by Eqs. (4.56) and (4.57) respectively. Instead, the internal Poincaré generators are $\bar{M}_{em}$, $\vec{P}_{em} \approx 0$, $\vec{\bar{S}}_{em}$, and $\vec{K}_{em}$. The natural gauge fixing for $\vec{P}_{em} \approx 0$ is $\vec{K}_{em} \approx 0$: It gives $\vec{R}_+ \approx \vec{q}_+ \approx 0$ and $\vec{\lambda}(\tau) = 0$.

One can choose the Fokker–Pryce center of inertia as the inertial observer origin of the 3-coordinates. A canonical set of collective and relative variables for the electromagnetic field (also in the presence of charged particles) after the elimination of the internal center of mass has been determined in Ref. [101] by means of the method used for the Klein–Gordon case.

4.3 Relativistic Atomic Physics

Standard atomic physics [190, 191] is a semi-relativistic treatment of quantum electrodynamics (QED) in which the matter fields are approximated by scalar (or spinning) particles, the relevant energies are below the threshold of pair production, and the electromagnetic field is described in the Coulomb gauge at the order $1/c$.

In Refs. [98–101, 149–151] a fully relativistic formulation of classical atomic physics in the rest-frame instant form of dynamics was given with the electromagnetic field in the radiation gauge and with the electric charges Q_i of the positive-energy particles being Grassmann-valued ($Q_i^2 = 0$, $Q_i Q_j = Q_j Q_i$ for $i \neq j$) to regularize the electromagnetic self-energies on the world-lines of particles. In the language of QED this is both an ultraviolet regularization (no loop contributions) and an infrared one (no bremsstrahlung), so that only the one-photon exchange diagram contributes and its static and non-static effects are replaced by potentials in a formulation based on the Cauchy problem.

Therefore, the starting point is a parametrized Minkowski theory with the electromagnetic field and N charged positive-energy particles.

The description of N charged scalar particles with the pseudo-classical description of the electric charge shown in Appendix B is done by replacing Eq. (2.11) with the following action:

$$
\begin{aligned}
S &= \int d\tau d^3\sigma\, \mathcal{L}(\tau, \sigma^u) = \int d\tau L(\tau),\\
\mathcal{L}(\tau, \sigma^u) &= \frac{i}{2} \sum_{i=1}^N \delta^3(\sigma^u - \eta_i^u(\tau)) \Big[\theta_i^*(\tau)\dot{\theta}_i(\tau) - \dot{\theta}_i^*(\tau)\theta_i(\tau)\Big] - \sum_{i=1}^N \delta^3(\sigma^u - \eta_i^u(\tau))\\
&\quad \Big[m_i\, c \sqrt{{}^4g_{\tau\tau}(\tau, \sigma^u) + 2\, {}^4g_{\tau r}(\tau, \sigma^u)\, \dot{\eta}_i^r(\tau) + {}^4g_{rs}(\tau, \sigma^u)\, \dot{\eta}_i^r(\tau)\, \dot{\eta}_i^s(\tau)}\\
&\quad + \frac{Q_i(\tau)}{c} \Big(A_\tau(\tau, \sigma^u) + A_r(\tau, \sigma^u)\, \dot{\eta}_i^r(\tau)\Big)\Big]\\
&\quad - \frac{1}{4\, c} \sqrt{g(\tau, \sigma^u)}\, {}^4g^{AC}(\tau, \sigma^u)\, {}^4g^{BD}(\tau, \sigma^u) F_{AB}(\tau, \sigma^u) F_{CD}(\tau, \sigma^u),\\
Q_i(\tau) &= e\theta_i^*(\tau)\theta_i(\tau).
\end{aligned} \tag{4.59}
$$

In this action, the configuration variables are $z^\mu(\tau, \sigma^u)$, $\eta_i^r(\tau)$, $\theta_i(\tau)$, and $A_A(\tau, \sigma^u)$. The sign of the energy of the particles is $\eta_i = \pm$. The electric charge of the particles is $Q_i = e_i\, \theta_i^*\, \theta_i$.

The canonical momenta are $\rho_\mu(\tau, \sigma^u)$, $\kappa_{ir}(\tau)$, $\pi^A(\tau, \sigma^u)$, plus those of the Grassmann variables given in Appendix B:

$$
\begin{aligned}
\rho_\mu(\tau, \sigma^u) &= -\frac{\partial \mathcal{L}(\tau, \sigma^u)}{\partial z_\tau^\mu(\tau, \sigma^u)} = \sum_{i=1}^N \delta^3(\sigma^u - \eta_i^u(\tau))\, \eta_i\, m_i c\\
&\quad \frac{z_{\tau\mu}(\tau, \sigma^u) + z_{\check{r}\mu}(\tau, \sigma^u)\, \dot{\eta}_i^r(\tau)}{\sqrt{{}^4g_{\tau\tau}(\tau, \sigma^u) + 2\, {}^4g_{\tau r}(\tau, \sigma^u)\, \dot{\eta}_i^r(\tau) + {}^4g_{rs}(\tau, \sigma^u)\, \dot{\eta}_i^r(\tau)\, \dot{\eta}_i^s(\tau)}}\\
&\quad + \frac{\sqrt{g(\tau, \sigma^u)}}{4} \Big(({}^4g^{\tau\tau}\, z_{\tau\mu} + {}^4g^{\tau r}\, z_{r\mu})\, {}^4g^{AC\, 4}g^{BD}\, F_{AB}\, F_{CD}\\
&\quad - 2\Big[z_{\tau\mu}\, ({}^4g^{A\tau\, 4}g^{\tau C\, 4}g^{BD} + {}^4g^{AC\, 4}g^{B\tau\, 4}g^{\tau D})\\
&\quad + z_{r\mu}\, ({}^4g^{Ar\, 4}g^{\tau C} + {}^4g^{A\tau\, 4}g^{rC})\, {}^4g^{BD}\Big]\, F_{AB}\, F_{CD}\Big)(\tau, \sigma^u),
\end{aligned}
$$

$$
\begin{aligned}
\pi^\tau(\tau,\sigma^u) &= \frac{\partial L}{\partial\partial_\tau\, A_\tau(\tau,\sigma^u)} = 0,\\
\pi^r(\tau,\sigma^u) &= \frac{\partial L}{\partial\partial_\tau\, A_r(\tau,\sigma^u)} = -\frac{\gamma(\tau,\sigma^u)}{\sqrt{g(\tau,\sigma^u)}}\, {}^3g^{rs}(\tau,\sigma^u)\,\Big(F_{\tau s} - {}^4g_{\tau v}\, {}^3g^{vu}\, F_{us}\Big)(\tau,\sigma^u)\\
&= \frac{\gamma(\tau,\sigma^u)}{\sqrt{g(\tau,\sigma^u)}}\, {}^3g^{rs}(\tau,\sigma^u)\,\Big(E_s(\tau,\sigma^u) + {}^4g_{\tau v}(\tau,\sigma^u)\, {}^3g^{vu}(\tau,\sigma^u)\,\epsilon_{ust}\, B_t(\tau,\sigma^u)\Big),\\
\kappa_{ir}(\tau) &= -\frac{\partial L(\tau)}{\partial \dot\eta^r_i(\tau)}\\
&= \eta_i\, m_i c\, \frac{{}^4g_{\tau r}(\tau,\eta^u_i(\tau)) + {}^3g_{rs}(\tau,\eta^u_i(\tau))\,\dot\eta^s_i(\tau)}{\sqrt{{}^4g_{\tau\tau}(\tau,\eta^u_i(\tau)) + 2\,{}^4g_{\tau r}(\tau,\eta^u_i(\tau))\,\dot\eta^r_i(\tau) + {}^3g_{rs}(\tau,\eta^u_i(\tau))\,\dot\eta^r_i(\tau)\,\dot\eta^s_i(\tau)}}\\
&\quad + e_i\,\theta^*_i(\tau)\,\theta_i(\tau)\, A_r(\tau,\eta^u_i(\tau)). \qquad (4.60)
\end{aligned}
$$

The energy–momentum tensor $T^{AB}(\tau,\sigma^u) = -\Big(\frac{1}{\sqrt{g}}\,\frac{\delta S}{\delta\, {}^4g_{AB}(\tau,\sigma^u)}\Big)(\tau,\sigma^u)$ is the energy–momentum tensor of the isolated system of particles plus the electromagnetic field given by Eqs. (2.14) and (4.42), plus interaction terms.

The primary and secondary first-class constraints are

$$
\begin{aligned}
&\mathcal{H}^\mu(\tau,\sigma^u) = \rho^\mu(\tau,\sigma^u) - l^\mu(\tau,\sigma^u)\, T^{\tau\tau}(\tau,\sigma^u) - z^\mu_r(\tau,\sigma^u)\, T^{\tau r}(\tau,\sigma^u) \approx 0,\\
&\pi^\tau(\tau,\sigma^u) \approx 0,\\
&\Gamma(\tau,\sigma^u) = \partial_r\, \pi^r(\tau,\sigma^u) - \sum_{i=1}^N Q_i\, \delta^3(\sigma^u - \eta^u_i(\tau)) \approx 0. \qquad (4.61)
\end{aligned}
$$

The Grassmann momenta are determined by the second-class constraint $\pi_{\theta\, i}(\tau) - \frac{\partial L(\tau)}{\partial\dot\theta_i(\tau)} = -\frac{i}{2}\,\theta^*_i(\tau) \approx 0$, $\pi_{\theta^*\, i}(\tau) - \frac{\partial L(\tau)}{\partial\dot\theta^*_i(\tau)} = -\frac{i}{2}\,\theta_i(\tau) \approx 0$, which allows replacing the Poisson brackets $\{\theta_i(\tau), \pi_{\theta\, j}(\tau)\} = \{\theta^*_i(\tau), \pi_{\theta^*\, j}(\tau)\} = -\delta_{ij}$ with Dirac brackets $\{\theta_i(\tau), \theta_j(\tau)\}^* = \{\theta^*_i(\tau), \theta^*_j(\tau)\}^* = 0$, $\{\theta_i(\tau), \theta^*_j(\tau)\}^* = -i\,\delta_{ij}$. From now on we will use $\{.,.\}$ to denote these Dirac brackets.

The Dirac Hamiltonian is like Eq. (4.45).

For the configurations with $\epsilon\, P^2 > 0$ we can formulate the rest-frame instant form with Wigner 3-spaces (with Cartesian 3-coordinates $\vec\sigma$; we go on to use $\{\dots\}$ for the Dirac brackets implied by the gauge fixings for the embeddings).

We can eliminate the electromagnetic gauge freedom by going to the Wigner-covariant radiation gauge (Eq. 4.55). However, now the particle momenta and the Grassmann variables are not electromagnetic DOs because we have $\{\kappa^r_i(\tau), \bar\Gamma(\tau,\vec\sigma)\}^* = e_i\,\theta^*_i\,\theta_i\,\frac{\partial}{\partial\eta^r_i}\,\delta^3(\vec\sigma - \vec\eta_i(\tau))$, $\{\theta_i(\tau), \bar\Gamma(\tau,\vec\sigma)\}^* = i\, e_i\,\theta_i(\tau)\,\delta^3(\vec\sigma - \vec\eta_i(\tau))$. The electromagnetic DOs of the particles, with vanishing Poisson brackets with Gauss's law, turn out to be $\vec\eta_i(\tau)$ and

$$
\begin{aligned}
\tilde{\vec\kappa}_i(\tau) &= \vec\kappa_i(\tau) - e_i\,\theta^*_i\,\theta_i\,\vec\partial\,\eta_{em}(\tau,\vec\eta_i(\tau)),\\
\tilde\theta_i(\tau) &= e^{i\,\eta_{em}(\tau,\vec\eta_i(\tau))}\,\theta_i(\tau),\\
\tilde\theta^*_i(\tau) &= e^{-i\,\eta_{em}(\tau,\vec\eta_i(\tau))}\,\theta^*_i(\tau). \qquad (4.62)
\end{aligned}
$$

We have $\tilde{Q}_i = \tilde{\theta}_i^* \tilde{\theta}_i = \theta_i^* \theta_i = Q_i$, $\vec{\kappa}_i(\tau) - e_i \theta_i^* \theta_i \vec{A}(\tau, \vec{\eta}_i(\tau)) = \tilde{\vec{\kappa}}_i(\tau) - e_i \tilde{\theta}_i^* \tilde{\theta}_i \vec{A}_\perp(\tau, \vec{\eta}_i(\tau))$, $\{\eta_i^r(\tau), \tilde{\kappa}_j^s(\tau)\} = \delta_{ij}\,\delta^{rs}$, $\{\tilde{\kappa}_i^r(\tau), \tilde{\theta}_j(\tau)\} = 0$, $\{\tilde{\kappa}_i^r(\tau), \tilde{\theta}_j^*(\tau)\} = 0$.

The electromagnetic DOs of the particles $\vec{\eta}_i(\tau)$, $\tilde{\vec{\kappa}}_i(\tau)$ describe scalar particles dressed with a Coulomb cloud.

The electromagnetic potential satisfies the following equations ($P_\perp^{rs}(\vec{\sigma}) = \delta^{rs} + \frac{\partial^r\,\partial^s}{\triangle}$, $\triangle = -\vec{\partial}^2$, $\partial^r = -\partial_r = -\partial/\partial\,\sigma^r$):

$$A_\tau(\tau, \vec{\sigma}) \approx \sum_{i=1}^N \frac{Q_i}{4\pi\,|\vec{\sigma} - \vec{\eta}_i(\tau)|},$$

$$\Box\, A_\perp^r(\tau, \vec{\sigma}) \stackrel{\circ}{=} \sum_{i=1}^N Q_i\, P_\perp^{rs}(\vec{\sigma})\, \dot{\eta}_i^s(\tau)\, \delta^3(\vec{\sigma} - \vec{\eta}_i(\tau) = j_\perp^r(\tau, \vec{\sigma}),$$

$$\{A_\perp^r(\tau, \vec{\sigma}), \pi_\perp^s(\tau, \vec{\sigma}_1)\} = -c\, P_\perp^{rs}(\vec{\sigma})\, \delta^3(\vec{\sigma} - \vec{\sigma}_1), \tag{4.63}$$

so that $\vec{\partial} \cdot \vec{A}_\perp(\tau, \vec{\sigma}) = 0$.

The solution of the equation $\Box\, A_\perp^r(\tau, \vec{\sigma}) \stackrel{\circ}{=} j_\perp^r(\tau, \vec{\sigma})$ gives the Lienard–Wiechert transverse electromagnetic potential and the electric and magnetic fields (see Ref. [151]):

$$\vec{A}_{\perp S}(\tau, \vec{\sigma}) \stackrel{\circ}{=} \sum_{i=1}^N Q_i \vec{A}_{\perp Si}(\vec{\sigma} - \vec{\eta}_i(\tau), \vec{\kappa}_i(\tau)),$$

$$\vec{A}_{\perp Si}(\vec{\sigma} - \vec{\eta}_i, \vec{\kappa}_i) = \frac{1}{4\pi|\vec{\sigma} - \vec{\eta}_i|}\, \frac{1}{\sqrt{m_i^2\, c^2 + \vec{\kappa}_i^2} + \sqrt{m_i^2\, c^2 + (\vec{\kappa}_i \cdot \frac{\vec{\sigma} - \vec{\eta}_i}{|\vec{\sigma} - \vec{\eta}_i|})^2}} \times \left[\vec{\kappa}_i + \frac{[\vec{\kappa}_i \cdot (\vec{\sigma} - \vec{\eta}_i)]\, (\vec{\sigma} - \vec{\eta}_i)}{|\vec{\sigma} - \vec{\eta}_i|^2}\, \frac{\sqrt{m_i^2\, c^2 + \vec{\kappa}_i^2}}{\sqrt{m_i^2\, c^2 + (\vec{\kappa}_i \cdot \frac{\vec{\sigma} - \vec{\eta}_i}{|\vec{\sigma} - \vec{\eta}_i|})^2}} \right], \tag{4.64}$$

$$\begin{aligned}
\vec{E}_{\perp S}(\tau, \vec{\sigma}) &= \vec{\pi}_{\perp S}(\tau, \vec{\sigma}) = -\frac{\partial \vec{A}_{\perp S}(\tau, \vec{\sigma})}{\partial \tau} \\
&= \sum_{i=1}^N Q_i\, \vec{\pi}_{\perp Si}(\vec{\sigma} - \vec{\eta}_i(\tau), \vec{\kappa}_i(\tau)) \\
&= \sum_{i=1}^N Q_i\, \frac{\vec{\kappa}_i(\tau) \cdot \vec{\partial}_\sigma}{\sqrt{m_i^2\, c^2 + \vec{\kappa}_i^2(\tau)}}\, \vec{A}_{\perp Si}(\vec{\sigma} - \vec{\eta}_i(\tau), \vec{\kappa}_i(\tau)) \\
&= -\sum_{i=1}^N Q_i \times \frac{1}{4\pi|\vec{\sigma} - \vec{\eta}_i(\tau)|^2} \left[\vec{\kappa}_i(\tau)\, [\vec{\kappa}_i(\tau) \cdot \frac{\vec{\sigma} - \vec{\eta}_i(\tau)}{|\vec{\sigma} - \vec{\eta}_i(\tau)|} \right] \\
&\quad \frac{\sqrt{m_i^2\, c^2 + \vec{\kappa}_i^2(\tau)}}{[m_i^2\, c^2 + (\vec{\kappa}_i(\tau) \cdot \frac{\vec{\sigma} - \vec{\eta}_i(\tau)}{|\vec{\sigma} - \vec{\eta}_i(\tau)|})^2]^{3/2}}
\end{aligned}$$

$$+\frac{\vec{\sigma}-\vec{\eta}_i(\tau)}{|\vec{\sigma}-\vec{\eta}_i(\tau)|}\left(\frac{\vec{\kappa}_i^2(\tau)+(\vec{\kappa}_i(\tau)\cdot\frac{\vec{\sigma}-\vec{\eta}_i(\tau)}{|\vec{\sigma}-\vec{\eta}_i(\tau)|})^2}{\vec{\kappa}_i^2(\tau)-(\vec{\kappa}_i(\tau)\cdot\frac{\vec{\sigma}-\vec{\eta}_i(\tau)}{|\vec{\sigma}-\vec{\eta}_i(\tau)|})^2}\right.$$

$$\left(\frac{\sqrt{m_i^2\,c^2+\vec{\kappa}_i^2(\tau)}}{\sqrt{m_i^2\,c^2+(\vec{\kappa}_i(\tau)\cdot\frac{\vec{\sigma}-\vec{\eta}_i(\tau)}{|\vec{\sigma}-\vec{\eta}_i(\tau)|})^2}}-1\right)$$

$$\left.\left.+\frac{(\vec{\kappa}_i(\tau)\cdot\frac{\vec{\sigma}-\vec{\eta}_i(\tau)}{|\vec{\sigma}-\vec{\eta}_i(\tau)|})^2\sqrt{m_i^2\,c^2+\vec{\kappa}_i^2(\tau)}}{[m_i^2\,c^2+(\vec{\kappa}_i(\tau)\cdot\frac{\vec{\sigma}-\vec{\eta}_i(\tau)}{|\vec{\sigma}-\vec{\eta}_i(\tau)|})^2]^{3/2}}\right)\right], \tag{4.65}$$

$$\begin{aligned}\vec{B}_S(\tau,\vec{\sigma})&=\vec{\partial}\times\vec{A}_{\perp S}(\tau,\vec{\sigma})=\sum_{i=1}^N Q_i\,\vec{B}_{Si}(\vec{\sigma}-\vec{\eta}_i(\tau),\vec{\kappa}_i(\tau))\\&=\sum_{i=1}^N Q_i\,\frac{1}{4\pi|\vec{\sigma}-\vec{\eta}_i(\tau)|^2}\,\frac{m_i^2\,c^2\,\vec{\kappa}_i(\tau)\times\frac{\vec{\sigma}-\vec{\eta}_i(\tau)}{|\vec{\sigma}-\vec{\eta}_i(\tau)|}}{[m_i^2\,c^2+(\vec{\kappa}_i(\tau)\cdot\frac{\vec{\sigma}-\vec{\eta}_i(\tau)}{|\vec{\sigma}-\vec{\eta}_i(\tau)|})^2]^{3/2}}.\end{aligned} \tag{4.66}$$

In the absence of particles, the electromagnetic field in the radiation gauge is described by the conjugate variables (the DOs) $\vec{A}_\perp(\tau,\vec{\sigma})$, $\vec{\pi}_\perp(\tau,\vec{\sigma})$, as shown in the canonical basis (Eq. 4.55).

In ref. [149] (see also Ref. [100], where there is also the non-relativistic limit of the rest-frame instant form), it is shown that the energy–momentum tensor of the system "N charged particles plus electromagnetic field" in the inertial rest-frame has the following form (the Grassmann-valued electric charges eliminate diverging self-energies) in the radiation gauge:

$$\begin{aligned}T^{\tau\tau}(\tau,\vec{\sigma})&=\sum_{i=1}^N\delta^3(\vec{\sigma}-\vec{\eta}_i(\tau))\sqrt{m_i^2\,c^2+[\vec{\kappa}_i(\tau)-\frac{Q_i}{c}\,\vec{A}_\perp(\tau,\vec{\eta}_i(\tau))]^2}\\&\quad+\frac{1}{2\,c}\left[\left(\vec{\pi}_\perp+\sum_{i=1}^N Q_i\,\frac{\vec{\partial}}{\triangle}\delta^3(\vec{\sigma}-\vec{\eta}_i(\tau))\right)^2+\vec{B}^2\right](\tau,\vec{\sigma})\\&=T^{\tau\tau}_{matter}(\tau,\vec{\sigma})+T^{\tau\tau}_{em}(\tau,\vec{\sigma}),\\T^{\tau\tau}_{em}(\tau,\vec{\sigma})&=\frac{1}{2\,c}\left(\vec{\pi}_\perp^2+\vec{B}^2\right)(\tau,\vec{\sigma}),\\T^{r\tau}(\tau,\vec{\sigma})&=\sum_{i=1}^N\delta^3(\vec{\sigma}-\vec{\eta}_i(\tau))\left[\kappa_i^r(\tau)-\frac{Q_i}{c}\,A_\perp^r(\tau,\vec{\eta}_i(\tau))\right]\\&\quad+\frac{1}{c}\left[\left(\vec{\pi}_\perp+\sum_{i=1}^N Q_i\,\frac{\vec{\partial}}{\triangle}\,\delta^3(\vec{\sigma}-\vec{\eta}_i(\tau))\right)\times\vec{B}\right](\tau,\vec{\sigma})\\&=T^{r\tau}_{matter}(\tau,\vec{\sigma})+T^{r\tau}_{em}(\tau,\vec{\sigma}),\\T^{r\tau}_{em}(\tau,\vec{\sigma})&=\frac{1}{c}\left(\vec{\pi}_\perp\times\vec{B}\right)(\tau,\vec{\sigma}),\\T^{rs}(\tau,\vec{\sigma})&=\sum_{i=1}^N\delta^3(\vec{\sigma}-\vec{\eta}_i(\tau))\,\frac{[\kappa_i^r(\tau)-\frac{Q_i}{c}\,A_\perp^r(\tau,\vec{\eta}_i(\tau))]\,[\kappa_i^s(\tau)-\frac{Q_i}{c}\,A_\perp^s(\tau,\vec{\eta}_i(\tau))]}{\sqrt{m_i^2\,c^2+[\,\vec{\kappa}_i(\tau)-\frac{Q_i}{c}\,\vec{A}_\perp(\tau,\vec{\eta}_i(\tau))]^2}}\end{aligned}$$

$$
\begin{aligned}
&-\frac{1}{c}\left[\frac{1}{2}\delta^{rs}\left[\left(\vec{\pi}_\perp+\sum_{i=1}^N Q_i\frac{\vec{\partial}}{\triangle}\delta^3(\vec{\sigma}-\vec{\eta}_i(\tau))\right)^2+\vec{B}^2\right]\right.\\
&-\left[\left(\vec{\pi}_\perp+\sum_{i=1}^N Q_i\frac{\vec{\partial}}{\triangle}\delta^3(\vec{\sigma}-\vec{\eta}_i(\tau))\right)^r\left(\vec{\pi}_\perp+\sum_{i=1}^N Q_i\frac{\vec{\partial}}{\triangle}\delta^3(\vec{\sigma}-\vec{\eta}_i(\tau))\right)^s\right.\\
&\left.\left.+B^r\,B^s\right]\right](\tau,\vec{\sigma})\\
&= T^{rs}_{matter}(\tau,\vec{\sigma})+T^{rs}_{em}(\tau,\vec{\sigma}),\\
&T^{rs}_{em}(\tau,\vec{\sigma})=-\frac{1}{c}\left[\frac{1}{2}\delta^{rs}\left(\vec{\pi}_\perp^2+\vec{B}^2\right)-\left(\pi^r_\perp\,\pi^s_\perp+B^r\,B^s\right)\right](\tau,\vec{\sigma}). \qquad (4.67)
\end{aligned}
$$

The internal Poincaré generators have the expression ($\vec{B}=\vec{\partial}\times\vec{A}_\perp$, $c(\vec{\sigma})=-1/4\pi\,|\vec{\sigma}|$):

$$
\begin{aligned}
\mathcal{E}_{(int)}&=\mathcal{P}_{(int)}=M\,c^2=c\sum_{i=1}^N\sqrt{m_i^2\,c^2+\left(\vec{\kappa}_i(\tau)-\frac{Q_i}{c}\,\vec{A}_\perp(\tau,\vec{\eta}_i(\tau))\right)^2}\\
&+\sum_{i\neq j}\frac{Q_i\,Q_j}{4\pi\ \mid\vec{\eta}_i(\tau)-\vec{\eta}_j(\tau)\mid}+\frac{1}{2}\int d^3\sigma\,[\vec{\pi}_\perp^2+\vec{B}^2](\tau,\vec{\sigma}),\\
\vec{\mathcal{P}}_{(int)}&=\sum_{i=1}^N\vec{\kappa}_i(\tau)+\frac{1}{c}\int d^3\sigma\,[\vec{\pi}_\perp\times\vec{B}](\tau,\vec{\sigma})\approx 0,\\
\bar{S}^r&=\sum_{i=1}^N\left(\vec{\eta}_i(\tau)\times\vec{\kappa}_i(\tau)\right)^r+\frac{1}{c}\int d^3\sigma(\vec{\sigma}\times\left([\vec{\pi}_\perp\times\vec{B}]\right)^r(\tau,\vec{\sigma}),\\
\mathcal{K}^r_{(int)}&=-\sum_{i=1}^N\eta_i^r(\tau)\sqrt{m_i^2\,c^2+\left(\,\vec{\kappa}_i(\tau)-\frac{Q_i}{c}\,\vec{A}_\perp(\tau,\ \vec{\eta}_i(\tau))\right)^2}\\
&+\frac{1}{c}\sum_{i=1}^N\left[\sum_{j\neq i}^{1\dots N}Q_i\,Q_j\left[\frac{1}{\triangle_{\vec{\eta}_j}}\frac{\partial}{\partial\eta^r_j}c(\vec{\eta}_i(\tau)-\vec{\eta}_j(\tau))\right.\right.\\
&\left.-\ \eta^r_j(\tau)\,c(\,\vec{\eta}_i(\tau)-\vec{\eta}_j(\tau))\right]\\
&\left.+\ Q_i\int d^3\sigma\,\pi^r_\perp(\tau,\vec{\sigma})\,c(\vec{\sigma}-\ \vec{\eta}_i(\tau))\right]-\frac{1}{2c}\int d^3\sigma\,\sigma^r\,(\vec{\pi}_\perp^2+\vec{B}^2)(\tau,\vec{\sigma}).
\end{aligned}
\qquad (4.68)
$$

In Ref. [149] it is shown that the use of the Lienard–Wiechert solution (see Ref. [151]) with "no incoming radiation field" allows one to arrive at a description of N charged particles dressed with a Coulomb cloud and mutually interacting through the Coulomb potential augmented with the full relativistic Darwin potential. This happens independently from the choice of the Green's function (retarded, advanced, symmetric, etc.) due to the Grassmann regularization. The quantization allows one to recover the standard instantaneous approximation for relativistic bound states, which till now had only been obtained starting from

QED (either in the instantaneous approximations of the Bethe–Salpeter equation or in the quasi-potential approach). In the case of spinning particles [150] the relativistic Salpeter potential was identified.

Moreover, in Ref. [100] it is shown that by using the previous results one can find a canonical transformation from the canonical basis $\vec{\eta}_i(\tau)$, $\tilde{\vec{\kappa}}_i(\tau)$, $\vec{A}_\perp(\tau,\sigma^r)$, $\vec{\pi}_\perp(\tau,\sigma^r)$, to a new canonical basis $\hat{\vec{\eta}}_i(\tau)$, $\hat{\vec{\kappa}}_i(\tau)$, $\vec{A}_{\perp rad}(\tau,\sigma^r)$, $\vec{\pi}_{\perp rad}(\tau,\sigma^r)$ so that in the rest-frame there is a decoupled free radiation transverse field (like the one in the radiation gauge (Eq. 4.55) in absence of particles) and a system of charged particles mutually interacting with Coulomb plus Darwin potentials.

The new canonical variables are

$$\begin{aligned}
\hat{\eta}^r_i(\tau) &= e^{\{\cdot,S\}}\,\eta^r_i(\tau),\\
\hat{\kappa}_{ir}(\tau) &= e^{\{\cdot,S\}}\,\kappa_{ir}(\tau)^r_i(\tau),\\
\vec{A}_{\perp rad}(\tau,\vec{\sigma}) &= e^{\{\cdot,S\}}\,\vec{A}_\perp(\tau,\vec{\sigma}),\\
\vec{\pi}_{\perp rad}(\tau,\vec{\sigma}) &= e^{\{\cdot,S\}}\,\vec{\pi}_\perp(\tau,\vec{\sigma}),
\end{aligned} \tag{4.69}$$

where $S = \frac{1}{c}\sum_i^{1...N} Q_i \int d^3\sigma \left[\vec{\pi}_\perp\cdot\vec{A}_{\perp Si} - \vec{A}_\perp\cdot\vec{\pi}_{\perp Si}\right](\tau,\vec{\sigma})$ is the generating function of the canonical transformation and $e^{\{.,S\}}\,A = A + \{A,S\} + \frac{1}{2}\{\{A,S\},S\}$.

The new internal Poincaré generators in the $N=2$ case are ($\hat{\vec{A}}_{\perp Si}$ and $\hat{\vec{\pi}}_{\perp Si}$ are the same functions of Eqs. (4.64) and (4.65) with $\vec{\eta}_i$ and $\vec{\pi}_i$ replaced by $\hat{\vec{\eta}}_i$ and $\vec{\pi}_i$):

$$\begin{aligned}
\mathcal{E}_{(int)} = M\,c^2 &= c\sum_{i=1}^{2}\sqrt{m_i^2\,c^2 + \hat{\vec{\kappa}}_i^2(\tau)}\\
&+\frac{Q_1\,Q_2}{4\pi\,|\,\hat{\vec{\eta}}_1(\tau) - \hat{\vec{\eta}}_2(\tau)|}\\
&+V_{DARWIN}(\hat{\vec{\kappa}}_1(\tau),\hat{\vec{\kappa}}_2(\tau),\hat{\vec{\eta}}_1(\tau) - \hat{\vec{\eta}}_2(\tau))\\
&+\frac{1}{2}\int d^3\sigma\left(\vec{\pi}^2_{\perp rad} + \vec{B}^2_{rad}\right)(\tau,\vec{\sigma}) = \mathcal{E}_{matter} + \mathcal{E}_{rad},
\end{aligned}$$

$$\begin{aligned}
V_{DARWIN}(\hat{\vec{\eta}}_1(\tau) - \hat{\vec{\eta}}_2(\tau);\hat{\vec{\kappa}}_i(\tau)) &= \sum_{i\neq j}^{1...N} Q_i\,Q_j\left(\frac{\hat{\vec{\kappa}}_i\cdot\hat{\vec{A}}_{\perp Sj}(\tau,\hat{\vec{\eta}}_i(\tau))}{\sqrt{m_i^2\,c^2 + \hat{\vec{\kappa}}_i^2}}\right.\\
&+\int d^3\sigma\left[\frac{1}{2}\left(\hat{\vec{\pi}}_{\perp Si}\cdot\hat{\vec{\pi}}_{\perp Sj} + \hat{\vec{B}}_{Si}\cdot\,\hat{\vec{B}}_{Sj}\right)\right.\\
&+\left(\frac{\hat{\vec{\kappa}}_i}{\sqrt{m_i^2\,c^2 + \hat{\vec{\kappa}}_i^2}}\cdot\frac{\partial}{\partial\,\hat{\vec{\eta}}_i}\right)\left(\hat{\vec{A}}_{\perp Si}\cdot\hat{\vec{\pi}}_{\perp Sj}\right.\\
&\left.\left.\left.-\hat{\vec{\pi}}_{\perp Si}\cdot\hat{\vec{A}}_{\perp Sj}\right)\right](\tau,\vec{\sigma})\right),
\end{aligned}$$

$$\vec{\mathcal{P}}_{(int)} = \sum_{i=1}^{2} \hat{\vec{\kappa}}_i(\tau) + \frac{1}{c} \int d^3\sigma \Big(\vec{\pi}_{\perp rad} \times \vec{B}_{rad}\Big)(\tau, \vec{\sigma}) = \vec{\mathcal{P}}_{matter} + \vec{\mathcal{P}}_{rad} \approx 0,$$

$$\vec{\bar{S}} = \sum_i \hat{\vec{\eta}}_i \times \hat{\vec{\kappa}}_i + \frac{1}{c} \int d^3\sigma\, \vec{\sigma} \times \Big(\vec{\pi}_{\perp rad} \times \vec{B}_{rad}\Big)(\tau, \vec{\sigma}) = \vec{\bar{S}}_{matter} + \vec{\bar{S}}_{rad},$$

$$\begin{aligned}
\vec{\mathcal{K}}_{(int)} &= - \sum_{i=1}^{2} \hat{\vec{\eta}}_i \sqrt{m_i^2\, c^2 + \hat{\vec{\kappa}}_i^2} \\
&- \frac{1}{2} \frac{Q_1\, Q_2}{c} \Bigg[\hat{\vec{\eta}}_1 \frac{\hat{\vec{\kappa}}_1 \cdot \Big(\frac{1}{2} \frac{\partial\, \hat{\mathcal{K}}_{12}(\hat{\vec{\kappa}}_1, \hat{\vec{\kappa}}_2, \hat{\vec{\rho}}_{12})}{\partial\, \hat{\vec{\rho}}_{12}} - 2\, \vec{A}_{\perp S2}(\hat{\vec{\kappa}}_2, \hat{\vec{\rho}}_{12})\Big)}{\sqrt{m_1^2\, c^2 + \hat{\vec{\kappa}}_1^2}} \\
&+ \hat{\vec{\eta}}_2 \frac{\hat{\vec{\kappa}}_2 \cdot \Big(\frac{1}{2} \frac{\partial\, \hat{\mathcal{K}}_{12}(\hat{\vec{\kappa}}_1, \hat{\vec{\kappa}}_2, \hat{\vec{\rho}}_{12})}{\partial\, \hat{\vec{\rho}}_{12}} - 2\, \vec{A}_{\perp S1}(\hat{\vec{\kappa}}_1, \hat{\vec{\rho}}_{12})\Big)}{\sqrt{m_2^2\, c^2 + \hat{\vec{\kappa}}_2^2}} \Bigg] \\
&- \frac{1}{2} \frac{Q_1\, Q_2}{c} \Big(\sqrt{m_1^2\, c^2 + \hat{\vec{\kappa}}_1^2}\, \frac{\partial}{\partial\, \hat{\vec{\kappa}}_1} + \sqrt{m_2^2\, c^2 + \hat{\vec{\kappa}}_2^2}\, \frac{\partial}{\partial\, \hat{\vec{\kappa}}_2} \Big) \hat{\mathcal{K}}_{12}(\hat{\vec{\kappa}}_1, \hat{\vec{\kappa}}_2, \hat{\vec{\rho}}_{12}) \\
&- \frac{Q_1\, Q_2}{4\pi\, c} \int d^3\sigma \Big(\frac{\hat{\vec{\pi}}_{\perp S1}(\vec{\sigma} - \hat{\vec{\eta}}_1, \hat{\vec{\kappa}}_1)}{|\vec{\sigma} - \hat{\vec{\eta}}_2|} + \frac{\hat{\vec{\pi}}_{\perp S2}(\vec{\sigma} - \hat{\vec{\eta}}_2, \hat{\vec{\kappa}}_2)}{|\vec{\sigma} - \hat{\vec{\eta}}_1|} \Big) \\
&- \frac{Q_1\, Q_2}{c} \int d^3\sigma\, \vec{\sigma}\, \Big(\hat{\vec{\pi}}_{\perp S1}(\vec{\sigma} - \hat{\vec{\eta}}_1, \hat{\vec{\kappa}}_1) \cdot \hat{\vec{\pi}}_{\perp S2}(\vec{\sigma} - \hat{\vec{\eta}}_2, \hat{\vec{\kappa}}_2) \\
&+ \hat{\vec{B}}_{S1}(\vec{\sigma} - \hat{\vec{\eta}}_1, \hat{\vec{\kappa}}_1) \cdot \hat{\vec{B}}_{S2}(\vec{\sigma} - \hat{\vec{\eta}}_2, \hat{\vec{\kappa}}_2) \Big) \\
&- \frac{1}{2\, c} \int d^3\sigma\, \vec{\sigma}\, \Big(\vec{\pi}_{\perp rad}^2 + \vec{B}_{rad}^2 \Big)(\tau, \vec{\sigma}) = \vec{\mathcal{K}}_{matter} + \vec{\mathcal{K}}_{rad} \approx 0. \qquad (4.70)
\end{aligned}$$

The only restriction on the two decoupled systems is the determination and elimination of their overall internal 3-center of mass $\vec{q}_+$ of Eq. (3.11) inside the Wigner 3-spaces with the gauge fixing $\vec{q}_+ \approx 0$. Therefore, at the classical level there is a way out from the Haag theorem forbidding the existence of the interaction picture in QED, so that there is no unitary evolution based on interpolating fields from the "in" states to the "out" ones in scattering processes. While the extension of these results to the non-inertial rest-frame is done in Refs. [98, 99], the quantization of this framework is still to be done.

In conclusion, the pseudo-classical description of the electric charge and the Wigner-covariant rest-frame instant form of dynamics in the radiation gauge is able:

1. to extract the static action-at-a-distance Coulomb potential from the electromagnetic field and to get differential equations of motion with a well-defined Cauchy problem;
2. to regularize the classical electromagnetic self-energies (evaluated from the pseudo-classical ones with the Berezin–Marinov distribution function of Appendix B) producing the $\sum_{i\neq j}$ rule and to eliminate troubles like the

runaway solutions, the pre-acceleration and the Schott term of the Abraham–Lorentz-Dirac equation (see Refs. [192–194]);

3. to recover the asymptotic Larmor formula in the presence of more than one charged particle with only the $Q_i Q_j$, $i \neq j$, terms in accordance with macroscopic experimental facts;
4. to get a unique Lienard–Wiechert solution, because the pseudo-classical rule $Q_i^2 = 0$ forces the standard retarded and advanced solutions to coincide, avoiding the problems of the standard Feynman–Wheeler theory [195, 196];
5. to extract terms proportional to Q_i (due to $Q_i^2 = 0$) from the square roots of the particle energies and to find the expression of the action-at-a-distance Darwin potential; and
6. to describe the radiation electromagnetic field in the presence of charged particles.

See Refs. [171, 172] for the coupling of charged spinning particles to the electromagnetic field in the rest-frame instant form.

In Ref. [98] there is the extension of these results to the non-inertial frames of the 3+1 approach. Then, in Ref. [99] there is the 3+1 description of the rotating disk, of the Sagnac effect, and of the Faraday rotation in astrophysics.

4.4 The Dirac Field

In this section we shall study the 3+1 approach to the pseudo-classical, namely Grassmann-valued, Dirac field interacting with the electromagnetic field from the point of view of constraint theory. We shall describe its second-class constraints and how to find its DOs. Similar considerations hold for the one-particle Dirac equation coupled to a classical electromagnetic field. Even if this is not a classical or pseudo-classical description of fermions but a first quantized one, it is used to describe the asymptotic states of the scattering of fermions in quantum field theory, starting from the Dirac quantum field, which corresponds to the quantization of the pseudo-classical Dirac field. As shown by Berezin [197], Grassmann-valued Dirac fields go into anti-commuting Dirac quantum fields. The results of this section have not been previously published.

The pseudo-classical fermionic Dirac field [27] $\psi(x) = \{\psi_\alpha(x)\}$ is a complex 4-component spinor ($\alpha = 1, \ldots, 4$), which carries a unitary representation of the gauge group[1] $U(1)$ ($T^o = -i$ is the anti-Hermitian generator). The classical Dirac field is recovered with the Berezin distribution function of Appendix B.

The fields $\psi(x)$ together with their complex conjugated $\bar{\psi}(x) = \psi^\dagger(x)\gamma_o$ form an 8-dimensional Grassmann algebra:

[1] The theory of gauge transformation and the asymptotic behavior at spatial infinity will be studied in Section 4.5 in the context of Yang–Mills theory.

$$\begin{aligned}\psi_\alpha(x)\,\psi_\beta(y) + \psi_\beta(y)\,\psi_\alpha(x) &= \psi_\alpha(x)\,\bar{\psi}_\beta(y) + \bar{\psi}_\beta(y)\,\psi_\alpha(x)\\ &= \bar{\psi}_\alpha(x)\,\bar{\psi}_\beta(y) + \bar{\psi}_\beta(y)\,\bar{\psi}_\alpha(x) = 0,\\ &\Rightarrow\ [\bar{\psi}(x)\,\psi(x)]^2 = \sum_{\alpha\neq\beta} \bar{\psi}_\alpha(x)\,\psi_\alpha(x)\,\bar{\psi}_\beta(x)\,\psi_\beta(x),\end{aligned} \tag{4.71}$$

and are assumed to behave as $\psi(x^o, \vec{x}) \rightarrow \chi\, r^{-3/2+\epsilon} + O(r^{-2})$ for $r = |\vec{x}| \rightarrow \infty$. The derivatives with respect to the Grassmann-valued fields are always right derivatives.

The 4-dimensional Dirac matrices γ_μ are defined by

$$\begin{aligned}&\gamma_\mu\,\gamma_\nu + \gamma_\nu\,\gamma_\mu = 2\,\eta_{\mu\nu}, \qquad \gamma^\dagger_\mu = \gamma_o\,\gamma_\mu\,\gamma_o,\\ &\gamma_5 = \gamma^5 = -i\,\gamma_o\,\gamma_1\,\gamma_2\,\gamma_3, \qquad \gamma_5\,\gamma_\mu + \gamma_\mu\,\gamma_5 = 0.\end{aligned} \tag{4.72}$$

If we introduce the notation $\beta = \gamma_o$ and $\vec{\alpha} = \gamma_o\,\vec{\gamma}$, we have $\alpha_i\,\alpha_j + \alpha_j\,\alpha_i = \alpha_i\,\beta + \beta\,\alpha_i = 0$ $(i \neq j)$, $\alpha^2 = \beta^2 = 1$.

The Grassmann-valued Dirac field, satisfying Eq. (4.71), on the 3-spaces Σ_τ of the 3+1 approach is

$$\tilde{\psi}(\tau, \sigma^u) \equiv \psi(z(\tau, \sigma^u)), \qquad \tilde{\bar{\psi}}(\tau, \sigma^u) \equiv \bar{\psi}(z(\tau, \sigma^u)) = \psi^\dagger(\tau, \sigma^u)\,\gamma^o. \tag{4.73}$$

The Lagrangian of parametrized Minkowski theory for Dirac fields coupled to the electromagnetic field on the space-like hyper-surfaces Σ_τ is

$$\begin{aligned}\mathcal{L}(\tau, \sigma^u) = N(\tau, \sigma^u)\sqrt{\gamma(\tau, \sigma^u)}\,\Big[&\frac{i}{2}\tilde{\bar{\psi}}(\tau, \sigma^u)\gamma^\mu\, z^A_\mu(\tau, \sigma^u)\Big(\partial_A - i\,e\,A_A(\tau, \sigma^u)\Big)\tilde{\psi}(\tau, \sigma^u)\\ &- \frac{i}{2}\Big(\partial_A + i\,e\,A_A(\tau, \sigma^u)\Big)\,\tilde{\bar{\psi}}(\tau, \sigma^u)\, z^A_\mu(\tau, \sigma^u)\,\gamma^\mu\,\tilde{\psi}(\tau, \sigma^u)\\ &- m\,\tilde{\bar{\psi}}(\tau, \sigma^u)\,\tilde{\psi}(\tau, \sigma^u)\Big]\\ &- \frac{N(\tau, \sigma^u)\,\sqrt{\gamma(\tau, \sigma^u)}}{4}\, g^{AC}(\tau, \sigma^u)\, g^{BD}(\tau, \sigma^u)\, F_{AB}(\tau, \sigma^u)\, F_{CD}(\tau, \sigma^u).\end{aligned} \tag{4.74}$$

It is convenient to take as Lagrangian variables

$$\tilde{\psi} \rightarrow \psi = \sqrt[4]{\gamma}\,\tilde{\psi}, \qquad \tilde{\bar{\psi}} \rightarrow \bar{\psi} = \sqrt[4]{\gamma}\,\tilde{\bar{\psi}}. \tag{4.75}$$

Since on space-like hyper-planes one has $\gamma(\tau, \sigma^u) = 1$, there we shall recover $\psi = \tilde{\psi}$.

Eq. (4.74) becomes ($l^\mu(\tau, \sigma^u)$ which is normal to the 3-space Σ_τ in the point with 3-coordinates σ^u):

$$\begin{aligned}\mathcal{L}(\tau, \sigma^u) = N(\tau, \sigma^u)\,\Big[&\frac{i}{2}\,\bar{\psi}(\tau, \sigma^u)\,\gamma^\mu\, z^A_\mu(\tau, \sigma^u)\,\Big(\partial_A - i\,e\,A_A(\tau, \sigma^u)\Big)\,\psi(\tau, \sigma^u)\\ &- \frac{i}{2}\Big(\partial_A + i\,e\,A_A(\tau, \sigma^u)\Big)\,\bar{\psi}(\tau, \sigma^u)\, z^A_\mu(\tau, \sigma^u)\,\gamma^\mu\,\psi(\tau, \sigma^u)\\ &- m\,\bar{\psi}(\tau, \sigma^u)\,\psi(\tau, \sigma^u)\Big]\end{aligned}$$

$$
\begin{aligned}
& - \frac{N(\tau,\sigma^u)\,\sqrt{\gamma(\tau,\sigma^u)}}{4}\, {}^4g^{AC}(\tau,\sigma^u)\, {}^4g^{BD}(\tau,\sigma^u)\, F_{AB}(\tau,\sigma^u)\, F_{CD}(\tau,\sigma^u) \\
&= \int d\tau d^3\sigma \Big[\frac{i}{2}\, l_\mu \Big(\bar\psi\, \gamma^\mu\, (\partial_\tau - i\, e\, A_\tau)\, \psi - (\partial_\tau + i\, e\, A_\tau)\, \bar\psi\, \gamma^\mu\, \psi \Big) \\
& - \frac{i}{2}\, N^r\, l_\mu \Big(\bar\psi\, \gamma^\mu\, (\partial_r - i\, e\, A_r)\, \psi - (\partial_r + i\, e\, A_r)\, \bar\psi\, \gamma^\mu\, \psi \Big) \\
& + \frac{i}{2}\, N\, {}^3g^{rs}\, z_{s\mu} \Big(\bar\psi\, \gamma^\mu\, (\partial_r - i\, e\, A_r)\, \psi \\
& - (\partial_r + i\, e\, A_r)\, \bar\psi\, \gamma^\mu\, \psi \Big) - N\, mc\, \bar\psi\, \psi \Big](\tau,\sigma^u) \\
& - \Big[\frac{\sqrt{\gamma}}{2\, N} \Big(F_{\tau r} - N^u\, F_{ur} \Big)\, {}^3g^{rs} \Big(F_{\tau s} - N^v\, F_{vs} \Big) \\
& + \frac{N\, \sqrt{\gamma}}{4}\, {}^3g^{rs}\, {}^3g^{uv}\, F_{ru}\, F_{sv} \Big](\tau,\sigma^u).
\end{aligned} \tag{4.76}
$$

The canonical momenta are

$$
\begin{aligned}
\pi_\alpha(\tau,\sigma^u) &= \frac{\partial \mathcal{L}(\tau,\sigma^u)}{\partial\, (\partial_\tau\, \psi_\alpha)} = - \frac{i}{2} \Big(\bar\psi(\tau,\sigma^u)\, \gamma^\mu \Big)_\alpha\, l_\mu(\tau,\sigma^u), \\
\bar\pi_\alpha(\tau,\sigma^u) &= \frac{\partial \mathcal{L}(\tau,\sigma^u)}{\partial\, (\partial_\tau\, \bar\psi_\alpha)} = - \frac{i}{2} \Big(\gamma^\mu\, \psi(\tau,\sigma^u) \Big)_\alpha\, l_\mu(\tau,\sigma^u), \\
\pi^\tau(\tau,\sigma) &= \frac{\partial^L \mathcal{L}(\tau,\sigma^u)}{\partial\, (\partial_\tau\, A_\tau)} = 0, \\
\pi^r(\tau,\sigma^u) &= \frac{\partial^L \mathcal{L}(\tau,\sigma^u)}{\partial\, (\partial_\tau\, A_r)} = - \frac{\gamma(\tau,\sigma^u)}{\sqrt{g(\tau,\sigma^u)}}\, {}^3g^{rs}(\tau,\sigma^u) \Big(F_{\tau s} - {}^4g_{\tau v}\, {}^3g^{vu}\, F_{us} \Big)(\tau,\sigma^u) \\
&= \frac{\gamma(\tau,\sigma^u)}{\sqrt{g(\tau,\sigma^u)}}\, {}^3g^{rs}(\tau,\sigma^u) \Big(E_s(\tau,\sigma^u) \\
& + {}^4g_{\tau v}(\tau,\sigma^u)\, {}^3g^{vu}(\tau,\sigma^u)\, \epsilon_{ust}\, B_t(\tau,\sigma^u) \Big),
\end{aligned}
$$

$$
\begin{aligned}
\rho_\mu(\tau,\sigma^u) &= - \frac{\partial^L \mathcal{L}(\tau,\sigma^u)}{\partial z^\mu_\tau} \\
&= l_\mu(\tau,\sigma^u) \Big(- \frac{i}{2}\, {}^3g^{rs}(\tau,\sigma^u)\, z_{\nu s}(\tau,\sigma^u) \Big[\bar\psi(\tau,\sigma^u)\, \gamma^\nu\, \partial_r\, \psi(\tau,\sigma^u) \\
& - \partial_r\, \bar\psi(\tau,\sigma^u)\, \gamma^\nu\, \psi(\tau,\sigma^u) \Big] + mc\, \bar\psi(\tau,\sigma^u)\, \psi(\tau,\sigma^u) \\
& - \frac{1}{2\, \sqrt{\gamma(\tau,\sigma^u)}}\, \pi^r(\tau,\sigma^u)\, {}^3g_{rs}(\tau,\sigma^u)\, \pi^s(\tau,\sigma^u) \\
& + \frac{\sqrt{\gamma(\tau,\sigma^u)}}{4}\, {}^3g^{rs}(\tau,\sigma^u)\, {}^3g^{uv}(\tau,\sigma^u)\, F_{ru}(\tau,\sigma^u)\, F_{sv}(\tau,\sigma^u) \\
& - e\, {}^3g^{rs}(\tau,\sigma^u)\, z_{\nu s}(\tau,\sigma^u)\, A_r(\tau,\sigma^u)\, \bar\psi(\tau,\sigma^u)\, \gamma^\nu\, \psi(\tau,\sigma^u) \Big) \\
& + z_{\mu s}(\tau,\sigma^u)\, {}^3g^{rs}(\tau,\sigma^u) \Big(\frac{i}{2}\, l_\nu(\tau,\sigma^u)
\end{aligned}
$$

$$\Big[\bar{\psi}(\tau,\sigma^u)\,\gamma^\nu\,\partial_r\,\psi(\tau,\sigma^u)\partial_r\,\bar{\psi}(\tau,\sigma^u)\,\gamma^\nu\,\psi(\tau,\sigma^u)\Big]$$
$$+F_{ru}(\tau,\sigma^u)\,\pi^u(\tau,\sigma^u)+e\,A_r(\tau,\sigma^u)\,\bar{\psi}(\tau,\sigma^u)\,\gamma^\nu\,l_\nu(\tau,\sigma^u)\,\psi(\tau,\sigma^u)\Big). \tag{4.77}$$

These satisfy the Poisson brackets:

$$\begin{aligned}
\{\psi_\alpha(\tau,\sigma^u),\pi_\beta(\tau,\sigma'^u)\} &= \{\pi_\beta(\tau,\sigma'^u),\psi_\alpha(\tau,\sigma^u)\} = -\delta_{\alpha\beta}\,\delta^3(\sigma^u-\sigma'^u),\\
\{\bar{\psi}_\alpha(\tau,\sigma^u),\bar{\pi}_\beta(\tau,\sigma'^u)\} &= \{\bar{\pi}_\beta(\tau,\sigma'^u),\bar{\psi}_\alpha(\tau,\sigma^u)\} = -\delta_{\alpha\beta}\,\delta^3(\sigma^u-\sigma'^u),\\
\{z^\mu(\tau,\sigma^u),\rho_\nu(\tau,\sigma'^u)\} &= -{}^4\eta^\mu_\nu\,\delta^3(\sigma^u-\sigma'^u),\\
\{A_A(\tau,\sigma^u),\pi^B(\tau,\sigma'^u)\} &= {}^4\eta^B_A\,\delta^3(\sigma^u-\sigma'^u).
\end{aligned} \tag{4.78}$$

The primary constraints are

$$\begin{aligned}
&\tilde{\chi}_\alpha(\tau,\sigma^u)\equiv\pi_\alpha(\tau,\sigma^u)+\frac{i}{2}\,(\bar{\psi}(\tau,\sigma^u)\,\gamma^\mu)_\alpha\,l_\mu(\tau,\sigma^u)\approx 0,\\
&\tilde{\tilde{\chi}}_\alpha(\tau,\sigma^u)\equiv\bar{\pi}_\alpha(\tau,\sigma^u)+\frac{i}{2}\,(\gamma^\mu\,\psi(\tau,\sigma^u))_\alpha\,l_\mu(\tau,\sigma^u)\approx 0,\\
&\pi^\tau(\tau,\sigma^u)\approx 0,
\end{aligned}$$

$$\begin{aligned}
\bar{\mathcal{H}}_\mu(\tau,\sigma^u)\equiv\rho_\mu(\tau,\sigma^u)-l_\mu(\tau,\sigma^u)\Big(&-\frac{i}{2}\,{}^3g^{rs}(\tau,\sigma^u)\,z_{\nu s}(\tau,\sigma^u)\,\Big[\bar{\psi}(\tau,\sigma^u)\,\gamma^\nu\,\partial_r\,\psi(\tau,\sigma^u)\\
&-\partial_r\,\bar{\psi}(\tau,\sigma^u)\,\gamma^\nu\,\psi(\tau,\sigma^u)\Big]+m\,\bar{\psi}(\tau,\sigma^u)\,\psi(\tau,\sigma^u)\\
&-\frac{1}{2\,\sqrt{\gamma(\tau,\sigma^u)}}\,\pi^r(\tau,\sigma^u)\,{}^3g_{rs}(\tau,\sigma^u)\,\pi^s(\tau,\sigma^u)\\
&+\frac{\sqrt{\gamma(\tau,\sigma^u)}}{4}\,{}^3g^{rs}(\tau,\sigma^u)\,{}^3g^{uv}(\tau,\sigma^u)\,F_{ru}(\tau,\sigma^u)\,F_{sv}(\tau,\sigma^u)\\
&-e\,{}^3g^{rs}(\tau,\sigma^u)\,z_{\nu s}(\tau,\sigma^u)\,A_r(\tau,\sigma^u)\,\bar{\psi}(\tau,\sigma^u)\,\gamma^\nu\,\psi(\tau,\sigma^u)\Big)\\
&+{}^3g^{rs}(\tau,\sigma^u)\,z_{\mu s}(\tau,\sigma^u)\,\Big(\frac{i}{2}\,l_\nu(\tau,\sigma^u)\,\Big[\bar{\psi}(\tau,\sigma^u)\,\gamma^\nu\,\partial_r\,\psi(\tau,\sigma^u)\\
&-\partial_r\,\bar{\psi}(\tau,\sigma^u)\,\gamma^\nu\,\psi(\tau,\sigma^u)\Big]\\
&+F_{ru}(\tau,\sigma^u)\,\pi^u(\tau,\sigma^u)+e\,A_r(\tau,\sigma^u)\bar{\psi}(\tau,\sigma^u)\,\gamma^\nu\,l_\nu(\tau,\sigma^u)\,\psi(\tau,\sigma^u)\Big)\approx 0.
\end{aligned} \tag{4.79}$$

The canonical and Dirac Hamiltonians are

$$\begin{aligned}
\bar{H}_c=\int d^3\sigma\,\Big[&-\pi_\alpha(\tau,\sigma^u)\,\partial_\tau\,\psi_\alpha(\tau,\sigma^u)-\bar{\pi}_\alpha(\tau,\sigma^u)\,\partial_\tau\,\bar{\psi}_\alpha(\tau,\sigma^u)\\
&+\pi^A(\tau,\sigma^u)\,\partial_\tau\,A_A(\tau,\sigma^u)\\
&-\rho_\mu(\tau,\sigma^u)\,z^\mu_\tau(\tau,\sigma^u)-\mathcal{L}(\tau,\sigma^u)\Big]=-\int d^3\sigma\,\bar{\Gamma}(\tau,\sigma^u)\,A_\tau(\tau,\sigma^u),
\end{aligned}$$

$$\bar{H}_D = \int d^3\sigma \Big[- A_\tau(\tau,\sigma^u)\,\bar{\Gamma}(\tau,\sigma^u) + \lambda^\mu(\tau,\sigma^u)\,\bar{\mathcal{H}}_\mu(\tau,\sigma^u) + \bar{a}_\alpha(\tau,\sigma^u)\,\bar{\chi}_\alpha(\tau,\sigma^u) \\ + \tilde{\bar{\chi}}_\alpha(\tau,\sigma^u)\,a_\alpha(\tau,\sigma^u) + \mu_\tau(\tau,\sigma^u)\,\pi^\tau(\tau,\sigma^u)\Big], \tag{4.80}$$

where λ^μ, μ_τ, a_α are Dirac multipliers (a_α are odd multipliers).

The constraints $\bar{\chi}_\alpha \approx 0$, $\tilde{\bar{\chi}}_\alpha \approx 0$, are second class, because they satisfy the Poisson brackets

$$\{\bar{\chi}_\alpha(\tau,\sigma^u), \tilde{\bar{\chi}}_\beta(\tau,\sigma'^u)\} = -i\,(\gamma^\mu\,l_\mu(\tau,\sigma^u))_{\beta\alpha}\,\delta^3(\sigma^u - \sigma'^u). \tag{4.81}$$

It is not convenient to go to Dirac brackets with respect to them at this stage, because the fundamental variables would not have any more diagonal Dirac brackets; the elimination of these constraints will be delayed till when the theory will be restricted to space-like hyper-planes.

Since the constraints $\bar{\mathcal{H}}_\mu(\tau,\sigma^u) \approx 0$ do not have zero Poisson brackets with $\bar{\chi}_\alpha$, $\tilde{\bar{\chi}}_\alpha$,

$$\begin{aligned}
\{\bar{\mathcal{H}}_\perp(\tau,\sigma^u), \bar{\chi}_\alpha(\tau,\sigma\prime u)\} &= i\,{}^3g^{rs}(\tau,\vec{\sigma})\,z_{\mu s}(\tau,\sigma^u)\Big(\partial_r\,\bar{\psi}(\tau,\sigma^u)\,\gamma^\mu\Big)_\alpha\,\delta^3(\sigma^u - \sigma'^u)\\
&+ \frac{i}{2}\Big(\bar{\psi}(\tau,\sigma^u)\,\gamma^\mu\Big)_\alpha\,\partial_r\Big({}^3g^{rs}\,z_{\mu s}\Big)(\tau,\sigma^u)\,\delta^3(\sigma^u - \sigma'^u)\\
&+ m\,\bar{\psi}_\alpha(\tau,\sigma^u)\,\delta^3(\sigma^u - \sigma'^u)\\
&- e\,{}^3g^{rs}(\tau,\sigma^u)\,z_{\mu s}(\tau,\sigma^u)\,A_r(\tau,\sigma^u)\\
&\quad \Big(\bar{\psi}(\tau,\sigma^u)\,\gamma^\mu\Big)_\alpha\,\delta^3(\sigma^u - \sigma'^u),\\
\{\bar{\mathcal{H}}_\perp(\tau,\sigma^u), \tilde{\bar{\chi}}_\alpha(\tau,\sigma'^u)\} &= i\,{}^3g^{rs}(\tau,\sigma^u)\,z_{\mu s}(\tau,\sigma^u)\Big(\gamma^\mu\,\partial_r\,\psi(\tau,\sigma^u)\Big)_\alpha\,\delta^3(\sigma^u - \sigma'^u)\\
&+ \frac{i}{2}\Big(\gamma^\mu\,\psi(\tau,\sigma^u)\Big)_\alpha\,\partial_r\Big({}^3g^{rs}\,z_{\mu s}\Big)(\tau,\sigma^u)\,\delta^3(\sigma^u - \sigma'^u)\\
&- mc\,\psi_\alpha(\tau,\sigma^u)\,\delta^3(\sigma^u - \sigma'^u)\\
&+ e\,{}^3g^{rs}(\tau,\sigma^u)\,z_{\mu s}(\tau,\sigma^u)\,A_r(\tau,\sigma^u)\\
&\quad \Big(\gamma^\mu\,\psi(\tau,\sigma^u)\Big)_\alpha\,\delta^3(\sigma^u - \sigma'^u),\\
\{\bar{\mathcal{H}}_r(\tau,\sigma^u), \bar{\chi}_\alpha(\tau,\sigma'^u)\} &= -\,i\,\partial_r\Big(\bar{\psi}(\tau,\sigma^u)\,\gamma^\mu\Big)_\alpha\,l_\mu(\tau,\sigma^u)\,\delta^3(\sigma^u - \sigma'^u)\\
&+ \frac{i}{2}\Big(\bar{\psi}(\tau,\sigma'^u)\,\gamma^\mu\Big)_\alpha\,l_\mu(\tau,\sigma'^u)\,\partial_r\,\delta^3(\sigma^u - \sigma'^u)\\
&+ e\,A_r(\tau,\sigma^u)\Big(\bar{\psi}(\tau,\sigma^u)\,\gamma^\mu\Big)_\alpha\,l_\mu(\tau,\sigma^u)\,\delta^3(\sigma^u - \sigma'^u),\\
\{\bar{\mathcal{H}}_r(\tau,\sigma^u), \tilde{\bar{\chi}}_\alpha(\tau,\sigma'^u)\} &= -\,i\,\partial_r\Big(\gamma^\mu\,\psi(\tau,\sigma^u)\Big)_\alpha\,l_\mu(\tau,\sigma^u)\,\delta^3(\sigma^u - \sigma'^u)\\
&+ \frac{i}{2}\Big(\gamma^\mu\,\psi(\tau,\sigma'^u)\Big)_\alpha\,l_\mu(\tau,\sigma'^u)\,\partial_r\,\delta^3(\sigma^u - \sigma'^u)\\
&- e\,A_r(\tau,\sigma^u)\Big(\gamma^\mu\,\psi(\tau,\sigma^u)\Big)_\alpha\,l_\mu(\tau,\sigma^u)\,\delta^3(\sigma^u - \sigma'^u),
\end{aligned} \tag{4.82}$$

where

$$\bar{\mathcal{H}}_\perp(\tau,\sigma^u) \equiv l^\mu(\tau,\sigma^u)\,\bar{\mathcal{H}}_\mu(\tau,\sigma^u) \approx 0, \qquad \bar{\mathcal{H}}_r(\tau,\sigma^u) \equiv z^\mu_r(\tau,\sigma^u)\,\bar{\mathcal{H}}_\mu(\tau,\sigma^u) \approx 0. \tag{4.83}$$

It is convenient to introduce the new constraints,

$$\begin{aligned}
\bar{\mathcal{H}}^*_\mu(\tau,\sigma^u) &= \bar{\mathcal{H}}_\mu(\tau,\sigma^u) - \int d^3v\,\{\bar{\mathcal{H}}_\mu(\tau,\sigma^u),\\
&\quad \tilde{\bar{\chi}}_\beta(\tau,v^u)\}\, i\left(\gamma^\mu\, l_\mu(\tau,v^u)\right)_{\alpha\beta} \bar{\chi}_\alpha(\tau,v^u)\\
&\quad - \int d^3v\,\{\bar{\mathcal{H}}_\mu(\tau,\sigma^u), \bar{\chi}_\beta(\tau,v^u)\}\, i\left(\gamma^\mu\, l_\mu(\tau,v^u)\right)_{\beta\alpha}\\
&\quad \tilde{\bar{\chi}}_\alpha(\tau,u^v) \approx \bar{\mathcal{H}}_\mu(\tau,\sigma^u) \approx 0,\\
\{\bar{\mathcal{H}}^*_\mu(\tau,\sigma^u), \bar{\chi}_\alpha(\tau,\sigma'^u)\} &\approx 0, \qquad \{\bar{\mathcal{H}}^*_\mu(\tau,\sigma^u), \tilde{\bar{\chi}}_\alpha(\tau,\sigma'^u)\} \approx 0,\\
\{\bar{\mathcal{H}}^*_\mu(\tau,\sigma^u), \bar{\mathcal{H}}^*_\nu(\tau,\sigma'^u)\} &\approx \Big(\Big[l_\mu(\tau,\sigma^u)\, z_{r\nu}(\tau,\sigma^u)\\
&\quad - l_\nu(\tau,\sigma^u)\, z_{r\mu}(\tau,\sigma^u)\Big]\, \frac{\pi^r(\tau,\sigma^u)}{\sqrt{\gamma(\tau,\sigma^u)}}\\
&\quad - z_{u\mu}(\tau,\sigma^u)\, {}^3g^{ur}(\tau,\sigma^u)\, F_{rs}(\tau,\sigma^u)\, {}^3g^{sv}(\tau,\sigma^u)\\
&\quad z_{v\nu}(\tau,\sigma^u)\Big)\bar{\Gamma}(\tau,\sigma^u)\,\delta^3(\sigma^u - \sigma'^u) \approx 0,
\end{aligned} \tag{4.84}$$

where $\bar{\Gamma}(\tau,\sigma^u) \approx 0$, the Gauss's law constraint, will be defined in the next equation. Therefore, the constraints $\bar{\mathcal{H}}^*_\mu(\tau,\sigma^u) \approx 0$ are constants of the motion and are first class.

The time constancy of $\pi^\tau(\tau,\sigma^u) \approx 0$, i.e., $\partial_\tau\, \pi^\tau(\tau,\sigma^u) \mathrel{\circ\!\!=} \{\pi^\tau(\tau,\sigma^u), \bar{H}_D\} \approx 0$, gives the Gauss's law secondary constraint:

$$\bar{\Gamma}(\tau,\sigma^u) = \partial_r\, \pi^r(\tau,\sigma^u) + e\,\bar{\psi}(\tau,\sigma^u)\,\gamma^\mu\, l_\mu(\tau,\sigma^u)\,\psi(\tau,\sigma^u) \approx 0. \tag{4.85}$$

Since we have

$$\begin{aligned}
\{\bar{\Gamma}(\tau,\sigma^u), \bar{\chi}_\alpha(\tau,\sigma'^u)\} &= -e\,\bar{\psi}_\beta(\tau,\sigma^u)\left(\gamma^\mu\, l_\mu(\tau,\sigma^u)\right)_{\beta\alpha} \delta^3(\sigma^u - \sigma'^u),\\
\{\bar{\Gamma}(\tau,\sigma^u), \tilde{\bar{\chi}}_\alpha(\tau,\sigma'^u)\} &= e\left(\gamma^\mu\, l_\mu(\tau,\sigma^u)\right)_{\alpha\beta} \psi_\beta(\tau,\sigma^u)\,\delta^3(\sigma^u - \sigma'^u),
\end{aligned} \tag{4.86}$$

let us define

$$\bar{\Gamma}^*(\tau,\sigma^u) \equiv \bar{\Gamma}(\tau,\sigma^u) + i\,e\,[\bar{\chi}_\alpha\,\psi_\alpha + \bar{\psi}_\alpha\,\tilde{\bar{\chi}}_\alpha](\tau,\sigma^u) \approx 0. \tag{4.87}$$

This new constraint satisfies

$$\begin{aligned}
&\{\bar{\Gamma}^*(\tau,\sigma^u), \bar{\chi}_\alpha(\tau,\sigma'^u)\} = -i\,e\,\bar{\chi}_\alpha(\tau,\sigma)\,\delta^3(\sigma^u - \sigma'^u),\\
&\{\bar{\Gamma}^*(\tau,\sigma^u), \tilde{\bar{\chi}}_\alpha(\tau,\sigma'^u)\} = i\,e\,\tilde{\bar{\chi}}_\alpha(\tau,\sigma^u)\,\delta^3(\sigma^u - \sigma'^u),\\
&\{\bar{\Gamma}^*(\tau,\sigma^u), \bar{\mathcal{H}}^*_\mu(\tau,\sigma'^u)\} \approx 0,\\
&\{\bar{\Gamma}^*(\tau,\sigma^u), \bar{\Gamma}^*(\tau,\sigma'^u)\} \approx 0, \qquad \{\bar{\Gamma}^*(\tau,\sigma^u), \pi^\tau(\tau,\sigma'^u)\} = 0.
\end{aligned} \tag{4.88}$$

Therefore, $\bar{\Gamma}^*(\tau,\sigma^u) \approx 0$ is a constant of motion and a first-class constraint like $\pi^\tau(\tau,\sigma^u) \approx 0$.

The time constancy of $\bar{\chi}_\alpha$, $\tilde{\bar{\chi}}_\alpha$

$$\begin{aligned}\partial_\tau \bar{\chi}_\alpha(\tau,\sigma^u) &\stackrel{\circ}{=} \{\bar{\chi}_\alpha(\tau,\sigma^u), \bar{H}_D\} = -e\, A_\tau(\tau,\sigma^u) \left(\bar{\psi}(\tau,\sigma^u)\,\gamma^\mu\right)_\alpha l_\mu(\tau,\sigma^u) \\ &\quad + i\, \bar{a}_\beta(\tau,\sigma^u) \left(\gamma^\mu\, l_\mu(\tau,\sigma^u)\right)_{\beta\alpha} \approx 0, \\ \partial_\tau \tilde{\bar{\chi}}_\alpha(\tau,\sigma) &\stackrel{\circ}{=} \{\tilde{\bar{\chi}}_\alpha(\tau,\sigma^u), \bar{H}_D\} = e\, A_\tau(\tau,\sigma^u)\, l_\mu(\tau,\sigma^u) \left(\gamma^\mu\, \psi(\tau,\sigma^u)\right)_\alpha \\ &\quad - i\, a_\beta(\tau,\sigma^u) \left(\gamma^\mu\, l_\mu(\tau,\sigma^u)\right)_{\alpha\beta} \approx 0, \end{aligned} \tag{4.89}$$

gives

$$\bar{a}_\alpha(\tau,\sigma^u)\, \bar{\chi}_\alpha(\tau,\sigma^u) + a_\alpha(\tau,\sigma^u)\, \tilde{\bar{\chi}}_\alpha(\tau,\sigma^u) \approx -i\, e\, A_\tau(\tau,\sigma^u) \left(\bar{\chi}_\alpha\, \psi_\alpha + \bar{\psi}_\alpha\, \tilde{\bar{\chi}}_\alpha\right)(\tau,\sigma^u), \tag{4.90}$$

so that the final Dirac Hamiltonian is

$$\bar{H}_D = \int d^3\sigma \left[-A_\tau(\tau,\sigma^u)\, \bar{\Gamma}^*(\tau,\sigma^u) + \lambda^\mu(\tau,\sigma^u)\, \bar{\mathcal{H}}^*_\mu(\tau,\sigma^u) + \mu_\tau(\tau,\sigma^u)\, \pi^\tau(\tau,\sigma^u)\right], \tag{4.91}$$

in which only the first-class constraints $\bar{\mathcal{H}}^*_\mu$, π^τ, $\bar{\Gamma}^*$ appear.

One can show that the external Poincaré generators

$$\begin{aligned} P_\mu &= \int d^3\sigma\, \rho_\mu(\tau,\sigma^u), \\ J^{\mu\nu} &= \int d^3\sigma \left[z^\mu(\tau,\sigma^u)\, \rho^\nu(\tau,\sigma^u) - z^\nu(\tau,\sigma^u)\, \rho^\mu(\tau,\sigma^u)\right] \\ &\quad + \frac{i}{2} \int d^3\sigma \left[\pi(\tau,\sigma^u)\, \sigma^{\mu\nu}\, \psi(\tau,\sigma^u) + \bar{\psi}(\tau,\sigma^u)\, \sigma^{\mu\nu}\, \bar{\pi}(\tau,\sigma^u)\right], \end{aligned} \tag{4.92}$$

and the electric charge

$$Q = -e \int d^3\sigma \left[\bar{\psi}(\tau,\sigma^u)\, \gamma^\mu\, l_\mu(\tau,\sigma^u)\, \psi(\tau,\sigma^u)\right], \tag{4.93}$$

are constants of the motion.

Let us note that, like γ^o, the matrix $\gamma^\mu\, l_\mu(\tau,\sigma^u)$ satisfies $\left[\gamma^\mu\, l_\mu(\tau,\sigma^u)\right]^2 = I$ and that we have $\{z^\mu(\tau,\sigma^u), p_s{}^\nu\} = -{}^4\eta^{\mu\nu}$.

After the restriction to space-like hyper-planes $\Sigma_{H\tau}$ ($\{.,.\}^*$ are the Dirac brackets implied by the gauge fixings on the embedding $z^\mu(\tau,\sigma^u)$; $\vec{\sigma}$ are Cartesian 3-coordinates), the angular momentum tensor becomes

$$\begin{aligned} J^{\mu\nu} &= L^{\mu\nu}_s + S^{\mu\nu}_s + S^{\mu\nu}_\psi, \\ L^{\mu\nu}_s &= x^\mu_s(\tau)\, p^\nu_s - x^\nu_s(\tau)\, p^\mu_s, \\ S^{\mu\nu}_s &= b^\mu_r(\tau) \int d^3\sigma\, \sigma^r\, \rho^\nu(\tau,\vec{\sigma}) - b^\nu_r(\tau) \int d^3\sigma\, \sigma^r\, \rho^\mu(\tau,\vec{\sigma}), \end{aligned}$$

$$S_\psi^{\mu\nu} = \frac{i}{2}\int d^3\sigma\,\Big[\pi(\tau,\vec{\sigma})\,\sigma^{\mu\nu}\,\psi(\tau,\vec{\sigma}) + \bar{\psi}(\tau,\vec{\sigma})\,\sigma^{\mu\nu}\,\bar{\pi}(\tau,\vec{\sigma})\Big],$$

$$\begin{aligned} \{J^{\mu\nu}, J^{\alpha\beta}\}^* &= C^{\mu\nu\alpha\beta}_{\gamma\delta}\,J^{\gamma\delta}, & \{L_s^{\mu\nu}, L_s^{\alpha\beta}\}^* &= C^{\mu\nu\alpha\beta}_{\gamma\delta}\,L_s^{\gamma\delta},\\ \{S_s^{\mu\nu}, S_s^{\alpha\beta}\}^* &= C^{\mu\nu\alpha\beta}_{\gamma\delta}\,S_s^{\gamma\delta}, & \{S_\xi^{\mu\nu}, S_\xi^{\alpha\beta}\}^* &= C^{\mu\nu\alpha\beta}_{\gamma\delta}\,S_\xi^{\gamma\delta}. \end{aligned} \tag{4.94}$$

The relations

$$\{S_\psi^{\mu\nu}, \psi_\alpha(\tau,\vec{\sigma})\}^* = \frac{i}{2}\Big(\sigma^{\mu\nu}\,\psi(\tau,\vec{\sigma})\Big)_\alpha, \qquad \{S_\psi^{\mu\nu}, \bar{\psi}_\alpha(\tau,\vec{\sigma})\}^* = -\frac{i}{2}\Big(\bar{\psi}(\tau,\vec{\sigma})\,\sigma^{\mu\nu}\Big)_\alpha, \tag{4.95}$$

show that $S_\psi^{\mu\nu}$ is the generator of the symplectic action of the Lorentz transformations on the Dirac fields.

The Dirac Hamiltonian becomes

$$\begin{aligned} \bar{H}_D^F &= \tilde{\lambda}^\mu(\tau)\,\bar{\tilde{\mathcal{H}}}^*_\mu(\tau) - \frac{1}{2}\,\tilde{\lambda}^{\mu\nu}(\tau)\,\bar{\tilde{\mathcal{H}}}^*_{\mu\nu}(\tau)\\ &\quad + \int d^3\sigma\,\Big[-A_\tau(\tau,\vec{\sigma})\,\bar{\Gamma}^*(\tau,\vec{\sigma}) + \mu_\tau(\tau,\vec{\sigma})\,\pi^\tau(\tau,\vec{\sigma})\Big],\\ \bar{\tilde{\mathcal{H}}}^*_\mu(\tau) &= \int d^3\sigma\,\bar{\mathcal{H}}^*_\mu(\tau,\vec{\sigma}) \approx 0,\\ \bar{\tilde{\mathcal{H}}}^*_{\mu\nu}(\tau) &= b_{\mu r}(\tau)\int d^3\sigma\,\sigma^r\,\bar{\mathcal{H}}^*_\nu(\tau,\vec{\sigma}) - b_{\nu r}(\tau)\int d^3\sigma\,\sigma^r\,\bar{\mathcal{H}}^*_\mu(\tau,\vec{\sigma}) \approx 0. \end{aligned} \tag{4.96}$$

On $\Sigma_{H\tau}$ the second-class constraints have the form (from Eq. (2.17) we have $b^\mu_\tau = l^\mu$ on space-like hyper-planes):

$$\begin{aligned} \bar{\chi}_\alpha(\tau,\vec{\sigma}) &= \pi_\alpha(\tau,\vec{\sigma}) + \frac{i}{2}\Big(\bar{\psi}(\tau,\vec{\sigma})\,\gamma^\mu\Big)_\alpha\,b_{\mu\tau} \approx 0,\\ \bar{\tilde{\chi}}_\alpha(\tau,\vec{\sigma}) &= \bar{\pi}_\alpha(\tau,\vec{\sigma}) + \frac{i}{2}\Big(\gamma^\mu\,\psi(\tau,\vec{\sigma})\Big)_\alpha\,b_{\mu\tau} \approx 0,\\ \{\bar{\chi}_\alpha(\tau,\vec{\sigma}), \bar{\tilde{\chi}}_\beta(\tau,\vec{\sigma}')\}^* &= -i\Big(\gamma^\mu\,b_{\mu\tau}\Big)_{\beta\alpha}\,\delta^3(\vec{\sigma}-\vec{\sigma}'), \qquad \Big[\Big(\gamma^\mu\,b_{\mu\tau}\Big)^2 = 2\Big], \end{aligned} \tag{4.97}$$

and can be eliminated by introducing the Dirac brackets:

$$\begin{aligned} \{\bar{A},\bar{B}\}^*_D = \{\bar{A},\bar{B}\}^* &- \int d^3u\,\{\bar{A}, \bar{\chi}_\alpha(\tau,\vec{u})\}^*\,i\Big(\gamma^\mu\,b_{\mu\tau}\Big)_{\alpha\beta}\,\{\bar{\tilde{\chi}}_\beta(\tau,\vec{u}), \bar{B}\}^*\\ &- \int d^3u\,\{\bar{A}, \bar{\tilde{\chi}}_\alpha(\tau,\vec{u})\}^*\,i\Big(\gamma^\mu\,b_{\mu\tau}\Big)_{\beta\alpha}\,\{\bar{\chi}_\beta(\tau,\vec{u}), \bar{B}\}^*. \end{aligned} \tag{4.98}$$

Now we get $\bar{\tilde{\mathcal{H}}}^*_\mu(\tau) \equiv \bar{\mathcal{H}}_\mu(\tau) = \int d^3\sigma\,\bar{\mathcal{H}}_\mu(\tau,\vec{\sigma})$, $\bar{\tilde{\mathcal{H}}}^*_{\mu\nu}(\tau) \equiv \bar{\tilde{\mathcal{H}}}_{\mu\nu}(\tau) = b_{\mu r}(\tau)\int d^3\sigma\,\sigma^r\,\bar{\mathcal{H}}^*_\nu(\tau,\vec{\sigma}) - b_{\nu r}(\tau)\int d^3\sigma\,\sigma^r\,\bar{\mathcal{H}}^*_\mu(\tau,\vec{\sigma})$,

$$S_\psi^{\mu\nu} \equiv \frac{1}{4}\int d^3\sigma\,b_{\rho\tau}\,\bar{\psi}(\tau,\vec{\sigma})\,\Big[\gamma^\rho, \sigma^{\mu\nu}\Big]_+\psi(\tau,\vec{\sigma}), \tag{4.99}$$

and

$$\{\psi_\alpha(\tau,\vec{\sigma}), \bar{\psi}_\beta(\tau,\vec{\sigma}')\}^*_D = -i \left(\gamma^\mu\, b_{\mu\tau}\right)_{\alpha\beta} \delta^3(\vec{\sigma}-\vec{\sigma}'), \tag{4.100}$$

while the Dirac brackets of the variables x^μ_s (the inertial observer), P^μ, b^μ_A, A_A, π^A are left unaltered by the new brackets. Now only the total spin,

$$S^{\mu\nu} = S^{\mu\nu}_s + S^{\mu\nu}_\psi, \tag{4.101}$$

satisfies a Lorentz algebra, since we have

$$\begin{aligned}
&\{S^{\mu\nu}, S^{\alpha\beta}\}^*_D = C^{\mu\nu\alpha\beta}_{\gamma\delta}\, S^{\gamma\delta},\\
&\{S^{\mu\nu}, \psi_\alpha(\tau,\vec{\sigma})\}^*_D = \frac{i}{2}\left(\sigma^{\mu\nu}\, \psi(\tau,\vec{\sigma})\right)_\alpha, \quad \{S^{\mu\nu}, \bar{\psi}_\alpha(\tau,\vec{\sigma})\}^*_D = -\frac{i}{2}\left(\bar{\psi}(\tau,\vec{\sigma})\, \sigma^{\mu\nu}\right)_\alpha,\\
&\{L^{\mu\nu}_s, L^{\alpha\beta}_s\}^*_D = \{L^{\mu\nu}_s, L^{\alpha\beta}_s\}^* = C^{\mu\nu\alpha\beta}_{\gamma\delta}\, L^{\gamma\delta}_s, \qquad \{L^{\mu\nu}_s, S^{\alpha\beta}\}^*_D = \{L^{\mu\nu}_s, S^{\alpha\beta}\}^* = 0,\\
&\{J^{\mu\nu}, J^{\alpha\beta}\}^*_D = \{J^{\mu\nu}, J^{\alpha\beta}\}^* = C^{\mu\nu\alpha\beta}_{\gamma\delta}\, J^{\gamma\delta}.
\end{aligned} \tag{4.102}$$

The Dirac Hamiltonian becomes

$$\begin{aligned}
\bar{H}_D &= \tilde{\lambda}^\mu(\tau)\, \tilde{\bar{\mathcal{H}}}_\mu(\tau) - \frac{1}{2}\tilde{\lambda}^{\mu\nu}(\tau)\, \tilde{\bar{\mathcal{H}}}_{\mu\nu}(\tau)\\
&\quad + \int d^3\sigma \Big[- A_\tau(\tau,\vec{\sigma})\, \bar{\Gamma}(\tau,\vec{\sigma}) + \mu_\tau(\tau,\vec{\sigma})\, \pi^\tau(\tau,\vec{\sigma})\Big],
\end{aligned} \tag{4.103}$$

and contains all the first-class constraints:

$$\begin{aligned}
&\bar{\Gamma}(\tau,\vec{\sigma}) = \partial_r\, \pi^r(\tau,\vec{\sigma}) + e\, \bar{\psi}(\tau,\vec{\sigma})\, \gamma^\mu\, b_{\mu\tau}\, \psi(\tau,\vec{\sigma}) \approx 0,\\
&\pi^\tau(\tau,\vec{\sigma}) \approx 0,\\
&\tilde{\bar{\mathcal{H}}}_\mu(\tau) = P_\mu - b_{\mu\tau} \int d^3\sigma \Big[\frac{i}{2}\, b_{\nu r}(\tau) \Big(\bar{\psi}(\tau,\vec{\sigma})\, \gamma^\nu\, \partial_r\, \psi(\tau,\vec{\sigma})\\
&\qquad - \partial_r\, \bar{\psi}(\tau,\vec{\sigma})\, \gamma^\nu\, \psi(\tau,\vec{\sigma})\Big) + mc\, \bar{\psi}(\tau,\vec{\sigma})\, \psi(\tau,\vec{\sigma}) + \frac{\vec{\pi}^2(\tau,\vec{\sigma}) + \vec{B}^2(\tau,\vec{\sigma})}{2}\\
&\qquad + e\, b_{\nu r}(\tau)\, A_r(\tau,\vec{\sigma})\, \bar{\psi}(\tau,\vec{\sigma})\, \gamma^\nu\, \psi(\tau,\vec{\sigma})\Big]\\
&\qquad + b_{\mu r}(\tau) \int d^3\sigma \Big[\frac{i}{2}\, b_{\nu\tau} \Big(\bar{\psi}(\tau,\vec{\sigma})\, \gamma^\nu\, \partial_r\, \psi(\tau,\vec{\sigma})\\
&\qquad - \partial_r\, \bar{\psi}(\tau,\vec{\sigma})\, \gamma^\nu\, \psi(\tau,\vec{\sigma})\Big) + \Big(\vec{\pi}(\tau,\vec{\sigma}) \times \vec{B}(\tau,\vec{\sigma})\Big)_r\\
&\qquad + e\, A_r(\tau,\vec{\sigma})\, \bar{\psi}(\tau,\vec{\sigma})\, \gamma^\nu\, b_{\nu\tau}\, \psi(\tau,\vec{\sigma})\Big] \approx 0,
\end{aligned}$$

$$\begin{aligned}
&\tilde{\bar{\mathcal{H}}}^{\mu\nu}(\tau) = S^{\mu\nu}(\tau) - S^{\mu\nu}_\psi(\tau)\\
&- \Big(b^\mu_r(\tau)\, b^\nu_\tau - b^\nu_r(\tau)\, b^\mu_\tau\Big) \int d^3\sigma\, \sigma^r \Big[\frac{i}{2}\, b_{\nu s}(\tau) \Big(\bar{\psi}(\tau,\vec{\sigma})\, \gamma^\nu\, \partial_s\, \psi(\tau,\vec{\sigma})\\
&- \partial_s\, \bar{\psi}(\tau,\vec{\sigma})\, \gamma^\nu\, \psi(\tau,\vec{\sigma})\Big) + m\, \bar{\psi}(\tau,\vec{\sigma})\, \psi(\tau,\vec{\sigma}) + \frac{\vec{\pi}^2(\tau,\vec{\sigma}) + \vec{B}^2(\tau,\vec{\sigma})}{2}
\end{aligned}$$

$$+e\, b_{\nu s}(\tau)\, A_s(\tau,\vec{\sigma})\, \bar{\psi}(\tau,\vec{\sigma})\, \gamma^\nu\, \psi(\tau,\vec{\sigma})\Big]$$
$$+\Big(b^\mu_r(\tau)\, b^\nu_s(\tau) - b^\nu_r(\tau)\, b^\mu_s(\tau)\Big) \int d^3\sigma\, \sigma^r \Big[\frac{i}{2}\, b_{\rho\tau}\, \Big(\bar{\psi}(\tau,\vec{\sigma})\, \gamma^\rho\, \partial_s\, \psi(\tau,\vec{\sigma})$$
$$-\partial_s\, \bar{\psi}(\tau,\vec{\sigma})\, \gamma^\rho\, \psi(\tau,\vec{\sigma})\Big) + \Big(\vec{\pi}(\tau,\vec{\sigma}) \times \vec{B}(\tau,\vec{\sigma})\Big)_s$$
$$+e\, A_s(\tau,\vec{\sigma})\, \bar{\psi}(\tau,\vec{\sigma})\, \gamma^\rho\, b_{\rho\tau}\, \psi(\tau,\vec{\sigma})\Big] \approx 0,$$

$$\{\bar{\mathcal{H}}^*_\mu(\tau,\vec{\sigma}), \bar{\mathcal{H}}^*_\nu(\tau,\vec{\sigma}')\} = \{\bar{\mathcal{H}}^*_\mu(\tau,\vec{\sigma}), \bar{\mathcal{H}}^*_\nu(\tau,\vec{\sigma}')\}^* \approx \{\bar{\mathcal{H}}_\mu(\tau,\vec{\sigma}), \bar{\mathcal{H}}_\nu(\tau,\vec{\sigma}')\}^*_D \approx 0. \tag{4.104}$$

For the configurations of the isolated system which are time-like, namely with $P^2 > 0$, we can boost at rest with the standard Wigner boost $L^\mu_{\cdot\nu}(p, \overset{o}{p})$ of Eq. (A.7)[2] for time-like Poincaré orbits all the variables of the non-canonical basis $x^\mu(\tau)$, P^μ, $b^\mu_A(\tau)$, $S^{\mu\nu}_s(\tau)$, $A_A(\tau,\vec{\sigma})$, $\pi^A(\tau,\vec{\sigma})$, $\psi(\tau,\vec{\sigma})$, $\bar{\psi}(\tau,\vec{\sigma})$ with Lorentz indices (to make it canonical $x^\mu(\tau)$ must be replaced with $\tilde{x}^\mu(\tau)$ of Eq. (3.3)). This is done with a canonical transformation $e^{\{.,\bar{\mathcal{F}}\}}$ of this basis with the generating function

$$\bar{\mathcal{F}}(p_s) = \frac{1}{2}\, \omega(\vec{p}_s)\, I_{\mu\nu}(p_s)\, S^{\mu\nu}, \tag{4.105}$$

containing the total spin $S^{\mu\nu}$ of Eq. (4.102) and not only $S^{\mu\nu}_s$, as we have for scalar fields. We get

$$\begin{aligned}
\overset{o}{\psi}(\tau,\vec{\sigma}) &= \exp\{\bar{\mathcal{F}}, .\}^*_D\, \psi(\tau,\vec{\sigma}) \\
&= \psi(\tau,\vec{\sigma}) + \{\bar{\mathcal{F}}, \psi(\tau,\vec{\sigma})\}^*_D + \frac{1}{2}\, \{\bar{\mathcal{F}}, \{\bar{\mathcal{F}}, \psi(\tau,\vec{\sigma})\}^*_D\}^*_D + \cdots \\
&= \psi(\tau,\vec{\sigma}) + \frac{i}{4}\, \omega(\vec{p}_s)\, I_{\mu\nu}(p_s)\, \sigma^{\mu\nu}\, \psi(\tau,\vec{\sigma}) + \cdots \\
&= \exp\Big[\frac{i}{4}\, \omega(\vec{p}_s)\, I_{\mu\nu}(p_s)\, \sigma^{\mu\nu}\Big]\, \psi(\tau,\vec{\sigma}).
\end{aligned} \tag{4.106}$$

This shows that this canonical transformation implements on the Dirac fields the action of Wigner boosts realized by using the standard representation of Lorentz transformations in terms of Dirac matrices [198, 199]:

$$S(L(\overset{o}{p}, p)) = \exp\Big[\frac{i}{4}\, \omega(\vec{p})\, I_{\mu\nu}(p)\, \sigma^{\mu\nu}\Big]. \tag{4.107}$$

In this way (see Section A.5 of Appendix A) we get the non-canonical basis $\tilde{x}^\mu(\tau)$, p^μ, $b^A_A(\tau)$, $\tilde{S}^{\mu\nu}_s(\tau)$, $A_A(\tau,\vec{\sigma})$, $\pi^A(\tau,\vec{\sigma})$, and

$$\overset{o}{\psi}(\tau,\vec{\sigma}) = S(L(\overset{o}{p}, p))\, \psi(\tau,\vec{\sigma}), \qquad \overset{o}{\bar{\psi}}(\tau,\vec{\sigma}) = \bar{\psi}(\tau,\vec{\sigma})\, S^{-1}(L(\overset{o}{p}, p)). \tag{4.108}$$

[2] From now on we will use the notation $P^\mu = p^\mu = L^\mu_\nu(p, \overset{o}{p})\, \overset{o}{p}^\nu$, $\overset{o}{p}^\mu = mc\, \epsilon^4 \eta^{\mu\nu}$ used in Appendix A.

Since we have

$$
\begin{aligned}
\{\tilde{x}^\mu, \overset{o}{\psi}(\tau,\vec{\sigma})\}^*_D &= \{\tilde{x}^\mu, S(L(\overset{o}{p},p))\,\psi(\tau,\vec{\sigma})\}^*_D \\
&= -\frac{\partial S(L(\overset{o}{p},p))}{\partial p_\mu}\, S^{-1}(L(\overset{o}{p},p)) \circ \psi(\tau,\vec{\sigma}) \\
&\quad -\frac{i}{4}\,\epsilon^A_\nu(u(p))\,{}^4\eta_{AB}\,\frac{\partial \epsilon^B_\rho(u(p))}{\partial p_\mu}\, S(L(\overset{o}{p},p))\,\sigma^{\nu\rho}\, S^{-1}(L(\overset{o}{p},p))\;\overset{o}{\psi}(\tau,\vec{\sigma}) \\
&= \Big[-\frac{\partial S(L(\overset{o}{p},p))}{\partial p_\mu}\, S^{-1}(L(\overset{o}{p},p)) \\
&\quad -\frac{i}{4}\,{}^4\eta_{\sigma B}\,\frac{\partial \epsilon^B_\rho(u(p))}{\partial p_\mu}\, L^\rho_\eta(p,\overset{o}{p})\,\sigma^{\sigma\eta}\Big]\;\overset{o}{\psi}(\tau,\vec{\sigma}),
\end{aligned}
\tag{4.109}
$$

to verify if $\{\tilde{x}^\mu, \overset{o}{\psi}(\tau,\vec{\sigma})\}^*_D = 0$, we need the evaluation of $\frac{\partial S(L(\overset{o}{p},p))}{\partial p_\mu}$. In Ref. [200] there is the following formula:

$$
\frac{\partial e^{B(\lambda)}}{\partial \lambda} = \int_0^1 dx\, e^{x\,B(\lambda)}\,\frac{\partial B(\lambda)}{\partial \lambda}\, e^{-x\,B(\lambda)}\, e^{B(\lambda)},
\tag{4.110}
$$

giving the derivative with respect to a continuous parameter λ of the exponential of an operator $B(\lambda)$. If we put

$$
A(p) \equiv \frac{i}{4}\,\omega(\vec{p})\, I_{\mu\nu}(p)\,\sigma^{\mu\nu} = -\frac{i}{2}\,\frac{\omega(\vec{p})}{|\vec{p}|}\, p_i\,\sigma^{0i}, \qquad S(L(\overset{o}{p},p)) = e^{A(p)},
\tag{4.111}
$$

and if we suppose $p^\mu = p^\mu(\lambda)$, we have

$$
\begin{aligned}
&1.\quad \frac{\partial e^{A(p(\lambda))}}{\partial \lambda} = \frac{\partial p_\mu(\lambda)}{\partial \lambda}\,\frac{\partial e^{A(p(\lambda))}}{\partial p_\mu}, \\
&2.\quad \frac{\partial e^{A(p(\lambda))}}{\partial \lambda} = \int_0^1 dx\, e^{x\,A(p(\lambda))}\,\frac{\partial A(p(\lambda))}{\partial \lambda}\, e^{-x\,A(p(\lambda))}\, e^{A(p(\lambda))} \\
&\qquad = \frac{\partial p_\mu(\lambda)}{\partial \lambda}\int_0^1 dx\, e^{x\,A(p(\lambda))}\,\frac{\partial A(p(\lambda))}{\partial p_\mu}\, e^{-x\,A(p(\lambda))}\, e^{A(p(\lambda))}.
\end{aligned}
\tag{4.112}
$$

This implies

$$
\frac{\partial S(L(\overset{o}{p},p))}{\partial p_\mu} = \frac{\partial e^{A(p)}}{\partial p_\mu} = \int_0^1 dx\, e^{x\,A(p)}\,\frac{\partial A(p)}{\partial p_\mu}\, e^{-x\,A(p)}\, e^{A(p)}.
\tag{4.113}
$$

Following Ref. [200], the solution of this equation is

$$
\frac{\partial e^{A(p)}}{\partial p_\mu} = \sum_{n=0}^{\infty} \frac{\Big[A^n(p),\, \frac{\partial A(p)}{\partial p_\mu}\Big]}{(n+1)!}\, e^{A(p)},
\tag{4.114}
$$

where $[A^n, B]$ means

$$
[A^0, B] = B, \qquad [A^1, B] = A\,B - B\,A, \qquad [A^{n+1}, B] = [A, [A^n, B]\,].
\tag{4.115}
$$

From the following commutators of Dirac matrices,

$$[\gamma^\mu, \gamma^\nu] = -2\, i\, \sigma^{\mu\nu}, \quad [\sigma^{\mu\nu}, \gamma^\rho] = 2\, i\, (\gamma^\mu\, \eta^{\nu\rho} - \gamma^\nu\, \eta^{\mu\rho}), \quad [\sigma^{\mu\nu}, \sigma^{\alpha\beta}] = 2\, i\, C^{\mu\nu\alpha\beta}_{\gamma\delta}\, \sigma^{\gamma\delta}, \tag{4.116}$$

and using Eq. (A.7), we get ($e_p = \eta\, \sqrt{p^2}$ with $\eta = \pm$)

$$\begin{aligned}
\frac{\partial S(L(\overset{o}{p}, p))}{\partial p_\mu} &= \Big[\frac{i}{2}\, \frac{p_i\, \sigma^{0i}}{e_p^2\, (p_o + e_p)}\, \Big(p^\mu + 2\, e_p\, \eta^\mu_o\Big) - \frac{i}{2}\, \frac{\sigma^{0\mu}}{e_p} \\
&\quad + \frac{i}{2}\, \frac{p_i\, \sigma^{i\mu}}{e_p\, (p_o + e_p)}\Big]\, S(L(\overset{o}{p}, p)) \\
&= \Big[-\frac{i}{4}\, \eta_{\sigma B}\, \frac{\partial \epsilon^B_\rho(u(p))}{\partial p_\mu}\, L^\rho_{.\eta}(p, \overset{o}{p})\, \sigma^{\sigma\eta}\Big]\, S(L(\overset{o}{p}, p)).
\end{aligned} \tag{4.117}$$

In conclusion, the new variables have the following Dirac brackets:

$$\begin{aligned}
\{\tilde x^\mu, p^\nu\}^*_D &= -\eta^{\mu\nu}, \\
\{\tilde x^\mu, \overset{o}{\psi}(\tau, \vec\sigma)\}^*_D &= \{\tilde x^\mu, \overset{o}{\bar\psi}(\tau, \vec\sigma)\}^*_D = \{\tilde x^\mu, \tilde x^\nu\}^*_D = \{\tilde x^\mu, b^A_{\breve B}\}^*_D = 0, \\
\{\tilde S^{0i}, b^r_A\}^*_D &= \frac{\delta^{is}\, (p^r\, b^s_A - p^s\, b^r_A)}{(p^0 + \eta\, \sqrt{p^2})}, \\
\{\tilde S^{ij}, b^r_A\}^*_D &= (\delta^{ir}\, \delta^{js} - \delta^{is}\, \delta^{jr})\, b^s_A, \\
\{\tilde S^{\mu\nu}, \tilde S^{\alpha\beta}\}^*_D &= C^{\mu\nu\alpha\beta}_{\gamma\delta}\, \tilde S^{\gamma\delta}, \\
\{\tilde x^\mu, \tilde S^{0i}\}^*_D &= -\frac{1}{(p^0 + \eta\, \sqrt{p^2})}\, \Big[{}^4\eta^{\mu j}\, \tilde S^{ji} + \frac{(p^\mu}{} \\
&\quad + \eta^{\mu 0}\, \eta\, \sqrt{p^2})\, \tilde S^{ik}\, p^k\, \eta\, \sqrt{p^2}\, (p^0 + \eta\, \sqrt{p^2})\Big], \\
\{\tilde x^\mu, \tilde S^{ij}\}^*_D &= 0, \\
\{\overset{o}{\psi}_\alpha(\tau, \vec\sigma), \overset{o}{\bar\psi}_\beta(\tau, \vec\sigma')\}^*_D &= -i\, b^A_\tau(\tau)\, {}^4\eta_{A\sigma}\, (\gamma^\sigma)_{\alpha\beta}\, \delta^3(\vec\sigma - \vec\sigma'), \\
\{\tilde S^{\mu\nu}, \overset{o}{\psi}(\tau, \vec\sigma)\}^*_D &= \frac{i}{2}\, \Big[L^\mu_{.\alpha}(p, \overset{o}{p})\, L^\nu_{.\beta}(p, \overset{o}{p}) - \frac{1}{2}\, \epsilon^A_\rho(u(p))\, {}^4\eta_{AB}\, \Big(\frac{\partial \epsilon^B_\sigma(u(p))}{\partial p_\mu}\, p^\nu \\
&\quad - \frac{\partial\, \epsilon^B_\sigma(u(p))}{\partial p_\nu}\, p^\mu\Big)\, L^\rho_{.\alpha}(p, \overset{o}{p})\, L^\sigma_{.\beta}(p, \overset{o}{p})\Big]\, \sigma^{\alpha\beta}\, \overset{o}{\psi}(\tau, \vec\sigma), \\
\{\tilde S^{\mu\nu}, \overset{o}{\bar\psi}(\tau, \vec\sigma)\}^*_D &= -\frac{i}{2}\, \overset{o}{\bar\psi}(\tau, \vec\sigma)\, \Big[L^\mu_{.\alpha}(p, \overset{o}{p})\, L^\nu_{.\beta}(p, \overset{o}{p}) - \frac{1}{2}\, \epsilon^A_\rho(u(p))\, {}^4\eta_{AB} \\
&\quad \cdot \Big(\frac{\partial \epsilon^B_\sigma(u(p))}{\partial p_\mu}\, p^\nu - \frac{\partial \epsilon^B_\sigma(u(p))}{\partial p_\nu}\, p^\mu\Big)\, L^\rho_{.\alpha}(p, \overset{o}{p})\, L^\sigma_{.\beta}(p, \overset{o}{p})\Big]\, \sigma^{\alpha\beta}.
\end{aligned} \tag{4.118}$$

Now the Lorentz generators are $J^{\mu\nu} = \tilde L^{\mu\nu} + \tilde S^{\mu\nu} = \tilde x^\mu\, p^\nu - \tilde x^\nu\, p^\mu + \tilde S^{\mu\nu}$. The *rest-frame or Thomas spin tensor* is $\bar S^{AB} = \epsilon^A_\mu(u(p_s))\, \epsilon^A_\nu(u(p_s))\, S^{\mu\nu}$.

Moreover, we have

$$\{\bar{S}^{AB}, \bar{S}^{CD}\}^*_D = C^{ABCD}_{EF}\, \bar{S}^{EF},$$
$$\{\bar{S}^{AB}, \overset{\circ}{\psi}(\tau,\vec{\sigma})\}^*_D = \frac{i}{2}\,\delta^A_\alpha\,\delta^B_\beta\,\sigma^{\alpha\beta}\,\overset{\circ}{\psi}(\tau,\vec{\sigma}), \qquad \{\bar{S}^{AB}, \overset{\circ}{\bar{\psi}}(\tau,\vec{\sigma})\}^*_D$$
$$= -\frac{i}{2}\,\overset{\circ}{\bar{\psi}}(\tau,\vec{\sigma})\,\delta^A_\alpha\,\delta^B_\beta\,\sigma^{\alpha\beta}. \tag{4.119}$$

The new canonical origin $\tilde{x}^\mu(\tau)$ is not covariant, since under a Poincaré transformation (a,Λ) it transforms as

$$\tilde{x}^\mu \overset{(a,\Lambda)}{\longrightarrow} \tilde{x}'^\mu = \Lambda^\mu_{.\nu}\left[\tilde{x}^\nu + \frac{1}{2}\,\bar{S}_{rs}\,R^r_{.k}(\Lambda,p)\,\frac{\partial}{\partial p_\nu}\,R^s_{.k}(\Lambda,p)\right] + a^\mu. \tag{4.120}$$

After the restriction to Wigner hyper-planes, the remaining variables form a canonical basis, with Section A.5 of Appendix A implying the Dirac brackets

$$\{\tilde{x}^\mu(\tau), p^\nu\}^{**}_D = -{}^4\eta^{\mu\nu}, \qquad \{A_A(\tau,\vec{\sigma}), \pi^B(\tau,\vec{\sigma}^{'})\}^{**}_D = {}^4\eta^B_A\,\delta^3(\vec{\sigma}-\vec{\sigma}^{'}),$$
$$\{\overset{\circ}{\psi}_\alpha(\tau,\vec{\sigma}), \overset{\circ}{\bar{\psi}}_\beta(\tau,\vec{\sigma}')\}^{**}_D = -i\,(\gamma^{\bar{o}})_{\alpha\beta}\,\delta^3(\vec{\sigma}-\vec{\sigma}'). \tag{4.121}$$

Now $\gamma^{\bar{o}} = \gamma^\tau$ is a Lorentz-scalar matrix, while γ^r forms a Wigner spin-1 3-vector. Therefore, we have a Wigner-covariant realization of Dirac matrices:

$$\gamma^A = \Big(\gamma^{\bar{o}}; \{\gamma^r\}, r=1,2,3\Big), \qquad \left[\gamma^A, \gamma^B\right]_+ = 2\,\eta^{AB}, \tag{4.122}$$

like in the Chakrabarti representation [201]. Under a Lorentz transformation Λ, the bilinear in the Dirac field transforms with the associated Wigner rotation $R(\Lambda,p)$. For instance, we have

$$\overset{\circ}{\bar{\psi}}(\tau,\vec{\sigma})\,\gamma^A\,\overset{\circ}{\psi}(\tau,\vec{\sigma}) \overset{\Lambda}{\longrightarrow} R^A_{.B}(\Lambda,p)\,\overset{\circ}{\bar{\psi}}(\tau,\vec{\sigma})\,\gamma^B\,\overset{\circ}{\psi}(\tau,\vec{\sigma}),$$

$$\overset{\circ}{\bar{\psi}}(\tau,\vec{\sigma})\,\gamma^\tau\,\overset{\circ}{\psi}(\tau,\vec{\sigma}) \overset{\Lambda}{\longrightarrow} \overset{\circ}{\bar{\psi}}(\tau,\vec{\sigma})\,\gamma^\tau\,\overset{\circ}{\psi}(\tau,\vec{\sigma}) \qquad \text{(scalar)},$$
$$\overset{\circ}{\bar{\psi}}(\tau,\vec{\sigma})\,\gamma^r\,\overset{\circ}{\psi}(\tau,\vec{\sigma}) \overset{\Lambda}{\longrightarrow} R^r_{.s}(\Lambda,p_s)\,\overset{\circ}{\bar{\psi}}(\tau,\vec{\sigma})\,\gamma^s\,\overset{\circ}{\psi}(\tau,\vec{\sigma}) \quad \text{(Wigner 3-vector)}, \tag{4.123}$$

and the induced spinorial transformation on Dirac fields will be $(S(\Lambda) \mapsto \overset{\circ}{S}(R(\Lambda,p))$, which could be evaluated by using the last of the next formulas)

$$\overset{\circ}{\psi}(\tau,\vec{\sigma}) \overset{\Lambda}{\longrightarrow} \overset{\circ}{\psi}{}'(\tau,\vec{\sigma}) = \overset{\circ}{S}(R(\Lambda,p))\,\overset{\circ}{\psi}(\tau,\vec{\sigma}),$$
$$\overset{\circ}{\bar{\psi}}(\tau,\vec{\sigma}) \overset{\Lambda}{\longrightarrow} \overset{\circ}{\bar{\psi}}{}'(\tau,\vec{\sigma}) = \overset{\circ}{\bar{\psi}}(\tau,\vec{\sigma})\,\overset{\circ}{S}{}^{-1}(R(\Lambda,p)),$$

$$\overset{\circ}{S}{}^{-1}(R(\Lambda,p))\;\gamma^A\;\overset{\circ}{S}(R(\Lambda,p)) = R^A_{.B}(\Lambda,p)\,\gamma^B. \tag{4.124}$$

The original variables $z^\mu(\tau,\vec\sigma)$, $\rho_\mu(\tau,\vec\sigma)$ are reduced only to $\tilde x^\mu(\tau)$, p^μ on the Wigner hyper-plane Σ_τ. On it there remain only six first-class constraints:

$$\begin{aligned}
&\pi^\tau(\tau,\vec\sigma)\approx 0,\\
&\bar\Gamma(\tau,\vec\sigma)=\partial_r\,\pi^r(\tau,\vec\sigma)+e\;\bar{\overset{o}{\psi}}(\tau,\vec\sigma)\,\gamma^\tau\;\overset{o}{\psi}(\tau,\vec\sigma)\approx 0,\\
&\tilde{\bar{\mathcal{H}}}^\mu(\tau)=p^\mu-[u^\mu(p)\,\bar M(\tau)+\epsilon^\mu_r(u(p))\,\bar{\mathcal{H}}_{pr}(\tau)]\\
&\qquad\;\;=u^\mu(p)\,\bar{\mathcal{H}}(\tau)+\epsilon^\mu_r(u(p))\,\bar{\mathcal{H}}_{pr}(\tau)\approx 0,
\end{aligned}$$

or

$$\begin{aligned}
\bar{\mathcal{H}}(\tau)&=\eta\,\sqrt{p^2}-\bar M(\tau)c\\
&=\eta_s\,\sqrt{p^2}-\int d^3\sigma\,\Big[\frac{i}{2}\Big(\bar{\overset{o}{\psi}}(\tau,\vec\sigma)\,\gamma_r\,\partial_r\;\overset{o}{\psi}(\tau,\vec\sigma)\\
&\quad-\partial_r\;\bar{\overset{o}{\psi}}(\tau,\vec\sigma)\,\gamma_r\;\overset{o}{\psi}(\tau,\vec\sigma)\Big)+m\;\bar{\overset{o}{\psi}}(\tau,\vec\sigma)\;\overset{o}{\psi}(\tau,\vec\sigma)\\
&\quad+\frac{1}{2}\Big(\vec\pi^2+\vec B^2\Big)(\tau,\vec\sigma)+e\,A_r(\tau,\vec\sigma)\;\bar{\overset{o}{\psi}}(\tau,\vec\sigma)\,\gamma_r\;\overset{o}{\psi}(\tau,\vec\sigma)\Big]\approx 0,\\
\bar{\mathcal{H}}_{pr}(\tau)&=\int d^3\sigma\,\Big[\frac{i}{2}\Big(\bar{\overset{o}{\psi}}(\tau,\vec\sigma)\,\gamma^\tau\,\partial_r\;\overset{o}{\psi}(\tau,\vec\sigma)-\partial_r\;\bar{\overset{o}{\psi}}(\tau,\vec\sigma)\,\gamma^\tau\;\overset{o}{\psi}(\tau,\vec\sigma)\Big)\\
&\quad+\Big(\vec\pi\times\vec B\Big)_r(\tau,\vec\sigma)+e\,A_r(\tau,\vec\sigma)\;\bar{\overset{o}{\psi}}(\tau,\vec\sigma)\,\gamma^\tau\;\overset{o}{\psi}(\tau,\vec\sigma)\Big]\approx 0,
\end{aligned}$$

$$\begin{aligned}
\{\tilde{\bar{\mathcal{H}}}^\mu(\tau),\tilde{\bar{\mathcal{H}}}^\nu(\tau)\}^{**}_D&=\int d^3\sigma\,\Big[\Big(u^\mu(p)\,\epsilon^\nu_r(u(p))-u^\nu(p)\,\epsilon^\mu_r(u(p))\Big)\,\pi^r(\tau,\vec\sigma)\\
&\quad-\epsilon^\mu_r(u(p))\,F_{rs}(\tau,\vec\sigma)\,\epsilon^\nu_s(u(p))\Big]\,\bar\Gamma(\tau,\vec\sigma)\approx 0,\\
\{\bar\Gamma(\tau,\vec\sigma),\tilde{\bar{\mathcal{H}}}^\mu(\tau)\}^{**}_D&=\{\bar\Gamma(\tau,\vec\sigma),\bar\Gamma(\tau,\vec\sigma')\}^{**}_D=\{\bar\Gamma(\tau,\vec\sigma),\pi^\tau(\tau,\vec\sigma')\}^{**}_D\\
&=\{\pi^\tau(\tau,\vec\sigma),\tilde{\bar{\mathcal{H}}}^\mu(\tau)\}^{**}_D=\{\pi^\tau(\tau,\vec\sigma),\pi^\tau(\tau,\vec\sigma')\}=0.
\end{aligned}\tag{4.125}$$

On $\Sigma_{W\tau}$ the Dirac Hamiltonian becomes

$$\bar H_D=\lambda(\tau)\,\bar{\mathcal{H}}(\tau)-\vec\lambda(\tau)\,\vec{\bar{\mathcal{H}}}_p(\tau)+\int d^3\sigma\,\Big[\mu_\tau(\tau,\vec\sigma)\,\pi^\tau(\tau,\vec\sigma)-A_\tau(\tau,\vec\sigma)\,\bar\Gamma(\tau,\vec\sigma)\Big].\tag{4.126}$$

Since $\tilde{\bar{\mathcal{H}}}^{\mu\nu}(\tau)\equiv 0$ implies

$$\begin{aligned}
S^{\mu\nu}&=S^{\mu\nu}_\psi+\Big(\epsilon^\mu_r(u(p))\,u^\nu(p)\\
&\quad-\epsilon^\nu_r(u(p))\,u^\mu(p)\Big)\int d^3\sigma\,\sigma^r\,\Big[\frac{i}{2}\Big(\bar{\overset{o}{\psi}}(\tau,\vec\sigma)\,\gamma_s\,\partial_s\;\overset{o}{\psi}(\tau,\vec\sigma)\\
&\quad-\partial_s\;\bar{\overset{o}{\psi}}(\tau,\vec\sigma)\,\gamma_s\;\overset{o}{\psi}(\tau,\vec\sigma)\Big)+m\;\bar{\overset{o}{\psi}}(\tau,\vec\sigma)\;\overset{o}{\psi}(\tau,\vec\sigma)\\
&\quad+\frac{1}{2}\Big(\vec\pi^2+\vec B^2\Big)(\tau,\vec\sigma)+e\,A_s(\tau,\vec\sigma)\;\bar{\overset{o}{\psi}}(\tau,\vec\sigma)\,\gamma_s\;\overset{o}{\psi}(\tau,\vec\sigma)\Big]
\end{aligned}$$

$$-\Big(\epsilon^\mu_r(u(p))\,\epsilon^\nu_s(u(p)) - \epsilon^\nu_r(u(p))\,\epsilon^\mu_s(p)\Big) \int d^3\sigma\, \sigma^r \Big[\frac{i}{2}\Big(\overset{o}{\bar\psi}(\tau,\vec\sigma)\,\gamma^\tau\,\partial_s\,\overset{o}{\psi}(\tau,\vec\sigma)$$
$$-\partial_s\,\overset{o}{\bar\psi}(\tau,\vec\sigma)\,\gamma^\tau\,\overset{o}{\psi}(\tau,\vec\sigma)\Big) + \Big(\vec\pi\times\vec B\Big)_s(\tau,\vec\sigma)$$
$$+e\,A_s(\tau,\vec\sigma)\,\overset{o}{\bar\psi}(\tau,\vec\sigma)\,\gamma^\tau\,\overset{o}{\psi}(\tau,\vec\sigma)\Big], \tag{4.127}$$

with

$$S^{\mu\nu}_\psi = \frac{1}{4}\,L^\mu_{.\alpha}(p,\overset{o}{p})\,L^\nu_{.\beta}(p,\overset{o}{p}) \int d^3\sigma\,\overset{o}{\bar\psi}(\tau,\vec\sigma)\,\big[\gamma^\tau,\sigma^{\alpha\beta}\big]_+\,\overset{o}{\psi}(\tau,\vec\sigma), \tag{4.128}$$

we get the following expression for the rest-frame spin tensor:

$$\bar S^{AB} = \bar S^{AB}_\psi + \Big[\delta^A_r\,\delta^B_\tau - \delta^B_r\,\delta^A_\tau\Big] \int d^3\sigma\,\sigma^r \Big[\frac{i}{2}\Big(\overset{o}{\bar\psi}(\tau,\vec\sigma)\,\gamma_s\,\partial_s\,\overset{o}{\psi}(\tau,\vec\sigma)$$
$$-\partial_s\,\overset{o}{\bar\psi}(\tau,\vec\sigma)\,\gamma_s\,\overset{o}{\psi}(\tau,\vec\sigma)\Big) + m\,\overset{o}{\bar\psi}(\tau,\vec\sigma)\,\overset{o}{\psi}(\tau,\vec\sigma)$$
$$+\frac{1}{2}\Big(\vec\pi^2 + \vec B^2\Big)(\tau,\vec\sigma) + e\,A_s(\tau,\vec\sigma)\,\overset{o}{\bar\psi}(\tau,\vec\sigma)\,\gamma_s\,\overset{o}{\psi}(\tau,\vec\sigma)\Big]$$
$$-\Big[\delta^A_r\,\delta^B_s - \delta^B_r\,\delta^A_s\Big] \int d^3\sigma\,\sigma^r \Big[\frac{i}{2}\Big(\overset{o}{\bar\psi}(\tau,\vec\sigma)\,\gamma^\tau\,\partial_s\,\overset{o}{\psi}(\tau,\vec\sigma)$$
$$-\partial_s\,\overset{o}{\bar\psi}(\tau,\vec\sigma)\,\gamma^\tau\,\overset{o}{\psi}(\tau,\vec\sigma)\Big) + \Big(\vec\pi\times\vec B\Big)_s(\tau,\vec\sigma)$$
$$+e\,A_s(\tau,\vec\sigma)\,\overset{o}{\bar\psi}(\tau,\vec\sigma)\,\gamma^\tau\,\overset{o}{\psi}(\tau,\vec\sigma)\Big], \tag{4.129}$$

where

$$\bar S^{AB}_\psi = \frac{1}{4}\int d^3\sigma\,\overset{o}{\bar\psi}(\tau,\vec\sigma)\,\big[\gamma^\tau,\sigma^{AB}\big]_+\,\overset{o}{\psi}(\tau,\vec\sigma), \tag{4.130}$$

is the component connected with the Dirac field.

On $\Sigma_{W\tau}$ the external Poincaré generators are

$$p^\mu, \qquad J^{\mu\nu} = \hat x^\mu\,p^\nu - \hat x^\nu\,p^\mu + \tilde S^{\mu\nu},$$
$$\tilde S^{\mu\nu} \equiv \tilde S^{\mu\nu}_s + \tilde S^{\mu\nu}_\xi, \qquad \tilde S^{0i} = -\frac{\delta^{ir}\,\bar S^{rs}\,p^s}{p^0 + \eta\sqrt{p^2}}, \qquad \tilde S^{ij} = \delta^{ir}\,\delta^{js}\,\bar S^{rs}, \tag{4.131}$$

because one can express $\tilde S^{\mu\nu}$ in terms of $\bar S^{AB} = \epsilon^A_\mu(u(p))\,\epsilon(u(p))\,S^{\mu\nu}$. Only the Thomas spin $\bar S_r = \frac{1}{2}\,\epsilon_{rst}\,\bar S_{st}$ contributes to them.

The electric charge takes the form

$$Q = -e\int d^3\sigma\,\overset{o}{\bar\psi}(\tau,\vec\sigma)\,\gamma^\tau\,\overset{o}{\psi}(\tau,\vec\sigma), \tag{4.132}$$

and is the weak Noether charge of the conserved 4-current $j^A(\tau,\vec\sigma) = -e\,\overset{o}{\bar\psi}(\tau,\vec\sigma)\,\gamma^A\,\overset{o}{\psi}(\tau,\vec\sigma)$, $\partial_A\,j^A(\tau,\vec\sigma)\,\overset{\circ}{=}\,0$.

Therefore, the rest-frame Wigner-covariant instant form of the system Dirac plus electromagnetic fields (pseudo-classical electrodynamics) formally coincides with the standard non-covariant Hamiltonian formulation of the system on the hyper-planes $z^o(\tau,\vec{\sigma}) = \text{const.}$: Only the covariance properties of the objects are different and, moreover, there are the first-class constraints $\vec{\mathcal{H}}_p(\tau) \approx 0$ defining the rest-frame.

The DOs of the electromagnetic field are given in Eq. (4.55). Since we have

$$\{\overset{\circ}{\psi}(\tau,\vec{\sigma}), \bar{\Gamma}(\tau,\vec{\sigma}^{'})\}^{**}_D = -i\, e\, \gamma^\tau\, \psi(\tau,\vec{\sigma})\, \delta^3(\vec{\sigma}-\vec{\sigma}^{'}), \tag{4.133}$$

we see that the Dirac field is not gauge-invariant. It turns out that its DOs are

$$\begin{aligned}
\check{\overset{\circ}{\psi}}(\tau,\vec{\sigma}) = e^{-i\, e\, \eta_{em}(\tau,\vec{\sigma})}\, \overset{\circ}{\psi}(\tau,\vec{\sigma}), \qquad &\check{\overset{\circ}{\bar{\psi}}}(\tau,\vec{\sigma}) = \overset{\circ}{\bar{\psi}}(\tau,\vec{\sigma})\, e^{i\, e\, \eta_{em}(\tau,\vec{\sigma})},\\
\{\check{\overset{\circ}{\psi}}{}^a(\tau,\vec{\sigma}), \check{\overset{\circ}{\bar{\psi}}}_b(\tau,\vec{\sigma}')\}^{**}_D &= -i\, (\gamma^\tau)^a_{.b}\, \delta^3(\vec{\sigma}-\vec{\sigma}'),
\end{aligned} \tag{4.134}$$

representing Dirac fields dressed with a Coulomb cloud.

Since we get

$$\begin{aligned}
\int d^3\sigma\, \vec{\pi}^2(\tau,\vec{\sigma}) &= \int d^3\sigma\, \vec{\pi}^2_\perp(\tau,\vec{\sigma})\\
&\quad + e^2 \int\!\!\int d^3\sigma d^3\sigma'\, \frac{\Big[\overset{\circ}{\bar{\psi}}(\tau,\vec{\sigma})\, \gamma^\tau\, \overset{\circ}{\psi}(\tau,\vec{\sigma})\Big]\Big[\overset{\circ}{\bar{\psi}}(\tau,\vec{\sigma}')\, \gamma^\tau\, \overset{\circ}{\psi}(\tau,\vec{\sigma}')\Big]}{4\pi\, \mid \vec{\sigma}-\vec{\sigma}' \mid}\\
&= \int d^3\sigma\, \vec{\pi}^2_\perp(\tau,\vec{\sigma})\\
&\quad + e^2 \int\!\!\int d^3\sigma d^3\sigma'\, \frac{\Big[\check{\overset{\circ}{\bar{\psi}}}(\tau,\vec{\sigma})\, \gamma^\tau\, \check{\overset{\circ}{\psi}}(\tau,\vec{\sigma})\Big]\Big[\check{\overset{\circ}{\bar{\psi}}}(\tau,\vec{\sigma}')\, \gamma^\tau\, \check{\overset{\circ}{\psi}}(\tau,\vec{\sigma}')\Big]}{4\pi\, \mid \vec{\sigma}-\vec{\sigma}' \mid}
\end{aligned} \tag{4.135}$$

in the radiation gauge, where $A_\tau(\tau,\vec{\sigma}) = \pi^\tau(\tau,\vec{\sigma}) = \eta_{em}(\tau,\vec{\sigma}) = \bar{\Gamma}(\tau,\vec{\sigma}) = 0$, we have

$$\begin{aligned}
\bar{\mathcal{H}}(\tau) &= e_p - \bar{M}c\\
&= \eta\, \sqrt{p^2} - \int d^3\sigma\, \Big[\frac{i}{2}\Big(\check{\overset{\circ}{\bar{\psi}}}(\tau,\vec{\sigma})\, \gamma_r\, \partial_r\, \check{\overset{\circ}{\psi}}(\tau,\vec{\sigma}) -\\
&\quad - \partial_r\, \check{\overset{\circ}{\bar{\psi}}}(\tau,\vec{\sigma})\, \gamma_r\, \check{\overset{\circ}{\psi}}(\tau,\vec{\sigma})\Big) + m\, \check{\overset{\circ}{\bar{\psi}}}(\tau,\vec{\sigma})\, \check{\overset{\circ}{\psi}}(\tau,\vec{\sigma})\\
&\quad + \frac{1}{2}\Big(\vec{\pi}^2_\perp + \vec{B}^2\Big)(\tau,\vec{\sigma}) + e\, A_{\perp r}(\tau,\vec{\sigma})\, \check{\overset{\circ}{\bar{\psi}}}(\tau,\vec{\sigma})\, \gamma_r\, \check{\overset{\circ}{\psi}}(\tau,\vec{\sigma})\\
&\quad + \frac{e^2}{2}\, \check{\overset{\circ}{\bar{\psi}}}(\tau,\vec{\sigma})\, \gamma^\tau\, \check{\overset{\circ}{\psi}}(\tau,\vec{\sigma}) \int d^3\sigma'\, \frac{\check{\overset{\circ}{\bar{\psi}}}(\tau,\vec{\sigma}')\, \gamma^\tau\, \check{\overset{\circ}{\psi}}(\tau,\vec{\sigma}')}{4\pi\, \mid \vec{\sigma}-\vec{\sigma}' \mid}\Big] \approx 0.
\end{aligned} \tag{4.136}$$

The last term is the non-renormalizable Coulomb self-interaction of the Dirac field in the rest-frame. The constraints identifying the rest-frame take the interaction-independent form expected in an instant form of dynamics:

$$\bar{\mathcal{H}}_{pr}(\tau) = \int d^3\sigma \left[\frac{1}{2}\left(\bar{\check{\overset{\circ}{\psi}}}(\tau,\vec{\sigma})\,\gamma^\tau\,\partial_r\check{\overset{\circ}{\psi}}(\tau,\vec{\sigma}) - \partial_r\,\bar{\check{\overset{\circ}{\psi}}}(\tau,\vec{\sigma})\,\gamma^\tau\,\check{\overset{\circ}{\psi}}(\tau,\vec{\sigma})\right)\right.$$
$$\left. + \left(\vec{\pi}_\perp \times \vec{B}\right)_r(\tau,\vec{\sigma}) \approx 0. \qquad (4.137)$$

By adding the gauge fixing $T_s - \tau \approx 0$, whose time constancy implies $\lambda(\tau) = -1$, we get the Dirac Hamiltonian:

$$\hat{\bar{H}}_D = \bar{M}c - \vec{\lambda}(\tau)\cdot\vec{\bar{\mathcal{H}}}_p(\tau),$$
$$\bar{M}c = \int d^3\sigma \left[\frac{i}{2}\left(\bar{\check{\overset{\circ}{\psi}}}(\tau,\vec{\sigma})\,\gamma_r\,\partial_r\,\check{\overset{\circ}{\psi}}(\tau,\vec{\sigma}) - \partial_r\,\bar{\check{\overset{\circ}{\psi}}}(\tau,\vec{\sigma})\,\gamma_r\,\check{\overset{\circ}{\psi}}(\tau,\vec{\sigma})\right)\right.$$
$$\left. + m\,\bar{\check{\overset{\circ}{\psi}}}(\tau,\vec{\sigma})\,\check{\overset{\circ}{\psi}}(\tau,\vec{\sigma}) + \frac{1}{2}\,(\vec{\pi}_\perp^2 + \vec{B}^2)(\tau,\vec{\sigma})\right]$$
$$+ \frac{e^2}{2}\int\int d^3\sigma d^3\sigma'\,\frac{\left[\bar{\check{\overset{\circ}{\psi}}}(\tau,\vec{\sigma})\,\gamma^\tau\,\check{\overset{\circ}{\psi}}(\tau,\vec{\sigma})\right]\left[\bar{\check{\overset{\circ}{\psi}}}(\tau,\vec{\sigma}')\,\gamma^\tau\,\check{\overset{\circ}{\psi}}(\tau,\vec{\sigma}')\right]}{4\pi\,\mid\vec{\sigma}-\vec{\sigma}'\mid}$$
$$+ e\int d^3\sigma\,A_{r\perp}(\tau,\vec{\sigma})\,\bar{\check{\overset{\circ}{\psi}}}(\tau,\vec{\sigma})\,\gamma_r\,\check{\overset{\circ}{\psi}}(\tau,\vec{\sigma}). \qquad (4.138)$$

In the gauge $\vec{\lambda}(\tau) = 0$, the Dirac field has the following Hamilton equation:

$$\partial_\tau\,\check{\overset{\circ}{\psi}}(\tau,\vec{\sigma}) \overset{\circ}{=} \{\check{\overset{\circ}{\psi}}(\tau,\vec{\sigma}), \hat{\bar{H}}_D\}^{**}_D$$
$$= \gamma^\tau\,\gamma_r\left[\partial_r - i\,e\,A_{\perp r}(\tau,\vec{\sigma})\right]\check{\overset{\circ}{\psi}}(\tau,\vec{\sigma}) - i\,m\,\gamma^\tau\,\check{\overset{\circ}{\psi}}(\tau,\vec{\sigma})$$
$$+ i\,e^2\int d^3\sigma'\,\frac{\bar{\check{\overset{\circ}{\psi}}}(\tau,\vec{\sigma}')\,\gamma^\tau\,\check{\overset{\circ}{\psi}}(\tau,\vec{\sigma}')}{4\pi\,\mid\vec{\sigma}-\vec{\sigma}'\mid}\,\check{\overset{\circ}{\psi}}(\tau,\vec{\sigma}), \qquad (4.139)$$

which can be rewritten in the standard form:

$$\{i\,\gamma^A\,[\partial_A - i\,e\,\tilde{A}_A(\tau,\vec{\sigma})] - m\}\,\check{\overset{\circ}{\psi}}(\tau,\vec{\sigma}) \overset{\circ}{=} 0, \qquad (4.140)$$

with

$$\tilde{A}_r(\tau,\vec{\sigma}) \equiv A_{\perp r}(\tau,\vec{\sigma}), \qquad \tilde{A}_\tau(\tau,\vec{\sigma}) \equiv -e\int d^3\sigma'\,\frac{\bar{\check{\overset{\circ}{\psi}}}(\tau,\vec{\sigma}')\,\gamma^\tau\,\check{\overset{\circ}{\psi}}(\tau,\vec{\sigma}')}{4\pi\,\mid\vec{\sigma}-\vec{\sigma}'\mid} = \tilde{A}^\dagger_\tau(\tau,\vec{\sigma}). \qquad (4.141)$$

Analogously, we get

$$\bar{\check{\overset{\circ}{\psi}}}(\tau,\vec{\sigma})\,\{-i\,[\overleftarrow{\partial}_A + i\,e\,\tilde{A}_A(\tau,\vec{\sigma})]\,\gamma^A - m\} \overset{\circ}{=} 0. \qquad (4.142)$$

Eqs. (4.140) and (4.142) are non-local and non-linear due to the reduction to the rest-frame radiation gauge.

For the transverse electromagnetic fields $\vec{A}_\perp(\tau,\vec{\sigma})$, $\vec{\pi}_\perp(\tau,\vec{\sigma})$, we get the Hamilton equations:

$$\partial_\tau\, A^r_\perp(\tau,\vec{\sigma}) \circeq \{A^r_\perp(\tau,\vec{\sigma}), \hat{\hat{H}}_D\}^{**}_D = -\pi^r_\perp(\tau,\vec{\sigma}),$$
$$\partial_\tau\, \pi^r_\perp(\tau,\vec{\sigma}) \circeq \{\pi^r_\perp(\tau,\vec{\sigma}), \hat{\hat{H}}_D\}^{**}_D = \triangle_\sigma\, A^r_\perp(\tau,\vec{\sigma}) + e\, P^{rs}(\vec{\sigma})\, \Big[\check{\overset{\circ}{\bar{\psi}}}(\tau,\vec{\sigma})\, \gamma^s\, \check{\overset{\circ}{\psi}}(\tau,\vec{\sigma})\Big], \tag{4.143}$$

implying the equation

$$\Box_\sigma\, A^r_\perp(\tau,\vec{\sigma}) \circeq P^{rs}(\vec{\sigma})\, j^s(\tau,\vec{\sigma}) \equiv j^r_\perp(\tau,\vec{\sigma}), \tag{4.144}$$

with

$$\Box_\sigma \equiv \partial_A\, \partial^A, \qquad P^{rs}(\vec{\sigma}) \equiv \delta^{rs} + \frac{\partial^r\, \partial^s}{\triangle_\sigma}, \qquad j^s(\tau,\vec{\sigma}) = -e\, \check{\overset{\circ}{\bar{\psi}}}(\tau,\vec{\sigma})\, \gamma^s\, \check{\overset{\circ}{\psi}}(\tau,\vec{\sigma}). \tag{4.145}$$

The wave equation is actually an integro-differential equation due to the projector appearing in the transverse fermionic Wigner spin-1 3-current.

These results are valid for massive Dirac fields and for all the massive ($P^2 > 0$) configurations of massless Dirac fields. They also hold for the massive configurations of chiral fields simply by replacing everywhere ψ with $\psi_\pm = \frac{1}{2}\,(1 \pm \gamma_5)\,\psi$. Instead, the massless ($P^2 = 0$) and infrared ($P^\mu = 0$) configurations of either massless or chiral fermion fields have to be treated separately, because they require the reformulation of the front form of dynamics in the instant form, like for massless spinning particles.

As a final remark, let us note that the search for Wigner-covariant DOs and the definition of the rest-frame instant form of dynamics requires the Wigner 3-spaces orthogonal to the total time-like 4-momentum P^μ of the isolated system. Therefore, the pseudo-classical Dirac field must always be accompanied by fields like the Klein–Gordon or electromagnetic ones to avoid having P^μ bilinear in the Grassmann variables so that there could be differential geometrical and algebraic problems in the definition of the 3-spaces.

4.5 Yang–Mills Fields

The rest-frame formulation of the Yang–Mills field was done in Ref. [169] in the framework of the quark model (with scalar quarks). This paper puts in Wigner-covariant form the results of previous papers on the use of constraint theory for the description of Yang–Mills theory with fermions [93, 94], of Higgs models [183, 184], and of the $SU(3) \times SU(2) \times U(1)$ model [185, 186].

In Yang–Mills theory in the case of the trivial principal bundle $M^4 \times G$ over Minkowski space-time (see Refs. [198, 199] for a review of the theory and

Ref. [200] for the needed background in differential geometry), the gauge potentials are $A_\mu(x) = A^a_\mu(x) T_a$, where T_a are the generators of the Lie algebra g of the Lie group G with $T^\dagger_a = -T_a$, $\mathrm{Tr}(T_a T_b) = -\delta_{ab}$, $[T_a, T_b] = c_{abc} T_c$ and with the center of G denoted Z_G. G is a compact, simple (or semi-simple), connected, simply connected Lie group with compact real simple Lie algebra. The field strengths are $F_{\mu\nu}(x) = F^a_{\mu\nu}(x)\, T_a = \partial_\mu\, A_\nu(x) - \partial_\nu A_\mu(x) + [A_\mu(x), A_\nu(x)]$ and the Lagrangian is $\mathcal{L}(x^o, \vec{x}) = \frac{1}{2}\, Tr(F^{\mu\nu}(x^o, \vec{x})\, F_{\mu\nu}(x^o, \vec{x}))$.

The canonical momenta are $\pi^{\mu a}(x^o, \vec{x}) = F^{\mu o a}(x^o, \vec{x})$. The Euler–Lagrange equations are $L^\mu(x^o, \vec{x}) = L^{\mu\, a}(x^o, \vec{x})\, T_a = D_\nu\, F^{\nu\mu}(x^o, \vec{x}) = \partial_\nu\, F^{\nu\mu}(x^o, \vec{x}) + [A_\nu(x^o, \vec{x}), F^{\nu\mu}(x^o, \vec{x})] \stackrel{\circ}{=} 0$ and the Bianchi identities are $(D_\alpha\, F_{\beta\gamma} + D_\beta\, F_{\gamma\alpha} + D_\gamma\, F_{\alpha\beta})(x^o, \vec{x}) \equiv 0$. The non-Abelian electric and magnetic fields are $E^k_a(x^o, \vec{x}) = F^{ko}_a(x^o, \vec{x})$ and $B^k_a(x^o, \vec{x}) = -\frac{1}{2}\, \epsilon^{kij}\, F^{ij}_k(x^o, \vec{x})$.

The primary first-class constraints are $\pi^o_a(x^o, \vec{x}) \approx 0$, while the secondary ones are Gauss's laws $\Gamma_a(x^o, \vec{x}) = \partial_r \pi^r_a(x^o, \vec{x}) + c_{abc}\, A_{br}(x^o, \vec{x})\, \pi^r_c(x^o, \vec{x}) \approx 0$ $(\{\Gamma_a(x^o, \vec{x}), \Gamma_b(x^o, \vec{y})\} = c_{abc}\, \Gamma_c(x^o, \vec{x})\, \delta^3(\vec{x} - \vec{y}))$.

The infinitesimal gauge transformations under which $\delta\mathcal{L}(x^o, \vec{x}) \equiv 0$ are $\delta A_{\mu a}(x^o, \vec{x}) = \partial_\mu\, \epsilon_a(x^o, \vec{x}) + c_{abc}\, A_{\mu b}(x^o, \vec{x})\, \epsilon_c(x^o, \vec{x})$, $\delta\, F_{a\mu\nu}(x^o, \vec{x}) = c_{abc}\, F_{b\mu\nu}(x^o, \vec{x})\, \epsilon_c(x^o, \vec{x})$, while the finite ones are $A_\mu(x) \rightarrow A^U_\mu(x) = U^{-1}(x)\, A_\mu(x)\, U(x) + U^{-1}(x)\, \partial_\mu\, U(x)$, $F_{\mu\nu}(x) \rightarrow U^{-1}(x)\, F_{\mu\nu}(x)\, U(x)$ with $U(x) = e^{\epsilon_a(x) T^a}$ an element of the group G.

While in the electromagnetic case we can find the canonical basis of Eq. (4.55) containing both the DOs and a set of canonical gauge variables, in general gauge field theories the situation is more complicated, because some of the constraints are non-linear partial differential equations (PDEs) and the theorems ensuring the existence of the Shanmugadhasan canonical transformation have not been extended to the infinite-dimensional case so one must use heuristic extrapolations of them. In Yang–Mills theory some of the constraints are elliptic PDEs and they can have zero modes. Let us consider the stratum $\epsilon\, P^2 > 0$ of free Yang–Mills theory as a prototype and its first-class constraints. The problem of the zero modes will appear as a singularity structure of the gauge foliation of the allowed strata, in particular of the stratum $\epsilon\, P^2 > 0$. This phenomenon was discovered in Refs. [202, 203] by studying the space of solutions of Yang–Mills and Einstein equations, which can be mapped onto the constraint manifold of these theories in their Hamiltonian description. It turns out that the space of solutions has a cone-over-cone structure of singularities: If we have a line of solutions with a certain number of symmetries, in each point of this line there is a cone of solutions with one less symmetry.

Other possible sources of singularities of the gauge foliation of Yang–Mills theory in the stratum $\epsilon\, P^2 > 0$ may be: (1) different classes of gauge potentials identified by different values of the field invariants; and (2) the orbit structure of the rest-frame (or Thomas) spin $\vec{S}$, identified by the Pauli–Lubanski Casimir $W^2 = -\epsilon\, P^2\, \vec{S}^2$ of the Poincaré group. The final outcome of this structure of singularities is that the reduced phase-space, i.e., the space of the gauge orbits, is in general a stratified manifold with singularities [202, 203]. In the stratum $\epsilon P^2 > 0$ of the Yang–Mills theory these singularities survive the Wick rotation to the Euclidean formulation, and it is not clear how the ordinary path integral approach and the associated BRS method can take them into account (they are zero measure effects).

In the Yang–Mills case there is also the problem of the gauge symmetries of a gauge potential $A_\mu(x) = A^a_\mu(x) T_a$, which are connected with the generators of its stability group, i.e., with the subgroup of those special gauge transformations that leave invariant the gauge potential. This is the "Gribov ambiguity" for gauge potentials [204–206]. There is also a more general Gribov ambiguity for field strengths, the "gauge copies" problem due to those gauge transformations leaving invariant the field strengths. For all these problems see Refs. [93, 94] and their bibliograpies.

The function $G_{gt}[\lambda_{ao}, A_{ao}] = \int d^3x\, [\lambda_{ao}\, \pi^o_a - A_{ao}\, \Gamma_a](x^o, \vec{x})$ is the generator of the gauge transformations ($\lambda_{ao}(x^o, \vec{x})$ are Dirac multipliers). However, if $\lambda_{ao}(x^o, \vec{x})$ and $A_{ao}(x^o, \vec{x})$ do not have suitable boundary conditions at spatial infinity, it becomes the generator of "improper" gauge transformations, which have to be eliminated.

Moreover, for a gauge potential with a non-trivial stability group those combinations of Gauss's laws corresponding to the generators of the stability group cannot be first-class constraints, since they do not generate effective gauge transformations but special symmetry transformations. This problematic remains to be clarified, but it seems that in this case these components of Gauss's laws become third-class constraints, which are not generators of true gauge transformations. This new kind of constraint was introduced in the finite-dimensional case (see Section 9.2) as a result of the study of some examples, in which the Jacobi equations (the linearization of the Euler–Lagrange equations) are singular, i.e., some of their solutions are not infinitesimal deviations between two neighboring extremals of the Euler–Lagrange equations. This interpretation seems to be confirmed by the fact that the singularity structure discovered in Refs. [202, 203] follows from the existence of singularities of the linearized Yang–Mills and Einstein equations. Due to the Gribov ambiguity, to fix univocally a gauge one has to give topological numbers identifying a stratum besides the ordinary gauge fixing constraints.

If it is not possible to eliminate the Gribov ambiguity by using suitable weighted Sobolev spaces (assuming that it is only a mathematical obstruction without any hidden physics like in the approach to QCD confinement reviewed in Ref. [207]), the existence of global DOs for Yang–Mills theory is very problematic.

The non-Abelian charges are $Q_a = -g^{-2}\, c_{abc} \int d^3x\, E^k_b(x^o, \vec{x})\, A_{ck}(x^o, \vec{x})$, $\frac{d\,Q_a}{dx^o} = 0$ (see Chapter 9 for their equality with the strong and weak Noether charges) and their covariance requires that the gauge parameters vanish at spatial infinity in an angle-independent way [198, 199].

As shown in Ref. [93, 94], all these problems can be avoided at the Hamiltonian level by assuming the following boundary conditions ($r = |\vec{x}|$):

$$
\begin{aligned}
&U(x^o, \vec{x}) \quad r \overrightarrow{\rightarrow \infty} \quad U_\infty + O(r^{-1}), \qquad U_\infty = \text{const.}\\
&\vec{\partial} U(x^o, \vec{x}) \quad r \overrightarrow{\rightarrow \infty} \quad (\vec{\partial} U)_\infty = \frac{\vec{u}}{r^{2+\epsilon}} + O(r^{-3}),\\
&\partial_o U(x^o, \vec{x}) \quad r \overrightarrow{\rightarrow \infty} \quad (\partial_o U)_\infty = \frac{u_o}{r^{1+\epsilon}} + O(r^{-2}),\\
&A_{ao}(x^o, \vec{x}) \quad r \overrightarrow{\rightarrow \infty} \quad \frac{a_{ao}}{r^{1+\epsilon}} + O(r^{-2}),\\
&\lambda_{ao}(x^o, \vec{x}) \quad r \overrightarrow{\rightarrow \infty} \quad \frac{\lambda^\infty_{ao}}{r^{1+\epsilon}} + O(r^{-2}),\\
&\pi^o_a(x^o, \vec{x}) \quad r \overrightarrow{\rightarrow \infty} \quad \frac{p^o_a}{r^{1+\epsilon}},\\
&A^i_a(x^o, \vec{x}) \quad r \overrightarrow{\rightarrow \infty} \quad \frac{a^i_a}{r^{2+\epsilon}} + O(r^{-3}),\\
&A^{Ui}_a(x^o, \vec{x}) \quad r \overrightarrow{\rightarrow \infty} \quad U^{-1}_\infty (\partial^i U)_\infty + U^{-1}_\infty A^{(\infty)i}_a U_\infty = \frac{a^{Ui}}{r^{2+\epsilon}} + O(r^{-3}),\\
&B^i_a(x^o, \vec{x}) \quad r \overrightarrow{\rightarrow \infty} \quad \frac{b^i_a}{r^{3+\epsilon}} + O(r^{-4}),\\
&\pi^i_a(x^o, \vec{x}) = E^i_a(x^o, \vec{x})/g^2 \quad r \overrightarrow{\rightarrow \infty} \quad \frac{e^i_a}{r^{2+\epsilon}} + O(r^{-3}),\\
&\alpha_a(x^o, \vec{x}) \quad r \overrightarrow{\rightarrow \infty} \quad \frac{\alpha^\infty_a}{r^{3+\epsilon}} + O(r^{-4}),\\
&\bar{\Gamma}_a(x^o, \vec{x}) \quad r \overrightarrow{\rightarrow \infty} \quad \frac{\bar{\Gamma}^\infty_a}{r^{3+\epsilon}} + O(r^{-4}).
\end{aligned}
\tag{4.146}
$$

In the 3+1 approach on the hyper-surface Σ_τ, we describe the non-Abelian potential and field strength with Lorentz-scalar variables $A_{a\check{A}}(\tau, \sigma^u)$ and $F_{a\check{A}\check{B}}(\tau, \sigma^u)$ respectively: They contain the embedding $\Sigma(\tau) \to M^4$ and are defined by

$$
\begin{aligned}
A_{aA}(\tau, \sigma^u) &= z^\mu_A(\tau, \sigma^u)\, A_{a\mu}(x = z(\tau, \sigma^u)),\\
F_{aAB}(\tau, \sigma^u) &= \partial_A\, A_{aB}(\tau, \sigma^u) - \partial_B\, A_{aA}(\tau, \sigma^u) + c_{abc}\, A_{bA}(\tau, \sigma^u)\, A_{bB}(\tau, \sigma^u)\\
&= z^\mu_A(\tau, \sigma^u)\, z^\nu_B(\tau, \sigma^u)\, F_{a\mu\nu}(z(\tau, \sigma^u))\\
&= z^\mu_A(\tau, \sigma^u)\, z^\nu_B(\tau, \sigma^u) \Big[\partial_\mu\, A_{a\nu}(z(\tau, \sigma^u))\\
&\quad - \partial_\nu\, A_{a\mu}(z(\tau, \sigma^u)) + c_{abc}\, A_{b\mu}(z(\tau, \sigma^u))\, A_{c\nu}(z(\tau, \sigma^u))\Big].
\end{aligned}
\tag{4.147}
$$

The action of the parametrized Minkowski theory is

$$
\begin{aligned}
S &= \int d\tau d^3\sigma\, \mathcal{L}(\tau, \sigma^u) = \int d\tau\, L(\tau), \qquad L(\tau) = \int d^3\sigma\, \mathcal{L}(\tau, \sigma^u),\\
\mathcal{L}(\tau, \sigma^u) &= -\frac{1}{4g_s^2} \sqrt{g(\tau, \sigma^u)}\, {}^4g^{AC}(\tau, \sigma^u)\, {}^4g^{BD}(\tau, \sigma^u) \sum_a F_{aAB}(\tau, \sigma^u)\, F_{aCD}(\tau, \sigma^u),
\end{aligned}
\tag{4.148}
$$

where the configuration variables are $z^\mu(\tau,\sigma^u)$ $A_{aA}(\tau,\sigma^u)$. The explicit form of $\mathcal{L}(\tau,\sigma^u)$ is like the electromagnetic one of Eq. (4.38). The action is invariant under separate τ- and σ-reparametrizations, since $A_{a\tau}(\tau,\sigma^u)$ transforms as a τ-derivative; moreover, it is invariant under the odd phase transformations $\delta\,\theta_{i\alpha}\mapsto \alpha_a\,(T^a)_{\alpha\beta}\,\theta_{i\beta}$.

The canonical momenta are

$$
\begin{aligned}
\rho_\mu(\tau,\sigma^u) &= -\frac{\partial\mathcal{L}(\tau,\sigma^u)}{\partial z^\mu_\tau(\tau,\sigma^u)} = \frac{\sqrt{g(\tau,\sigma^u)}}{4}\Big[({}^4g^{\tau\tau}\,z_{\tau\mu}\\
&\quad +{}^4g^{\tau r}\,z_{r\mu})(\tau,\sigma^u)\,{}^4g^{AC}(\tau,\sigma^u)\,{}^4g^{BD}(\tau,\sigma^u)\sum_a F_{aAB}(\tau,\sigma^u)\,F_{aCD}(\tau,\sigma^u)\\
&\quad -2\,[z_{\tau\mu}(\tau,\sigma^u)\,({}^4g^{A\tau}\,{}^4g^{\tau C}\,{}^4g^{BD}+{}^4g^{AC}\,{}^4g^{B\tau}\,{}^4g^{\tau D})(\tau,\sigma^u)\\
&\quad +z_{r\mu}(\tau,\sigma^u)\,({}^4g^{Ar}\,{}^4g^{\tau C}+{}^4g^{A\tau}\,{}^4g^{rC})(\tau,\sigma^u)\,{}^4g^{BD}(\tau,\sigma^u)]\\
&\quad \sum_a F_{aAB}(\tau,\sigma^u)\,F_{aCD}(\tau,\sigma^u)\Big]\\
&= [(\rho_\nu l^\nu)\,l_\mu + (\rho_\nu\,z^\nu_r)\,{}^3g^{rs}\,z_{s\mu}](\tau,\sigma^u),\\
\pi^\tau_a(\tau,\sigma^u) &= \frac{\partial L}{\partial\partial_\tau\,A_{a\tau}(\tau,\sigma^u)} = 0,\\
\pi^r_a(\tau,\sigma^u) &= \frac{\partial L}{\partial\partial_\tau\,A_{ar}(\tau,\sigma^u)}\\
&= -g_s^{-2}\,\frac{\gamma(\tau,\sigma^u)}{\sqrt{g(\tau,\sigma^u)}}\,{}^3g^{rs}(\tau,\sigma^u)\,(F_{a\tau s}+{}^4g_{\tau v}\,{}^3g^{vu}\,F_{aus})(\tau,\sigma^u)\\
&= g_s^{-2}\,\frac{\gamma(\tau,\sigma^u)}{\sqrt{g(\tau,\sigma^u)}}\,{}^3g^{rs}(\tau,\sigma^u)\,(E_{as}(\tau,\sigma^u)\\
&\quad +{}^4g_{\tau v}(\tau,\sigma^u)\,{}^3g^{vu}(\tau,\sigma^u)\,\epsilon_{ust}\,B_{at}(\tau,\sigma^u)),
\end{aligned}
\tag{4.149}
$$

and the following Poisson brackets are assumed:

$$
\begin{aligned}
\{z^\mu(\tau,\sigma^u),\rho_\nu(\tau,\sigma'^u\} &= -\eta^\mu_\nu\,\delta^3(\sigma^u-\sigma'^u), \qquad \{A_{aA}(\tau,\sigma^u),\pi^B_b(\tau,\sigma'^u)\}\\
&= {}^4\eta^B_A\,\delta_{ab}\,\delta^3(\sigma^u-\sigma'^u).
\end{aligned}
\tag{4.150}
$$

The four primary constraints are (see Eq. (2.14) for the energy–momentum tensor $T_{AB}(\tau,\sigma^u)$ and for the external Poincaré generators P^μ, $J^{\mu\nu}$)

$$
\begin{aligned}
\bar{\mathcal{H}}_\mu(\tau,\sigma^u) &= \rho_\mu(\tau,\sigma^u) - l_\mu(\tau,\sigma^u)\,T_{\tau\tau}(\tau,\sigma^u)\\
&\quad -z_{r\mu}(\tau,\sigma^u)\,{}^3g^{rs}(\tau,\sigma^u)\Big(-T_{\tau s}(\tau,\sigma^u)\Big)\approx 0,\\
T_{\tau\tau}(\tau,\sigma^u) &= -\frac{1}{2}\sum_a\Big(\frac{g_s^2}{\sqrt{\gamma}}\,\pi^r_a\,{}^3g_{rs}\,\pi^s_a - \frac{\sqrt{\gamma}}{2\,g_s^2}\,{}^3g^{rs}\,{}^3g^{uv}\,F_{aru}\,F_{asv}\Big)(\tau,\sigma^u),\\
T_{\tau s}(\tau,\sigma^u) &= -\sum_a F_{ast}(\tau,\sigma^u)\,\pi^t_a(\tau,\sigma^u) = -\epsilon_{stu}\sum_a \pi^t_a(\tau,\sigma^u)\,B_{au}(\tau,\sigma^u)\\
&= \sum_a\,[\vec{\pi}_a(\tau,\sigma^u)\times\vec{B}_a(\tau,\sigma^u)]_s.
\end{aligned}
\tag{4.151}
$$

Since the canonical Hamiltonian is $\bar{H}_c = -\int d^3\sigma \sum_a A_{a\tau}(\tau,\sigma^u)\,\bar{\Gamma}_a(\tau,\sigma^u)$ with $\bar{\Gamma}_a(\tau,\sigma^u) = \partial_r\,\pi^r_a(\tau,\sigma^u) + c_{abc}\,A_{br}(\tau,\sigma^u)\,\pi^r_c(\tau,\sigma^u) = -\hat{\vec{D}}^{(A)}_{ab}(\tau,\sigma^a)\cdot\vec{\pi}_b(\tau,\sigma^u)$, we have the Dirac Hamiltonian ($\lambda^\mu(\tau,\sigma^u)$ and $\lambda_\tau(\tau,\sigma^u)$ are Dirac's multipliers):

$$\bar{H}_D = \int d^3\sigma \Big[\lambda^\mu(\tau,\sigma^u)\,\bar{\mathcal{H}}_\mu(\tau,\sigma^u) + \sum_a \lambda_{a\tau}(\tau,\sigma^u)\,\pi^\tau_a(\tau,\sigma^u)$$
$$-\sum_a A_{a\tau}(\tau,\sigma^u)\,\bar{\Gamma}_a(\tau,\sigma^u)\Big]. \tag{4.152}$$

The time constancy of the constraints $\pi^\tau_a(\tau,\sigma^u) \approx 0$ will produce the only secondary constraints (Gauss's laws):

$$\bar{\Gamma}_a(\tau,\sigma^u) \approx 0. \tag{4.153}$$

The six constraints $\bar{\mathcal{H}}_\mu(\tau,\sigma^u) \approx 0$, $\pi^\tau_a(\tau,\sigma^u) \approx 0$, $\bar{\Gamma}_a(\tau,\sigma) \approx 0$ are first class with the only non-vanishing Poisson brackets,

$$\{\bar{\mathcal{H}}_\mu(\tau,\sigma),\quad \bar{\mathcal{H}}_\nu(\tau,\sigma^{'u})\}$$
$$= \Big([l_\mu(\tau,\sigma^u)\,z_{r\nu}(\tau,\sigma^u) - l_\nu(\tau,\sigma^u)\,z_{r\mu}(\tau,\sigma^u)]\,\frac{\pi^r(\tau,\sigma^u)}{\sqrt{\gamma(\tau,\sigma^u)}}$$
$$-z_{u\mu}(\tau,\sigma^u)\,{}^3g^{ur}(\tau,\sigma^u)\sum_a F_{ars}(\tau,\sigma^u)\,{}^3g^{sv}(\tau,\sigma^u)\,z_{v\nu}(\tau,\sigma^u)\Big)$$
$$\bar{\Gamma}_a(\tau,\sigma)\,\delta^3(\sigma-\sigma^{'u}) \approx 0. \tag{4.154}$$

On space-like hyper-planes the Dirac Hamiltonian becomes $\bar{H}_D = \tilde{\lambda}^\mu(\tau)\,\tilde{\bar{\mathcal{H}}}_\mu(\tau) - \frac{1}{2}\,\tilde{\lambda}^{\mu\nu}(\tau)\,\tilde{\bar{\mathcal{H}}}_{\mu\nu}(\tau)$ and only the variables x^μ_s, p^μ_s, b^μ_A, $S^{\mu\nu}_s$, A_A, π^A and the following first-class constraints are left ($\vec{\sigma}$ are Cartesian 3-coordinates):

$$\tilde{\bar{\mathcal{H}}}^\mu(\tau) = p^\mu_s - l_\mu\,\frac{1}{2}\int d^3\sigma \sum_a [g^2_s\,\vec{\pi}^2_a(\tau,\vec{\sigma}) + g_s^{-2}\,\vec{B}^2_a(\tau,\vec{\sigma})]$$
$$-b^\mu_r(\tau)\int d^3\sigma \sum_a\,[\vec{\pi}_a(\tau,\vec{\sigma})\times\vec{B}_a(\tau,\vec{\sigma})]_r \approx 0,$$
$$\tilde{\bar{\mathcal{H}}}^{\mu\nu}(\tau) = S^{\mu\nu}_s(\tau) - [b^\mu_r(\tau)\,b^\nu_\tau - b^\nu_r(\tau)\,b^\mu_\tau]\,\frac{1}{2}\int d^3\sigma\,\sigma^r \sum_a [g^2_s\,\vec{\pi}^2_a(\tau,\vec{\sigma}) + g_s^{-2}\,\vec{B}^2_a(\tau,\vec{\sigma})]$$
$$+[b^\mu_r(\tau)\,b^\nu_s(\tau) - b^\nu_r(\tau)\,b^\mu_s(\tau)]\,[\int d^3\sigma\,\sigma^r \sum_a\,[\vec{\pi}_a(\tau,\vec{\sigma})\times\vec{B}_a(\tau,\vec{\sigma})]_s \approx 0,$$
$$\pi^\tau_a(\tau,\vec{\sigma}) \approx 0,$$
$$\bar{\Gamma}_a(\tau,\vec{\sigma}) \approx 0. \tag{4.155}$$

On the Wigner hyper-planes of the rest-frame instant form for the configurations with $\epsilon\,P^2 > 0$, the only remaining first-class constraints are ($\{.,.\}^{**}$ are the final Dirac brackets after the gauge fixings on the embedding):

$$\tilde{\bar{\mathcal{H}}}^{\mu}(\tau) = P^{\mu} - u^{\mu}(u(P))\,\frac{1}{2}\int d^3\sigma \sum_a [g_s^2\,\vec{\pi}_a^2(\tau,\vec{\sigma}) + g_s^{-2}\,\vec{B}_a^2(\tau,\vec{\sigma})]$$
$$-\epsilon^{\mu}_r(u(P))\,[\Big(\int d^3\sigma \sum_a [\vec{\pi}_a(\tau,\vec{\sigma}) \times \vec{B}_a(\tau,\vec{\sigma})]^r \approx 0,$$
$$\pi^{\tau}_a(\tau,\vec{\sigma}) \approx 0,$$
$$\bar{\Gamma}_a(\tau,\vec{\sigma}) \approx 0,$$

$$\{\tilde{\bar{\mathcal{H}}}^{\mu}, \tilde{\bar{\mathcal{H}}}^{\nu}\}^{**} = \sum_a \int d^3\sigma \Big([\epsilon^{\mu}_{\tau}(u(p_s))\,\epsilon^{\nu}_r(u(p_s)) - \epsilon^{\nu}_{\tau}(u(p_s))\,\epsilon^{\mu}_r(u(p_s))]\,\pi_{ar}(\tau,\vec{\sigma}) + \epsilon^{\mu}_r(u(p_s))\,F_{ars}(\tau,\vec{\sigma})\,\epsilon^{\nu}_s(u(p_s))\Big)\,\bar{\Gamma}_a(\tau,\vec{\sigma}), \tag{4.156}$$

or

$$\bar{\mathcal{H}}(\tau) = \epsilon_s - \frac{1}{2}\int d^3\sigma \sum_a [g_s^2\,\vec{\pi}_a^2(\tau,\vec{\sigma}) + g_s^{-2}\,\vec{B}_a^2(\tau,\vec{\sigma})] \approx 0,$$
$$\vec{\bar{\mathcal{H}}}_p(\tau) = \int d^3\sigma \sum_a \vec{\pi}_a(\tau,\vec{\sigma}) \times \vec{B}_a(\tau,\vec{\sigma}) \approx 0,$$
$$\pi^{\tau}_a(\tau,\vec{\sigma}) \approx 0,$$
$$\bar{\Gamma}_a(\tau,\vec{\sigma}) \approx 0. \tag{4.157}$$

The discussion now proceeds as in the electromagnetic case with the determination of the internal Poincaré generators and the elimination of the internal center of mass.

As shown in Ref. [94],[3] the search of a non-Abelian Shanmugadhasan canonical basis is highly non-trivial and there are global results only in the case of a trivial principal bundle (for non-trivial bundles there are only local results) by using methods from group theory and from differential geometry [198, 199, 208–212]. It is possible to find the DOs of Yang–Mills theory with a trivial principal bundle and the non-Abelian analogue of the Abelian Shanmugadhasan canonical transformation (Eq. 4.55) in the inertial frames and in particular in the rest-frames of Minkowski space-time.

With a trivial bundle in each point $(\tau,\vec{\sigma})$ of the Wigner 3-space there is a copy of the group manifold of the group G, namely at each τ there is the structure $\Sigma_\tau \times G$. In each point $(\tau,\vec{\sigma})$ of Σ_τ the fiber is a copy of the group manifold of G, which is parametrized with canonical coordinates of the first kind $\eta_a(\tau,\vec{\sigma})$ ($\eta_a = 0$ is the origin of G), satisfying $\mathcal{A}_{ab}(\eta_c(\tau,\vec{\sigma}))\,\eta_b(\tau,\vec{\sigma}) = \eta_a(\tau,\vec{\sigma})$, where $\mathcal{A}_{ab}(\eta_c(\tau,\vec{\sigma}))$ is a matrix satisfying the Maurer–Cartan equations $\partial_{\eta_b}\mathcal{A}_{ac}(\eta_u) - \partial_{\eta_c}\mathcal{A}_{ab}(\eta_u) = -c_{amn}\,\mathcal{A}_{mb}(\eta_u)\,\mathcal{A}_{nc}(\eta_u)$.

[3] In its appendices there is a review of the parametrization of principal G-bundles.

If we define the modified Gauss's laws $\tilde{\Gamma}_a(\tau,\vec{\sigma}) = \bar{\Gamma}_b(\tau,\vec{\sigma})\,\mathcal{A}_{ba}(\eta_u(\tau,\vec{\sigma})) \approx 0$, we find that the second canonical pair of gauge variables besides $A^\tau_a(\tau,\vec{\sigma})$, $\pi^\tau(\tau,\vec{\sigma}) \approx 0$ (see Eq. (4.150)) is $\eta_a(\tau,\vec{\sigma})$, $\tilde{\Gamma}_a(\tau,\vec{\sigma})$, whose non-zero Poisson brackets are

$$\{\eta_a(\tau,\vec{\sigma}), \tilde{\Gamma}_b(\tau,\vec{\sigma}^{\,\prime})\} = -\delta_{ab}\,\delta^3(\vec{\sigma}-\vec{\sigma}^{\,\prime}). \tag{4.158}$$

Let $\Omega(\eta_u) = \Omega_a(\eta_u)\,T^a$ be the canonical one-form on G in the adjoint representation, satisfying $d^*\Omega_a(\eta_u) = \mathcal{A}_{ab}(\eta_u)\,d\eta_b$ (d^* is the exterior derivative along the preferred line defining the canonical coordinates of the first kind).

It can be shown that the vector potential admits the following decomposition:

$$A_{ai}(\tau,\vec{\sigma}) = \mathcal{A}_{ab}(\eta_u(\tau,\vec{\sigma}))\,\frac{\partial\,\eta_b(\tau,\vec{\sigma})}{\partial\,\sigma^i} + \Big(P\,e^{\Omega(\eta_u(\tau,\vec{\sigma}))}\Big)_{ab}\,A_{bi\perp}(\tau,\vec{\sigma}), \tag{4.159}$$

where P denotes the path ordering along the preferred line and $\vec{A}_{a\perp}(\tau,\vec{\sigma})$ is a transverse vector potential: $\vec{\partial}\cdot\vec{A}_{a\perp}(\tau,\vec{\sigma}) = 0$.

For the conjugated momenta one has to use the decomposition[4]

$$\begin{aligned}
\pi^i_a(\tau,\vec{\sigma}) &= \pi^i_{a,D\perp}(\tau,\vec{\sigma}) + \int d^3\sigma^{\prime}\,\mathcal{G}^{(A)}_{ab}(\tau;\vec{\sigma},\vec{\sigma}^{\,\prime})\,\tilde{\Gamma}_b(\tau,\vec{\sigma}^{\,\prime}),\\
\pi^i_{a,D\perp}(\tau,\vec{\sigma}) &= \int d^3\sigma^{\prime}\,\Big[\delta^{ij}\,\delta_{av}\,\delta^3(\vec{\sigma}-\vec{\sigma}^{\,\prime})\\
&\quad -\frac{\partial^i_\sigma}{\triangle_\sigma}\,\vec{\partial}_\sigma\cdot\vec{\mathcal{G}}^{(A)}_{ub}(\tau;\vec{\sigma},\vec{\sigma}^{\,\prime})\,c_{buv}\,A^j_u(\tau,\vec{\sigma}^{\,\prime})\Big]\,\pi^j_{v,\perp}(\tau,\vec{\sigma}^{\,\prime}),\\
&\quad \vec{D}^{(A)}_{ab}(\tau,\vec{\sigma})\cdot\vec{\pi}_{b,D\perp}(\tau,\vec{\sigma}) = 0,
\end{aligned} \tag{4.160}$$

where

$$\vec{\mathcal{G}}^{(A)}_{ab}(\tau;\vec{\sigma},\vec{\sigma}^{\,\prime}) = \frac{\vec{\sigma}-\vec{\sigma}^{\,\prime}}{4\pi\,|\vec{\sigma}-\vec{\sigma}^{\,\prime}|^3}\,\Big(P\,e^{\int_{\vec{\sigma}^{\,\prime}}^{\vec{\sigma}} d^3\sigma^{\prime}\,\vec{A}_c(\tau,\vec{\sigma})\,T^c}\Big)_{ab} \tag{4.161}$$

is the Green's function of the covariant derivative $D^{(A)}_{abu}(\tau,\vec{\sigma}) = \delta_{ab}\,\frac{\partial}{\partial\,\sigma^u} + c_{abc}\,A_{du}(\tau,\vec{\sigma})$.

The final step is to introduce the following vector potentials:

$$\begin{aligned}
\hat{\pi}^i_{a,D\perp}(\tau,\vec{\sigma}) &= \Big[\Big(P\,e^{\Omega(\eta_u(\tau,\vec{\sigma}))}\Big)^{-1}\Big]_{ab}\,\pi^i_{b,D\perp}(\tau,\vec{\sigma}),\\
\hat{\pi}^i_{a\perp}(\tau,\vec{\sigma}) &= P^{ij}_\perp(\vec{\sigma})\,\hat{\pi}^j_{a,D\perp}(\tau,\vec{\sigma}).
\end{aligned} \tag{4.162}$$

The DOs of Yang–Mills theory with the trivial principal bundle turn out to be $A^i_{a\perp}(\tau,\vec{\sigma})$ and $\hat{\pi}_{a\perp}(\tau,\vec{\sigma})$, because it can be shown that they have zero Poisson bracket with the two pairs of canonical gauge variables $A^\tau_a(\tau,\vec{\sigma})$,

[4] $\pi^i_{a\perp}(\tau,\vec{\sigma}) = P^{ij}_\perp(\vec{\sigma})\,\pi^j_a(\tau,\vec{\sigma})$, $\vec{\partial}\cdot\vec{\pi}_{a\perp} = 0$ has non-vanishing Poisson brackets with Gauss's laws; $P^{ij}_\perp(\vec{\sigma}) = \delta^{ij} + \frac{\partial^i\partial^j}{\nabla}$, $\nabla = -\vec{\partial}^2$.

$\pi^\tau(\tau,\vec{\sigma}) \approx 0$, $\eta_a(\tau,\vec{\sigma})$, $\tilde{\Gamma}_a(\tau,\vec{\sigma})$. Moreover, their only non-vanishing Poisson brackets are

$$\{A^i_{a\perp}(\tau,\vec{\sigma}), \hat{\pi}^j_{b\perp}(\tau,\vec{\sigma}'\} = -\delta_{ab}\, P^{ij}_\perp(\vec{\sigma})\, \delta^3(\vec{\sigma}-\vec{\sigma}'). \tag{4.163}$$

These variables form the non-Abelian analogue of the Shanmugadhasan canonical basis (Eq. 4.55) of the electromagnetic field.

The global DOs are extremely complex, non-local, and with poor global control due to the need for the Green's function of the covariant divergence (it requires the use of path-dependent non-integrable phases; see Refs. [185, 186]). The non-local and non-polynomial (due to the presence of classical Wilson lines along flat geodesics) physical Hamiltonian has been obtained: It is non-local but without any kind of singularities and it has the correct Abelian limit if the structure constants are turned off.

Other models studied with the described technology are as follows:

1. In Ref. [185] there is the formulation in the rest-frame instant form of the relativistic quark model in the radiation gauge for the $SU(3)$ Yang–Mills fields with scalar quarks having Grassmann-valued color charges. While in equation 101 of that paper there is the rest-frame condition, in equation 97 there is the invariant mass M for a quark–antiquark system. In it the electromagnetic Coulomb potential is replaced with a potential, given in equation 95, depending on the color transverse vector potential through the Green's function of the $SU(3)$ covariant divergence. The non-linearity of the problem does not allow evaluation of a Lienard–Wiechert solution. Instead, in Ref. [186], there is the study of Yang–Mills theory with Grassmann-valued fermion fields in the case of a trivial principal bundle with suitable boundary conditions at the Hamiltonian level so as to exclude monopole and instanton solutions. The fermion fields turn out to be dressed with Yang–Mills (gluonic) clouds.

2. $SU(3)$ Yang–Mills theory with scalar particles with Grassmann-valued color charges [93, 94] for the regularization of self-energies. It is possible to show that in this relativistic scalar quark model the Dirac Hamiltonian expressed as a function of DOs has the property of asymptotic freedom.

3. The Abelian and non-Abelian $SU(2)$ Higgs models with fermion fields [183, 184], where the symplectic decoupling is a refinement of the concept of unitary gauge. There is an ambiguity in the solutions of Gauss's law constraints, which reflects the existence of disjoint sectors of solutions of the Euler–Lagrange equations of Higgs models. The physical Hamiltonian and Lagrangian of the Higgs phase have been found; the self-energy turns out to be local and contains a local four-fermion interaction.

4. The standard $SU(3)\times SU(2)\times U(1)$ model of elementary particles [185, 186] with Grassmann- valued fermion fields. The final reduced Hamiltonian contains non-local self-energies for the electromagnetic and color interactions, but "local

ones" for the weak interactions, implying the non-perturbative emergence of four-fermion interactions.

4.6 Relativistic Fluids, Relativistic Micro-Canonical Ensemble, and Steps toward Relativistic Statistical Mechanics

In this section we shall apply the 3+1 approach to non-inertial frames to relativistic fluids [213, 214] (whose extension to GR in the case of dust is done in Ref. [104]) and to the problems of how to define the relativistic micro-canonical ensemble [215] (in Ref. [216] there is an extended version of this paper) and relativistic statistical mechanics [217].

4.6.1 The Relativistic Perfect Fluid

The stability of stellar models for rotating stars, gravity–fluid models, neutron stars, accretion discs around compact objects, collapse of stars, and merging of compact objects are only some of the many topics in astrophysics and cosmology in which relativistic hydrodynamics is the basic underlying theory. This theory is also needed in heavy-ion collisions.

As shown in Ref. [218], there are many ways to describe relativistic perfect fluids by means of action functionals, both in SR and GR. Usually, besides the thermo-dynamical variables $\hat{n}$ (particle number density), ρ (energy density), p (pressure), T (temperature), s (entropy per particle), which are space-time scalar fields whose values represent measurements made in the rest-frame of the fluid (Eulerian observers), one characterizes the fluid motion by its unit time-like 4-velocity vector field U^μ.[5]

However, these variables are constrained due to ($;\mu$ denotes a covariant derivative):

1. particle number conservation, $(\hat{n}U^\mu)_{;\mu} = 0$;
2. absence of entropy exchange between neighboring flow lines, $(\hat{n}sU^\mu)_{;\mu} = 0$; and
3. the requirement that the fluid flow lines should be fixed on the boundary.

Therefore, one needs Lagrange multipliers to incorporate 1 and 2 into the action and this leads to using Clebsch (or velocity-potential) representations of the 4-velocity and action functionals depending on many redundant variables, generating first- and second-class constraints at the Hamiltonian level, which are reviewed in Appendix C.

[5] See Appendix C for a review of the relations among the local thermo-dynamical variables and for a review of covariant relativistic thermodynamics following Ref. [219].

Following Refs. [220, 221], the previous constraint 3 may be enforced by replacing the unit 4-velocity U^μ with a set of space-time scalar fields $\tilde{\alpha}^i(z)$, $i = 1, 2, 3$, interpreted as Lagrangian (or comoving) coordinates for the fluid labeling the fluid flow lines (physically determined by the average particle motions) passing through the points inside the boundary.[6] This requires the choice of an arbitrary space-like hyper-surface on which the α^i are the 3-coordinates. A similar point of view is contained in the concept of material space of Refs. [222–224], describing the collection of all the idealized points of the material; besides non-dissipative isentropic fluids, the scheme can be applied to isotropic elastic media and anisotropic (crystalline) materials, namely to an arbitrary non-dissipative relativistic continuum [225]. See Ref. [226] for the study of the transformation from Eulerian to Lagrangian coordinates (in the non-relativistic framework of the Euler–Newton equations).

Notice that the use of Lagrangian (comoving) coordinates in place of Eulerian quantities allows the use of standard Poisson brackets in the Hamiltonian description, avoiding the formulation with *Lie–Poisson brackets* of Refs. [227, 228], which could be recovered by a so-called Lagrangian to Eulerian map.

Following Ref. [213] we shall reformulate the theory of fluids on arbitrary space-like hyper-surfaces either of Minkowski space-time or of a curved globally hyperbolic space-time M^4 of GR, whose points have local coordinates z^μ. Let ${}^4g_{\mu\nu}(z)$ be its 4-metric with determinant ${}^4g = |\det {}^4g_{\mu\nu}|$. Given a perfect fluid with Lagrangian coordinates $\tilde{\alpha}(z) = \{\tilde{\alpha}^i(z)\}$, unit 4-velocity vector field $U^\mu(z)$ and particle number density $n(z)$, let us introduce the number flux vector

$$\hat{n}(z)\, U^\mu(z) = \frac{J^\mu(\tilde{\alpha}^i(z))}{\sqrt{{}^4g(z)}}, \tag{4.164}$$

and the densitized fluid number flux vector or material current:[7]

$$\begin{aligned}
J^\mu(\tilde{\alpha}^i(z)) &= -\sqrt{{}^4g}\, \epsilon^{\mu\nu\rho\sigma}\, \eta_{123}(\tilde{\alpha}^i(z))\, \partial_\nu\, \tilde{\alpha}^1(z)\, \partial_\rho\, \tilde{\alpha}^2(z)\, \partial_\sigma\, \tilde{\alpha}^3(z), \\
&\Rightarrow \hat{n}(z) = \frac{|J(\tilde{\alpha}^i(z))|}{\sqrt{{}^4g}} = \eta_{123}(\tilde{\alpha}^i(z))\, \frac{\sqrt{\epsilon\, {}^4g_{\mu\nu}(z)\, J^\mu(\tilde{\alpha}^i(z))\, J^\nu(\tilde{\alpha}^i(z))}}{\sqrt{{}^4g(z)}}, \\
&\Rightarrow \partial_\mu\, J^\mu(\tilde{\alpha}^i(z)) = \sqrt{{}^4g}\, [\hat{n}(z)\, U^\mu(z)]_{;\mu} = 0, \\
&\Rightarrow J^\mu(\tilde{\alpha}^i(z))\, \partial_\mu\, \tilde{\alpha}^i(z) = [\sqrt{{}^4g}\, \hat{n}\, U^\mu](z)\, \partial_\mu\, \tilde{\alpha}^i(z) = 0.
\end{aligned} \tag{4.165}$$

This shows that the fluid flow lines, whose tangent vector field is the fluid 4-velocity time-like vector field U^μ, are identified by $\tilde{\alpha}^i$ = const. and that the particle number conservation is automatic. Moreover, if the entropy for particles is a function only of the fluid Lagrangian coordinates, $s = s(\tilde{\alpha}^i)$, the assumed

[6] On the boundary they are fixed: Either the $\tilde{\alpha}^i(z^o, \vec{z})$ have a compact boundary $V_\alpha(z^o)$ or they have assigned boundary conditions at spatial infinity.

[7] $\epsilon^{0123} = 1/\sqrt{{}^4g}$; $\partial_\alpha\, (\sqrt{{}^4g}\, \epsilon^{\mu\nu\rho\sigma}) = 0$; $\eta_{123}(\tilde{\alpha}^i)$ describes the orientation of the volume in the material space.

form of J^μ also implies automatically the absence of entropy exchange between neighboring flow lines, $(\hat{n}\, s\, U^\mu)_{,\mu} = 0$. Since $U^\mu\, \partial_\mu\, s(\tilde{\alpha}^i) = 0$, the perfect fluid is locally adiabatic; instead, for an isentropic fluid we have $\partial_\mu\, s = 0$, namely $s = \text{const}$.

Even if in general the time-like vector field $U^\mu(z)$ is not surface forming,[8] in each point z we can consider the space-like hyper-surface orthogonal to the fluid flow line in that point[9] and consider $\frac{1}{3!}\, [U^\mu\, \epsilon_{\mu\nu\rho\sigma}\, dz^\nu \wedge dz^\rho \wedge dz^\sigma](z)$ as the infinitesimal 3-volume on it at z. Then the 3-form

$$\eta[z] = [\eta_{123}(\tilde{\alpha})\, d\tilde{\alpha}^1 \wedge d\tilde{\alpha}^2 \wedge d\tilde{\alpha}^3](z) = \frac{1}{3!}\, \hat{n}(z)\, [U^\mu\, \epsilon_{\mu\nu\rho\sigma}\, dz^\nu \wedge dz^\rho \wedge dz^\sigma](z) \quad (4.166)$$

may be interpreted as the number of particles in this 3-volume. If V is a volume around z on the space-like hyper-surface, then $\int_V \eta$ is the number of particle in V and $\int_V s\, \eta$ is the total entropy contained in the flow lines included in the volume V. Note that locally η_{123} can be set to unity by an appropriate choice of coordinates.

In Ref. [218] it is shown that the action functional

$$S[{}^4g_{\mu\nu}, \tilde{\alpha}] = - \int d^4z\, \sqrt{{}^4g(z)}\, \rho\Big(\frac{|J(\tilde{\alpha}^i(z))|}{\sqrt{{}^4g(z)}},\; s(\tilde{\alpha}^1(z))\Big) \quad (4.167)$$

has a variation with respect to the 4-metric, which gives rise to the correct stress tensor $T^{\mu\nu} = (\rho + p)\, U^\mu\, U^\nu - \epsilon\, p\, {}^4g^{\mu\nu}$ with $p = n\, \frac{\partial \rho}{\partial n}|_s - \rho$ for a perfect fluid (see Appendix C).

The Euler–Lagrange equations associated with the variation of the Lagrangian coordinates are [218] ($V_\mu = \mu\, U_\mu$, the Taub current – see Appendix C):

$$\frac{1}{\sqrt{{}^4g}}\, \frac{\delta S}{\delta \tilde{\alpha}^i} = \frac{1}{2}\, \epsilon^{\mu\nu\rho\sigma}\, V_{\mu;\nu}\, \eta_{ijk}\, \partial_\rho\, \tilde{\alpha}^j\, \partial_\sigma\, \tilde{\alpha}^k - \hat{n}\, T\, \frac{\partial s}{\partial \tilde{\alpha}^i} = 0,$$

$$\frac{1}{\sqrt{{}^4g}}\, \frac{\delta S}{\delta \tilde{\alpha}^i} \partial_\mu \tilde{\alpha}^i = \frac{1}{2}\, \epsilon^{\alpha\beta\gamma\delta}\, V_{\alpha;\beta}\, U^\nu\, \epsilon_{\nu\mu\gamma\delta} - T\, \partial_\mu\, s = 2\, V_{[\mu;\nu]}\, U^\nu - T\, \partial_\mu s = 0. \quad (4.168)$$

As shown in Appendix C, these equations together with the entropy exchange constraint imply the Euler equations, which are produced from the conservation of the stress–energy–momentum tensor.

Therefore, with this description the conservation laws are automatically satisfied and the Euler–Lagrange equations are equivalent to the Euler equations. In Minkowski space-time the conserved particle number is $\mathcal{N} = \int_{V_\alpha(z^o)} d^3z\, \hat{n}(z)\, U^o(z) = \int_{V_\alpha(z^o)} d^3z\, J^o(\tilde{\alpha}^i(z))$, while the conserved entropy per particle is $\int_{V_\alpha(z^o)} d^3z\, s(z)\, \hat{n}(z)\, U^o(z) = \int_{V_\alpha(z^o)} d^3z\, s(z)\, J^o(\tilde{\alpha}^i)$. Moreover, the conservation

[8] Namely has a non-vanishing vorticity.
[9] Namely we split the tangent space T_zM^4 at z in the $U^\mu(z)$ direction and in the orthogonal complement.

laws $T^{\mu\nu}_{,\nu} = 0$ will generate the conserved 4-momentum and angular momentum of the fluid.

However, in Ref. [218] there are only some comments on the Hamiltonian description implied by this particular action.

To define parametrized Minkowski theory for the relativistic perfect fluid with equation of state $\rho = \rho(\hat{n}, s)$, we have only to replace the external 4-metric ${}^4g_{\mu\nu}$ with ${}^4g_{AB}(\tau, \sigma^u)$ and scalar fields with $\alpha^i(\tau, \sigma^u) = \tilde{\alpha}^i(z(\tau, \sigma^u))$ for the Lagrangian coordinates. Either the $\alpha^i(\tau, \sigma^u)$ have a compact boundary $V_\alpha(\tau) \subset \Sigma_\tau$ or have boundary conditions at spatial infinity. For each value of τ, we could invert $\alpha^i = \alpha^i(\tau, \sigma^u)$ to $\sigma^u = \sigma^u(\tau, \alpha^i)$ and use the α^i as a special coordinate system on Σ_τ inside the support $V_\alpha(\tau) \subset \Sigma_\tau$: $z^\mu(\tau, \sigma^u(\tau, \alpha^i)) = \check{z}^\mu(\tau, \alpha^i)$.

By going to Σ_τ-adapted coordinates such that $\eta_{123}(\alpha) = 1$ we get ($\gamma = |\det g_{rs}|$; $\sqrt{g} = \sqrt{{}^4g} = \sqrt{|\det {}^4g_{AB}|} = N\sqrt{\gamma}$; $N = 1 + n$):

$$\begin{aligned}
J^A(\alpha^i(\tau, \sigma^u)) &= [N\sqrt{\gamma}\,\hat{n}\, U^A](\tau, \sigma^u),\\
J^\tau(\alpha^i(\tau, \sigma^u)) &= [-\epsilon^{ruv}\,\partial_r\,\alpha^1\,\partial_u\,\alpha^2\,\partial_v\,\alpha^3](\tau, \sigma^u) = -\det(\partial_r\,\alpha^i)(\tau, \sigma^u),\\
J^r(\alpha^i(\tau, \sigma^u)) &= \Big[\sum_{i=1;i,j,k\,cyclic}^{3} \partial_\tau\,\alpha^i\,\epsilon^{ruv}\,\partial_u\,\alpha^j\,\partial_v\,\alpha^k\Big](\tau, \sigma^u)\\
&= \frac{1}{2}\,\epsilon^{ruv}\,\epsilon_{ijk}\,[\partial_\tau\,\alpha^i\,\partial_u\,\alpha^j\,\partial_v\,\alpha^k](\tau, \sigma^u),\\
\Rightarrow \hat{n}(\tau, \sigma^u) &= \frac{|J|}{N\sqrt{\gamma}}(\tau, \sigma^u) = \frac{\sqrt{\epsilon\,{}^4g_{AB}\,J^A\,J^B}}{N\sqrt{\gamma}}(\tau, \sigma^u),
\end{aligned} \tag{4.169}$$

with $\mathcal{N} = \int_{V_\alpha(\tau)} d^3\sigma\, J^\tau(\alpha^i(\tau, \sigma^u))$ giving the conserved particle number and $\int_{V_\alpha(\tau)} d^3\sigma\,(s\,J^\tau)(\tau, \sigma^u)$ giving the conserved entropy per particle.

The action becomes

$$\begin{aligned}
S &= \int d\tau d^3\sigma\, L(z^\mu(\tau, \vec{\sigma}), \alpha^i(\tau, \vec{\sigma}))\\
&= -\int d\tau d^3\sigma\,(N\sqrt{\gamma})(\tau, \vec{\sigma})\,\rho\Big(\frac{|J(\alpha^i(\tau, \vec{\sigma}))|}{(N\sqrt{\gamma})(\tau\vec{\sigma})},\, s(\alpha^i(\tau, \vec{\sigma}))\Big)\\
&= -\int d\tau d^3\sigma\,(N\sqrt{\gamma})(\tau, \vec{\sigma})\\
&\quad \rho\Big(\frac{1}{\sqrt{\gamma(\tau, \vec{\sigma})}}\sqrt{\Big[(J^\tau)^2 - {}^3g_{\check{u}\check{v}}\,\frac{J^{\check{u}} + N^{\check{u}}\,J^\tau}{N}\,\frac{J^{\check{v}} + N^{\check{v}}\,J^\tau}{N}\Big](\tau, \vec{\sigma}; \alpha^i(\tau, \vec{\sigma}))},\, s(\alpha^i(\tau, \vec{\sigma}))\Big),
\end{aligned} \tag{4.170}$$

with $N = N_{[z](flat)}$, $N^{\check{r}} = N^{\check{r}}_{[z](flat)}$.

This is the form of the action whose Hamiltonian formulation has been studied in Ref. [213] and reformulated in terms of generalized Eulerian coordinates in Ref. [214].

Let us now consider the relativistic dust.

Let us consider first the simplest case of an isentropic perfect fluid, a dust with $p = 0$, $s = \text{const.}$, and equation of state $\rho = \mu\,\hat{n}$. In this case the chemical potential μ is the rest mass-energy of a fluid particle: $\mu = m$ (see Appendix C).

Eq. (4.170) implies that the Lagrangian density is

$$\begin{aligned} L(\alpha^i, z^\mu) &= -\sqrt{g}\,\rho = -\mu\,\sqrt{g}\,\hat{n} = -\mu\,\sqrt{\epsilon\, g_{\check{A}\check{B}}\, J^{\check{A}}\, J^{\check{B}}} \\ &= -\mu\, N\,\sqrt{(J^\tau)^2 - {}^3g_{\check{r}\check{s}}\, Y^{\check{r}}\, Y^{\check{s}}} = -\mu\, N\, X, \\ Y^{\check{r}} &= \frac{1}{N}\,(J^{\check{r}} + N^{\check{r}}\, J^\tau), \\ X &= \sqrt{(J^\tau)^2 - {}^3g_{\check{r}\check{s}}\, Y^{\check{r}}\, Y^{\check{s}}} = \frac{\sqrt{g}}{N}\,\hat{n} = \sqrt{\gamma}\,\hat{n}, \end{aligned} \tag{4.171}$$

with J^τ, $J^{\check{r}}$ given in Eqs. (4.169).

The momentum conjugate to α^i is

$$\begin{aligned} \Pi_i &= \frac{\partial L}{\partial \partial_\tau \alpha^i} = \mu\, \frac{Y^{\check{t}}\, {}^3g_{\check{t}\check{r}}\, \epsilon^{\check{r}\check{u}\check{v}}\, \partial_{\check{u}}\, \alpha^j\, \partial_{\check{v}}\, \alpha^k}{X}\Big|_{i,j,k \quad cyclic} \\ &= \mu\, \frac{Y^{\check{t}}}{2\,X}\, {}^3g_{\check{t}\check{r}}\, \epsilon^{\check{r}\check{u}\check{v}}\, \epsilon_{ijk}\, \partial_{\check{u}}\, \alpha^j\, \partial_{\check{v}}\, \alpha^k = \mu\, \frac{Y^{\check{r}}}{X}\, T_{\check{r}i}, \\ T_{\check{t}i} &\overset{def}{=} \frac{1}{2}\, g_{\check{t}\check{r}}\, \epsilon^{\check{r}\check{u}\check{v}}\, \epsilon_{ijk}\, \partial_{\check{u}}\, \alpha^j\, \partial_{\check{v}}\, \alpha^k = g_{\check{t}\check{r}}\, (ad\, J_{i\check{r}}), \end{aligned} \tag{4.172}$$

where $ad\, J_{i\check{r}} = (\det J)\, J^{-1}_{i\check{r}}$ is the adjoint matrix of the Jacobian $J = (J_{i\check{r}} = \partial_{\check{r}}\, \alpha^i)$ of the transformation from the Lagrangian coordinates $\alpha^i(\tau, \sigma^u)$ to the Eulerian ones σ^u on Σ_τ.

The momentum conjugate to z^μ is

$$\rho_\mu(\tau, \sigma^u) = -\frac{\partial L}{\partial z^\mu_\tau}(\tau, \sigma^u) = \Big[\mu\, l_\mu\, \frac{(J^\tau)^2}{X} + \mu\, z_{r\mu}\, J^\tau\, \frac{Y^r}{X}\Big](\tau, \sigma^u). \tag{4.173}$$

The following Poisson brackets are assumed:

$$\begin{aligned} \{z^\mu(\tau, \sigma^u), \rho_\nu(\tau, \sigma^{'u}\} &= -\epsilon\, {}^4\eta^\mu_\nu\, \delta^3(\sigma^u - \sigma^{'u}), \\ \{\alpha^i(\tau, \sigma^u), \Pi_j(\tau, \sigma^{'u})\} &= \delta^i_j\, \delta^3(\sigma^u - \sigma^{'u}). \end{aligned} \tag{4.174}$$

We can express $Y^{\check{r}}/X$ in terms of Π_i with the help of the inverse $(T^{-1})^{\check{r}i}$ of the matrix $T_{\check{t}i}$:

$$\frac{Y^{\check{r}}}{X} = \frac{1}{\mu}\, (T^{-1})^{\check{r}i}\Pi_i, \tag{4.175}$$

where

$$(T^{-1})^{ri} = \frac{{}^3g^{rs}\, \partial_s\, \alpha^i}{\det\, (\partial_u \alpha^k)}. \tag{4.176}$$

From the definition of X, we find

$$X = \frac{\mu\, J^\tau}{\sqrt{\mu^2 + {}^3g_{uv}\, (T^{-1})^{ui}\, (T^{-1})^{vj}\, \Pi_i\, \Pi_j}}. \tag{4.177}$$

Consequently, we can get the expression of the velocities of the Lagrangian coordinates in terms of the momenta

$$\partial_\tau \alpha^i = -\frac{J^r \partial_r \alpha^i}{J^\tau} = \frac{(N^r J^\tau - N Y^r) \partial_r \alpha^i}{J^\tau}, \tag{4.178}$$

namely

$$\partial_\tau \alpha^i = \partial_r \alpha^i \left[N^r - N (T^{-1})^{ri} \Pi_i \sqrt{\mu^2 c^2 + {}^3g_{uv} (T^{-1})^{ui} (T^{-1})^{vj} \Pi_i \Pi_j}\right]. \tag{4.179}$$

Now ρ_μ can be expressed as a function of the z, α, and Π:

$$\rho_\mu = l_\mu J^\tau \sqrt{\mu^2 + {}^3g_{uv} (T^{-1})^{ui} \Pi_i (T^{-1})^{vj} \Pi_j} + z_{r\mu} J^\tau (T^{-1})^{ri} \Pi_i. \tag{4.180}$$

The Dirac Hamiltonian and the primary constraints are

$$\begin{aligned}
\bar{H}_D &= \int d^3\sigma \, \lambda^\mu(\tau, \sigma^u) \, \bar{\mathcal{H}}_\mu(\tau, \sigma^u), \\
\bar{\mathcal{H}}_\mu(\tau, \sigma^u) &= \left[\rho_\mu - l_\mu \bar{\mathcal{M}} + z_{r\mu} \bar{\mathcal{M}}^r\right](\tau, \sigma^u) \approx 0, \\
\bar{\mathcal{M}} &= T^{\tau\tau} = J^\tau \sqrt{\mu^2 c^2 + {}^3g_{uv} (T^{-1})^{ui} \Pi_i (T^{-1})^{vj} \Pi_j}, \\
\bar{\mathcal{M}}^r &= T^{\tau r} = J^\tau (T^{-1})^{ri} \Pi_i.
\end{aligned} \tag{4.181}$$

On arbitrary space-like hyper-planes, the following ten first-class constraints remain ($\vec{\sigma}$ are Cartesian 3-coordinates):

$$\begin{aligned}
\tilde{\bar{\mathcal{H}}}^\mu(\tau) &= P^\mu \\
&\quad - \int d^3\sigma \left(J^\tau \left[l^\mu \sqrt{\mu^2 c^2 + \delta_{uv} (T^{-1})^{ui} \Pi_i (T^{-1})^{vj} \Pi_j}\right.\right. \\
&\quad \left.\left. + b^\mu_r (T^{-1})^{rl} \Pi_l\right]\right)(\tau, \vec{\sigma}) \approx 0, \\
\tilde{\bar{\mathcal{H}}}^{\mu\nu}(\tau) &= S^{\mu\nu}_s(\tau) \\
&\quad - [b^\mu_r(\tau) b^\nu_\tau - b^\nu_r(\tau) b^\mu_\tau] \int d^3\sigma \, \sigma^r \\
&\quad \left(J^\tau \sqrt{\mu^2 c^2 + \delta_{uv} (T^{-1})^{ui} \Pi_i (T^{-1})^{vj} \Pi_j}\right)(\tau, \vec{\sigma}) \\
&\quad + [b^\mu_r(\tau) b^\nu_s(\tau) - b^\nu_r(\tau) b^\mu_s(\tau)] \int d^3\sigma \, \sigma^r \left(J^\tau (T^{-1})^{sl} \Pi_l\right)(\tau, \vec{\sigma}) \approx 0.
\end{aligned} \tag{4.182}$$

On the Wigner hyper-planes of the rest-frame instant form the relevant spin tensor is

$$\begin{aligned}
\bar{S}^{AB}_s &\equiv ({}^4\eta^A_r \, {}^4\eta^B_\tau - {}^4\eta^B_r \, {}^4\eta^A_\tau) \, \bar{S}^{\tau r}_s - ({}^4\eta^A_r \, {}^4\eta^B_s - {}^4\eta^B_r \, {}^4\eta^A_s) \, \bar{S}^{rs}_s, \\
\bar{S}^{rs}_s &\equiv \int d^3\sigma \left(J^\tau [\sigma^r (T^{-1})^{sl} \Pi_l - \sigma^s (T^{-1})^{rl} \Pi_l]\right)(\tau, \vec{\sigma}),
\end{aligned}$$

$$\bar{S}_s^{\tau r} \equiv -\bar{S}_s^{r\tau} = -\int d^3\sigma \left(J^\tau\, \sigma^r\, \sqrt{\mu^2 + \delta_{uv}\,(T^{-1})^{ui}\,\Pi_i\,(T^{-1})^{vj}\,\Pi_j} \right)(\tau, \vec{\sigma}), \tag{4.183}$$

and the Dirac Hamiltonian is $\bar{H}_D = \tilde{\lambda}^\mu(\tau)\,\tilde{\bar{\mathcal{H}}}_\mu(\tau)$ with the four constraints (from Eqs. (4.176) and (4.169) we have $J^\tau = -\det(\partial_r\,\alpha^i)$, $(T^{-1})^{ri} = \delta^{rs}\,\partial_s\,\alpha^i/\det(\partial_u\alpha^k)$):

$$\begin{aligned}
\tilde{\bar{\mathcal{H}}}^\mu(\tau) &= P^\mu - \int d^3\sigma \Big(J^\tau \Big[u^\mu(P)\,\sqrt{\mu^2 + \delta_{uv}\,(T^{-1})^{ui}\,\Pi_i\,(T^{-1})^{vj}\,\Pi_j} \\
&\quad - \epsilon^\mu_r(u(P))\,\mu\,(T^{-1})^{rl}\,\Pi_l \Big] \Big)(\tau, \vec{\sigma}) \approx 0, \\
&\Downarrow \\
\bar{\mathcal{H}}(\tau) &= u^\mu(P)\,\tilde{\bar{\mathcal{H}}}_\mu(\tau) = \epsilon_s - \bar{M}_{sys} \approx 0, \\
&\quad \bar{M}_{sys} c = \int d^3\sigma\, \bar{\mathcal{M}}(\tau, \vec{\sigma}) c \\
&= \int d^3\sigma \left(J^\tau\, \sqrt{\mu^2 c^2 + \delta_{uv}\,(T^{-1})^{ui}\,\Pi_i\,(T^{-1})^{vj}\,\Pi_j} \right)(\tau, \vec{\sigma}) \\
&= -\int d^3\sigma \left[\det(\partial_r\,\alpha^k)\,\sqrt{\mu^2 c^2 + \delta^{uv}\,\frac{\partial_u\,\alpha^i\,\partial_v\,\alpha^j}{[\det(\partial_r\,\alpha^k)]^2}\,\Pi_i\,\Pi_j} \right](\tau, \vec{\sigma}), \\
\bar{\mathcal{H}}^r_p(\tau) &\overset{def}{=} P^r_{sys} = \int d^3\sigma\, \bar{\mathcal{M}}^r(\tau, \vec{\sigma}) \\
&= \int d^3\sigma\, \mu \left(J^\tau\,(T^{-1})^{rl}\,\Pi_l \right)(\tau, \vec{\sigma}) = -\int d^3\sigma\, \mu \left[\delta^{rs}\,\partial_s\,\alpha^i\,\Pi_i \right](\tau, \vec{\sigma}) \approx 0,
\end{aligned} \tag{4.184}$$

where $\bar{M}_{sys}$ is the invariant mass of the fluid. The first one gives the mass spectrum of the isolated system, while the other three say that the total 3-momentum of the N particles on the hyper-plane $\Sigma_{\tau W}$ vanishes.

The Hamilton equations for $\alpha^i(\tau, \vec{\sigma})$ in the Wigner-covariant rest-frame instant form are equivalent to the hydrodynamical Euler equations:

$$\begin{aligned}
\partial_\tau\,\alpha^i(\tau, \vec{\sigma}) &\quad \{\alpha^i(\tau, \vec{\sigma}), \bar{H}_D\} \\
&= -\Big(\frac{\delta^{uv}\,\partial_u\,\alpha^i\,\partial_v\,\alpha^j\,\Pi_j}{\det(\partial_r\,\alpha^k)\,\sqrt{\mu^2 c^2 + \delta^{uv}\,\frac{\partial_u\,\alpha^m\,\partial_v\,\alpha^n}{[\det(\partial_r\,\alpha^k)]^2}\,\Pi_m\,\Pi_n}} \Big)(\tau, \vec{\sigma}) \\
&\quad + \lambda^r(\tau)\,\partial_r\,\alpha^i(\tau), \\
\partial_\tau\,\Pi_i(\tau, \vec{\sigma}) &= \{\Pi_i(\tau, \vec{\sigma}), \bar{H}Dd\} = \qquad (ijk\, \text{cyclic}) \\
&= \frac{\partial}{\partial\sigma^s} \Bigg[\epsilon^{suv}\,\partial_u\,\alpha^j\,\partial_v\,\alpha^k\,\sqrt{\mu^2 c^2 + \delta^{uv}\,\frac{\partial_u\,\alpha^m\,\partial_v\,\alpha^n}{[\det(\partial_r\,\alpha^k)]^2}\,\Pi_m\,\Pi_n}
\end{aligned}$$

$$+\frac{(\delta_i^m\,\delta_s^u - \frac{\delta^{uv}\,\partial_v\,\alpha^m}{\det(\partial_r\,\alpha^l)}\,\epsilon^{slt}\,\epsilon_{ipq}\,\partial_l\,\alpha^p\,\partial_t\,\alpha^q)\frac{\partial_u\,\alpha^n}{\det(\partial_r\,\alpha^l)}\,\Pi_m\,\Pi_n}{\sqrt{\mu^2c^2+\delta^{uv}\,\frac{\partial_u\,\alpha^m\,\partial_v\,\alpha^n}{[\det(\partial_r\,\alpha^k)]^2}\,\Pi_m\,\Pi_n}}\Bigg](\tau,\vec{\sigma})$$

$$+\lambda^r(\tau)\,\frac{\partial}{\partial\sigma^s}\Big[\epsilon^{suv}\,\partial_u\,\alpha^j\,\partial_v\,\alpha^k\,\frac{\partial_r\,\alpha^m}{\det(\partial_t\,\alpha^l)}\,\Pi_m$$

$$+(\delta_i^m\,\delta_{rs} - \frac{\partial_r\,\alpha^m}{\det(\partial_t\,\alpha^l)}\,\epsilon^{slt}\,\epsilon_{ipq}\,\partial_l\,\alpha^p\,\partial_t\,\alpha^q)\,\Pi_m\Big](\tau,\vec{\sigma}). \qquad (4.185)$$

In this special gauge we have $b^\mu_A \equiv L^\mu_A(p_s, \overset{\circ}{p}_s)$ (the Wigner boost) and $S_s^{\mu\nu} \equiv S_{sys}^{\mu\nu}$ ($S^r_{sys} = \epsilon^{ruv}\int d^3\sigma\,\sigma^u\Big(J^\tau\,(T^{-1})^{vs}\,\Pi_s\Big)(\tau,\vec{\sigma})$), and the only remaining canonical variables are the non-covariant Newton–Wigner-like canonical *external* canonical non-covariant 3-center-of-mass $\vec{z}_s$ and $\vec{k}_s$.

While the *external* realization of the Poincaré group is standard, the generators of the *internal* Poincaré group, built using the invariant mass M_{sys} and the spin tensor $\bar{S}_s^{AB}$, are

$$\begin{aligned}
P^\tau &= \bar{M}_{sys}c = \int d^3\sigma\,\Big[J^\tau\,\sqrt{\mu^2c^2+\delta_{uv}\,(T^{-1})^{ui}\,(T^{-1})^{vj}\,\Pi_i\,\Pi_j}\Big](\tau,\vec{\sigma})\\
&= \int d^3\sigma\,\sqrt{\mu^2c^2\,[\det(\partial_u\,\alpha^k)]^2+\delta^{uv}\,(\partial_u\,\alpha^i)\,(\partial_v\,\alpha^j)\,\Pi_i\,\Pi_j}(\tau,\vec{\sigma}),\\
P^r &= P^r_{sys} = -\int d^3\sigma\,\Big[J^\tau\,(T^{-1})^{ri}\,\Pi_i\Big](\tau,\vec{\sigma})\\
&= -\mu\int d^3\sigma\,\delta^{rs}\,[\partial_s\,\alpha^i\,\Pi_i](\tau,\vec{\sigma}) \approx 0,\\
K^r &= J^{\tau r} = \bar{S}_s^{\tau r} \equiv \int d^3\sigma\,\sigma^r\,\Big[J^\tau\,\sqrt{\mu^2c^2+\delta_{uv}\,(T^{-1})^{ui}\,(T^{-1})^{vj}\,\Pi_i\,\Pi_j}\Big](\tau,\vec{\sigma})\\
&= \int d^3\sigma\,\sigma^r\,\sqrt{\mu^2c^2\,[\det(\partial_u\,\alpha^k)]^2+\delta^{uv}\,(\partial_u\,\alpha^i)\,(\partial_v\,\alpha^j)\,\Pi_i\,\Pi_j}(\tau,\vec{\sigma}),\\
J^r &= S^r_{sys} = \frac{1}{2}\epsilon^{ruv}\,\bar{S}_s^{uv} \equiv \epsilon^{ruv}\int d^3\sigma\,\sigma^u\,\Big[J^\tau\,(T^{-1})^{vi}\,\Pi_i\Big](\tau,\vec{\sigma})\\
&= \mu\,\epsilon^{ruv}\int d^3\sigma\,\sigma^u\,\delta^{vs}\,[\partial_s\,\alpha^i\,\Pi_i](\tau,\vec{\sigma}).
\end{aligned} \qquad (4.186)$$

The natural gauge fixing to the rest-frame conditions $\vec{P}_{sys} \approx 0$ are $\vec{q}_+ \approx 0$, with the internal canonical 3-center of mass determined by the internal 3-center of energy: $\vec{q}_+ \approx \vec{R}_+ = -\frac{\vec{K}}{P^\tau} = -\frac{1}{2\,P^\tau}\int d^3\sigma\,\vec{\sigma}\,\Big[J^\tau\,\sqrt{\mu^2+\delta_{uv}\,(T^{-1})^{ui}\,(T^{-1})^{vj}\,\Pi_i\,\Pi_j}\Big]$ $(\tau,\vec{\sigma})$.

In general, the dust 4-velocity is different from both of the 4-velocities of the two congruences of observers connected with the 3+1 formalism.

4.6.2 The Relativistic Micro-Canonical Ensemble

In non-relativistic physics, the description of systems of interacting N point particles in the Euclidean 3-spaces of the inertial frames of Galilei space-time can be done without problems due to the existence of an absolute time. The non-relativistic center of mass and the generators of the Galilei group are well defined. As a consequence, there is a well-defined formulation of kinetic theory and statistical mechanics in non-relativistic inertial frames [229, 230]. However, the extension to non-inertial frames is mainly restricted to the region near the axis of rotating frames without a universally accepted formulation in arbitrarily moving frames, where the inertial forces are always long-range forces.

As a consequence, it was not possible to define a consistent relativistic kinetic theory and then relativistic statistical mechanics in the inertial frames of Minkowski space-time. Therefore, in Ref. [231], after a review of the quoted problems, it is said that it is only possible to define a relativistic kinetic theory of world-lines, but not of relativistic particles. All of the existing approaches [230, 232–239] (see also the review in Ref. [240]) either consider only free particles or make ad hoc ansatzs whose validity is out of control.

The 3+1 approach with parametrized Minkowski theory allowed one for the first time to define the relativistic micro-canonical ensemble in the Hamiltonian framework for systems of N particles interacting through either short- or long-range forces both in inertial frames and the non-inertial rest-frames of Minkowski space-time [215, 217] (see Ref. [216] for an extended version with extra results) and a "Lorentz-scalar micro-canonical temperature" $T_{(mc)}$[10] extending the non-relativistic results of Ref. [247] for the extended distribution function of the micro-canonical ensemble[11] in the presence of long-range forces.[12] When the forces in inertial frames are short range and the thermodynamical limit ($N, V \to \infty$ with N/V = const.) exists, one can define the relativistic canonical ensemble with a Lorentz-scalar temperature [253–261] (see these papers for the debate on the three possible transformation properties of the temperature under Lorentz transformations and for the existing definitions of relativistic thermodynamics ($T = T_{rest}$, $T = T_{rest}\,(1 - v^2/c^2)^{1/2}$, $T = T_{rest}\,(1 - v^2/c^2)^{-1/2}$)).

Furthermore, the Liouville operators for single particles and the one-particle distribution function in relativistic kinetic theory can be defined in the relativistic frames of the 3+1 approach.

Due to the non-covariance of the canonical relativistic center of mass, one must decouple it, and one must reformulate the dynamics in the relativistic inertial frames only in terms of canonical Wigner-covariant relative variables $\vec{\rho}_a(\tau)$, $\vec{\pi}_a(\tau)$, $a = 1, \ldots, N-1$. As a consequence, the distribution function of the relativistic micro-canonical ensemble is defined only after the decoupling

10 For the problems in the definition of this temperature, see Refs. [241–245] in the presence of short-range forces and Ref. [246] for the case of long-range forces.

11 See Ref. [248] for the case of the ideal relativistic quantum gas in relativistic inertial frames.

12 See Refs. [249–252] for the non-equivalence of micro-canonical and canonical ensembles in the presence of long-range forces.

of the center of mass and depends on the ten internal Poincaré generators, which are functions only of the relative variables. In relativistic non-inertial frames, the explicit form of the relative variables is not known, but action-at-a-distance potentials can be described by using the Synge world-function, like in GR [262]. As a consequence, the explicit form of the micro-canonical distribution in arbitrary non-inertial frames is not known explicitly.

In the non-relativistic limit, it is possible to get the (ordinary and extended) Newtonian micro-canonical ensembles both in inertial [263] and non-inertial rest-frames of Galilei space-time. They are functions of the generators of the Galilei group without a dependence on the center of mass in the Hamilton–Jacobi framework. However, now one can reintroduce the motion of the center of mass and recover the known definition of the distribution function.

Since in the non-inertial rest-frames both the Galilei or Poincaré generators are asymptotic constants of motion at spatial infinity, also in them the micro-canonical distribution function is time-independent, as in the inertial frames: Therefore, the standard passive viewpoint[13] on equilibrium in inertial frames can be extended also to non-inertial rest-frames, notwithstanding the fact that the inertial forces are long-range independent of the type of interparticle interactions.

To get a consistent relativistic formulation in the presence of every type of interaction, one has to use the 3+1 splitting method to eliminate the non-covariant non-local (therefore, non-measurable) 4-center of mass (described by the Jacobi data $\vec{z}$, $\vec{h}$) and to describe the particles with the Wigner 3-vectors $\vec{\eta}_i(\tau)$ and $\vec{\kappa}_i(\tau)$, $i = 1, \ldots, N$, as fundamental canonical variables together with the rest-frame conditions eliminating the internal 3-center of mass inside the instantaneous Wigner 3-spaces of the inertial rest-frame (centered on the Fokker–Pryce external 4-center of inertia with Jacobi 3-velocity $\vec{h} = 0$) and reformulating the dynamics in terms of relative variables $\vec{\rho}_a$ and $\vec{\pi}_a$.

The relativistic micro-canonical partition function is defined in terms of the internal Poincaré generators (Eq. 3.10) living inside the Wigner 3-spaces Σ_τ of the inertial rest-frame. It is a function of the internal energy $\mathcal{P}^\tau = Mc$, of the rest spin $\vec{\bar{S}}$ in the rest-frame and of the volume V. The natural volume V would be a spherical box centered on the Fokker–Pryce center of inertia in the Wigner three-space ($|\vec{\eta}_i(\tau)| \leq R$) identified by a characteristic function $\chi(V) = \prod_i \theta(R - |\vec{\eta}_i(\tau)|)$. Since the internal center of mass is eliminated, the characteristic function depends only on the relative variables, namely $\chi(V) = \prod_a \theta(2R - |\vec{\rho}_a|)$.

[13] This is a different problem from how to describe equilibrium in GR [264], where there are physical tidal degrees of freedom of the gravitational field, and the equivalence principle forbids the existence of global inertial frames.

The extended and ordinary partition functions of the micro-canonical ensemble are:

$$\tilde{Z}(\mathcal{E}, \vec{\mathcal{S}}, V, N) = \frac{1}{N!} \int \prod_i^{1...N} d^3\eta_i\, \chi(V) \int \prod_j^{1...N} d^3\kappa_j\, \delta(M\,c - \mathcal{E})$$

$$\delta^3(\vec{\bar{\mathcal{S}}} - \vec{\mathcal{S}})\, \delta^3(\vec{\mathcal{P}}_{(int)})\, \delta^3(\frac{\vec{\mathcal{K}}_{(int)}}{Mc})$$

$$= \frac{1}{N!} \int d^3\eta \prod_{a=1}^{N-1} d^3\rho_a\, \chi(V) \int \prod_b^{1...N-1}$$

$$d^3\pi_b\, J(\vec{\rho}_a, \vec{\pi}_a)\, \delta^3(\vec{\eta} - \vec{\eta}_+(\vec{\rho}_a, \vec{\pi}_a))$$

$$\delta(\hat{M}(\vec{\rho}_a, \vec{\pi}_a)\, c - \mathcal{E})\, \delta^3(\sum_{a=1}^{N-1} \vec{\rho}_a \times \vec{\pi}_a - \vec{\mathcal{S}}),$$

$$\tilde{Z}(\mathcal{E}, V, N) = \int d^3\mathcal{S}\, \tilde{Z}(\mathcal{E}, \vec{\mathcal{S}}, V, N)$$

$$= \frac{1}{N!} \int \prod_i^{1...N} d^3\eta_i\, \chi(V) \int \prod_j^{1...N}$$

$$d^3\kappa_j\, \delta(M\,c - \mathcal{E})\, \delta^3(\vec{\mathcal{P}}_{(int)})\, \delta^3(\frac{\vec{\mathcal{K}}_{(int)}}{Mc})$$

$$= \frac{1}{N!} \int d^3\eta \prod_{a=1}^{N-1} d^3\rho_a\, \chi(V) \int \prod_b^{1...N-1}$$

$$d^3\pi_b\, J(\vec{\rho}_a, \vec{\pi}_a)\, \delta^3(\vec{\eta} - \vec{\eta}_+(\vec{\rho}_a, \vec{\pi}_a))$$

$$\delta(\hat{M}(\vec{\rho}_a, \vec{\pi}_a)\, c - \mathcal{E}). \tag{4.187}$$

Their explicit form is not known except in the case of free particles.

Since the 3-vectors $\vec{\eta}_i(\tau)$, $\vec{\kappa}_i(\tau)$, $\vec{\rho}_a(\tau)$, $\vec{\pi}_a(\tau)$ are Wigner spin-1 3-vectors, the invariant mass Mc is a Lorentz scalar. $\vec{\bar{\mathcal{S}}}$, $\vec{\mathcal{P}}_{(int)}$, and $\vec{\mathcal{K}}_{(int)}/Mc$ are Wigner spin-1 3-vectors: Under a Lorentz transformation Λ, they undergo a Wigner rotation $R(\Lambda)$, so that expressions like $\delta^3(\vec{\mathcal{P}}_{(int)})$ are Lorentz scalars. Furthermore, the volume V is a Lorentz scalar because both $|\vec{\eta}_i(\tau)|$ and $|\vec{\rho}_a(\tau)|$ are Lorentz scalars. In an ordinary inertial frame centered on an inertial observer A with Cartesian 4-coordinates x^μ, the world-lines of the particles would be $\tilde{x}_i^\mu(x^o) = (x^o; \vec{\tilde{x}}_i(x^o))$, and a volume $\tilde{V}$ would be defined as a spherical box centered on A in the 3-spaces $\Sigma_{x^o=\text{const.}}$: $|\vec{\tilde{x}}_i(x^o)| < R$. The spherical box $\tilde{V}$ centered on A is the standard non-Lorentz-invariant type of volume. It is extremely complicated to make explicit the transition between the two formulations.

As a consequence, $\tilde{Z}(\mathcal{E}, V, N)$ is a Lorentz scalar, while one gets $\tilde{Z}(\mathcal{E}, \vec{\mathcal{S}}, V, N) \mapsto \tilde{Z}(\mathcal{E}, R(\Lambda)^{-1}\vec{\mathcal{S}}, V, N)$ under a Lorentz transformation.

For the relativistic ordinary and extended distribution functions, one has:

$$f_{(mc)}(\vec{\rho}_1, \ldots, \vec{\pi}_{N-1} | \mathcal{E}, V, N) = \tilde{Z}^{-1}(\mathcal{E}, V, N)\, \frac{\chi(V)}{N!}$$

$$\delta(M\,c - \mathcal{E})\, \delta^3(\vec{\mathcal{P}}_{(int)})\, \delta^3\left(\frac{\vec{\mathcal{K}}_{(int)}}{M_N c}\right),$$

$$\tilde{f}_{(mc)}(\vec{\rho}_1, \ldots, \vec{\pi}_{N-1} | \mathcal{E}, \vec{\mathcal{S}}, V, N) = \tilde{Z}^{-1}(\mathcal{E}, \vec{\mathcal{S}}, V, N)\, \frac{\chi(V)}{N!}$$

$$\delta(M\,c - \mathcal{E})\, \delta^3(\vec{\bar{\mathcal{S}}} - \vec{\mathcal{S}}) \tag{4.188}$$

$$\delta^3(\vec{\mathcal{P}}_{(int)})\, \delta^3\left(\frac{\vec{\mathcal{K}}_{(int)}}{Mc}\right). \tag{4.189}$$

$f_{(mc)}$ satisfies the Liouville theorem with $H = Mc$. Moreover, it is time-independent, $\partial_\tau f_{(mc)} = 0$, so that it is an equilibrium distribution function in relativistic statistical mechanics.

A manifestly covariant micro-canonical distribution function of the type $F_{(mc)}(\vec{\tilde{x}}_i(x^o), \vec{\tilde{p}}_i(x^o)|\mathcal{E}, V, N)$, i.e., depending on the world-lines and their momenta in an arbitrary Lorentz frame (like in all of the existing approaches) does not exist due to the non-covariance of the Jacobi data $\vec{z}$ of the canonical external 4-center of mass. In group-theoretical terms, the basic obstruction to get this type of distribution function is that the Poincaré energy cannot be written as the center-of-mass energy plus an internal energy, like in the case of the Galilei group.

In Refs. [214–216] there is the definition of the two micro-canonical partition functions in the relativistic non-inertial rest-frame and it is shown how to make the non-relativistic limit to the non-relativistic non-inertial rest-frame. The definition of the non-rest non-inertial frames is possible, but extremely complicated. In all of these cases, the system of particles is isolated inside the Lorentz-scalar non-dynamical volume V, determined only by the range of the forces. In going to non-inertial frames, only the variables defining the volume are passively changed with a mathematical transformation. This formulation cannot be used for a gas in a box: In this case, the isolated system should be composed by the gas plus a dynamically described box.

In the relativistic inertial rest-frame, the micro-canonical temperature $T_{(mc)}$, defined by $\frac{1}{k_B\, T_{(mc)}} = \frac{1}{\tilde{Z}(\mathcal{E},V,N)} \frac{\partial\, \tilde{Z}(\mathcal{E},V,N)}{\partial\, \mathcal{E}}|_{V,N} = \frac{\partial S}{\partial \mathcal{E}}|_{V,N}$, where $S = \frac{1}{N}\ln\tilde{Z}$ is the entropy. It is a Lorentz scalar, because the relativistic internal energy $\mathcal{E}$ is a Lorentz scalar like the internal energy Mc, and also the non-dynamical volume V is a frame-independent Lorentz scalar.

4.6.3 Steps towards Relativistic Statistical Mechanics

In Ref. [229] on the non-relativistic kinetic theory of diluted gas, one can introduce the (non-equilibrium) one-particle distribution function in terms of the particle positions and momenta. It satisfies the Boltzmann transport equation, which has the Maxwell Boltzmann solution in the case of free particles. The non-relativistic Boltzmann equation can be derived as an approximation starting from the coupled equation of motion for the s-particle distribution functions (the so-called Bogoliubov–Born–Green–Kirkwood–Yvon (BBGKY) hierarchy) by using the Liouville theorem. At the relativistic level, lacking a consistent theory for N interacting particles, the Boltzmann equation is postulated.

As shown in Ref. [217] with the 3+1 approach, one can define the relativistic s-particle distribution functions and their coupled equations of motion in a consistent way in the Wigner-covariant rest-frame instant form.

However, these equations do not produce a relativistic BBGKY hierarchy, from which the relativistic Boltzmann equation implied by the 3+1 approach with a decoupled external relativistic center of mass could be defined in the absence of external forces. This is due to the presence of potentials under the square-roots of the particle energies. As a consequence, a relativistic Boltzmann equation emerging from this framework is expected to be more complex than the standard form [266–269]: This is one of the main open problems in relativistic statistical mechanics.

Moreover, in Ref. [217] there is a review of the papers concerning a hydrodynamical description of relativistic kinetic theory by means of an effective dissipative fluid, and there is a sketch of how this description can be formulated in terms of relativistic dissipative fluids described by an action principle in the relativistic inertial rest-frame.

Part II

General Relativity

Globally Hyperbolic Einstein Space-Times

In the years 1913–1916, Einstein developed general relativity (GR), relying on the equivalence principle (equality of inertial and gravitational masses of bodies in free fall) forbidding the existence of global inertial frames (see Refs. [4–11, 14–17]). It suggested to him the impossibility of distinguishing a uniform gravitational field from the effects of constant acceleration by means of local experiments in sufficiently small regions where the effects of tidal forces are negligible. This led to the geometrization of the gravitational interaction and to the replacement of Minkowski space-time with a pseudo-Riemannian 4-manifold M^4 with non-vanishing curvature Riemann tensor. The principle of general covariance (see Ref. [270] for a review), as the basis of the tensorial nature of Einstein's equations, has the two following consequences:

1. invariance of the Hilbert action under passive diffeomorphisms (the coordinate transformations in M^4), so that the second Noether theorem implies the existence of first-class constraints at the Hamiltonian level (see Chapter 9);
2. mapping of solutions of Einstein's equations among themselves under the action of active diffeomorphisms of M^4 extended to the tensors over M^4 (dynamical symmetries of Einstein's equations).

The basic field of metric gravity is the 4-metric tensor with components ${}^4g_{\mu\nu}(x)$ in an arbitrary coordinate system of M^4. The peculiarity of gravity is that the 4-metric field, differently from the fields of electromagnetic, weak, and strong interactions and from the matter fields, has a double role:

1. it is the mediator of the gravitational interaction (in analogy to all the other gauge fields);
2. it determines the chrono-geometric structure of the space-time M^4 in a dynamical way through the line element $ds^2 = {}^4g_{\mu\nu}(x)\, dx^\mu\, dx^\nu$.

Let us make a comment about the two main existing approaches to the quantization of gravity.

1. *Effective quantum field theory and string theory* [147]. This approach contains the standard model of elementary particles and its extensions. However, since the quantization, namely the definition of the Fock space, requires a background space-time where it is possible to define creation and annihilation operators, one must use the splitting ${}^4g_{\mu\nu} = {}^4\eta^{(B)}_{\mu\nu} + {}^4h_{\mu\nu}$ and quantize only the perturbation ${}^4h_{\mu\nu}$ of the background 4-metric $\eta^{(B)}_{\mu\nu}$ (usually B is either Minkowski or DeSitter space-time). In this way property 2 is lost (one uses the fixed non-dynamical chrono-geometrical structure of the background space-time), gravity is replaced by a field of spin-2 over the background (and passive diffeomorphisms are replaced by gauge transformations acting in an inner space), and the only difference among gravitons, photons, and gluons lies in their quantum numbers.
2. *Loop quantum gravity* [18, 19]. This approach never introduces a background space-time, but being inequivalent to a Fock space, it has problems incorporating particle physics. It uses a fixed 3+1 splitting of the space-time M^4 and is a quantization of the associated instantaneous 3-spaces Σ_τ (quantum geometry). However, there is no known way to implement a consistent unitary evolution (the problem of the super-Hamiltonian constraint) and, since it is usually formulated in spatially compact space-times without boundary, there is no notion of a Poincaré group (and therefore no extra dimensions) and a problem of time (frozen picture without evolution).

Let us remark that in all known formulations, particle and nuclear physics are a chapter of the theory of representations of the Poincaré group in inertial frames in the spatially non-compact Minkowski space-time. As a consequence, if one looks at GR from the point of view of particle physics, the main problem getting a unified theory is how to reconcile the Poincaré group (the kinematical group of transformations connecting inertial frames) with the diffeomorphism group, implying the non-existence of global inertial frames in GR (special relativity [SR] holds only in a small neighborhood of a body in free fall).

While in SR Minkowski space-time is an absolute notion, unifying the absolute notions of time and 3-space of the non-relativistic Galilei space-time, in Einstein GR also the space-time is a dynamical object [271–274] and the gravitational field is described by the metric structure of the space-time, namely by the ten dynamical fields ${}^4g_{\mu\nu}(x)$ (x^μ are world 4-coordinates) satisfying Einstein equations.

The ten dynamical fields ${}^4g_{\mu\nu}(x)$ are not only a (pre)potential for the gravitational field (like the electromagnetic and Yang–Mills fields are the potentials for electromagnetic and non-Abelian forces) but as already said they determine the "chrono-geometrical structure" of space-time through the line element. Therefore, the 4-metric teaches relativistic causality to the other fields: It says to massless particles like photons and gluons what are the allowed world-lines in each point of space-time. The ACES mission of ESA [275–277] will give the first precision measurement of the gravitational red-shift of the geoid, namely of the $1/c^2$ deformation of Minkowski light-cone caused by the geo-potential.

5

Hamiltonian Gravity in Einstein Space-Times

In this chapter I will show which is the class of space-times admitting 3+1 splittings so that one can formulate ADM canonical gravity [278, 279] and then reformulate it in the non-inertial rest-frames of the 3+1 instant form of dynamics [95–97, 280–287] and [48] (see Refs. [143–146, 288] for reviews), having the same rest-frames as limit when the Newton constant is switched off.[1] For the background in differential geometry, see Refs. [292, 293].

5.1 Global 3+1 Splittings of Globally Hyperbolic Space-Times without Super-Translations and Asymptotically Minkowskian at Spatial Infinity Admitting a Hamiltonian Formulationof Gravity

We shall restrict ourselves to the simplest class of space-times allowing a consistent 3+1 approach. These space-times are time-oriented, orientable, topologically trivial (with the leaves of each 3+1 splitting diffeomorphic to R^3, so that they admit global coordinate charts), torsion-free, pseudo-Riemannian or Lorentzian 4-manifold $(M^4, {}^4g)$ with signature $\epsilon\,(+---)$ $(\epsilon = \pm 1)$ with a choice of time orientation (i.e., there exists a continuous, non-vanishing time-like vector field which is used to separate the non-space-like vectors at each point of M^4 in either future- or past-directed vectors).

Our space-times are assumed to be the following:

1. Globally hyperbolic 4-manifolds, i.e., topologically they are $M^4 = R \times \Sigma$, so have a well-posed Cauchy problem (with Σ the abstract model of Cauchy surface) at least till no singularity develops in M^4. Therefore, these space-times admit regular foliations with orientable, complete, non-intersecting space-like

[1] This is the deparametrization problem of general relativity (GR), only partially solved in Refs. [289–291] by using coordinate gauge conditions.

3-manifolds: The leaves of the foliation are the embeddings i_τ: $\Sigma \to \Sigma_\tau \subset M^4$, $(\tau, \sigma^u) \mapsto x^\mu = z^\mu(\tau, \sigma^u)$, where σ^r, $r = 1, 2, 3$, are local coordinates in a chart of the C^∞-atlas of the abstract 3-manifold Σ, x^μ are local coordinates in M^4, and τ: $M^4 \to R$, $z^\mu \mapsto \tau(z^\mu)$ is a global time-like future-oriented function labeling the leaves (surfaces of simultaneity), namely the 3-spaces Σ_τ. In this way, one obtains 3+1 splittings of M^4 and the possibility of a Hamiltonian formulation.

2. Asymptotically flat at spatial infinity, so as to have the possibility to define asymptotic Poincaré charges [294–300], which should reduce to the ten Poincaré generators of the isolated system when the Newton constant is switched off. They allow the definition of a Møller radius in GR (it opens the possibility of defining an intrinsic ultraviolet cutoff in canonical quantization) and are a bridge towards a future soldering with the theory of elementary particles in Minkowski space-time defined as irreducible representation of its kinematical, globally implemented Poincaré group according to Wigner. We will not compactify space infinity at a point like in the spatial infinity (SPI) approach of Ref. [301].
3. Since we want to be able to introduce Dirac fermion fields, our space-times M^4 must admit a spinor (or spin) structure [4]. Since we consider non-compact space- and time- and space-orientable space-times, spinors can be defined if and only if they are parallelizable [302]. This means that we have a trivial principal frame bundle $L(M^4) = M^4 \times GL(4, R)$ with $GL(4, R)$ as the structure group and a trivial orthonormal frame bundle $F(M^4) = M^4 \times SO(3, 1)$; the fibers of $F(M^4)$ are the disjoint union of four components and $F_o(M^4) = M^4 \times L_+^\uparrow$ (with projection π: $F_o(M^4) \to M^4$) corresponds to the proper subgroup $L_+^\uparrow \subset SO(3, 1)$ of the Lorentz group. Therefore, global frames (tetrads) and coframes (cotetrads) exist. A spin structure for $F_o(M^4)$ is, in this case, the trivial spin principal $SL(2, C)$-bundle $S(M^4) = M^4 \times SL(2, C)$ (with projection π_s: $S(M^4) \to M^4$) and a map λ: $S(M^4) \to F_o(M^4)$ such that $\pi(\lambda(p)) = \pi_s(p) \in M^4$ for all $p \in S(M^4)$ and $\lambda(pA) = \lambda(p)\Lambda(A)$ for all $p \in S(M^4)$, $A \in SL(2, C)$, with Λ: $SL(2, C) \to L_+^\uparrow$ the universal covering homomorphism. Then, Dirac fields are defined as cross-sections of a bundle associated with $S(M^4)$ [303]. $F(\Sigma_\tau) = \Sigma_\tau \times SO(3)$ is the trivial orthonormal frame $SO(3)$-bundle and, since $\pi_1(SO(3)) = \pi_1(L_+^\uparrow) = Z_2$, one can define $SU(2)$ spinors on Σ_τ (see Refs. [304–306] for the 3+1 decomposition of $SL(2, C)$ spinors of M^4 in terms of $SU(2)$ (the covering group of $SO(3)$) spinors of Σ_τ).
4. The non-compact parallelizable simultaneity 3-manifolds (the Cauchy surfaces), i.e., the 3-spaces Σ_τ are assumed to be topologically trivial, geodesically complete (so that the Hopf–Rinow theorem [292, 293] assures metric completeness of the Riemannian 3-manifold $(\Sigma_\tau, {}^3g)$) and, finally, diffeomorphic to R^3 (so that they may have global charts). These 3-manifolds have the same manifold structure as Euclidean spaces [292, 293]: (a) the geodesic

exponential map $\exp_p : T_p\Sigma_\tau \to \Sigma_\tau$ is a diffeomorphism (Hadamard theorem); (b) the sectional curvature is less than or equal to zero everywhere; and (c) they have no conjugate locus (i.e., there are no pairs of conjugate Jacobi points [intersection points of distinct geodesics through them] on any geodesic) and no cut locus (i.e., no closed geodesics through any point). In these manifolds, two points determine a line, so that the static tidal forces in Σ_τ due to the 3-curvature tensor are repulsive; instead in M^4 the tidal forces due to the 4-curvature tensor are attractive, since they describe gravitation, which is always attractive, and this implies that the sectional 4-curvature of time-like tangent planes must be negative (this is the source of the singularity theorems) [292, 293]. In 3-manifolds not of this class one has to give a physical interpretation of static quantities like the two quoted loci.

5. Like in Yang–Mills case [93, 94], the 3-spin connection on the orthogonal frame $SO(3)$-bundle (and therefore triads and cotriads) will have to be restricted to suited weighted Sobolev spaces to avoid Gribov ambiguities. In turn, this implies the absence of isometries of the non-compact Riemannian 3-manifold $(\Sigma_\tau, {}^3g)$ (see, for instance, the review paper in Ref. [307]).
6. Diffeomorphisms on Σ_τ ($\mathrm{Diff}\,\Sigma_\tau$) will be interpreted in the passive way, following Refs. [308–311], in accord with the Hamiltonian point of view that infinitesimal diffeomorphisms are generated by taking the Poisson bracket with the first-class diffeomorphism (or super-momentum) constraints. The Lagrangian approach based on the Hilbert action connects general covariance with the invariance of the action under space-time diffeomorphisms ($\mathrm{Diff}\,M^4$). Therefore, the moduli space (or super-space or space of 4-geometries) is the space $\mathrm{Riem}\,M^4/\mathrm{Diff}\,M^4$ [312, 313], where $\mathrm{Riem}\,M^4$ is the space of Lorentzian 4-metrics. As shown in Refs. [314–320], super-space, in general, is not a manifold (it is a stratified manifold with singularities [321]) due to the existence (in Sobolev spaces) of 4-metrics and 4-geometries with isometries. See Ref. [322] for the study of great diffeomorphisms, which are connected with the existence of disjoint components of the diffeomorphism group.

   Instead, in the ADM Hamiltonian formulation of metric gravity [278, 279] space diffeomorphisms are replaced by $\mathrm{Diff}\,\Sigma_\tau$, while time diffeomorphisms are distorted to the transformations generated by the super-Hamiltonian first-class constraint [289–291, 323, 324].

   In the Lichnerowicz–ChoquetBruhat–York conformal approach to canonical reduction [325–334] (see Refs. [10, 307, 335] for reviews), one defines, in the case of closed 3-manifolds, the conformal super-space as the space of conformal 3-geometries (namely the space of conformal 3-metrics modulo $\mathrm{Diff}\,\Sigma_\tau$ or, equivalently, as $\mathrm{Riem}\,\Sigma_\tau$ (the space of Riemannian 3-metrics) modulo $\mathrm{Diff}\,\Sigma_\tau$ and conformal transformations ${}^3g \mapsto \phi^4\,{}^3g$ ($\phi > 0$)), because in this approach gravitational dynamics is regarded as the time evolution of conformal 3-geometry (the momentum conjugate to the conformal factor ϕ

is proportional to York time [328–334, 336], i.e., to the trace of the extrinsic curvature of Σ_τ).

7. It is known that at spatial infinity the group of asymptotic symmetries (direction-dependent asymptotic Killing vectors) is the infinite dimensional SPI group [298–300, 337–341]. Besides an invariant 4-dimensional subgroup of translations, it contains an infinite number of Abelian "super-translations." This forbids the identification of a unique Lorentz subgroup.[2] The presence of super-translations is an obstruction to the definition of angular momentum in GR [4, 342, 343] and there is no idea how to measure this infinite number of constants of motion if they are allowed to exist. Therefore, suitable boundary conditions at spatial infinity have to be assumed to kill the super-translations. In this way the SPI group is reduced to a well-defined asymptotic ADM Poincaré group. A convenient set of boundary conditions is obtained by assuming that the coordinate atlas of the space-time is restricted in such a way that the 4-metric always tends to the Minkowski metric in Cartesian coordinates at spatial infinity with the 3-metric on each space-like hyper-surface associated with the allowed 3+1 splittings becoming Euclidean at spatial infinity in a "direction-independent" way [280]. Then, this last property is assumed also for all the other Hamiltonian variables like the lapse and shift functions. These latter variables are assumed to be the sum of their asymptotic part (growing linearly in the 3-coordinates on the 3-space [294, 295]), plus a bulk part with the quoted property. The final result of all these requirements is a set of boundary conditions compatible with Christodoulou–Klainermann space-times [8]. As shown in Refs. [280, 282], these boundary conditions are ${}^3g_{rs}(\tau,\sigma^u) \to (1 + \frac{\text{const.}}{r}\,\delta_{rs} + O(r^{-3/2})$, $N\tau,\sigma^u) = 1 + n(\tau,\sigma^u) \to 1 + O(r^{-(2+\epsilon)})$, $N_r(\tau,\sigma^u) \to 0 + O(r^{-\epsilon})$ for $r = \sqrt{\sum_u(\sigma^u)^2} \to \infty$ and $\epsilon > 0$. See Section 8.3 for a discussion of all the problems in the definition and interpretation of gauge transformations in field theory and GR leading to this choice of boundary conditions.

As a consequence, the allowed 3+1 splittings of space-time have all the space-like leaves approaching Minkowski space-like hyper-planes at spatial infinity in a direction-independent way and, as shown in Refs. [280, 281], the absence of super-translations implies that these asymptotic hyper-planes are orthogonal to the weak (namely the volume form[3] of the) ADM 4-momentum, for those space-times for which it is time-like. Therefore, these hyper-planes reduce to the Wigner hyper-planes in Minkowski space-time when the Newton constant is switched off. To arrive at these results, Dirac's strategy [20, 21, 344] of adding ten extra degrees of freedom (tetrads) at spatial infinity and then adding ten

[2] Only an abstract Lorentz group appears from the quotient of the SPI group with respect to the invariant subgroup of all translations and super-translations.

[3] See Section 9.4 for the distinction between weak and strong charges.

first-class constraints so that the new degrees of freedom are gauge variables, will be followed.

In this way, the non-inertial rest-frame instant form of metric gravity may be defined. The weak ADM energy turns out to play the role of the canonical Hamiltonian for the evolution in the scalar mathematical time, labeling the leaves of the 3+1 splitting (consistently with Ref. [345]). There will be a point near spatial infinity playing the role of the decoupled canonical center of mass of the universe and which can be interpreted as a point-particle clock for the mathematical time. There will be three first-class constraints implying the vanishing of the weak ADM 3-momentum – they define the rest-frame of the universe. Therefore, at spatial infinity we have inertial observers whose unit 4-velocity is determined by the time-like ADM 4-momentum. Modulo 3-rotations, these observers carry an asymptotic tetrad adapted to the asymptotic space-like hyper-planes: After a conventional choice of the 3-rotation this tetrad defines the dynamical fixed stars (standard of non-rotation). By using Frauendiener's reformulation [346] of Sen–Witten equations [347–349] for the triads and the adapted tetrads on a space-like hyper-surface of this kind, one can determine preferred dynamical adapted tetrads[4] in each point of the hyper-surface (they are dynamical because the solution of Einstein's equations is needed to find them). Therefore, these special space-like hyper-surfaces can be named Wigner–Sen–Witten (WSW) hyper-surfaces: As already said, they reduce to the Wigner hyper-planes of the rest-frames of Minkowski space-time when the Newton constant is switched off. When the space-time is restricted to the Minkowski space-time with Cartesian coordinates, the rest-frame instant form of ADM canonical gravity in the presence of matter reduces to the Minkowski rest-frame instant form description of the same matter when the Newton constant is switched off.

5.2 The ADM Hamiltonian Formulation of Einstein Gravity and the Asymptotic ADM Poincaré Generators in the Non-Inertial Rest-Frames

Let us reformulate ADM canonical gravity in the non-inertial rest-frame identified in the 3+1 splitting of the space-times chosen in the previous section as a generalization of the non-inertial rest-frames of Minkowski space-time. See Refs. [4, 8–10] for the standard formulation of GR.

Let M^4 be foliated with space-like Cauchy hyper-surfaces Σ_τ (the instantaneous 3-space) through the embeddings $x^\mu = z^\mu(\tau, \sigma^u)$ (x^μ are local coordinates of M^4) of a 3-manifold Σ, assumed diffeomorphic to R^3, into M^4,

[4] They seem to be the natural realization of the non-flat preferred observers of Bergmann [308–311].

where $\sigma^A = (\tau, \sigma^u) = \sigma^A(x^\mu)$ are the radar 4-coordinates introduced in Section 2.1 for the 3+1 approach to Minkowski space-time (from now on indices α, u of the flat Minkowski space-time will be denoted with (α), (u)).

The Σ_τ-adapted holonomic coordinate bases $\partial_A = \frac{\partial}{\partial\sigma^A} \mapsto b^\mu_A(\sigma)\partial_\mu = \frac{\partial z^\mu(\sigma)}{\partial\sigma^A}\partial_\mu$ for vector fields, and $dx^\mu \mapsto d\sigma^A = b^A_\mu(\sigma)dx^\mu = \frac{\partial\sigma^A(z)}{\partial z^\mu}dx^\mu$ for differential 1-forms are used. In flat Minkowski space-time the transformation coefficients $b^A_\mu(\tau,\sigma^u)$ and $b^\mu_A(\tau,\sigma^u)$ become the flat orthonormal tetrads $\delta^{(\mu)}_\mu z^A_{(\mu)}(\sigma) = \frac{\partial\sigma^A(x)}{\partial x^\mu}|_{x=z(\sigma)}$ and cotetrads $\delta^\mu_{(\mu)} z^{(\mu)}_A(\sigma) = \frac{\partial z^\mu(\sigma)}{\partial\sigma^A}$ of Section 2.2 [105, 350].

The induced 4-metric and inverse 4-metric become in the new basis (${}^4g_{AB}(\tau,\sigma^u) = b^\mu_A(\tau,\sigma^u) b^\nu_B(\tau,\sigma^u)\, {}^4g_{\mu\nu}(z(\tau,\sigma^u))$; $l^\mu(\tau,\sigma^u)$ is the unit normal to the 3-space Σ_τ in the point (τ,σ^u))

$$
\begin{aligned}
{}^4g(x = z(\tau,\sigma^u)) &= {}^4g_{\mu\nu}(x = z(\tau,\sigma^u))dx^\mu \otimes dx^\nu = {}^4g_{AB}(\tau,\sigma^u)\, d\sigma^A \otimes d\sigma^B,\\
{}^4g_{\mu\nu}(x = z(\tau,\sigma^u)) &= (b^A_\mu\, {}^4g_{AB} b^B_\nu)(\tau,\sigma^u)\\
&= \Big[\epsilon\,(N^2 - {}^3g_{rs}N^rN^s)\partial_\mu\tau(x)\,\partial_\nu\tau(x)\\
&\quad -\epsilon\,{}^3g_{rs}N^s(\partial_\mu\tau(x)\,\partial_\nu\sigma^r(x) + \partial_\nu\tau(x)\,\partial_\mu\sigma^r(x))\\
&\quad -\epsilon\,{}^3g_{rs}\,\partial_\mu\sigma^r(x)\,\partial_\nu\sigma^s(x)\Big](\tau,\sigma^u)\\
&= \Big[\epsilon\, l_\mu l_\nu - \epsilon\,{}^3g_{rs}(\partial_\mu\sigma^r(x) + N^r\,\partial_\mu\tau(x))(\partial_\nu\sigma^s(\tau)\\
&\quad +N^s\,\partial_\nu\tau(x))\Big](\tau,\sigma^u),\\
&\Rightarrow {}^4g_{AB}(\tau,\sigma^u) = \Big({}^4g_{\tau\tau} = \epsilon(N^2 - {}^3g_{rs}N^rN^s);\, {}^4g_{\tau r}\\
&= -\epsilon\,{}^3g_{rs}\,N^s;\, {}^4g_{rs} = -\epsilon\,{}^3g_{rs}\Big)(\tau,\sigma^u)\\
&= \epsilon[l_Al_B - {}^3g_{rs}(\delta^r_A + N^r\delta^\tau_A)(\delta^s_B + N^s\delta^\tau_B)](\tau,\sigma^u),\\
{}^4g^{\mu\nu}(x = z(\tau,\sigma^u)) &= \Big(b^\mu_A\, {}^4g^{AB} b^\nu_B\Big)(\tau,\sigma^u)\\
&= \Big[\frac{\epsilon}{N^2}\partial_\tau z^\mu\partial_\tau z^\nu - \frac{\epsilon\, N^r}{N^2}(\partial_\tau z^\mu\partial_r z^\nu + \partial_\tau z^\nu\partial_r z^\mu)\\
&\quad -\epsilon\Big({}^3g^{rs} - \frac{N^rN^s}{N^2}\Big)\partial_r z^\mu\partial_s z^\nu\Big](\tau,\sigma^u)\\
&= \epsilon[\,l^\mu l^\nu - {}^3g^{rs}\partial_r z^\mu\partial_s z^\nu],\\
&\Rightarrow {}^4g^{AB}(\tau,\sigma^u) = \Big({}^4g^{\tau\tau} = \frac{\epsilon}{N^2};\, {}^4g^{\tau r} = -\frac{\epsilon\, N^r}{N^2};\, {}^4g^{rs}\\
&= -\epsilon\Big({}^3g^{rs} - \frac{N^rN^s}{N^2}\Big)\Big)(\tau,\sigma^u)\\
&= \epsilon[l^Al^B - {}^3g^{rs}\delta^A_r\delta^B_s](\tau,\sigma^u),\\
l^A(\tau,\sigma^u) &= \Big(l^\mu b^A_\mu\Big)(\tau,\sigma^u) = (N\,{}^4g^{A\tau})(\tau,\sigma^u) = \frac{\epsilon}{N(\tau,\sigma^u)}(1; -N^r(\tau,\sigma^u)),\\
l_A(\tau,\sigma^u) &= (l_\mu b^\mu_A)(\tau,\sigma^u) = N(\tau,\sigma^u)\,\delta^\tau_A,\\
b^\mu_\tau(\tau,\sigma^u) &= \Big(N\, l^\mu + N^r\, b^\mu_r\Big)(\tau,\sigma^u).
\end{aligned} \tag{5.1}
$$

Here, the 3-metric ${}^3g_{rs}(\tau,\sigma^u) = -\epsilon\,{}^4g_{rs}(\tau,\sigma^u)$, with signature $(+++)$, of Σ_τ was introduced. If ${}^4\gamma^{rs}(\tau,\sigma^u)$ is the inverse of the spatial part of the 4-metric $[{}^4\gamma^{ru}(\tau,\sigma^u)\,{}^4g_{us}(\tau,\sigma^u) = \delta^r_s]$, the inverse of the 3-metric is ${}^3g^{rs}(\tau,\sigma^u) = -\epsilon\,{}^4\gamma^{rs}(\tau,\sigma^u)$ $[{}^3g^{ru}(\tau,\sigma^u)\,{}^3g_{us}(\tau,\sigma^u) = \delta^r_s]$. ${}^3g_{rs}(\tau,\sigma^u)$ are the components of the first fundamental form of the Riemann 3-manifold $(\Sigma_\tau, {}^3g)$, and the line element of M^4 is

$$\begin{aligned} ds^2 &= {}^4g_{\mu\nu}dx^\mu dx^\nu \\ &= \epsilon\Big[N^2(\tau,\sigma^u)(d\tau)^2 - {}^3g_{rs}(\tau,\sigma^u)\,(d\sigma^r + N^r(\tau,\sigma^u)\,d\tau)\,(d\sigma^s + N^s(\tau,\sigma^u)\,d\tau)\Big]. \end{aligned} \tag{5.2}$$

It must be $\epsilon\,{}^4g_{oo} > 0$, $\epsilon\,{}^4g_{ij} < 0$, $\begin{vmatrix} {}^4g_{ii} & {}^4g_{ij} \\ {}^4g_{ji} & {}^4g_{jj} \end{vmatrix} > 0$, $\epsilon\det{}^4g_{ij} > 0$ for each value of $x^\mu = z^\mu(\tau,\sigma^u)$.

Defining $g(\tau,\sigma^u) = {}^4g(\tau,\sigma^u) = |\det({}^4g_{\mu\nu}(\tau,\sigma^u))|$ and $\gamma(\tau,\sigma^u) = {}^3g(\tau,\sigma^u) = |\det({}^3g_{rs}(\tau,\sigma^u))|$, the lapse and shift functions assume the following form:

$$\begin{aligned} N(\tau,\sigma^u) &= \sqrt{\frac{{}^4g(\tau,\sigma^u)}{{}^3g(\tau,\sigma^u)}} = \frac{1}{\sqrt{{}^4g^{\tau\tau}(\tau,\sigma^u)}} = \sqrt{\frac{g(\tau,\sigma^u)}{\gamma(\tau,\sigma^u)}} \\ &= \sqrt{({}^4g_{\tau\tau} - \epsilon\,{}^3g^{rs}\,{}^4g_{\tau r}{}^4g_{\tau s})(\tau,\sigma^u)}, \\ N^r(\tau,\sigma^u) &= -\epsilon\,{}^3g^{rs}(\tau,\sigma^u)\,{}^4g_{\tau s}(\tau,\sigma^u) = \frac{{}^4g^{\tau r}(\tau,\sigma^u)}{{}^4g^{\tau\tau}(\tau,\sigma^u)}, \\ N_r(\tau,\sigma^u) &= {}^3g_{rs}(\tau,\sigma^u)\,N^s(\tau,\sigma^u)(\tau,\sigma^u) \\ &= -\epsilon\,{}^4g_{rs}(\tau,\sigma^u)\,N^s(\tau,\sigma^u) = -\epsilon{}^4g_{\tau r}(\tau,\sigma^u). \end{aligned} \tag{5.3}$$

See Refs. [27, 351–356] for the 3+1 decomposition of 4-tensors on M^4. The horizontal projector ${}^3h^\nu_\mu(\tau,\sigma^u) = \delta^\nu_\mu - \epsilon\,l_\mu(\tau,\sigma^u)\,l^\nu(\tau,\sigma^u)$ on Σ_τ defines the 3-tensor fields on Σ_τ, starting from the 4-tensor fields on M^4.

In the standard non-Hamiltonian description of the 3+1 decomposition a Σ_τ-adapted non-holonomic non-coordinate basis $[\bar{A} = (l; r)]$ is used:

$$\begin{aligned} \hat{b}^\mu_{\bar{A}}(\tau,\sigma^u) &= \{\hat{b}^\mu_l(\tau,\sigma^u) = \epsilon\,l^\mu(\tau,\sigma^u) = N^{-1}(\tau,\sigma^u)[b^\mu_\tau(\tau,\sigma^u) \\ &\quad - N^r(\tau,\sigma^u)b^\mu_r(\tau,\sigma^u)];\, \hat{b}^\mu_r(\tau,\sigma^u) = b^\mu_r(\tau,\sigma^u)\}, \\ \hat{b}^{\bar{A}}_\mu(\tau,\sigma^u) &= \{\hat{b}^l_\mu(\tau,\sigma^u) = l_\mu(\tau,\sigma^u) = N(\tau,\sigma^u)b^\tau_\mu(\tau,\sigma^u) \\ &= N(\tau,\sigma^u)\partial_\mu\tau(z(\tau,\sigma^u));\, \hat{b}^r_\mu(\tau,\sigma^u) \\ &= b^r_\mu(\tau,\sigma^u) + N^r(\tau,\sigma^u)b^\tau_\mu(\tau,\sigma^u)\}, \\ &\quad \hat{b}^{\bar{A}}_\mu(\tau,\sigma^u)\hat{b}^\nu_{\bar{A}}(\tau,\sigma^u) = \delta^\nu_\mu, \quad \hat{b}^{\bar{A}}_\mu(\tau,\sigma^u)\hat{b}^\mu_{\bar{B}}(\tau,\sigma^u) = \delta^{\bar{A}}_{\bar{B}}, \\ {}^4\bar{g}_{\bar{A}\bar{B}}(z(\tau,\sigma^u)) &= \hat{b}^\mu_{\bar{A}}(\tau,\sigma^u){}^4g_{\mu\nu}(z(\tau,\sigma^u))\hat{b}^\nu_{\bar{B}}(\tau,\sigma^u) \\ &= \{{}^4\bar{g}_{ll}(\tau,\sigma^u) = \epsilon;\, {}^4\bar{g}_{lr}(\tau,\sigma^u) = 0;\, {}^4\bar{g}_{rs}(\tau,\sigma^u) \\ &= {}^4g_{rs}(\tau,\sigma^u) = -\epsilon\,{}^3g_{rs}(\tau,\sigma^u)\}, \\ {}^4\bar{g}^{\bar{A}\bar{B}}(\tau,\sigma^u) &= \{{}^4\bar{g}^{ll}(\tau,\sigma^u) = \epsilon;\, {}^4\bar{g}^{lr}(\tau,\sigma^u) = 0;\, {}^4\bar{g}^{rs}(\tau,\sigma^u) \end{aligned}$$

$$\begin{aligned}
&= {}^4\gamma^{rs}(\tau,\sigma^u) = -\epsilon {}^3g^{rs}(\tau,\sigma^u)\},\\
X_{\bar A} &= \hat b^\mu_{\bar A}(\tau,\sigma^u)\partial_\mu = \left\{X_l = \frac{1}{N(\tau,\sigma^u)}(\partial_\tau - N^r(\tau,\sigma^u)\,\partial_r); \partial_r\right\},\\
\theta^{\bar A} &= \hat b^{\bar A}_\mu(x(\tau,\sigma^u))\,dx^\mu = \{\theta^l(\tau,\sigma^u) = N(\tau,\sigma^u)\,d\tau; \theta^r(\tau,\sigma^u)\\
&= d\sigma^r + N^r(\tau,\sigma^u)\,d\tau\},\\
&\Rightarrow l_\mu(\tau,\sigma^u)\,b^\mu_r(\tau,\sigma^u) = 0, \quad l^\mu(\tau,\sigma^u) b^r_\mu(\tau,\sigma^u)\\
&= -N^r(\tau,\sigma^u)/N(\tau,\sigma^u),\\
l^{\bar A}(\tau,\sigma^u) &= l^\mu(\tau,\sigma^u)\hat b^{\bar A}_\mu(\tau,\sigma^u) = (\epsilon; l^r(\tau,\sigma^u) + N^r(\tau,\sigma^u)\,l^\tau(\tau,\sigma^u)) = (\epsilon;\vec 0),\\
l_{\bar A}(\tau,\sigma^u) &= l_\mu(\tau,\sigma^u)\,\hat b^\mu_{\bar A}(\tau,\sigma^u) = (1; l_r(\tau,\sigma^u)) = (1;\vec 0). \qquad (5.4)
\end{aligned}$$

One has ${}^3h_{\mu\nu}(\tau,\sigma^u) = {}^4g_{\mu\nu}(\tau,\sigma^u) - \epsilon\, l_\mu(\tau,\sigma^u)\, l_\nu(\tau,\sigma^u) = -\epsilon \Big[{}^3g_{rs}(b^r_\mu + N^r b^\tau_\mu)(b^s_\mu + N^s b^\tau_\mu)\Big](\tau,\sigma^u) = -\epsilon\, {}^3g_{rs}(\tau,\sigma^u)\,\hat b^r_\mu(\tau,\sigma^u)\,\hat b^s_\nu(\tau,\sigma^u)$. For a 4-vector ${}^4V^\mu = {}^4V^{\bar A}\hat b^\mu_{\bar A} = {}^4V^l l^\mu + {}^4V^r \hat b^\mu_r$ one gets ${}^3V^\mu = {}^3V^r \hat b^\mu_r = {}^3h^\mu_\nu\, {}^4V^\nu$, ${}^3V^r = {}^4V^r = \hat b^r_\mu\, {}^3V^\mu$ for each value of (τ,σ^u).

The non-holonomic basis in Σ_τ-adapted coordinates is

$$\begin{aligned}
\hat b^{\bar A}_A(\tau,\sigma^u) &= \hat b^{\bar A}_\mu(\tau,\sigma^u)\, b^\mu_A(\tau,\sigma^u) = \{\hat b^l_A = l_A;\ \hat b^r_A = \delta^r_A + N^r\delta^\tau_A\}(\tau,\sigma^u),\\
\hat b^A_{\bar A}(\tau,\sigma^u) &= \hat b^\mu_{\bar A}(\tau,\sigma^u)\, b^A_\mu(\tau,\sigma^u) = \{\hat b^A_l = \epsilon l^A;\ \hat b^A_r = \delta^A_r\}(\tau,\sigma^u). \qquad (5.5)
\end{aligned}$$

The 3-dimensional covariant derivative (denoted ${}^3\nabla$ or with the subscript "|"; $({}^3\nabla_\rho)^\alpha_\beta = \partial_\rho\,\delta^\alpha_\beta + {}^3\Gamma^\alpha_{\rho\beta}(\tau,\sigma^u))$ of a 3-dimensional tensor ${}^3T^{\mu_1\dots\mu_p}_{\nu_1\dots\nu_q}(x(\tau,\sigma^u))$ of rank (p,q) is the 3-dimensional tensor of rank (p,q+1) ${}^3\nabla_\rho\, {}^3T^{\mu_1\dots\mu_p}_{\nu_1\dots\nu_q}(\tau,\sigma^u) = {}^3T^{\mu_1\dots\mu_p}_{\nu_1\dots\nu_q|\rho}(\tau,\sigma^u) = \Big({}^3h^{\mu_1}_{\alpha_1}\cdots {}^3h^{\mu_p}_{\alpha_p}\, {}^3h^{\beta_1}_{\nu_1}\cdots {}^3h^{\beta_q}_{\nu_q}\, {}^3h^\sigma_\rho\, {}^4\nabla_\sigma\, {}^3T^{\alpha_1\dots\alpha_p}_{\beta_1\dots\beta_q}\Big)(\tau,\sigma^u)$, where ${}^4\nabla$ is the 4-dimensional covariant derivative.

The components of the second fundamental form of $(\Sigma_\tau, {}^3g)$ describe its extrinsic curvature ($\mathcal{L}_l$ is the Lie derivative along the normal)

$$\begin{aligned}
{}^3K_{\mu\nu}(x = z(\tau,\sigma^u)) &= {}^3K_{\nu\mu}(x = (\tau,\sigma^u)) = -\frac{1}{2}\mathcal{L}_l\, {}^3g_{\mu\nu}(x = z(\tau,\sigma^u))\\
&= \Big(\hat b^r_\mu \hat b^s_\nu\, {}^3K_{rs}\Big)(\tau,\sigma^u),\\
{}^3K_{rs}(\tau,\sigma^u) = {}^3K_{sr}(\tau,\sigma^u) &= \frac{1}{2N(\tau,\sigma^u)}\Big(N_{r|s}(\tau,\sigma^u)\\
&\quad + N_{s|r}(\tau,\sigma^u) - \frac{\partial\, {}^3g_{rs}(\tau,\sigma^u)}{\partial\tau}\Big). \qquad (5.6)
\end{aligned}$$

One has $({}^4\nabla_\rho\, l^\mu)(\tau,\sigma^u) = (\epsilon\, {}^3a^\mu l_\rho - {}^3K^\mu_\rho)(\tau,\sigma^u)$, with the acceleration ${}^3a^\mu(\tau,\sigma^u) = {}^3a^r(\tau,\sigma^u)\,\hat b^\mu_r(\tau,\sigma^u)$ of the observers traveling along the congruence of time-like curves with tangent vector $l^\mu(\tau,\sigma^u)$ given by ${}^3a_r(\tau,\sigma^u) = \partial_r \ln N(\tau,\sigma^u)$.

The information contained in the 20 independent components ${}^4R^\alpha_{\mu\beta\nu}(x = z(\tau,\sigma^u)) = \Big({}^4\Gamma^\alpha_{\beta\rho}\, {}^4\Gamma^\rho_{\nu\mu} - {}^4\Gamma^\alpha_{\nu\rho}\, {}^4\Gamma^\rho_{\beta\mu} + \partial_\beta\, {}^4\Gamma^\alpha_{\mu\nu} - \partial_\nu\, {}^4\Gamma^\alpha_{\beta\mu}\Big)(\tau,\sigma^u)$ (with the associated Ricci tensor ${}^4R_{\mu\nu} = {}^4R^\beta_{\mu\beta\nu}$) of the curvature Riemann tensor of M^4 is replaced by its three projections given by Gauss, Codazzi–Mainardi, and Ricci equations [354]. In the non-holonomic basis the Einstein tensor becomes ${}^4G_{\mu\nu}(x = z(\tau,\sigma^u)) = ({}^4R_{\mu\nu} - \frac{1}{2}\,{}^4g_{\mu\nu}\,{}^4R)(x = z(\tau,\sigma^u)) = (\epsilon\,{}^4\bar{G}_{ll} l_\mu l_\nu + \epsilon\,{}^4\bar{G}_{lr}(l_\mu \hat{b}^r_\nu + l_\nu \hat{b}^r_\mu) + {}^4\bar{G}_{rs}\hat{b}^r_\mu \hat{b}^s_\nu)(\tau,\sigma^u)$. The Bianchi identities ${}^4G^{\mu\nu}_{;\nu}(\tau,\sigma^u) \equiv 0$ imply the following four contracted Bianchi identities:

$$
\begin{aligned}
&\Big(\frac{1}{N}\partial_\tau\,{}^4\bar{G}_{ll} - \frac{N^r}{N}\partial_r\,{}^4\bar{G}_{ll} - {}^3K\,{}^4\bar{G}_{ll} + \partial_r\,{}^4\bar{G}_l{}^r \\
&+(2\,{}^3a_r + {}^3\Gamma^s_{sr}){}^4\bar{G}_l{}^r - {}^3K_{rs}\,{}^4\bar{G}^{rs}\Big)(\tau,\sigma^u) \equiv 0, \\
&\Big(\frac{1}{N}\partial_\tau\,{}^4\bar{G}_l{}^r - \frac{N^s}{N}\partial_s\,{}^4\bar{G}_l{}^r + {}^3a^r\,{}^4\bar{G}_{ll} - \Big(2\,{}^3K^r_s + \delta^r_s\,{}^3K + \frac{\partial_s N^r}{N}\Big)\,{}^4\bar{G}_l{}^s + \partial_s\,{}^4\bar{G}^{rs} \\
&+({}^3a_s + {}^3\Gamma^u_{us}){}^4\bar{G}^{rs}\Big)(\tau,\sigma^u) \equiv 0.
\end{aligned}
\tag{5.7}
$$

The vanishing of ${}^4\bar{G}_{ll}(\tau,\sigma^u)$, ${}^4\bar{G}_{lr}(\tau,\sigma^u)$ corresponds to the four secondary constraints (restrictions of Cauchy data) of the ADM Hamiltonian formalism. The four contracted Bianchi identities imply [4] that, if the restrictions of Cauchy data are satisfied initially and the spatial equations ${}^4G_{ij} \stackrel{\circ}{=} 0$ are satisfied everywhere, then the secondary constraints are satisfied also at later times (see Refs. [4, 307, 357, 358] for the initial value problem). The four contracted Bianchi identities plus the four secondary constraints imply that only two combinations of the Einstein equations contain the accelerations (second time derivatives) of the two (non-tensorial) independent degrees of freedom of the gravitational field and that these equations can be put in normal form (this was one of the motivations behind the discovery of the Shanmugadhasan canonical transformations [see Chapter 9]).

The intrinsic geometry of Σ_τ is defined by the Riemannian 3-metric ${}^3g_{rs}(\tau,\sigma^u)$ (it allows evaluating the length of space curves), the Levi–Civita affine connection, i.e., the Christoffel symbols ${}^3\Gamma^u_{rs}(\tau,\sigma^u)$, (for the parallel transport of 3-dimensional tensors on Σ_τ) and the curvature Riemann tensor ${}^3R^r_{stu}(\tau,\sigma^u)$ (for the evaluation of the holonomy and for the geodesic deviation equation). The extrinsic geometry of Σ_τ is defined by the lapse $N(\tau,\sigma^u)$ and shift $N^r(\tau,\sigma^u)$ functions (which describe the "evolution" of Σ_τ in M^4) and by the extrinsic curvature ${}^3K_{rs}$ (it is needed to evaluate how much a 3-dimensional vector goes outside Σ_τ under space-time parallel transport and to rebuild the space-time curvature from the 3-dimensional one).

Given an arbitrary 3+1 splitting of M^4, the ADM action [278, 279] expressed in terms of the independent Σ_τ-adapted variables $N(\tau,\sigma^u)$, $N_r(\tau,\sigma^u) = {}^3g_{rs}(\tau,\sigma^u)\,N^s(\tau,\sigma^u)$, ${}^3g_{rs}(\tau,\sigma^u)$ is

$$
\begin{aligned}
S_{ADM} &= \int d\tau\, L_{ADM}(\tau) = \int d\tau d^3\sigma \mathcal{L}_{ADM}(\tau,\sigma^u) \\
&= -\epsilon k \int_{\triangle\tau} d\tau \int d^3\sigma\, \{\sqrt{\gamma} N\,[{}^3R + {}^3K_{rs}\,{}^3K^{rs} - ({}^3K)^2]\}(\tau,\sigma^u),
\end{aligned}
\tag{5.8}
$$

where $k = \frac{c^3}{16\pi G}$, with G the Newton constant.

The Euler–Lagrange equations are

$$
\begin{aligned}
L_N(\tau,\sigma^u) &= \Big(\frac{\partial \mathcal{L}_{ADM}}{\partial N} - \partial_\tau \frac{\partial \mathcal{L}_{ADM}}{\partial \partial_\tau N} - \partial_r \frac{\partial \mathcal{L}_{ADM}}{\partial \partial_r N}\Big)(\tau,\sigma^u) \\
&= -\epsilon k \Big(\sqrt{\gamma}[{}^3R - {}^3K_{rs}\, {}^3K^{rs} + ({}^3K)^2]\Big)(\tau,\sigma^u) = -2\,\epsilon k\, {}^4\bar{G}_{ll}(\tau,\sigma^u) \,\mathring{=}\, 0, \\
L^r_{\vec{N}}(\tau,\sigma^u) &= \Big(\frac{\partial \mathcal{L}_{ADM}}{\partial N_r} - \partial_\tau \frac{\partial \mathcal{L}_{ADM}}{\partial \partial_\tau N_r} - \partial_s \frac{\partial \mathcal{L}_{ADM}}{\partial \partial_s N_r}\Big)(\tau,\sigma^u) \\
&= 2\epsilon k\, [\sqrt{\gamma}({}^3K^{rs} - {}^3g^{rs}\, {}^3K)]_{|s}(\tau,\sigma^u) = 2k\, {}^4\bar{G}_l{}^r(\tau,\sigma^u) \,\mathring{=}\, 0, \\
L^{rs}_g(\tau,\sigma^u) &= -\epsilon k \Big[\frac{\partial}{\partial \tau}[\sqrt{\gamma}({}^3K^{rs} - {}^3g^{rs}\, {}^3K)] - N\sqrt{\gamma}({}^3R^{rs} - \frac{1}{2}{}^3g^{rs}\, {}^3R) \\
&\quad + 2N\sqrt{\gamma}({}^3K^{ru}\, {}^3K_u{}^s - {}^3K\, {}^3K^{rs}) + \frac{1}{2}N\sqrt{\gamma}[({}^3K)^2 - {}^3K_{uv}\, {}^3K^{uv})^3g^{rs} \\
&\quad + \sqrt{\gamma}({}^3g^{rs} N^{|u}_{|u} - N^{|r|s})\Big](\tau,\sigma^u) = -\epsilon k \Big(N\sqrt{\gamma}\, {}^4\bar{G}^{rs}\Big)(\tau,\sigma^u) \,\mathring{=}\, 0,
\end{aligned}
\tag{5.9}
$$

and correspond to the Einstein equations in the form ${}^4\bar{G}_{ll}(\tau,\sigma^u) \,\mathring{=}\, 0$, ${}^4\bar{G}_{lr}(\tau,\sigma^u) \,\mathring{=}\, 0$, ${}^4\bar{G}_{rs}(\tau,\sigma^u) \,\mathring{=}\, 0$, respectively. The four contracted Bianchi identities imply that only two of the six equations $L^{rs}_g(\tau,\sigma^u) \,\mathring{=}\, 0$ are independent.

The canonical momenta (densities of weight -1) are

$$
\begin{aligned}
\tilde{\pi}^N(\tau,\sigma^u) &= \frac{\delta S_{ADM}}{\delta \partial_\tau N(\tau,\sigma^u)} = 0, \\
\tilde{\pi}^r_{\vec{N}}(\tau,\sigma^u) &= \frac{\delta S_{ADM}}{\delta \partial_\tau N_r(\tau,\sigma^u)} = 0, \\
{}^3\tilde{\Pi}^{rs}(\tau,\sigma^u) &= \frac{\delta S_{ADM}}{\delta \partial_\tau {}^3g_{rs}(\tau,\sigma^u)} = \epsilon k\, [\sqrt{\gamma}\, ({}^3K^{rs} - {}^3g^{rs}\, {}^3K)](\tau,\sigma^u), \\
&\Downarrow \\
{}^3K_{rs}(\tau,\sigma^u) &= \frac{\epsilon}{k\sqrt{\gamma(\tau,\sigma^u)}} \Big[{}^3\tilde{\Pi}_{rs} - \frac{1}{2}{}^3g_{rs}\, {}^3\tilde{\Pi}\Big](\tau,\sigma^u), \\
{}^3\tilde{\Pi}(\tau,\sigma^u) &= {}^3g_{rs}(\tau,\sigma^u)\, {}^3\tilde{\Pi}^{rs}(\tau,\sigma^u) = -2\epsilon k\sqrt{\gamma}(\tau,\sigma^u)\, {}^3K(\tau,\sigma^u),
\end{aligned}
\tag{5.10}
$$

and satisfy the Poisson brackets

$$
\begin{aligned}
\{N(\tau,\sigma^u), \tilde{\Pi}^N(\tau,\sigma'^u)\} &= \delta^3(\sigma^u,\sigma'^u), \\
\{N_r(\tau,\sigma^u), \tilde{\Pi}^s_{\vec{N}}(\tau,\sigma'^u)\} &= \delta^s_r\, \delta^3(\sigma^u,\sigma'^u), \\
\{{}^3g_{rs}(\tau,\sigma^u), {}^3\tilde{\Pi}^{uv}(\tau,\sigma'^u\} &= \frac{1}{2}(\delta^u_r\delta^v_s + \delta^v_r\delta^u_s)\delta^3(\sigma^u,\sigma'^u).
\end{aligned}
\tag{5.11}
$$

The Wheeler–De Witt super-metric [312, 313] is

$$
{}^3G_{rstw}(\tau,\sigma^u) = [{}^3g_{rt}\, {}^3g_{sw} + {}^3g_{rw}\, {}^3g_{st} - {}^3g_{rs}\, {}^3g_{tw}](\tau,\sigma^u). \tag{5.12}
$$

Its inverse is defined by the equations

$$\frac{1}{2}{}^3G_{rstw}(\tau,\sigma^u)\,\frac{1}{2}{}^3G^{twuv}(\tau,\sigma^u) = \frac{1}{2}(\delta^u_r\delta^v_s + \delta^v_r\delta^u_s),$$
$${}^3G^{twuv}(\tau,\vec{\sigma}) = [{}^3g^{tu}\,{}^3g^{wv} + {}^3g^{tv}\,{}^3g^{wu} - 2\,{}^3g^{tw}\,{}^3g^{uv}](\tau,\vec{\sigma}), \tag{5.13}$$

so that one gets

$${}^3\tilde{\Pi}^{rs}(\tau,\sigma^u) = \frac{1}{2}\epsilon k\sqrt{\gamma}(\tau,\sigma^u)\,{}^3G^{rsuv}(\tau,\sigma^u)\,{}^3K_{uv}(\tau,\sigma^u),$$
$${}^3K_{rs}(\tau,\sigma^u) = \frac{\epsilon}{2k\sqrt{\gamma(\tau,\sigma^u)}}\,{}^3G_{rsuv}(\tau,\sigma^u)\,{}^3\tilde{\Pi}^{uv}(\tau,\sigma^u),$$

$$\begin{aligned}
&[{}^3K^{rs}\,{}^3K_{rs} - ({}^3K)^2](\tau,\sigma^u)\\
&= k^{-2}[\gamma^{-1}({}^3\tilde{\Pi}^{rs}\,{}^3\tilde{\Pi}_{rs} - \frac{1}{2}({}^3\tilde{\Pi})^2](\tau,\vec{\sigma})\\
&= (2k)^{-1}[\gamma^{-1}\,{}^3G_{rsuv}\,{}^3\tilde{\Pi}^{rs}\,{}^3\tilde{\Pi}^{uv}](\tau,\sigma^u),
\end{aligned}$$
$$\partial_\tau\,{}^3g_{rs}(\tau,\sigma^u) = \left[N_{r|s} + N_{s|r} - \frac{\epsilon N}{k\sqrt{\gamma}}\,{}^3G_{rsuv}\,{}^3\tilde{\Pi}^{uv}\right](\tau,\sigma^u). \tag{5.14}$$

Since $\left({}^3\tilde{\Pi}^{rs}\partial_\tau\,{}^3g_{rs}\right)(\tau,\sigma^u) = \left({}^3\tilde{\Pi}^{rs}[N_{r|s} + N_{s|r} - \frac{\epsilon N}{k\sqrt{\gamma}}{}^3G_{rsuv}{}^3\tilde{\Pi}^{uv}]\right)(\tau,\sigma^u)$ $= -2\left(N_r{}^3\tilde{\Pi}^{rs}_{|s} - \frac{\epsilon N}{k\sqrt{\gamma}}\,{}^3G_{rsuv}\,{}^3\tilde{\Pi}^{rs}{}^3\tilde{\Pi}^{uv} + (2N_r\,{}^3\tilde{\Pi}^{rs})_{|s}\right)(\tau,\sigma^u)$, we obtain the canonical Hamiltonian[5]

$$\begin{aligned}
H_{(c)ADM} &= \int_S d^3\sigma\,[\tilde{\pi}^N\partial_\tau N + \tilde{\pi}^r_{\vec{N}}\partial_\tau N_r + {}^3\tilde{\Pi}^{rs}\partial_\tau\,{}^3g_{rs}](\tau,\sigma^u) - L_{ADM}\\
&= \int_S d^3\sigma\left[\epsilon N\left(k\sqrt{\gamma}\,{}^3R - \frac{1}{2k\sqrt{\gamma}}{}^3G_{rsuv}\,{}^3\tilde{\Pi}^{rs}{}^3\tilde{\Pi}^{uv}\right) - 2N_r\,{}^3\tilde{\Pi}^{rs}_{|s}\right](\tau,\sigma^u)\\
&\quad +2\int_{\partial S} d^2\Sigma_s[N_r\,{}^3\tilde{\Pi}^{rs}\,](\tau,\sigma^u).
\end{aligned}\tag{5.15}$$

In the following discussion, the surface term will be omitted.

The Dirac Hamiltonian is ($\lambda(\tau,\sigma^u)$ are arbitrary Dirac multipliers)

$$H_{(D)ADM} = H_{(c)ADM} + \int d^3\sigma\,[\lambda_N\,\tilde{\pi}^N + \lambda^{\vec{N}}_r\,\tilde{\pi}^r_{\vec{N}}](\tau,\sigma^u). \tag{5.16}$$

The τ-constancy of the primary constraints $[\partial_\tau\tilde{\pi}^N(\tau,\sigma^u) \overset{\circ}{=} \{\tilde{\pi}^N(\tau,\sigma^u), H_{(D)ADM}\} \approx 0,\ \partial_\tau\tilde{\pi}^r_{\vec{N}}(\tau,\sigma^u) \overset{\circ}{=} \{\tilde{\pi}^r_{\vec{N}}(\tau,\sigma^u), H_{(D)ADM}\} \approx 0]$ generates four secondary constraints (they are densities of weight -1), which correspond to the Einstein equations ${}^4\bar{G}_{ll}(\tau,\sigma^u) \overset{\circ}{=} 0$, ${}^4\bar{G}_{lr}(\tau,\sigma^u) \overset{\circ}{=} 0$:

[5] Since $N_r(\tau,\sigma^u)\,{}^3\tilde{\Pi}^{rs}(\tau,\sigma^u)$ is a vector density of weight -1, it holds that ${}^3\nabla_s(N_r\,{}^3\tilde{\Pi}^{rs})(\tau,\sigma^u) = \partial_s(N_r\,{}^3\tilde{\Pi}^{rs})(\tau,\sigma^u)$.

$$
\begin{aligned}
\tilde{\mathcal{H}}(\tau,\sigma^u) &= \epsilon\left[k\sqrt{\gamma}\,{}^3R - \frac{1}{2k\sqrt{\gamma}}\,{}^3G_{rsuv}\,{}^3\tilde{\Pi}^{rs}\,{}^3\tilde{\Pi}^{uv}\right](\tau,\sigma^u)\\
&= \epsilon\left[\sqrt{\gamma}\,{}^3R - \frac{1}{k\sqrt{\gamma}}\left({}^3\tilde{\Pi}^{rs}\,{}^3\tilde{\Pi}_{rs} - \frac{1}{2}({}^3\tilde{\Pi})^2\right)\right](\tau,\sigma^u)\\
&= \epsilon k\{\sqrt{\gamma}[{}^3R - ({}^3K_{rs}\,{}^3K^{rs} - ({}^3K)^2)]\}(\tau,\sigma^u) \approx 0,\\
{}^3\tilde{\mathcal{H}}^r(\tau,\sigma^u) &= -2\,{}^3\tilde{\Pi}^{rs}_{|s}(\tau,\sigma^u) = -2[\partial_s\,{}^3\tilde{\Pi}^{rs} + {}^3\Gamma^r_{su}\,{}^3\tilde{\Pi}^{su}](\tau,\sigma^u)\\
&= -2\epsilon k\{\partial_s[\sqrt{\gamma}({}^3K^{rs} - {}^3g^{rs}\,{}^3K)]\\
&\quad + {}^3\Gamma^r_{su}\sqrt{\gamma}({}^3K^{su} - {}^3g^{su}\,{}^3K)\}(\tau,\sigma^u) \approx 0,
\end{aligned}
\tag{5.17}
$$

so that the Hamiltonian becomes

$$
H_{(c)ADM} = \int d^3\sigma[N\,\tilde{\mathcal{H}} + N_r\,{}^3\tilde{\mathcal{H}}^r](\tau,\sigma^u) \approx 0,
\tag{5.18}
$$

with $\tilde{\mathcal{H}}(\tau,\sigma^u) \approx 0$ called the super-Hamiltonian constraint and ${}^3\tilde{\mathcal{H}}^r(\tau,\sigma^u) \approx 0$ the super-momentum constraints. In $\tilde{\mathcal{H}}(\tau,\sigma^u) \approx 0$, one can say that the term $-\epsilon k[\sqrt{\gamma}({}^3K_{rs}\,{}^3K^{rs} - {}^3K^2)](\tau,\sigma^u)$ is the kinetic energy and $\epsilon k(\sqrt{\gamma}\,{}^3R)(\tau,\sigma^u)$ the potential energy.

All the constraints are first class, because the only non-identically zero Poisson brackets correspond to the so-called universal Dirac algebra [20, 21]:

$$
\begin{aligned}
\{{}^3\tilde{\mathcal{H}}_r(\tau,\sigma^u), {}^3\tilde{\mathcal{H}}_s(\tau,\sigma'^u)\} &= {}^3\tilde{\mathcal{H}}_r(\tau,\sigma'^u)\,\frac{\partial\delta^3(\sigma^u,\sigma'^u)}{\partial\sigma^s} + {}^3\tilde{\mathcal{H}}_s(\tau,\sigma^u)\frac{\partial\delta^3(\sigma,\sigma'^u)}{\partial\sigma^r},\\
\{\tilde{\mathcal{H}}(\tau,\sigma^u), {}^3\tilde{\mathcal{H}}_r(\tau,\sigma'^u)\} &= \tilde{\mathcal{H}}(\tau,\sigma^u)\frac{\partial\delta^3(\sigma,\sigma'^u)}{\partial\sigma^r},\\
\{\tilde{\mathcal{H}}(\tau,\sigma^u), \tilde{\mathcal{H}}(\tau,\sigma'^u)\} &= [{}^3g^{rs}(\tau,\sigma^u){}^3\tilde{\mathcal{H}}_s(\tau,\sigma^u)\\
&\quad + {}^3g^{rs}(\tau,\sigma'^u){}^3\tilde{\mathcal{H}}_s(\tau,\sigma'^u)]\frac{\partial\delta^3(\sigma,\sigma')}{\partial\sigma^r},
\end{aligned}
\tag{5.19}
$$

with ${}^3\tilde{\mathcal{H}}_r(\tau,\sigma^u) = {}^3g_{rs}(\tau,\sigma^u)\,{}^3\tilde{\mathcal{H}}^r(\tau,\sigma^u)$ as the combination of the super-momentum constraints satisfying the algebra of 3-diffeomorphisms. In Ref. [359] it is shown that Eq. (5.19) is a sufficient condition for the embeddability of Σ_τ into M^4. In Ref. [361] it is shown that the last two lines of the Dirac algebra are the equivalent in phase space of the Bianchi identities ${}^4G^{\mu\nu}_{;\nu}(\tau,\sigma^u) \equiv 0$.

The Hamilton Dirac equations are [$\mathcal{L}$ is the notation for the Lie derivative]

$$
\begin{aligned}
\partial_\tau N(\tau,\sigma^u) &\overset{\circ}{=} \{N(\tau,\sigma^u), H_{(D)ADM}\} = \lambda_N(\tau,\sigma^u),\\
\partial_\tau N_r(\tau,\sigma^u) &\overset{\circ}{=} \{N_r(\tau,\sigma^u), H_{(D)ADM}\} = \lambda^{\vec{N}}_r(\tau,\sigma^u),\\
\partial_\tau\,{}^3g_{rs}(\tau,\sigma^u) &\overset{\circ}{=} \{{}^3g_{rs}(\tau,\sigma^u), H_{(D)ADM}\}\\
&= \left[N_{r|s} + N_{s|r} - \frac{2\epsilon N}{k\sqrt{\gamma}}\left({}^3\tilde{\Pi}_{rs} - \frac{1}{2}{}^3g_{rs}\,{}^3\tilde{\Pi}\right)\right](\tau,\sigma^u)\\
&= [N_{r|s} + N_{s|r} - 2N\,{}^3K_{rs}](\tau,\sigma^u),\\
\partial_\tau\,{}^3\tilde{\Pi}^{rs}(\tau,\sigma^u) &\overset{\circ}{=} \{{}^3\tilde{\Pi}^{rs}(\tau,\sigma^u), H_{(D)ADM}\} = \epsilon\left[N\,k\sqrt{\gamma}\left({}^3R^{rs} - \frac{1}{2}{}^3g^{rs}\,{}^3R\right)\right](\tau,\sigma^u)
\end{aligned}
$$

$$
\begin{aligned}
&-2\epsilon\Bigg[\frac{N}{k\sqrt{\gamma}}\left(\frac{1}{2}{}^3\tilde{\Pi}\,{}^3\tilde{\Pi}^{rs}-{}^3\tilde{\Pi}^r_u\,{}^3\tilde{\Pi}^{us}\right)\\
&-\frac{\epsilon N}{2}\frac{{}^3g^{rs}}{k\sqrt{\gamma}}\left(\frac{1}{2}{}^3\tilde{\Pi}^2-{}^3\tilde{\Pi}_{uv}\,{}^3\tilde{\Pi}^{uv}\right)\Bigg](\tau,\sigma^u)\\
&+\mathcal{L}_{\vec{N}}\,{}^3\tilde{\Pi}^{rs}(\tau,\sigma^u)+\epsilon[k\sqrt{\gamma}(N^{|r|s}-{}^3g^{rs}\,N^{|u}_{|u})](\tau,\sigma^u),\\
&\Downarrow
\end{aligned}
$$

$$
\begin{aligned}
\partial_\tau\,{}^3K_{rs}(\tau,\sigma^u)\circeq{}&\Big(N[{}^3R_{rs}+{}^3K\,{}^3K_{rs}-2\,{}^3K_{ru}\,{}^3K^u_s]\\
&-N_{|s|r}+N^u_{|s}\,{}^3K_{ur}+N^u_{|r}\,{}^3K_{us}+N^u\,{}^3K_{rs|u}\Big)(\tau,\sigma^u),
\end{aligned}
$$

with

$$
(\mathcal{L}_{\vec{N}}\,{}^3\tilde{\Pi}^{rs})(\tau,\sigma^u)=-\left(\sqrt{\gamma}\,{}^3\nabla_u\left(\frac{N^u}{\sqrt{\gamma}}{}^3\tilde{\Pi}^{rs}\right)+{}^3\tilde{\Pi}^{ur}\,{}^3\nabla_uN^s+{}^3\tilde{\Pi}^{us}\,{}^3\nabla_uN^r\right)(\tau,\sigma^u).
\tag{5.20}
$$

The equation for $\partial_\tau\,{}^3g_{rs}(\tau,\sigma^u)$ shows that the generator of space pseudo-diffeomorphisms[6] $\int d^3\sigma N_r(\tau,\sigma^u)\ {}^3\tilde{\mathcal{H}}^r(\tau,\sigma^u)$ produces a variation, tangent to Σ_τ, $\delta_{tangent}{}^3g_{rs}(\tau,\sigma^u)=\mathcal{L}_{\vec{N}}\,{}^3g_{rs}(\tau,\sigma^u)=N_{r|s}(\tau,\sigma^u)+N_{s|r}(\tau,\sigma^u)$ in accord with the infinitesimal pseudo-diffeomorphisms in Diff Σ_τ. Instead, the gauge transformations induced by the super-Hamiltonian generator $\int d^3\sigma N(\tau,\sigma^u)\ \tilde{\mathcal{H}}(\tau,\sigma^u)$ do not reproduce the infinitesimal diffeomorphisms in Diff M^4 normal to Σ_τ (see Ref. [323]). For the clarification of the connection between space-time diffeomorphisms and Hamiltonian gauge transformations, see Refs. [364, 365]. See section IX of Ref. [280] for more details on the interpretation of the action of the Hamiltonian gauge transformations of metric gravity and their comparison with the transformations induced by the space-time diffeomorphisms of the space-time (Diff M^4).

Finally, the canonical transformation $(\tilde{\pi}^N\,dN+\tilde{\pi}^r_{\vec{N}}\,dN_r+{}^3\tilde{\Pi}^{rs}\,d^3g_{rs})={}^4\tilde{\Pi}^{AB}\,d^4g_{AB}$ allows defining the following momenta conjugated to ${}^4g_{AB}$:

$$
\begin{aligned}
{}^4\tilde{\Pi}^{\tau\tau}(\tau,\sigma^u)&=\frac{\epsilon}{2N(\tau,\sigma^u)}\tilde{\pi}^N(\tau,\sigma^u),\\
{}^4\tilde{\Pi}^{\tau r}(\tau,\sigma^u)&=\frac{\epsilon}{2}\left(\frac{N^r}{N}\tilde{\pi}^N-\tilde{\pi}^r_{\vec{N}}\right)(\tau,\sigma^u),\\
{}^4\tilde{\Pi}^{rs}(\tau,\sigma^u)&=\epsilon\left(\frac{N^rN^s}{2N}\tilde{\pi}^N-{}^3\tilde{\Pi}^{rs}\right)(\tau,\sigma^u),\\
&\{{}^4g_{AB}(\tau,\sigma^u),{}^4\tilde{\Pi}^{CD}(\tau,\sigma'^u)\}=\frac{1}{2}(\delta^C_A\delta^D_B+\delta^D_A\delta^C_B)\delta^3(\sigma^u,\sigma'^u),\\
\tilde{\pi}^N(\tau,\sigma^u)&=\frac{2\epsilon}{\sqrt{\epsilon^4g^{\tau\tau}(\tau,\sigma^u)}}{}^4\tilde{\Pi}^{\tau\tau}(\tau,\sigma^u),
\end{aligned}
$$

[6] The Hamiltonian transformations generated by these constraints are the extension to the 3-metric of passive or pseudo- diffeomorphisms, namely changes of coordinate charts, of Σ_τ [Diff Σ_τ].

$$\tilde{\pi}^r_{\vec{N}}(\tau,\sigma^u) = 2\epsilon \frac{{}^4g^{\tau r}(\tau,\sigma^u)}{{}^4g^{\tau\tau}(\tau,\sigma^u)} {}^4\tilde{\Pi}^{\tau\tau}(\tau,\sigma^u) - 2\epsilon {}^4\tilde{\Pi}^{\tau r}(\tau,\sigma^u),$$
$$ {}^3\tilde{\Pi}^{rs}(\tau,\sigma^u) = \epsilon \frac{{}^4g^{\tau r}(\tau,\sigma^u){}^4g^{\tau s}(\tau,\sigma^u)}{({}^4g^{\tau\tau}(\tau,\sigma^u))^2} {}^4\tilde{\Pi}^{\tau\tau}(\tau,\sigma^u) - \epsilon {}^4\tilde{\Pi}^{rs}(\tau,\sigma^u), \qquad (5.21)$$

which would emerge if the ADM action were considered a function of ${}^4g_{AB}$ instead of N, N_r, and ${}^3g_{rs}$.

Let us add a comment on the structure of gauge fixings for metric gravity. As said in Refs. [169, 170, 366], in a system with only primary and secondary first-class constraints (like electromagnetism, Yang–Mills theory, and both metric and tetrad gravity), the Dirac Hamiltonian H_D contains only the arbitrary Dirac multipliers associated with the primary first-class constraints. The secondary first-class constraints are already contained in the canonical Hamiltonian with well-defined coefficients (the temporal components A_{ao} of the gauge potential in Yang–Mills theory; the lapse and shift functions in metric and tetrad gravity as evident from Eq. (5.18); in both cases, through the first half of the Hamilton equations, the Dirac multipliers turn out to be equal to the τ-derivatives of these quantities, which, therefore, inherit an induced arbitrariness). See Section 9.1 for a discussion of this point and for a refusal of Dirac's conjecture [20, 21] according to which also the secondary first-class constraints must have arbitrary Dirac multipliers.[7] In the standard cases, one must adopt the following gauge fixing strategy: (1) add gauge fixing constraints $\chi_a(\tau,\sigma^u) \approx 0$ to the secondary constraints; (2) their time constancy, $\partial_\tau \chi_a(\tau,\sigma^u) \stackrel{\circ}{=} \{\chi_a(\tau,\sigma^u), H_D\} = g_a(\tau,\sigma^u) \approx 0$ implies the appearance of gauge fixing constraints $g_a(\tau,\sigma^u) \approx 0$ for the primary constraints; (3) the time constancy of the constraints $g_a(\tau,\sigma^u) \approx 0$, $\partial_\tau g_a(\tau,\sigma^u) \stackrel{\circ}{=} \{g_a(\tau,\sigma^u), H_D\} \approx 0$ determines the Dirac multipliers in front of the primary constraints (the $\lambda(\tau,\sigma^u)$ in Eq. (5.16)).

As shown in Ref. [170] for the electromagnetic case, this method works also with covariant gauge fixings: The electromagnetic Lorentz gauge $\partial^\mu A_\mu(x) \approx 0$ may be rewritten in phase space as a gauge fixing constraint depending upon the Dirac multiplier; its time constancy gives a multiplier-dependent gauge fixing for $A_o(x)$ and the time constancy of this new constraint gives the elliptic equation for the multiplier with the residual gauge freedom connected with the kernel of the elliptic operator.

In metric gravity, the covariant gauge fixings analogous to the Lorentz gauge are those determining the harmonic coordinates (harmonic or De Donder gauge): $\chi^B(\tau,\sigma^u) = \frac{1}{\sqrt{{}^4g(\tau,\sigma^u)}} \partial_A(\sqrt{{}^4g}\,{}^4g^{AB})(\tau,\sigma^u) \approx 0$ in the Σ_τ-adapted holonomic coordinate basis. More explicitly, they are:

1. for $B = \tau$: $\Big(N\partial_\tau\gamma - \gamma\partial_\tau N - N^2\partial_r(\frac{\gamma N^r}{N})\Big)(\tau,\sigma^u) \approx 0$;
2. for $B = s$: $\Big(NN^s\partial_\tau\gamma + \gamma(N\partial_\tau N^s - N^s\partial_\tau N) + N^2\partial_r[N\gamma({}^3g^{rs} - \frac{N^rN^s}{N^2})]\Big)$ $(\tau,\sigma^u) \approx 0$.

[7] In such a case, one does not recover the original Lagrangian by inverse Legendre transformation, and one obtains a different "off-shell" theory.

From the Hamilton Dirac equations we get

$$\begin{aligned}
\partial_\tau N(\tau,\sigma^u) &\stackrel{\circ}{=} \lambda_N(\tau,\sigma^u),\\
\partial_\tau N_r(\tau,\sigma^u) &\stackrel{\circ}{=} \lambda_r^{\vec{N}}(\tau,\sigma^u),\\
\partial_\tau \gamma(\tau,\sigma^u) &= \frac{1}{2}\gamma(\tau,\sigma^u)\,{}^3g^{rs}(\tau,\sigma^u)\partial_\tau\,{}^3g_{rs}(\tau,\sigma^u)\\
&\stackrel{\circ}{=} \frac{1}{2}\gamma(\tau,\sigma^u)\left[{}^3g^{rs}(N_{r|s}+N_{s|r}) - \frac{5\epsilon N}{k\sqrt{\gamma}}\,{}^3\tilde{\Pi}\right](\tau,\sigma^u). \qquad (5.22)
\end{aligned}$$

Therefore, in phase space the harmonic coordinate gauge fixings associated with the secondary super-Hamiltonian and super-momentum constraints take the form

$$\chi^B(\tau,\sigma^u) = \bar{\chi}^B\left[N, N_r, N_{r|s}, {}^3g_{rs}, {}^3\tilde{\Pi}^{rs}, \lambda_N, \lambda_r^{\vec{N}}\right](\tau,\sigma^u) \approx 0. \qquad (5.23)$$

The conditions $\partial_\tau\bar{\chi}^B(\tau,\sigma^u) \stackrel{\circ}{=} \{\bar{\chi}^B(\tau,\sigma^u), H_D\} = g^B(\tau,\sigma^u) \approx 0$ give the gauge fixings for the primary constraints $\tilde{\pi}^N(\tau,\sigma^u) \approx 0$, $\tilde{\pi}^r_{\vec{N}}(\tau,\sigma^u) \approx 0$.

The conditions $\partial_\tau g^B(\tau,\sigma^u) \stackrel{\circ}{=} \{g^B(\tau,\sigma^u), H_D\} \approx 0$ are partial differential equations for the Dirac multipliers $\lambda_N(\tau,\sigma^u)$, $\lambda_r^{\vec{N}}(\tau,\sigma^u)$, implying a residual gauge freedom as happens for the electromagnetic Lorentz gauge.

See sections 5 and 6 of Ref. [281] for the identification of the asymptotic ADM Poincaré generators as both weak and strong conserved charges.

The search for the DOs in metric gravity is a difficult, unsolved task [282, 367–372] because no one is able to identify which are the true unknowns contained in the super-Hamiltonian and super-momentum constraints and to solve the non-linear partial differential equations implied by the constraints for these unknowns.

6

ADM Tetrad Gravity and Its Constraints

In this chapter I replace ADM gravity with ADM tetrad gravity to be able to include fermions in the matter. We give its formulation in the non-inertial rest-frames identified in Section 5.1 of the previous chapter. Then we find a Shanmugadhasan canonical transformation (suggested by the York map) adapted to all the first-class constraints except the super-Hamiltonian and super-momentum ones [284]. This limitation is due to the problem that it is not known how to solve these non-linear partial differential equations (PDEs): When a solution appears it will be possible to find a canonical transformation adapted to all the constraints and to find the Dirac observables (DOs) of general relativity (GR).

Even if they are not the real DOs of Einstein gravity, I identify the tidal variables of the gravitational field (the gravitational waves of the linearized theory) and I give a metrological interpretation of the gauge inertial variables [288]. A class of non-harmonic 3-orthogonal Schwinger time gauges is clearly suggested by the York canonical basis.

6.1 ADM Tetrad Gravity, Its Hamiltonian Formulation, and Its First-Class Constraints

To take into account the coupling of fermions to the gravitational field, metric gravity has to be replaced with tetrad gravity [350]. This can be achieved in the same class of space-times by decomposing the 4-metric of Eq. (5.2) on cotetrad fields:

$$ {}^4g_{AB}(\tau,\sigma^u) = E^{(\alpha)}_A(\tau,\sigma^u)\, {}^4\eta_{(\alpha)(\beta)}\, E^{(\beta)}_B(\tau,\sigma^u), \tag{6.1} $$

by putting this expression into the ADM action and by considering the resulting action, a functional of the 16 fields $E^{(\alpha)}_A(\tau,\sigma^u)$, as the action for ADM tetrad gravity. In Eq. (6.1), (α) are flat indices and the cotetrad fields $E^{(\alpha)}_A(\tau,\sigma^u)$ are the inverse of the tetrad fields $E^A_{(\alpha)}(\tau,\sigma^u)(\tau,\sigma^u)$, which are connected to the world tetrad fields by $E^{\mu}_{(\alpha)}(x) = z^{\mu}_A(\tau,\sigma^u)\, E^A_{(\alpha)}(x = z(\tau,\sigma^u))$ by the embeddings of the 3+1 approach in the non-inertial rest-frames defined in Chapter 5.

This leads to an interpretation of gravity based on a congruence of time-like observers endowed with orthonormal tetrads: In each point of space-time the time-like axis is the unit 4-velocity of the observer, while the spatial axes are a (gauge) convention for the observer's gyroscopes. This framework was developed in Refs. [281, 282].

Even if the action of ADM tetrad gravity depends upon 16 fields, the counting of the physical degrees of freedom of the gravitational field does not change, because this action is invariant not only under the group of 4-diffeomorphisms but also under the $O(3,1)$ gauge group of the Newman–Penrose approach [9] (the extra gauge freedom acting on the tetrads in the tangent space of each point of space-time).

The cotetrads $E^{(\alpha)}_A(\tau,\sigma^u)$ are the new configuration variables. They are connected to cotetrads ${}^4\overset{\circ}{E}{}^{(\alpha)}_A(\tau,\sigma^u)$ adapted to the 3+1 splitting of space-time, namely such that the inverse adapted time-like tetrad ${}^4\overset{\circ}{E}{}^A_{(o)}(\tau,\sigma^u)$ is the unit normal $l^A(\tau,\sigma^u)$ to the 3-space Σ_τ, by a standard Wigner boost for time-like Poincaré orbits with parameters $\varphi_{(a)}(\tau,\sigma^u)$, $a=1,2,3$ (see Appendix A):

$$
\begin{aligned}
E^{(\alpha)}_A(\tau,\sigma^u) &= L^{(\alpha)}_{(\beta)}(\varphi_{(a)}(\tau,\sigma^u))\,\overset{o}{E}{}^{(\beta)}_A(\tau,\sigma^u),\\
{}^4g_{AB}(\tau,\sigma^u) &= {}^4\overset{\circ}{E}{}^{(\alpha)}_A(\tau,\sigma^u)\,{}^4\eta_{(\alpha)(\beta)}\,{}^4\overset{\circ}{E}{}^{(\beta)}_B(\tau,\sigma^u),\\
L^{(\alpha)}_{(\beta)}(\varphi_{(a)}(\tau,\sigma^u)) &\overset{def}{=} L^{(\alpha)}_{(\beta)}(V(z(\tau,\sigma^u));\,\overset{\circ}{V}) = \delta^{(\alpha)}_{(\beta)} + 2\epsilon\,V^{(\alpha)}(z(\tau,\sigma^u))\,\overset{\circ}{V}_{(\beta)}\\
&\quad -\epsilon\,\frac{(V^{(\alpha)}(z(\tau,\sigma^u)) + \overset{o}{V}{}^{(\alpha)})\,(V_{(\beta)}(z(\tau,\sigma^u)) + \overset{\circ}{V}_{(\beta)})}{1+V^{(o)}(z(\tau,\sigma^u))}.
\end{aligned}
\tag{6.2}
$$

In each tangent plane to a point of Σ_τ, this point-dependent standard Wigner boost sends the unit future-pointing time-like vector $\overset{o}{V}{}^{(\alpha)} = (1;0)$ into the unit time-like vector $V^{(\alpha)}(\tau,\sigma^u) = {}^4E^{(\alpha)}_A(\tau,\sigma^u)\,l^A(\tau,\sigma^u) = \Big(\sqrt{1+\sum_a \varphi^2_{(a)}(\tau,\sigma^u)};\ \varphi^{(a)}(\tau,\sigma^u) = -\epsilon\,\varphi_{(a)}(\tau,\sigma^u)\Big)$. As a consequence, the flat indices (a) of the adapted tetrads and cotetrads and of the triads and cotriads on Σ_τ transform as Wigner spin-1 indices under point-dependent $SO(3)$ Wigner rotations $R_{(a)(b)}(V(z(\tau,\sigma^u));\ \Lambda(z(\tau,\sigma^u)))$ associated with Lorentz transformations $\Lambda^{(\alpha)}_{(\beta)}(z(\tau,\sigma^u))$ in the tangent plane to the space-time in the given point of Σ_τ. Instead, the index (o) of the adapted tetrads and cotetrads is a local Lorentz scalar index.

The adapted tetrads and cotetrads have the expression

$$
\begin{aligned}
{}^4\overset{\circ}{E}{}^A_{(o)}(\tau,\sigma^u) &= \frac{1}{1+n(\tau,\sigma^u)}\left(1;-\sum_a n_{(a)}(\tau,\sigma^u)\,{}^3e^r_{(a)}(\tau,\sigma^u)\right) = l^A(\tau,\sigma^u),\\
{}^4\overset{\circ}{E}{}^A_{(a)}(\tau,\sigma^u) &= (0;{}^3e^r_{(a)}(\tau,\sigma^u)),
\end{aligned}
$$

$$
\begin{aligned}
&{}^4\overset{\circ}{E}{}^{(o)}_A(\tau,\sigma^u) = (1+n(\tau,\sigma^u))\,(1;\vec{0}) = \epsilon\, l_A(\tau,\sigma^u),\\
&{}^4\overset{\circ}{E}{}^{(a)}_A(\tau,\sigma^u) = (n_{(a)}(\tau,\sigma^u);{}^3e_{(a)r}(\tau,\sigma^u)),
\end{aligned}
\tag{6.3}
$$

where ${}^3e^r_{(a)}(\tau,\sigma^u)$ and ${}^3e_{(a)r}(\tau,\sigma^u)$ are triads and cotriads on Σ_τ and $n_{(a)}(\tau,\sigma^u) = n_r(\tau,\sigma^u)\,{}^3e^r_{(a)}(\tau,\sigma^u) = n^r(\tau,\sigma^u)\,{}^3e_{(a)r}(\tau,\sigma^u)$[1] are adapted shift functions. In Eq. (6.3) $N(\tau,\sigma^u) = 1+n(\tau,\sigma^u) > 0$, with $n(\tau,\sigma^u)$ vanishing at spatial infinity, so that $N(\tau,\sigma^u)\,d\tau$ is positive from Σ_τ to $\Sigma_{\tau+d\tau}$, is the lapse function; $N^r(\tau,\sigma^u) = n^r(\tau,\sigma^u)$, vanishing at spatial infinity, are the shift functions. These are the boundary conditions given in Section 5.1 for the absence of super-translations. For the cotriads, one has ${}^3e_{(a)r}(\tau,\sigma^u) \rightarrow (1+\frac{\text{const.}}{2r})\,\delta_{ar} + O(r^{-3/2})$.

The adapted tetrads ${}^4\overset{\circ}{E}{}^A_{(a)}(\tau,\sigma^u)$ are defined modulo $SO(3)$ rotations ${}^4\overset{\circ}{E}{}^A_{(a)}(\tau,\sigma^u) = \sum_b R_{(a)(b)}(\alpha_{(e)}(\tau,\sigma^u))\,{}^4\overset{\circ}{\bar{E}}{}^A_{(b)}(\tau,\sigma^u)$, ${}^3e^r_{(a)}(\tau,\sigma^u) = \sum_b R_{(a)(b)}(\alpha_{(e)}(\tau,\sigma^u))\,{}^3\bar{e}^r_{(b)}(\tau,\sigma^u)$, where $\alpha_{(a)}(\tau,\sigma^u)$ are three point-dependent Euler angles. After choosing an arbitrary point-dependent origin $\alpha_{(a)}(\tau,\sigma^u) = 0$, one arrives at the following adapted tetrads and cotetrads: $[\bar{n}_{(a)}(\tau,\sigma^u) = \sum_b n_{(b)}(\tau,\sigma^u)\,R_{(b)(a)}(\alpha_{(e)}(\tau,\sigma^u))$, $\sum_a n_{(a)}(\tau,\sigma^u)\,{}^3e^r_{(a)}(\tau,\sigma^u) = \sum_a \bar{n}_{(a)}(\tau,\sigma^u)\,{}^3\bar{e}^r_{(a)}(\tau,\sigma^u)]$,

$$
\begin{aligned}
{}^4\overset{\circ}{\bar{E}}{}^A_{(o)}(\tau,\sigma^u) &= {}^4\overset{\circ}{E}{}^A_{(o)}(\tau,\sigma^u)\\
&= \frac{1}{1+n(\tau,\sigma^u)}\left(1;-\sum_a \bar{n}_{(a)}(\tau,\sigma^u)\,{}^3\bar{e}^r_{(a)}(\tau,\sigma^u)\right) = l^A(\tau,\sigma^u),\\
{}^4\overset{\circ}{\bar{E}}{}^A_{(a)}(\tau,\sigma^u) &= (0;{}^3\bar{e}^r_{(a)}(\tau,\sigma^u)),\\
{}^4\overset{\circ}{\bar{E}}{}^{(o)}_A(\tau,\sigma^u) &= {}^4\overset{\circ}{E}{}^{(o)}_A(\tau,\sigma^u) = (1+n(\tau,\sigma^u))\,(1;\vec{0}) = \epsilon\, l_A(\tau,\sigma^u),\\
{}^4\overset{\circ}{\bar{E}}{}^{(a)}_A(\tau,\sigma^u) &= (\bar{n}_{(a)}(\tau,\sigma^u);{}^3\bar{e}_{(a)r}(\tau,\sigma^u)),
\end{aligned}
\tag{6.4}
$$

which we will use as a reference standard.

The expression for the general tetrad,

$$
\begin{aligned}
{}^4E^A_{(\alpha)}(\tau,\sigma^u) &= {}^4\overset{\circ}{E}{}^A_{(\beta)}(\tau,\sigma^u)\,L^{(\beta)}{}_{(\alpha)}(\varphi_{(a)}(\tau,\sigma^u)) = {}^4\overset{\circ}{\bar{E}}{}^A_{(o)}(\tau,\sigma^u)\,L^{(o)}{}_{(\alpha)}(\varphi_{(c)}(\tau,\sigma^u))\\
&\quad + \sum_{ab} {}^4\overset{\circ}{\bar{E}}{}^A_{(b)}(\tau,\sigma^u)\,R^T_{(b)(a)}(\alpha_{(c)}(\tau,\sigma^u))\,L^{(a)}{}_{(\alpha)}(\varphi_{(c)}(\tau,\sigma^u)),
\end{aligned}
\tag{6.5}
$$

[1] Since one uses the positive-definite 3-metric $\delta_{(a)(b)}$, one will use only lower flat spatial indices. Therefore for the cotriads one uses the notation ${}^3e^{(a)}_r \overset{def}{=} {}^3e_{(a)r}$, with $\delta_{(a)(b)} = {}^3e^r_{(a)}\,{}^3e_{(b)r}$.

shows that every point-dependent Lorentz transformation Λ in the tangent planes may be parametrized with the (Wigner) boost parameters $\varphi_{(a)}(\tau, \sigma^u)$ and the Euler angles $\alpha_{(a)}(\tau, \sigma^u)$, being the product $\Lambda = R\,L$ of a rotation and a boost.

The future-oriented unit normal to Σ_τ and the projector on Σ_τ are $l_A(\tau, \sigma^u) = \epsilon\,(1 + n(\tau, \sigma^u))\,\Big(1;\,0\Big)$, ${}^4g^{AB}(\tau, \sigma^u)\,l_A(\tau, \sigma^u)\,l_B(\tau, \sigma^u) = \epsilon$, $l^A(\tau, \sigma^u) = \epsilon\,(1 + n(\tau, \sigma^u))\,{}^4g^{A\tau}(\tau, \sigma^u) = \frac{1}{1+n(\tau,\sigma^u)}\Big(1;\,-n^r(\tau, \sigma^u)\Big) = \frac{1}{1+n(\tau,\sigma^u)}\Big(1;\,-\sum_a \bar{n}_{(a)}(\tau, \sigma^u)\,{}^3\bar{e}^r_{(a)}(\tau, \sigma^u)\Big)$, ${}^3h^B_A(\tau, \sigma^u) = \delta^B_A - \epsilon\,l_A(\tau, \sigma^u)\,l^B$.

The 4-metric has the following expression:

$$
\begin{aligned}
{}^4g_{\tau\tau}(\tau, \sigma^u) &= \epsilon\,[(1+n)^2 - {}^3g^{rs}\,n_r\,n_s](\tau, \sigma^u) = \epsilon\,[(1+n)^2 - \sum_a \bar{n}^2_{(a)}](\tau, \sigma^u),\\
{}^4g_{\tau r}(\tau, \sigma^u) &= -\epsilon\,n_r(\tau, \sigma^u) = -\epsilon \sum_a \bar{n}_{(a)}(\tau, \sigma^u)\,{}^3\bar{e}_{(a)r}(\tau, \sigma^u),\\
{}^4g_{rs}(\tau, \sigma^u) &= -\epsilon\,{}^3g_{rs}(\tau, \sigma^u) = -\epsilon \sum_a {}^3e_{(a)r}(\tau, \sigma^u)\,{}^3e_{(a)s}(\tau, \sigma^u)\\
&= -\epsilon \sum_a {}^3\bar{e}_{(a)r}(\tau, \sigma^u)\,{}^3\bar{e}_{(a)s}(\tau, \sigma^u),
\end{aligned}
$$

$$
\begin{aligned}
{}^4g^{\tau\tau}(\tau, \sigma^u) &= \frac{\epsilon}{(1+n(\tau, \sigma^u))^2},\\
{}^4g^{\tau r}(\tau, \sigma^u) &= -\epsilon\,\frac{n^r(\tau, \sigma^u)}{(1+n(\tau, \sigma^u))^2} = -\epsilon\,\frac{\sum_a {}^3\bar{e}^r_{(a)}(\tau, \sigma^u)\,\bar{n}_{(a)}(\tau, \sigma^u)}{(1+n(\tau, \sigma^u))^2},\\
{}^4g^{rs}(\tau, \sigma^u) &= -\epsilon\,({}^3g^{rs}(\tau, \sigma^u) - \frac{n^r(\tau, \sigma^u)\,n^s(\tau, \sigma^u)}{(1+n(\tau, \sigma^u))^2})\\
&= -\epsilon \sum_{ab} {}^3\bar{e}^r_{(a)}(\tau, \sigma^u)\,{}^3\bar{e}^s_{(b)}(\tau, \sigma^u)\,(\delta_{(a)(b)} - \frac{\bar{n}_{(a)}(\tau, \sigma^u)\,\bar{n}_{(b)}(\tau, \sigma^u)}{(1+n(\tau, \sigma^u))^2}),
\end{aligned}
$$

$$
\begin{aligned}
\sqrt{-g(\tau, \sigma^u)} &= \sqrt{|{}^4g(\tau, \sigma^u)|} = \frac{\sqrt{{}^3g(\tau, \sigma^u)}}{\sqrt{\epsilon\,{}^4g^{\tau\tau}(\tau, \sigma^u)}}\\
&= \sqrt{\gamma}(\tau, \sigma^u)\,(1+n(\tau, \sigma^u)) = {}^3e(\tau, \sigma^u)\,(1+n(\tau, \sigma^u)),\\
{}^3g(\tau, \sigma^u) &= \gamma(\tau, \sigma^u) = ({}^3e(\tau, \sigma^u))^2, \quad {}^3e(\tau, \sigma^u) = \det{}^3e_{(a)r}(\tau, \sigma^u).
\end{aligned}
\tag{6.6}
$$

The 3-metric ${}^3g_{rs}(\tau, \sigma^u)$ has the signature $(+++)$, so one may put all the flat 3-indices down. One has ${}^3g^{ru}(\tau, \sigma^u)\,{}^3g_{us}(\tau, \sigma^u) = \delta^r_s$.

The given parametrization of the cotetrad fields allows us to rewrite the action of ADM tetrad gravity in terms of the following 16 fields as configuration variables: three boost parameters $\varphi_{(a)}(\tau, \sigma^u)$; and the lapse $N(\tau, \sigma^u) = 1 + n(\tau, \sigma^u)$ and shift $n_{(a)}(\tau, \sigma^u)$ functions; and the nine components of cotriad fields ${}^3e_{(a)r}(\tau, \sigma^u)$ on the 3-spaces Σ_τ. As shown in Refs. [95, 281, 282, 284], the ADM action for the gravitational field has the expression

$$
\begin{aligned}
S_{grav} &= \frac{c^3}{16\pi\, G}\int d\tau d^3\sigma\, \Big[(1+n)\, {}^3e\, \epsilon_{(a)(b)(c)}\, {}^3e^r_{(a)}\, {}^3e^s_{(b)}\, {}^3\Omega_{rs(c)} \\
&+ \frac{{}^3e}{2(1+n)} ({}^3G_o^{-1})_{(a)(b)(c)(d)}\, {}^3e^r_{(b)} (n_{(a)|r} - \partial_\tau\, {}^3e_{(a)r}) \\
&\quad {}^3e^s_{(d)} (n_{(c)|s} - \partial_\tau\, {}^3e_{(c)\ s})\Big](\tau,\sigma^u).
\end{aligned} \tag{6.7}
$$

In it, ${}^3\Omega_{rs(a)}(\tau,\sigma^u) = \Big(\partial_r\, {}^3\omega_{s(a)} - \partial_s\, {}^3\omega_{r(a)} - \epsilon_{(a)(b)(c)}\, {}^3\omega_{r(b)}\, {}^3\omega_{s(c)}\Big)(\tau,\sigma^u)$ is the field strength associated with the 3-spin connection ${}^3\omega_{r(a)}(\tau,\sigma^u) = \frac{1}{2}\,\epsilon_{(a)(b)(c)}$ $\Big[{}^3e^u_{(b)}\, (\partial_r\, {}^3e_{(c)u} - \partial_u\, {}^3e_{(c)r}) + \frac{1}{2}\, {}^3e^u_{(b)}\, {}^3e^v_{(c)}\, {}^3e_{(d)r}\, (\partial_v\, {}^3e_{(d)u} - \partial_u\, {}^3e_{(d)v})\Big](\tau,\sigma^u)$ and $({}^3G_o^{-1})_{(a)(b)(c)(d)} = \delta_{(a)(c)}\delta_{(b)(d)} + \delta_{(a)(d)}\delta_{(b)(c)} - 2\delta_{(a)(b)}\delta_{(c)(d)}$ is the flat (with lower indices) inverse of the flat Wheeler–DeWitt super-metric ${}^3G_{o(a)(b)(c)(d)} = \delta_{(a)(c)}\,\delta_{(b)(d)} + \delta_{(a)(d)}\,\delta_{(b)(c)} - \delta_{(a)(b)}\,\delta_{(c)(d)}$, ${}^3G_{o(a)(b)(e)(f)}\, {}^3G^{-1}_{o(e)(f)(c)(d)} = 2\,(\delta_{(a)(c)}\, \delta_{(b)(d)} + \delta_{(a)(d)}\,\delta_{(b)(c)})$.

The canonical momenta $\pi_{\varphi_{(a)}}(\tau,\sigma^u)$, $\pi_n(\tau,\sigma^u)$, $\pi_{n_{(a)}}(\tau,\sigma^u)$, ${}^3\pi^r_{(a)}(\tau,\sigma^u)$, conjugate to the configuration variables satisfy 14 first-class constraints: the ten primary constraints (the last three constraints generate rotations on quantities with flat indices (a) like the cotriads),

$$
\begin{aligned}
&\pi_{\varphi_{(a)}}(\tau,\sigma^u) \approx 0, \qquad \pi_n(\tau,\sigma^u) \approx 0, \qquad \pi_{n_{(a)}}(\tau,\sigma^u) \approx 0, \\
&{}^3M_{(a)}(\tau,\sigma^u) = \epsilon_{(a)(b)(c)}\, {}^3e_{(b)r}(\tau,\sigma^u)\, {}^3\pi^r_{(c)}(\tau,\sigma^u) \approx 0,
\end{aligned} \tag{6.8}
$$

and the secondary super-Hamiltonian and super-momentum constraints,

$$
\begin{aligned}
\mathcal{H}(\tau,\sigma^u) &= \Big[\frac{c^3}{16\pi\, G}\, {}^3e\; \epsilon_{(a)(b)(c)}\, {}^3e^r_{(a)}\, {}^3e^s_{(b)}\, {}^3\Omega_{rs(c)} \\
&- \frac{2\pi\, G}{c^3\, {}^3e}\, {}^3G_{o(a)(b)(c)(d)}\, {}^3e_{(a)r}\, {}^3\pi^r_{(b)}\, {}^3e_{(c)s}\, {}^3\pi^s_{(d)}\Big](\tau,\sigma^u) + \mathcal{M}(\tau,\sigma^u) \approx 0, \\
\mathcal{H}_{(a)}(\tau,\sigma^u) &= \Big[\partial_r\, {}^3\pi^r_{(a)} - \epsilon_{(a)(b)(c)}\, {}^3\omega_{r(b)}\, {}^3\pi^r_{(c)} + {}^3e^r_{(a)}\mathcal{M}_r\Big](\tau,\sigma^u) \approx 0.
\end{aligned} \tag{6.9}
$$

The functions $\mathcal{M}(\tau,\sigma^u)$ and $\mathcal{M}_r(\tau,\sigma^u)$ describe the matter present in the space-time: $\mathcal{M}(\tau,\sigma^u)$ is the (matter- and metric-dependent) internal mass density, while $\mathcal{M}_r(\tau,\sigma^u)$ is the universal (metric-independent) internal momentum density. If the action of matter is added to Eq. (6.7), one can evaluate the energy–momentum tensor $T^{AB}(\tau,\sigma^u) = -\Big[\frac{2}{\sqrt{-{}^4g}}\,\frac{\delta S_{matter}}{\delta\, {}^4g_{AB}}\Big](\tau,\sigma^u)$ of the matter[2] and determine these functions from the following parametrization:

$$
\begin{aligned}
T^{\tau\tau}(\tau,\sigma^u) &= \frac{\mathcal{M}(\tau,\sigma^u)}{[{}^3e\,(1+n)^2](\tau,\sigma^u)}, \\
T^{\tau r}(\tau,\sigma^u) &= \Big(\frac{{}^3e^r_{(a)}\,\Big[(1+n)\, {}^3e^s_{(a)}\,\mathcal{M}_s - n_{(a)}\,\mathcal{M}\Big]}{{}^3e\,(1+n)^2}\Big)(\tau,\sigma^u).
\end{aligned} \tag{6.10}
$$

[2] The Hamilton equations imply ${}^4\nabla_A\, T^{AB}(\tau,\sigma^u) \equiv 0$ in accord with Einstein's equations and the Bianchi identity.

The extrinsic curvature tensor of the 3-spaces Σ_τ as 3-manifolds embedded into the space-time has the following expression in terms of the barred cotriads of Eq. (6.4) and their conjugate barred momenta:

$$
\begin{aligned}
{}^3K_{rs}(\tau,\sigma^u) = -\frac{4\pi\,G}{c^3\,{}^3\bar{e}(\tau,\sigma^u)} \sum_{abu} \Big[& \Big({}^3\bar{e}_{(a)r}\,{}^3\bar{e}_{(b)s} + {}^3\bar{e}_{(a)s}\,{}^3\bar{e}_{(b)r}\Big)\,{}^3\bar{e}_{(a)u}\,\bar{\pi}^u_{(b)} \\
& - {}^3\bar{e}_{(a)r}\,{}^3\bar{e}_{(a)s}\,{}^3\bar{e}_{(b)u}\,\bar{\pi}^u_{(b)}\Big](\tau,\sigma^u).
\end{aligned} \tag{6.11}
$$

Therefore, the basis of canonical variables for this formulation of tetrad gravity, naturally adapted to 7 of the 14 first-class constraints, is

$\varphi_{(a)}(\tau,\sigma^u)$	$n(\tau,\sigma^u)$	$n_{(a)}(\tau,\sigma^u)$	${}^3e_{(a)r}(\tau,\sigma^u)$
$\pi_{\varphi_{(a)}}(\tau,\sigma^u)\approx 0$	$\pi_n(\tau,\sigma^u)\approx 0$	$\pi_{n_{(a)}}(\tau,\sigma^u)\approx 0$	${}^3\pi^r_{(a)}(\tau,\sigma^u)$

(6.12)

The behavior of some of these fields at spatial infinity, compatible with the absence of super-translations, is given after Eq. (6.3); for the others there is the following behavior for $r=\sqrt{\sum_r(\sigma^r)^2}\rightarrow\infty$ and $\epsilon>0$; ${}^3\pi^r_{(a)}(\tau,\sigma^u)\rightarrow O(r^{-5/2})$, $\varphi_{(a)}(\tau,\sigma^u)\rightarrow O(r^{1+\epsilon})$, $\pi_{\varphi_{(a)}}(\tau,\sigma^u)\rightarrow O(r^{-2})$, $\pi_n(\tau,\sigma^u)\rightarrow O(r^{-3})$, $\pi_{n_{(a)}}(\tau,\sigma^u)\rightarrow O(r^{-3})$

From the action (Eq. 6.7), after having added the matter action, one can obtain the standard non-Hamiltonian ADM equations ($|r$ denotes the 3 covariant derivative in the 3-space Σ_τ with 3-metric ${}^3g_{rs}$; ${}^3R_{rs}$ is the 3-Ricci tensor of Σ_τ):

$$
\begin{aligned}
\partial_\tau\,{}^3g_{rs}(\tau,\sigma^u) &\stackrel{\circ}{=} \Big(n_{r|s}+n_{s|r}-2\,(1+n)\,{}^3K_{rs}\Big)(\tau,\sigma^u),\\
\partial_\tau\,{}^3K_{rs}(\tau,\sigma^u) &\stackrel{\circ}{=} (1+n(\tau,\sigma^u))\,\Big({}^3R_{rs}+{}^3K\,{}^3K_{rs}-2\,{}^3K_{ru}\,{}^3K^u_s\Big)(\tau,\sigma^u)\\
&\quad -\Big(n_{|s|r}+n^u_{|s}\,{}^3K_{ur}+n^u_{|r}\,{}^3K_{us}+n^u\,{}^3K_{rs|u}\Big)(\tau,\sigma^u)(\tau,\sigma^u),
\end{aligned} \tag{6.13}
$$

with the quantities appearing in these equations re-expressed in terms of the configurational variables of Eq. (6.12).

Instead, at the Hamiltonian level one can get the Hamilton equations for all the variables of the canonical basis (Eq. 6.12), as shown in Ref. [95], by using the Dirac Hamiltonian. As shown in Ref. [281], the Dirac Hamiltonian has the form (if the matter contains the electromagnetic field there are extra terms with the electromagnetic first-class constraints)

$$
\begin{aligned}
H_D &= \frac{1}{c}\,\hat{E}_{ADM} + \int d^3\sigma\,\Big[n\,\mathcal{H} - n_{(a)}\,\mathcal{H}_{(a)}\Big](\tau,\sigma^u)\\
&\quad + \int d^3\sigma\,\Big[\lambda_n\,\pi_n + \lambda_{n_{(a)}}\,\pi_{n_{(a)}} + \lambda_{\varphi_{(a)}}\,\pi_{\varphi_{(a)}} + \mu_{(a)}\,{}^3M_{(a)}\Big](\tau,\sigma^u),
\end{aligned} \tag{6.14}
$$

where $\hat{E}_{ADM}$ is the weak ADM energy and the λ are arbitrary Dirac multipliers.

From equations 2.22, 3.43, and 3.47 of Ref. [95] we get the following expression of the ten weak asymptotic ADM Poincaré generators:

$$\hat{P}^{\tau}_{ADM} = \frac{1}{c}\,\hat{E}_{ADM} = \int d^3\sigma \Big[-\frac{c^3}{16\pi\, G}\,\sqrt{\gamma}\; {}^3g^{rs}\left({}^3\Gamma^u_{rv}\,{}^3\Gamma^v_{su} - {}^3\Gamma^u_{rs}\,{}^3\Gamma^v_{vu}\right)$$
$$+ \frac{8\pi\, G}{c^3\,\sqrt{\gamma}}\,{}^3G_{rsuv}\,{}^3\Pi^{rs}\,{}^3\Pi^{uv} + \mathcal{M}\Big](\tau,\vec{\sigma}),$$
$$\hat{P}^{r}_{ADM} = -2\int d^3\sigma \left[{}^3\Gamma^r_{su}(\tau,\vec{\sigma})\,{}^3\Pi^{su} - \frac{1}{2}\,{}^3g^{rs}\,\mathcal{M}_s\right](\tau,\vec{\sigma}) \approx 0,$$

$$\hat{J}^{\tau r}_{ADM} = -\hat{J}^{r\tau}_{ADM} = \epsilon \int d^3\sigma \Big(\sigma^r$$
$$\Big[\frac{c^3}{16\pi\, G}\,\sqrt{\gamma}\;{}^3g^{ns}\,({}^3\Gamma^u_{nv}\,{}^3\Gamma^v_{su} - {}^3\Gamma^u_{ns}\,{}^3\Gamma^v_{vu}) - \frac{8\pi\, G}{c^3\,\sqrt{\gamma}}\,{}^3G_{nsuv}\,{}^3\Pi^{ns}\,{}^3\Pi^{uv} + \mathcal{M}\Big]$$
$$+ \frac{c^3}{16\pi\, G}\,\delta^r_u\,({}^3g_{vs} - \delta_{vs})\,\partial_n\left[\sqrt{\gamma}\,({}^3g^{ns}\,{}^3g^{uv} - {}^3g^{nu}\,{}^3g^{sv})\right]\Big)(\tau,\vec{\sigma}) \approx 0,$$

$$\hat{J}^{rs}_{ADM} = \int d^3\sigma \left[(\sigma^r\,{}^3\Gamma^s_{uv} - \sigma^s\,{}^3\Gamma^r_{uv})\,{}^3\Pi^{uv} - \frac{1}{2}\,(\sigma^r\,{}^3g^{su} - \sigma^s\,{}^3g^{ru})\,\mathcal{M}_u\right](\tau,\vec{\sigma}), \tag{6.15}$$

by using ${}^3g_{rs} = {}^3e_{(a)r}\,{}^3e_{(a)s}$, ${}^3\Pi^{rs} = \frac{1}{4}[{}^3e^r_{(a)}\,{}^3\pi^s_{(a)} + {}^3e^s_{(a)}\,{}^3\pi^r_{(a)}]$ and ${}^3\Gamma^u_{rs} = \frac{1}{2}\,{}^3g^{uv}\,(\partial_s\,{}^3g_{vr} + \partial_r\,{}^3g_{vs} - \partial_v\,{}^3g_{rs}) = \frac{1}{2}\,{}^3e^u_{(a)}\Big[\partial_r\,{}^3e_{(a)s} + \partial_s\,{}^3e_{(a)r} + {}^3e^v_{(a)}\Big({}^3e_{(b)r}(\partial_s\,{}^3e_{(b)v} - \partial_v\,{}^3e_{(b)s}) + {}^3e_{(b)s}(\partial_r\,{}^3e_{(b)v} - \partial_v\,{}^3e_{(b)r})\Big)\Big]$.

Since one is in a non-inertial rest-frame (due to the absence of super-translations), one has the rest-frame conditions $\hat{P}^r_{ADM} \approx 0$, as in special relativity (SR). Then, one has to add the conditions $\hat{\mathcal{K}}^r_{ADM} \approx 0$ to eliminate the internal 3-center of mass of the 3-universe, as in SR. Therefore, the 3-universe can be seen as a decoupled external canonical non-covariant center of mass carrying a pole–dipole structure: the invariant mass $M = \frac{1}{c^2}\,\hat{E}_{ADM}$ and the rest spin $\hat{J}^{rs}_{ADM}$. This view is in accord with an old suggestions of Dirac [20, 21]. As in SR, one can choose the Fokker–Pryce world-line, built in terms of the asymptotic ADM Poincaré generators, as the asymptotic inertial observer origin of the 3-coordinates σ^r. The connection with the dynamical observers [90] on the Earth can be faced only in the post-Newtonian approximation of GR. As in SR, the elimination of the external center of mass implies that the gravitational physics depends only on relative variables hidden inside the tetrads to be determined like in the case of the Klein–Gordon field in Section 4.1. See section 5.2 of Ref. [97] for the center-of-mass problem in GR. This relevance of relative variables in the relativistic dynamics fits with Rovelli's relational point of view [373, 374].

In Refs. [95–97] the study of ADM canonical tetrad gravity was done with the following type of matter: – charged scalar particles (described by the canonical variables $\eta_i^r(\tau)$, $\kappa_{ir}(\tau)$), and the electromagnetic field in the non-covariant radiation gauge (described by the canonical variables $A^r_\perp(\tau,\sigma^u)$, $\pi^r_\perp(\tau,\sigma^u)$ as shown in Section 4.2). The particles (described by an action like the one in Eq. (2.12)) have not only Grassmann-valued electric charges Q_i ($Q_i^2 = 0$, $Q_i\,Q_j = Q_j\,Q_i$ for $i \neq j$) to regularize the electromagnetic self-energies, but also Grassmann-valued signs of the energy ($\eta_i^2 = 0$, $\eta_i\,\eta_j = \eta_j\,\eta_i$ for $i \neq j$) to regularize the gravitational self-energies (see Appendix B).

Instead, in Ref. [104] the matter is the perfect fluid described in Section 4.6.

In the case of N particles the functions $\mathcal{M}$ and $\mathcal{M}_r$ have the expression (see Section 6.4 and Refs. [95–97] for their form in the presence of the electromagnetic field)

$$\mathcal{M}(\tau,\sigma^u) = \sum_{i=1}^N \delta^3(\sigma^u,\eta_i^u(\tau))\,\eta_i\,\sqrt{m_i^2\,c^2 + {}^3e^r_{(a)}(\tau,\sigma^u)\,\kappa_{ir}(\tau)\,{}^3e^s_{(a)}(\tau,\sigma^u)\,\kappa_{is}(\tau)},$$
$$\mathcal{M}_r(\tau,\sigma^u) = \sum_{i=1}^N \eta_i\,\kappa_{ir}(\tau), \tag{6.16}$$

while in the case of dust [104], described by canonical coordinates $\alpha^i(\tau,\sigma^u)$, $\Pi_i(\tau,\sigma^u)$, $i = 1,2,3$, they have the expression

$$\mathcal{M}(\tau,\sigma^u) = \sqrt{\mu^2\,[\det(\partial_s\,\alpha^j)]^2 + \tilde{\phi}^{-2/3}\sum_{arsij} Q_a^{-2}\,V_{ra}\,V_{sa}\,\partial_r\,\alpha^i\,\partial_s\,\alpha^j\,\Pi_i\,\Pi_j(\tau,\sigma^u)},$$
$$\mathcal{M}_r(\tau,\sigma^u) = \sum_i \partial_r\,\alpha^i(\tau,\sigma^u)\,\Pi_i(\tau,\sigma^u). \tag{6.17}$$

6.2 The Shanmugadhasan Canonical Transformation to the York Canonical Basis for the Search of the Dirac Observables of the Gravitational Field

The presence of 14 first-class constraints in the phase space having the 32 fields of Eq. (6.12) as a canonical basis implies that there are 14 gauge variables describing inertial effects and 2 canonical pairs of physical degrees of freedom describing the tidal effects of the gravitational field (namely gravitational waves in the weak field limit). To disentangle the inertial effects from the tidal ones, one needs a canonical transformation to a new canonical basis adapted to all the ten primary constraints (Eq. 6.8) and containing the barred variables defined in Eq. (6.4).

A canonical transformation adapted to the ten primary constraints (Eq. 6.8) was found in Ref. [284]. It implements the York map (see Section 8.3) in the cases in which the 3-metric ${}^3g_{rs}$ has three distinct eigenvalues and diagonalizes the York–Lichnerowicz approach (see Ref. [10] for a review of this approach).

As said before Eq. (6.4), one can decompose the cotriads on Σ_τ in the product of a rotation matrix, belonging to the subgroup $SO(3)$ of the tetrad gauge group and depending on three Euler angles $\alpha_{(a)}(\tau,\sigma^r)$, and of barred cotriads depending only on six independent fields. The canonical transformation Abelianizes the constraints ${}^3M_{(a)}(\tau,\sigma^u) \approx 0$ of Eq. (6.8), satisfying $\{{}^3M_{(a)}(\tau,\sigma^u), {}^3M_{(b)}(\tau,\sigma'^u)\} = \epsilon_{(a)(b)(c)}\, {}^3M_{(c)}(\tau,\sigma^u)\delta^3(\sigma^u,\sigma'^u)$, and replaces them with the vanishing of the three momenta $\pi^{(\alpha)}_{(a)}(\tau,\sigma^r) \approx 0$ conjugate to the Euler angles.

The new canonical basis, named York canonical basis, is ($a = 1,2,3$; $\bar{a} = 1,2$; all these canonical variables are functions of (τ,σ^u))

$\varphi_{(a)}$	$\alpha_{(a)}$	n	$\bar{n}_{(a)}$	θ^r	$\tilde{\phi}$	$R_{\bar{a}}$
$\pi_{\varphi_{(a)}} \approx 0$	$\pi^{(\alpha)}_{(a)} \approx 0$	$\pi_n \approx 0$	$\pi_{\bar{n}_{(a)}} \approx 0$	$\pi^{(\theta)}_r$	$\pi_{\tilde{\phi}} = \frac{c^3}{12\pi G}\, {}^3K$	$\Pi_{\bar{a}}$

(6.18)

In it the cotriads and the components of the 4-metric have the following expression:

$$
\begin{aligned}
{}^3e_{(a)r}(\tau,\sigma^u) &= \sum_b R_{(a)(b)}(\alpha_{(c)}(\tau,\sigma^u))\, {}^3\bar{e}_{(b)r}(\tau,\sigma^u)\\
&= \sum_b R_{(a)(b)}(\alpha_{(c)}(\tau,\sigma^u))\, V_{rb}(\theta^i(\tau,\sigma^u))\, \tilde{\phi}^{1/3}(\tau,\sigma^u)\, e^{\sum_{\bar{a}}^{1,2} \gamma_{\bar{a}a} R_{\bar{a}}(\tau,\sigma^u)},\\
{}^4g_{\tau\tau}(\tau,\sigma^u) &= \epsilon\left[(1+n)^2 - \sum_a \bar{n}^2_{(a)}\right](\tau,\sigma^u),\\
{}^4g_{\tau r}(\tau,\sigma^u) &= -\epsilon \sum_a n_{(a)}(\tau,\sigma^u)\, {}^3e_{(a)r}(\tau,\sigma^u)\\
&= -\epsilon \sum_a \bar{n}_{(a)}(\tau,\sigma^u)(\tau,\sigma^u)\, {}^3\bar{e}_{(a)r}(\tau,\sigma^u),\\
\tilde{\phi}(\tau,\sigma^u) &= \phi^6(\tau,\sigma^u) = \sqrt{\det {}^3g_{rs}(\tau,\sigma^u)},\\
{}^4g_{rs}(\tau,\sigma^u) &= -\epsilon\, {}^3g_{rs}(\tau,\sigma^u)\\
&= -\epsilon\, \tilde{\phi}^{2/3}(\tau,\sigma^u) \sum_a V_{ra}(\theta^i(\tau,\sigma^u))\, V_{sa}(\theta^i(\tau,\sigma^u))\, Q^2_a(\tau,\sigma^u),\\
Q_a(\tau,\sigma^u) &= e^{\sum_{\bar{a}}^{1,2} \gamma_{\bar{a}a} R_{\bar{a}}(\tau,\sigma^u)} = e^{\Gamma_a(\tau,\sigma^u)},
\end{aligned}
\tag{6.19}
$$

The set of numerical parameters $\gamma_{\bar{a}a}$ appearing in $Q_a(\tau,\sigma^u)$ satisfies [281] $\sum_u \gamma_{\bar{a}u} = 0$, $\sum_u \gamma_{\bar{a}u}\gamma_{\bar{b}u} = \delta_{\bar{a}\bar{b}}$, $\sum_{\bar{a}} \gamma_{\bar{a}u}\gamma_{\bar{a}v} = \delta_{uv} - \frac{1}{3}$. Each solution of these equations defines a different York canonical basis.

This canonical basis has been found due to the fact that the 3-metric ${}^3g_{rs}$ is a real symmetric 3×3 matrix, which may be diagonalized with an orthogonal matrix $V(\theta^r)$, $V^{-1} = V^T$ ($\sum_u V_{ua}V_{ub} = \delta_{ab}$, $\sum_a V_{ua}V_{va} = \delta_{uv}$, $\sum_{uv} \epsilon_{wuv} V_{ua} V_{vb} = \sum_c \epsilon_{abc} V_{cw}$), $\det V = 1$, depending on three parameters θ^i ($i = 1,2,3$),[3] whose conjugate momenta $\Pi^{(\theta)}_i$ are to be determined as solutions of

[3] Due to the positive signature of the 3-metric, one defines the matrix V with the following indices: V_{ru}. Since the choice of Shanmugadhasan canonical bases breaks manifest covariance, one will use the notation $V_{ua} = \sum_v V_{uv}\,\delta_{v(a)}$ instead of $V_{u(a)}$.

the super-momentum constraints. If one chooses these three gauge parameters to be Euler angles $\hat{\theta}^i(\tau, \sigma^u)$, one gets a description of the 3-coordinate systems on Σ_τ from a local point of view, because they give the orientation of the tangents to the three 3-coordinate lines through each point. However, for the calculations (see Refs. [95–97]) it is more convenient to choose the three gauge parameters as first-kind coordinates $\theta^i(\tau, \sigma^u)$ $(-\infty < \theta^i < +\infty)$ on the $O(3)$ group manifold, so that by definition one has $V_{ru}(\theta^i) = \left(e^{-\sum_i \hat{T}_i \theta^i}\right)_{ru}$, where $(\hat{T}_i)_{ru} = \epsilon_{rui}$ are the generators of the $O(3)$ Lie algebra in the adjoint representation, and the Euler angles may be expressed as $\hat{\theta}^i = f^i(\theta^n)$. The Cartan matrix has the form $A(\theta^n) = \frac{1-e^{-\sum_i \hat{T}_i \theta^i}}{\sum_i \hat{T}_i \theta^i}$ and satisfies $A_{ri}(\theta^n)\, \theta^i = \delta_{ri}\, \theta^i$; $B(\theta^i) = A^{-1}(\theta^i)$.

From now on, for the sake of notational simplicity, the symbol V will mean $V(\theta^i)$.

The extrinsic curvature tensor of the 3-space Σ_τ has the expression

$$
\begin{aligned}
{}^3K_{rs}(\tau, \sigma^u) = -\frac{4\pi\, G}{c^3} \Big[\tilde{\phi}^{-1/3} \Big(\sum_a Q_a^2\, V_{ra}\, V_{sa} \Big[2 \sum_{\bar{b}} \gamma_{\bar{b}a}\, \Pi_{\bar{b}} - \tilde{\phi}\, \pi_{\tilde{\phi}} \Big] \\
+ \sum_{ab} Q_a\, Q_b\, (V_{ra}\, V_{sb} + V_{rb}\, V_{sa}) \sum_{twi} \frac{\epsilon_{abt}\, V_{wt}\, B_{iw}\, \pi_i^{(\theta)}}{Q_b\, Q_a^{-1} - Q_a\, Q_b^{-1}} \Big) \Big] (\tau, \sigma^u).
\end{aligned}
\tag{6.20}
$$

This canonical transformation realizes a York map because the gauge variable $\pi_{\tilde{\phi}}(\tau, \sigma^u)$ (describing the freedom in the choice of the trace of the extrinsic curvature of the instantaneous 3-spaces Σ_τ) is proportional to York internal extrinsic time ${}^3K(\tau, \sigma^u)$. It is the only gauge variable among the momenta: This is a reflex of the Lorentz signature of space-time, because $\pi_{\tilde{\phi}}(\tau, \sigma^u)$ and $\theta^n(\tau, \sigma^u)$ can be used as a set of 4-coordinates for the space-time [271–274]. The York time describes the effect of gauge transformations producing a deformation of the shape of the 3-space along the 4-normal to the 3-space as a 3-sub-manifold of space-time.

Its conjugate variable, to be determined by the super-Hamiltonian constraint (interpreted as the Lichnerowicz equation [10]), is $\tilde{\phi}(\tau, \sigma^u) = \phi^6(\tau, \sigma^u) = {}^3\bar{e}(\tau, \sigma^u) = \sqrt{\det {}^3g_{rs}(\tau, \sigma^u)}$, which is proportional to Misner's internal intrinsic time; moreover $\tilde{\phi}(\tau, \sigma^u)$ is the 3-volume density on Σ_τ: $V_R = \int_R d^3\sigma\, \tilde{\phi}(\tau, \sigma^u)$, $R \subset \Sigma_\tau$. Since one has ${}^3g_{rs}(\tau, \sigma^u) = \tilde{\phi}^{2/3}(\tau, \sigma^u)\, {}^3\hat{g}_{rs}(\tau, \sigma^u)$ with $\det {}^3\hat{g}_{rs}(\tau, \sigma^u) = 1$, $\tilde{\phi}(\tau, \sigma^u)$ is also called the conformal factor of the 3-metric. Instead, the unknowns in the super-momentum constraints are the momenta $\pi_i^{(\theta)}(\tau, \sigma^u)$.

The two pairs of canonical variables $R_{\bar{a}}(\tau, \sigma^u)$, $\Pi_{\bar{a}}(\tau, \sigma^u)$, $\bar{a} = 1, 2$, describe the generalized tidal effects, namely the independent physical degrees of freedom of the gravitational field. They are 3-scalars on Σ_τ and the configuration tidal variables $R_{\bar{a}}(\tau, \sigma^u)$ parametrize the two eigenvalues of the 3-metric ${}^3\hat{g}_{rs}(\tau, \sigma^u)$ with unit determinant. They are DOs only with respect to the gauge transformations generated by 10 of the 14 first-class constraints. Let us remark that, if one fixes completely the gauge and one goes to Dirac brackets, then the only surviving dynamical variables $R_{\bar{a}}(\tau, \sigma^u)$ and $\Pi_{\bar{a}}(\tau, \sigma^u)$ become two pairs of non-canonical DOs for that gauge: The two pairs of canonical DOs have to be found as a Darboux basis of the copy of the reduced phase space identified by the gauge and they will be (in general non-local) functionals of the $R_{\bar{a}}(\tau, \sigma^u)$, $\Pi_{\bar{a}}(\tau, \sigma^u)$ variables. To find the real two pairs of canonical DOs will be possible only after having found a solution of the super-Hamiltonian and super-momentum constraints.

Therefore, the 14 arbitrary gauge variables are $\varphi_{(a)}(\tau, \sigma^u)$, $\alpha_{(a)}(\tau, \sigma^u)$, $n(\tau, \sigma^u)$, $\bar{n}_{(a)}(\tau, \sigma^u)$, $\theta^i(\tau, \sigma^u)$, $\pi_{\tilde{\phi}}(\tau, \sigma^u)$: They describe the following generalized inertial effects [284]:

1. $\alpha_{(a)}(\tau, \sigma^u)$ and $\varphi_{(a)}(\tau, \sigma^u)$ are the six configuration variables parametrizing the $O(3,1)$ gauge freedom in the choice of the tetrads in the tangent plane to each point of Σ_τ and describe the arbitrariness in the choice of a tetrad to be associated to a time-like observer, whose world-line goes through the point $(\tau, \vec{\sigma})$. They fix the unit 4-velocity of the observer and the conventions for the orientation of three gyroscopes and their transport along the world-line of the observer. The Schwinger time gauges [375] are defined by the gauge fixings $\alpha_{(a)}(\tau, \sigma^u) \approx 0$, $\varphi_{(a)}(\tau, \sigma^u) \approx 0$.
2. $\theta^i(\tau, \sigma^u)$ (depending only on the 3-metric) describe the arbitrariness in the choice of the 3-coordinates in the instantaneous 3-spaces Σ_τ of the chosen non-inertial frame centered on an arbitrary time-like observer. Their choice will induce a pattern of relativistic inertial forces for the gravitational field, whose potentials are the functions $V_{ra}(\theta^i)$ present in the weak ADM energy $\hat{E}_{ADM}$.
3. $\bar{n}_{(a)}(\tau, \sigma^u)$, the shift functions, describe which points on different instantaneous 3-spaces have the same numerical value of the 3-coordinates. They are the inertial potentials describing the effects of the non-vanishing off-diagonal components ${}^4g_{\tau r}(\tau, \sigma^u)$ of the 4-metric, namely they are the gravito-magnetic potentials[4] responsible for effects like the dragging of inertial frames (Lense–Thirring effect) in the post-Newtonian approximation. The shift functions are determined by the τ-preservation of the gauge fixings determining the gauge variables $\theta^i(\tau, \sigma^u)$.
4. $\pi_{\tilde{\phi}}(\tau, \sigma^u)$, i.e., the York time ${}^3K(\tau, \sigma^u)$, describes the non-dynamical arbitrariness in the choice of the convention for the synchronization of distant clocks, which remains in the transition from SR to GR. Since the York time is present in the Dirac Hamiltonian, it is a new inertial potential connected

[4] In the post-Newtonian approximation in harmonic gauges they are the counterpart of the electromagnetic vector potentials describing magnetic fields [10]: (1) $N = 1 + n$, $n \stackrel{def}{=} -\frac{4\epsilon}{c^2}\Phi_G$ with Φ_G the gravito-electric potential; (2) $n_r \stackrel{def}{=} \frac{2\epsilon}{c^2} A_{G\,r}$ with $A_{G\,r}$ the gravito-magnetic potential; and (3) $E_{G\,r} = \partial_r \Phi_G - \partial_\tau (\frac{1}{2} A_{G\,r})$ (the gravito-electric field) and $B_{G\,r} = \epsilon_{ruv}\, \partial_u\, A_{G\,v} = c\, \Omega_{G\,r}$ (the gravito-magnetic field). Let us remark that in arbitrary gauges the analogy with electromagnetism breaks down.

to the problem of the relativistic freedom in the choice of the shape of the instantaneous 3-space, which has no Newtonian analogue (in Galilei space-time, time is absolute and there is an absolute notion of Euclidean 3-space). Its effects are completely unexplored. Instead, the other components of the extrinsic curvature of Σ_τ are dynamically determined by the tidal variables and by the choice of the 3-coordinate system in the 3-space.

5. $1+n(\tau,\sigma^u)$, the lapse function appearing in the Dirac Hamiltonian describes the arbitrariness in the choice of the unit of proper time in each point of the simultaneity surfaces Σ_τ, namely how these surfaces are packed in the 3+1 splitting. The lapse function is determined by the τ-preservation of the gauge fixing for the gauge variable ${}^3K(\tau,\sigma^u)$.

As shown in Refs. [271–274], the dynamical nature of space-time implies that each solution (i.e., an Einstein 4-geometry) of Einstein's equations (or of the associated ADM Hamilton equations) dynamically selects a preferred 3+1 splitting of the space-time, namely in GR the instantaneous 3-spaces are dynamically determined modulo only one inertial gauge function (the gauge freedom in clock synchronization in GR). In the York canonical basis this function is the York time, namely the trace of the extrinsic curvature of the 3-space. Instead, in SR the gauge freedom in clock synchronization depends on four basic gauge functions: the embeddings $z^\mu(\tau,\sigma^r)$ and both the 4-metric and the whole extrinsic curvature tensor are derived inertial potentials. In GR the extrinsic curvature tensor of the 3-spaces is a mixture of dynamical (tidal) pieces and inertial gauge variables playing the role of inertial potentials.

6.3 The Non-Harmonic 3-Orthogonal Schwinger Time Gauges and the Metrological Interpretation of the Inertial Gauge Variables

As shown in Ref. [95], in the York canonical basis the Dirac Hamiltonian becomes (the λ are arbitrary Dirac multipliers; the Dirac multiplier $\lambda_r(\tau)$ implements the rest-frame condition $\hat{P}^r_{ADM}\approx 0$)

$$H_D=\frac{1}{c}\,\hat{E}_{ADM}+\int d^3\sigma\,\Big[n\,\mathcal{H}-n_{(a)}\,\mathcal{H}_{(a)}\Big](\tau,\sigma^u)+\lambda_r(\tau)\,\hat{P}^r_{ADM}$$
$$+\int d^3\sigma\,\Big[\lambda_n\,\pi_n+\lambda_{\bar{n}_{(a)}}\,\pi_{\bar{n}_{(a)}}+\lambda_{\varphi_{(a)}}\,\pi_{\varphi_{(a)}}+\lambda_{\alpha_{(a)}}\,\pi^{(\alpha)}_{(a)}\Big](\tau,\sigma^u), \tag{6.21}$$

with the following expression for the weak ADM energy:

$$\hat{E}_{ADM}=c\int d^3\sigma\,\Bigg[\check{\mathcal{M}}-\frac{c^3}{16\pi G}\,\mathcal{S}+\frac{2\pi G}{c^3}\,\tilde{\phi}^{-1}\,\Bigg(-3\,(\tilde{\phi}\,\pi_{\tilde{\phi}})^2+2\sum_{\bar{b}}\Pi^2_{\bar{b}}$$
$$+2\sum_{abtwiuvj}\frac{\epsilon_{abt}\,\epsilon_{abu}\,V_{wt}\,B_{iw}\,V_{vu}\,B_{jv}\,\pi_i^{(\theta)}\,\pi_j^{(\theta)}}{\Big[Q_a\,Q_b^{-1}-Q_b\,Q_a^{-1}\Big]^2}\Bigg)\Bigg](\tau,\sigma^u). \tag{6.22}$$

Here, $\mathcal{S}(\tau, \sigma^u)$ is a function of $\tilde{\phi}(\tau, \sigma^u)$, $\theta^i(\tau, \sigma^u)$, and $R_{\bar{a}}(\tau, \sigma^u)$ (given in equation B8 of Ref. [95]), which play the role of inertial potentials, depending on the choice of the 3-coordinates in the 3-space (it is the $\Gamma - \Gamma$ term in the scalar 3-curvature of the 3-space). B_{ij} is the inverse of the Cartan matrix (see after Eq. (6.19)).

Eq. (6.22) shows that the kinetic term, quadratic in the momenta, is not positive definite. While the kinetic energy of the tidal variables and the last term[5] are positive definite, there is the negative kinetic terms (vanishing only in the gauges ${}^3K(\tau, \sigma^u) = 0$) $-\frac{6\pi G}{c^2} \int d^3\sigma\, \tilde{\phi}(\tau, \sigma^u)\, \pi^2_{\tilde{\phi}}(\tau, \sigma^u) = -\frac{c^4}{24\pi G} \int d^3\sigma\, \tilde{\phi}(\tau, \sigma^u)\, {}^3K^2(\tau, \sigma^u)$. It is an inertial potential associated with the inertial gauge variable York time, which is a momentum due to the Lorentz signature of space-time. It was known that this quadratic form is not definite positive, but only in the York canonical basis can this be made explicit.

In the York canonical basis it is possible to follow the procedure for the fixation of a gauge natural from the point of view of constraint theory when there are chains of first-class constraints (see Chapter 8). This procedure implies that one has to add six gauge fixings to the primary constraints without secondaries ($\pi_{\varphi_{(a)}}(\tau, \sigma^u) \approx 0$, $\pi_{\alpha_{(a)}}(\tau, \sigma^u) \approx 0$) and four gauge fixings to the secondary super-Hamiltonian and super-momentum constraints. These ten gauge fixings must be preserved in time, namely their Poisson brackets with the Dirac Hamiltonian must vanish. The τ-preservation of the six gauge fixings determining the gauge variables $\alpha_{(a)}(\tau, \sigma^u)$ and $\varphi_{(a)}(\tau, \sigma^u)$ produces the equations determining the six Dirac multipliers $\lambda_{\varphi_{(a)}}(\tau, \sigma^u)$, $\lambda_{\alpha_{(a)}}(\tau, \sigma^u)$. The τ-preservation of the other four gauge fixings, determining the gauge variables $\theta^i(\tau, \sigma^u)$ and the York time ${}^3K(\tau, \sigma^u)$, produces four secondary gauge fixing constraints for the determination of the lapse and shift functions. Then, the τ-preservation of these secondary gauge fixings determines the four Dirac multipliers $\lambda_n(\tau, \sigma^u)$, $\lambda_{\bar{n}_{(a)}}(\tau, \sigma^u)$. Instead, in numerical gravity one gives independent gauge fixings for both the primary and secondary gauge variables in such a way as to minimize the computer time.

In section V of Ref. [95] there is a review of the gauges usually used in canonical gravity. It is shown that the commonly used family of the harmonic gauges is not natural according to the above procedure. The harmonic gauge fixings imply hyperbolic PDE for the lapse and shift functions, to be added to the hyperbolic PDE for the tidal variables. Therefore, in harmonic gauges both the tidal variables and the lapse and shift functions depend (in a retarded way) on the no-incoming radiation condition on the Cauchy surface in the past (so that the knowledge of ${}^3K$ from the initial time till today is needed).

Instead, the natural gauge fixings in the York canonical basis of ADM tetrad gravity are the family of Schwinger time gauges [375], where the $O(3,1)$

[5] It describes gravito-magnetic effects.

gauge freedom of the tetrads is eliminated with the gauge fixings (implying $\lambda_{\varphi_{(a)}}(\tau, \sigma^u) = \lambda_{\alpha_{(a)}}(\tau, \sigma^u) = 0$),

$$\alpha_{(a)}(\tau, \sigma^u) \approx 0, \qquad \varphi_{(a)}(\tau, \sigma^u) \approx 0, \tag{6.23}$$

and the subfamily of the 3-orthogonal gauges,

$$\theta^i(\tau, \sigma^u) \approx 0, \qquad {}^3K(\tau, \sigma^u) \approx F(\tau, \sigma^u) = \text{numerical function}, \tag{6.24}$$

in which the 3-coordinates are chosen in such a way that the 3-metric in the 3-spaces Σ_τ is diagonal. The τ-preservation of Eqs. (6.24) gives four coupled elliptic PDEs for the lapse and shift functions. Therefore, in these gauges only the tidal variables (the gravitational waves after linearization), and therefore only the eigenvalues of the 3-metric with unit determinant inside Σ_τ, depend (in a retarded way) on the no-incoming radiation condition. The solutions $\tilde{\phi}(\tau, \sigma^u)$ and $\pi_i^{(\theta)}(\tau, \sigma^u)$ of the constraints and the lapse $1 + n(\tau, \sigma^u)$ and shift $\bar{n}_{(a)}(\tau, \sigma^u)$ functions depend only on the 3-space Σ_τ with fixed τ. If the matter consists of positive-energy particles (with a Grassmann regularization of the gravitational self-energies) [95–97], these solutions will contain action-at-a-distance gravitational potentials (replacing the Newton ones) and gravito-magnetic potentials.

In the family of 3-orthogonal gauges the weak ADM energy and the super-Hamiltonian and super-momentum constraints (they are coupled elliptic PDEs for their unknowns) have the expression (see equation 3.47 of Ref. [95] for the other weak ADM Poincaré generators)

$$\begin{aligned}
\hat{E}_{ADM}|_{\theta^i=0} &= c \int d^3\sigma \Bigg[\mathcal{M}|_{\theta^i=0} - \frac{c^3}{16\pi\, G}\, \mathcal{S}|_{\theta^i=0} \\
&\quad + \frac{2\pi\, G}{c^3}\, \tilde{\phi}^{-1} \Bigg(-3\, (\tilde{\phi}\, \pi_{\tilde{\phi}})^2 + 2 \sum_{\bar{b}} \Pi_{\bar{b}}^2 \\
&\quad + 2 \sum_{abij} \frac{\epsilon_{abi}\, \epsilon_{abj}\, \pi_i^{(\theta)}\, \pi_j^{(\theta)}}{\Big[Q_a\, Q_b^{-1} - Q_b\, Q_a^{-1}\Big]^2} \Bigg) \Bigg] (\tau, \sigma^u),
\end{aligned}$$

$$\begin{aligned}
\mathcal{H}(\tau, \sigma^u)|_{\theta^i=0} &= \frac{c^3}{16\pi\, G}\, \tilde{\phi}^{1/6}(\tau, \sigma^u)\, [8\, \hat{\triangle}\, \tilde{\phi}^{1/6} - {}^3\hat{R}|_{\theta^i=0}\, \tilde{\phi}^{1/6}](\tau, \sigma^u) \\
&\quad + \mathcal{M}|_{\theta^i=0}(\tau, \sigma^u) + \frac{2\pi\, G}{c^3}\, \tilde{\phi}^{-1} \Bigg[-3\, (\tilde{\phi}\, \pi_{\tilde{\phi}})^2 + 2 \sum_{\bar{b}} \Pi_{\bar{b}}^2 \\
&\quad + 2 \sum_{abij} \frac{\epsilon_{abi}\, \epsilon_{abj}\, \pi_i^{(\theta)}\, \pi_j^{(\theta)}}{\Big[Q_a\, Q_b^{-1} - Q_b\, Q_a^{-1}\Big]^2} \Bigg] (\tau, \sigma^u) \approx 0,
\end{aligned}$$

$$\tilde{\bar{\mathcal{H}}}_{(a)}|_{\theta^i=0}(\tau,\sigma^u) = \phi^{-2}(\tau,\vec{\sigma}) \Bigg[\sum_{b\neq a}\sum_i \frac{\epsilon_{abi}\, Q_b^{-1}}{Q_b\, Q_a^{-1} - Q_a\, Q_b^{-1}}\, \partial_b\, \pi_i^{(\theta)}$$
$$+2\sum_{b\neq a}\sum_i \frac{\epsilon_{abi}\, Q_a^{-1}}{\left(Q_b\, Q_a^{-1} - Q_a\, Q_b^{-1}\right)^2} \sum_{\bar{c}} (\gamma_{\bar{c}a} - \gamma_{\bar{c}b})\, \partial_b\, R_{\bar{c}}\, \pi_i^{(\theta)}$$
$$+ Q_a^{-1}\Big(\phi^6\, \partial_a\, \pi_{\tilde{\phi}} + \sum_{\bar{b}} (\gamma_{\bar{b}a}\, \partial_a\, \Pi_{\bar{b}} - \partial_a\, R_{\bar{b}}\, \Pi_{\bar{b}}) + \mathcal{M}_a\Big)\Bigg] (\tau,\sigma^u) \approx 0,$$

$$\hat{\triangle} = \sum_r Q_r^{-2}\Big[\partial_r^2 + 2\sum_{\bar{a}} \gamma_{\bar{a}r}\, \partial_r\, R_{\bar{a}}(\tau,\sigma^u)\, \partial_r\Big],$$
$$\mathcal{S}_{\theta^i=0}(\tau,\sigma^u) = \Bigg(\tilde{\phi}^{1/3} \sum_a Q_a^{-2}\Bigg[\frac{2}{9}(\tilde{\phi}^{-1}\, \partial_a\, \tilde{\phi})^2$$
$$+ \sum_{\bar{b}}\Big(\sum_{\bar{c}} (2\,\gamma_{\bar{b}a}\,\gamma_{\bar{c}a} - \delta_{\bar{b}\bar{c}})\, \partial_a\, R_{\bar{c}} - \frac{2}{3}\gamma_{\bar{b}a}\, \tilde{\phi}^{-1}\, \partial_a\, \tilde{\phi}\Big)\partial_a\, R_{\bar{b}}\Bigg]\Bigg)(\tau,\sigma^u). \tag{6.25}$$

In Ref. [95] there is the explicit form of the Hamilton equations for all the canonical variables of the gravitational field and of the matter replacing the standard 12 ADM equations and the matter equations ${}^4\nabla_A\, T^{AB}(x) = 0$ in the Schwinger time gauges and their restriction to the 3-orthogonal gauges. They could also be obtained from the effective Dirac Hamiltonian of the 3-orthogonal gauges, which is evaluated by means of a τ-dependent canonical transformation sending the gauge momentum $\pi_{\tilde{\phi}}(\tau,\sigma^u)$ in the gauge fixing conditions $\pi'_{\tilde{\phi}}(\tau,\sigma^u) = \frac{c^3}{12\pi\, G}\left({}^3K(\tau,\sigma^u) - F(\tau,\sigma^u)\right) \approx 0$, and which is given in equation (4.39) of Ref. [96].

These Hamilton equations are divided into five groups:

A. The contracted Bianchi identities, namely the evolution equations for the solutions $\tilde{\phi}(\tau,\sigma^u)$ and $\pi_i^{(\theta)}(\tau,\sigma^u)$ of the super-Hamiltonian and super-momentum constraints: They are identities stating that, given a solution of the constraints on a Cauchy surface, it remains a solution also at later times.
B. The evolution equation for the four basic gauge variables $\theta^i(\tau,\sigma^u)$ and ${}^3K(\tau,\sigma^u)$ (the equation for the York time is the Raychaudhuri equation[6]): These equations determine the lapse and the shift functions once four gauge fixings for the basic gauge variables are given.

[6] This equation is relevant for studying the development of caustics in a congruence of time-like geodesics for converging values of the expansion θ and of singularities in Einstein space-times [4]. However, the boundary conditions of asymptotically Minkowskian space-times without super-translations should avoid the singularity theorems, as happens with their subfamily without matter in Ref. [8].

C. The equations $\partial_\tau\, n(\tau,\sigma^u) = \lambda_n(\tau,\sigma^u)$ and $\partial_\tau\, \bar{n}_{(a)}(\tau,\sigma^u) = \lambda_{\bar{n}_{(a)}}(\tau,\sigma^u)$. Once the lapse and shift functions of the chosen gauge have been found, they determine the associated Dirac multipliers.
D. The hyperbolic evolution PDE for the tidal variables $R_{\bar{a}}(\tau,\sigma^u)$, $\Pi_{\bar{a}}(\tau,\sigma^u)$. When the equations for $\partial_\tau\, R_{\bar{a}}(\tau,\sigma^u)$ are inverted to get $\Pi_{\bar{a}}(\tau,\sigma^u)$ in terms of $R_{\bar{a}}(\tau,\sigma^u)$ and its derivatives, then the Hamilton equations for $\Pi_{\bar{a}}(\tau,\sigma^u)$ become hyperbolic PDEs for the evolution of the physical tidal variable $R_{\bar{a}}(\tau,\sigma^u)$.
E. The Hamilton equations for matter, when present.

Given a solution of the super-momentum and super-Hamiltonian constraints and the Cauchy data for the tidal variables on an initial 3-space, one can find a solution to Einstein's equations in radar 4-coordinates adapted to a time-like observer in the chosen gauge.

As in SR, one can consider the congruence of the Eulerian observers with zero vorticity associated with the 3+1 splitting of space-time, whose properties are described by Eq. (2.7). In Ref. [95] it is shown that in ADM tetrad gravity the congruence has the following properties in each point (τ,σ^r):

1. The acceleration ${}^3a^A = l^B\, {}^4\nabla_B\, l^A = {}^4g^{AB}\, {}^3a_B$ has the components ${}^3a^\tau = 0$, ${}^3a^r = \epsilon\, \tilde{\phi}^{-2/3}\, Q_a^{-2}\, V_{ra}\, V_{sa}\, \partial_s \ln(1+n)$, ${}^3a_\tau = -\tilde{\phi}^{-1/3}\, Q_a^{-1}\, V_{ra}\, \bar{n}_{(a)}\, \partial_r \ln(1+n)$, ${}^3a_r = -\partial_r \ln(1+n)$.
2. The expansion[7] coincides with the York time:
$$\theta = {}^4\nabla_A\, l^A = -\epsilon\, {}^3K = -\epsilon\, \frac{12\pi\, G}{c^3}\, \pi_{\tilde{\phi}}. \tag{6.26}$$
In cosmology, the expansion is proportional to the Hubble constant and the dimensionless cosmological deceleration parameter is $q = 3\, l^A\, {}^4\nabla_A \frac{1}{\theta} - 1 = -3\, \theta^{-2}\, l^A\, \partial_A\, \theta - 1$.
3. By using Eq. (6.4) it can be shown that the shear[8] $\sigma_{AB} = \sigma_{BA} = -\frac{\epsilon}{2}\, ({}^3a_A\, l_B + {}^3a_B\, l_A) + \frac{\epsilon}{2}\, ({}^4\nabla_A\, l_B + {}^4\nabla_B\, l_A) - \frac{1}{3}\, \theta\, {}^3h_{AB} = \sigma_{(\alpha)(\beta)}\, {}^4\overset{\circ}{\bar{E}}{}^{(\alpha)}_A\, {}^4\overset{\circ}{\bar{E}}{}^{(\beta)}_B$ has the following components: $\sigma_{(o)(o)} = \sigma_{(o)(r)} = 0$, $\sigma_{(a)(b)} = \sigma_{(b)(a)} = ({}^3K_{rs} - \frac{1}{3}\, {}^3g_{rs}\, {}^3K)\, {}^3\bar{e}^r_{(a)}\, {}^3\bar{e}^s_{(b)}$, $\sum_a \sigma_{(a)(a)} = 0$. $\sigma_{(a)(b)}$ depends upon the canonical variables θ^r, $\tilde{\phi}$, $R_{\bar{a}}$, $\pi_i^{(\theta)}$, and $\Pi_{\bar{a}}$.

By using Eq. (6.20) for the extrinsic curvature tensor, one finds that the diagonal elements $\sigma_{(a)(a)}$ of the shear are also connected with the tidal momenta $\Pi_{\bar{a}}$,

[7] It measures the average expansion of the infinitesimally nearby world-lines surrounding a given world-line in the congruence.
[8] It measures how an initial sphere in the tangent space to the given world-line, which is Lie-transported along the world-line tangent l^μ (i.e., it has zero Lie derivative with respect to $l^\mu\, \partial_\mu$), is distorted toward an ellipsoid with principal axes given by the eigenvectors of σ^μ_ν, with the rate given by the eigenvalues of σ^μ_ν.

while the non-diagonal elements $\sigma_{(a)(b)}|_{a\neq b}$ are connected with the momenta $\pi_i^{(\theta)}$ (the unknowns in the super-momentum constraints):

$$
\begin{aligned}
\Pi_{\bar a}(\tau,\sigma^u) &= -\frac{c^3}{8\pi\, G}\,\tilde\phi(\tau,\sigma^u)\sum_a \gamma_{\bar aa}\,\sigma_{(a)(a)}(\tau,\sigma^u),\\
\pi_i^{(\theta)}(\tau,\sigma^u) &= \frac{c^3}{8\pi\, G}\,\tilde\phi(\tau,\sigma^u)\sum_{wtab}\Big(A_{wi}\,V_{wt}\,Q_a\,Q_b^{-1}\,\epsilon_{tab}\,\sigma_{(a)(b)}|_{a\neq b}\Big)(\tau,\sigma^u),
\end{aligned}
$$

$$
\begin{aligned}
{}^3K_{rs}(\tau,\sigma^u) &= \Big[\tilde\phi^{2/3}\sum_{ab}\Big(-\frac{\epsilon}{3}\,\theta\,\delta_{ab}+\sigma_{(a)(b)}\Big)\,Q_a\,Q_b\,V_{ra}\,V_{sb}\Big](\tau,\sigma^u)\\
&\rightarrow_{\theta^i\rightarrow 0}(\tau,\sigma^u)\Big[\tilde\phi^{2/3}\,Q_r\,Q_s\Big(-\frac{\epsilon}{3}\,\theta+\sigma_{(a)(b)}\Big)\Big](\tau,\sigma^u). \qquad (6.27)
\end{aligned}
$$

Therefore, the Eulerian observers associated to the 3+1 splitting of space-time induce a geometrical interpretation of some of the momenta of the York canonical basis:

1. Their expansion $\theta(\tau,\sigma^u)$ is the gauge variable York time ${}^3K(\tau,\sigma^u) = \frac{12\pi G}{c^3}\,\pi_{\tilde\phi}(\tau,\sigma^u)$ determining the non-dynamical gauge part of the shape of the instantaneous 3-spaces Σ_τ as a sub-manifold of space-time.
2. The diagonal elements of their shear describe the tidal momenta $\Pi_{\bar a}(\tau,\sigma^u)$, while the non-diagonal elements are connected to the variables $\pi_i^{(\theta)}(\tau,\sigma^u)$, determined by the super-momentum constraints.

In Eq. (6.25), valid in the 3-orthogonal gauges, the term quadratic in the momenta $\pi_i^{(\theta)}(\tau,\sigma^u)$ in the weak ADM energy and in the super-Hamiltonian constraint can be written as $\frac{c^3}{16\pi G}\,\tilde\phi(\tau,\sigma^u)\sum_{ab,a\neq b}\sigma^2_{(a)(b)}(\tau,\sigma^u)$, while the super-momentum constraints can be written in the form of a PDE for the non-diagonal elements of the shear:

$$
\begin{aligned}
\bar{\mathcal{H}}_{(a)}|_{\theta^i=0}(\tau,\sigma^u) &= -\frac{c^3}{8\pi\, G}\,\tilde\phi^{2/3}(\tau,\sigma^u)\Big(\sum_{b\neq a} Q_b^{-1}\Big[\partial_b\,\sigma_{(a)(b)}\\
&\quad+\Big(\tilde\phi^{-1}\,\partial_b\,\tilde\phi+\sum_{\bar b}(\gamma_{\bar ba}-\gamma_{\bar bb})\,\partial_b\,R_{\bar b}\Big)\,\sigma_{(a)(b)}\Big]\\
&\quad-\frac{8\pi\, G}{c^3}\,\tilde\phi^{-1}\,Q_a^{-1}\Big[\tilde\phi\,\partial_a\,\pi_{\tilde\phi}+\sum_{\bar b}(\gamma_{\bar ba}\,\partial_a\,\Pi_{\bar b}-\partial_a\,R_{\bar b}\,\Pi_{\bar b})\\
&\quad+\mathcal{M}_a\Big]\Big)(\tau,\sigma^u)\approx 0. \qquad (6.28)
\end{aligned}
$$

As a consequence, by using ${}^3g_{rs}(\tau,\sigma^u)$ of Eq. (6.19) and ${}^3K_{rs}(\tau,\sigma^u)$ of Eq. (6.27), the first-order non-Hamiltonian ADM equations (6.13) can be re-expressed in terms of the configurational variables $n(\tau,\sigma^u)$, $\bar n_{(a)}(\tau,\sigma^u)$, $\tilde\phi(\tau,\sigma^u)$, $\theta^i(\tau,\sigma^u)$, and $R_{\bar a}(\tau,\sigma^u)$, and of the expansion $\theta(\tau,\sigma^u)$ and shear $\sigma_{(a)(b)}(\tau,\sigma^u)$ of the Eulerian observers. Then, the 12 equations can be put in the form of equations determining $\partial_\tau\,\tilde\phi(\tau,\sigma^u)$, $\partial_\tau\,R_{\bar a}(\tau,\sigma^u)$, $\partial_\tau\,\theta^i(\tau,\sigma^u)$, $\partial_\tau\,\theta(\tau,\sigma^u)$, and

$\partial_\tau\, \sigma_{(a)(b)}(\tau,\sigma^u)$. In equation 2.17 of Ref. [95] this manipulation is explicitly done for the first six equations (6.13).

These results are important for extending the identification of the inertial and tidal variables of the gravitational field, achieved with the York canonical basis, to cosmological space-times. Since these space-times are only conformally asymptotically flat, the Hamiltonian formalism is not defined. However, they are globally hyperbolic and admit 3+1 splittings with the associated congruence of Eulerian observers. As a consequence, in them Einstein's equations are usually replaced with the non-Hamiltonian first-order ADM equations plus the super-Hamiltonian and super-momentum constraints. Our analysis implies that, since the 4-metric can always be put in the form of Eq. (6.19), the inertial gauge variables of the cosmological space-times are $n(\tau,\sigma^u)$, $\bar{n}_{(a)}(\tau,\sigma^u)$, $\theta^i(\tau,\sigma^u)$, and the expansion $\theta(\tau,\sigma^u) = -\epsilon\, {}^3K(\tau,\sigma^u)$, while the physical tidal variables are $R_{\bar{a}}(\tau,\sigma^u)$ and the diagonal components of the shear $\sigma_{(a)(a)}(\tau,\sigma^u)$ ($\sum_a \sigma_{(a)(a)}(\tau,\sigma^u) = 0$). The unknown in the super-Hamiltonian constraint is the conformal factor $\tilde{\phi}(\tau,\sigma^u)$ of the 3-metric in Σ_τ, while the unknowns in the super-Hamiltonian constraints are the non-diagonal components of the shear $\sigma_{(a)(b)}|_{a\neq b}(\tau,\sigma^u)$.

In Ref. [288] there is the Lorentz-scalar expression of the Riemann, Ricci, and Weyl tensors and of Ricci and Weyl scalars in the York canonical basis, and a comparison of DOs with the observables proposed by Bergmann [308 311].

Till now we have considered only space-times without Killing symmetries. In Ref. [287] it is shown that the existence of a Killing symmetry in a gauge theory is equivalent to the addition by hand of extra Hamiltonian constraints in its phase space formulation, which imply restrictions both on the DOs and on the gauge freedom. When there is a time-like Killing vector, only pure gauge electromagnetic fields survive in Maxwell theory in SR; in ADM, canonical gravity in asymptotically Minkowskian space-times, only inertial effects without gravitational waves survive.

6.4 Point Particles and the Electromagnetic Field as Matter

The tetrad ADM action for tetrad gravity (see Ref. [95]) plus the electromagnetic field and N charged scalar particles with Grassmann-valued electric charges Q_i and sign of the energy η_i, depending on the configuration variables $n(\tau,\sigma^u)$, $n_{(a)}(\tau,\sigma^u)$, $\varphi_{(a)}(\tau,\sigma^u)$, ${}^3e_{(a)r}(\tau,\sigma^u)$, $A_A(\tau,\sigma^u)$, $\vec{\eta}_i(\tau)$, $\theta_i(\tau)$, $\theta_i^{(Q)}(\tau)$, is

$$\begin{aligned}
S &= S_{grav} + S_{em} + S_{part} + S_{Grassmann}\\
&= \frac{c^3}{16\pi\, G}\int d\tau d^3\sigma\, \{(1+n)\, {}^3e\, \epsilon_{(a)(b)(c)}\, {}^3e^r_{(a)}\, {}^3e^s_{(b)}\, {}^3\Omega_{rs(c)}\\
&\quad + \frac{{}^3e}{2(1+n)}({}^3G_o^{-1})_{(a)(b)(c)(d)}\, {}^3e^r_{(b)}(n_{(a)|r}\\
&\quad - \partial_\tau\, {}^3e_{(a)r})\, {}^3e^s_{(d)}(n_{(c)|s} - \partial_\tau\, {}^3e_{(c)\ s})\}(\tau,\sigma^u)
\end{aligned}$$

$$
\begin{aligned}
&-\frac{1}{4}\int d\tau d^3\sigma\, {}^3e(\tau,\sigma^u)\Big[-\frac{2\,{}^3e^r_{(a)}\,{}^3e^s_{(a)}}{1+n}\,F_{\tau r}\,F_{\tau s}+\frac{4\,{}^3e^r_{(a)}\,n_{(a)}\,{}^3e^s_{(b)}\,{}^3e^u_{(b)}}{1+n}\,F_{\tau u}\,F_{rs}\\
&+{}^3e^r_{(a)}\,{}^3e^s_{(b)}\,{}^3e^u_{(c)}\,{}^3e^v_{(d)}\,[(1+n)\,\delta_{(a)(b)}\,\delta_{(c)(d)}\\
&-\frac{\delta_{(a)(b)}\,n_{(c)}\,n_{(d)}+\delta_{(c)(d)}\,n_{(a)}\,n_{(b)}}{1+n}]\,F_{ru}\,F_{sv}\Big](\tau,\sigma^u)\\
&-\sum_{i=1}^N\int d\tau d^3\sigma\,\delta^3(\vec{\sigma},\vec{\eta}_i(\tau))\Big(m_i\,c\\
&\eta_i\sqrt{\big(1+n(\tau,\sigma^u)\big)^2-\big({}^3e_{(a)r}(\tau,\sigma^u)\,\dot{\eta}^r_i(\tau)+n_{(a)}(\tau,\sigma^u)\big)\big({}^3e_{(a)s}(\tau,\sigma^u)\,\dot{\eta}^s_i(\tau)+n_{(a)}(\tau,\sigma^u)\big)}\\
&-\frac{\eta_i\,Q_i}{c}\Big[A_\tau(\tau,\sigma^u)+A_r(\tau,\sigma^u)\,\dot{\eta}^r_i(\tau)\Big]\Big)\\
&+\frac{i\hbar}{2}\int d\tau\sum_i[\theta^*_i(\tau)\,\dot{\theta}_i(\tau)-\dot{\theta}^*_i(\tau)\,\theta_i(\tau)]+\frac{i\hbar}{2}\int d\tau\sum_i[\theta^{(Q)*}_i(\tau)\,\dot{\theta}^{(Q)}_i(\tau)-\dot{\theta}^{(Q)*}_i(\tau)\,\theta^{(Q)}_i(\tau)],\\
&\eta_i=\theta^*_i\,\theta_i,\qquad Q_i=\theta^{(Q)*}_i\,\theta^{(Q)}_i.
\end{aligned}
\tag{6.29}
$$

While Q_i is the Grassmann-valued electric charge of particle "i", η_i is its Grassmann-valued sign of the energy (particles with negative energy have the opposite electric charge, $\eta_i\,Q_i$). See after Eq. (6.7) for the definition of ${}^3\Omega_{rs(a)}$ and $({}^3G_o^{-1})_{(a)(b)(c)(d)}$. For the momenta and the constraints connected with the Grassmann variables, see Appendix B.

The canonical momenta for the tetrad gravity variables and the electromagnetic field are (we are in the canonical basis (Eq. 6.12))

$$
\begin{aligned}
\pi_{\varphi_{(a)}}(\tau,\sigma^u)&=\frac{\delta S_{ADM}}{\delta\partial_\tau\varphi_{(a)}(\tau,\sigma^u)}=0,\\
\pi_n(\tau,\sigma^u)&=\frac{\delta S_{ADM}}{\delta\partial_\tau n(\tau,\sigma^u)}=0,\\
\pi_{n_{(a)}}(\tau,\sigma^u)&=\frac{\delta S_{ADM}}{\delta\partial_\tau n_{(a)}(\tau,\sigma^u)}=0,\\
{}^3\pi^r_{(a)}(\tau,\sigma^u)&=\frac{\delta S_{ADM}}{\delta\partial_\tau{}^3e_{(a)r}(\tau,\sigma^u)}\\
&=-\frac{c^3}{16\pi\,G}\left[\frac{{}^3e}{1+n}\,({}^3G_o^{-1})_{(a)(b)(c)(d)}\,{}^3e^r_{(b)}\,{}^3e^s_{(d)}\,\big(n_{(c)|s}-\partial_\tau\,{}^3e_{(c)s}\big)\right](\tau,\sigma^u)\\
&=-\frac{c^3}{8\pi\,G}\,[{}^3e({}^3K^{rs}-{}^3e^r_{(c)}\,{}^3e^s_{(c)}\,{}^3K){}^3e_{(a)s}](\tau,\sigma^u),\\
\pi^\tau(\tau,\sigma^u)&=\frac{\delta\,S_{ADM}}{\delta\,\partial_\tau\,A_\tau(\tau,\sigma^u)}=0,\\
\pi^r(\tau,\sigma^u)&=\frac{\delta\,S_{ADM}}{\delta\,\partial_\tau\,A_r(\tau,\sigma^u)}=\Big(\frac{\sqrt{\gamma}}{1+n}\,{}^3e^r_{(a)}\,{}^3e^s_{(a)}\,(F_{\tau s}-n^u\,F_{us})\Big)(\tau,\sigma^u),
\end{aligned}
$$

$$\{A_A(\tau,\sigma^u),\pi^B(\tau,\sigma'^u)\}=c\,\eta^B_A\,\delta^3(\sigma^u,\sigma'^u),$$

$$\{n(\tau,\sigma^u),\pi_n(\tau,\sigma'^u)\}=\delta^3(\sigma^u,\sigma'^u),$$

$$\{n_{(a)}(\tau,\sigma^u), \pi_{n_{(b)}}(\tau,\sigma'^u)\} = \delta_{(a)(b)}\delta^3(\sigma^u,\sigma'^u),$$
$$\{\varphi_{(a)}(\tau,\sigma^u), \pi_{\varphi_{(b)}}(\tau,\sigma'^u)\} = \delta_{(a)(b)}\delta^3(\sigma^u,\sigma'^u),$$
$$\{{}^3e_{(a)r}(\tau,\sigma^u), {}^3\pi^s_{(b)}(\tau,\sigma'^u)\} = \delta_{(a)(b)}\delta^s_r\delta^3(\sigma^u,\sigma'^u),$$

$$\{{}^3e^r_{(a)}(\tau,\sigma^u), {}^3\tilde{\pi}^s_{(b)}(\tau,\sigma'^u)\} = -{}^3e^r_{(b)}(\tau,\sigma^u)\,{}^3e^s_{(a)}(\tau,\sigma^u)\delta^3(\sigma^u,\sigma'^u),$$
$$\{{}^3e(\tau,\sigma^u), {}^3\tilde{\pi}^r_{(a)}(\tau,\sigma'^u)\} = {}^3e(\tau,\sigma^u)\,{}^3e^r_{(a)}(\tau,\sigma^u)\,\delta^3(\sigma^u,\sigma'^u), \tag{6.30}$$

while the particle momenta are

$$\tilde{\kappa}_{ir}(\tau) = \frac{\partial L_{ADM}(\tau)}{\partial \dot{\eta}^r_i(\tau)} \overset{def}{=} \eta_i\,\kappa_{ir}(\tau) = \frac{\eta_i\,Q_i}{c}\,A_r(\tau,\eta^u_i(\tau))$$
$$+ \frac{\eta_i\,m_i\,c\;{}^3e_{(a)r}(\tau,\eta^u_i(\tau))\,({}^3e_{(a)s}(\tau,\sigma^u)\,\dot{\eta}^s_i(\tau) + n_{(a)}(\tau,\sigma^u))}{\sqrt{(1+n(\tau,\sigma^u))^2 - ({}^3e_{(a)r}(\tau,\sigma^u)\,\dot{\eta}^r_i(\tau) + n_{(a)}(\tau,\sigma^u))\,({}^3e_{(a)s}(\tau,\sigma^u)\,\dot{\eta}^s_i(\tau) + n_{(a)}(\tau,\sigma^u))}}$$
$$|_{\sigma^u=\eta^u_i(\tau)},$$

$$\kappa_{ir}(\tau) = \int d\psi_i\,d\psi^*_i\,\tilde{\kappa}_{ir}(\tau), \qquad \{\eta^r_i(\tau), \kappa_{is}(\tau)\} = \delta_{ij}\,\delta^r_s,$$

$$\Downarrow$$

$$\sqrt{m_i^2c^2 + {}^3e^r_{(a)}(\tau,\sigma^u)\left(\kappa_{ir}(\tau) - \frac{Q_i}{c}\,A_r(\tau,\sigma^u)\right){}^3e^s_{(a)}(\tau,\sigma^u)\left(\kappa_{is}(\tau) - \frac{Q_i}{c}\,A_s(\tau,\sigma^u)\right)}$$
$$|_{\sigma^u=\eta^u_i(\tau)}$$
$$= \frac{m_i\,[1+n(\tau,\sigma^u)]}{\sqrt{(1+n(\tau,\sigma^u))^2 - ({}^3e_{(a)r}(\tau,\sigma^u)\,\dot{\eta}^r_i(\tau) + n_{(a)}(\tau,\sigma^u))\,({}^3e_{(a)s}(\tau,\sigma^u)\,\dot{\eta}^s_i(\tau) + n_{(a)}(\tau,\sigma^u))}}$$
$$|_{\sigma^u=\eta^u_i(\tau)},$$

$$\dot{\eta}^r_i(\tau) = {}^3e^r_{(a)}(\tau,\eta^u_i(\tau))\Bigg[\frac{(1+n)\,{}^3e^s_{(a)}\left(\kappa_{is}(\tau) - \frac{Q_i}{c}\,A_r\right)}{\sqrt{m_i^2c^2 + {}^3e^u_{(b)}\,{}^3e^v_{(b)}\left(\kappa_{iu}(\tau) - \frac{Q_i}{c}\,A_u\right)\left(\kappa_{iv}(\tau) - \frac{Q_i}{c}\,A_v\right)}}$$
$$-n_{(a)}\Bigg](\tau,\eta^u_i(\tau)). \tag{6.31}$$

The primary constraints are

$$\pi_{\varphi_{(a)}}(\tau,\sigma^u) \approx 0, \qquad \pi_n(\tau,\sigma^u) \approx 0, \qquad \pi_{n_{(a)}}(\tau,\sigma^u) \approx 0,$$
$${}^3M_{(a)}(\tau,\sigma^u) = \epsilon_{(a)(b)(c)}\,{}^3e_{(b)r}(\tau,\sigma^u)\,{}^3\pi^r_{(c)}(\tau,\sigma^u) \approx 0,$$
$$\pi^\tau(\tau,\sigma^u) \approx 0. \tag{6.32}$$

By evaluating the canonical Hamiltonian by Legendre transformation (see Ref. [282]) and by asking that the primary constraints are constants of the motion under the τ-evolution generated by it, we get the following secondary constraints:

$$
\begin{aligned}
\Gamma(\tau,\sigma^u) &= \partial_r\, \pi^r(\tau,\sigma^u) + \sum_i Q_i\, \eta_i\, \delta^3(\sigma^u, \eta_i^u(\tau)) \approx 0,\\
\mathcal{H}(\tau,\sigma^u) &= \Big[\frac{c^3}{16\pi\, G}\; {}^3e\; \epsilon_{(a)(b)(c)}\, {}^3e^r_{(a)}\, {}^3e^s_{(b)}\, {}^3\Omega_{rs(c)}\\
&\quad - \frac{2\pi\, G}{c^3\, {}^3e}\; {}^3G_{o(a)(b)(c)(d)}\, {}^3e_{(a)r}\, {}^3\pi^r_{(b)}\, {}^3e_{(c)s}\, {}^3\pi^s_{(d)}\Big](\tau,\sigma^u) + \mathcal{M}(\tau,\sigma^u) \approx 0,\\
\mathcal{H}_{(a)}(\tau,\sigma^u) &= \Big[\partial_r\, {}^3\pi^r_{(a)} - \epsilon_{(a)(b)(c)}\, {}^3\omega_{r(b)}\, {}^3\pi^r_{(c)} + {}^3e^r_{(a)}\mathcal{M}_r\Big](\tau,\sigma^u)\\
&= -{}^3e^r_{(a)}(\tau,\sigma^u)[{}^3\Theta_r - {}^3\omega_{r(b)}\, {}^3M_{(b)}](\tau,\sigma^u) \approx 0,\\
&\Downarrow\\
{}^3\Theta_r(\tau,\sigma^u) &= [{}^3\pi^s_{(a)}\partial_r\, {}^3e_{(a)s} - \partial_s({}^3e_{(a)r}\, {}^3\pi^s_{(a)})](\tau,\sigma^u) - \mathcal{M}_r(\tau,\sigma^u) \approx 0. \qquad (6.33)
\end{aligned}
$$

In Eq. (6.33) the following notation has been introduced for the matter terms (the mass density $\mathcal{M}(\tau,\sigma^u)$ and the mass current density $\mathcal{M}_r(\tau,\sigma^u)$) to conform with the treatment of the same matter in the non-inertial rest-frames of Minkowski space-time of Chapter 2:

$$
\begin{aligned}
\mathcal{M}(\tau,\sigma^u) &= \sum_{i=1}^N \delta^3(\sigma^u, \eta_i^u(\tau))\, M_i(\tau,\sigma^u)\, c + {}^3e\, T^{(em)}_{\perp\perp}(\tau,\sigma^u),\\
M_i(\tau,\sigma^u)\, c &= \eta_i \sqrt{m_i^2\, c^2 + {}^3e^r_{(a)}\left(\kappa_{ir}(\tau) - \frac{Q_i}{c}\, A_r\right) {}^3e^s_{(a)}\left(\kappa_{is}(\tau) - \frac{Q_i}{c}\, A_s\right)}(\tau,\sigma^u),\\
\mathcal{M}_r(\tau,\sigma^u) &= \sum_{i=1}^N \eta_i\left(\kappa_{ir}(\tau) - \frac{Q_i}{c}\, A_r(\tau,\sigma^u)\right)\delta^3(\sigma^u, \eta_i^u(\tau)) - {}^3e\, T^{(em)}_{\perp r}(\tau,\sigma^u),\\
T^{(em)}_{\perp\perp}(\tau,\sigma^u) &= \frac{1}{2\, c\, {}^3e}\left(\frac{1}{{}^3e}\, {}^3e_{(a)r}\, {}^3e_{(a)s}\, \pi^r\, \pi^s\right.\\
&\quad \left. + \frac{{}^3e}{2}\, {}^3e^r_{(a)}\, {}^3e^s_{(a)}\, {}^3e^u_{(b)}\, {}^3e^v_{(b)}\, F_{ru}\, F_{sv}\right)(\tau,\sigma^u),\\
T^{(em)}_{\perp r}(\tau,\sigma^u) &= \frac{1}{c\, {}^3e}\, F_{rs}(\tau,\sigma^u)\, \pi^s(\tau,\sigma^u). \qquad (6.34)
\end{aligned}
$$

Let us remark that the mass current density $\mathcal{M}_r(\tau,\sigma^u)$ does not depend upon the 4-metric.

In Eq. (6.33), $\Gamma(\tau,\sigma^u) \approx 0$ is the electromagnetic Gauss's law; $\mathcal{H}(\tau,\sigma^u) \approx 0$ and $\mathcal{H}_{(a)}(\tau,\sigma^u) \approx 0$ are the super-Hamiltonian and super-momentum constraints respectively; and ${}^3\Theta(\tau,\sigma^u) \approx 0$ are the generators of 3-diffeomorphisms on Σ_τ. In the constraint $\mathcal{H}(\tau,\sigma^u) \approx 0$ we have $\left({}^3e\, \epsilon_{(a)(b)(c)}\, {}^3e^r_{(a)}\, {}^3e^s_{(b)}\, \Omega_{rs(c)}\right)(\tau,\sigma^u) = \left({}^3e\, {}^3R\right)(\tau,\sigma^u)$, with ${}^3R(\tau,\sigma^u)$ the scalar 3-curvature of the instantaneous 3-space Σ_τ.

One can check that the constraints are all first class with the following algebra:

$$
\begin{aligned}
\{{}^3M_{(a)}(\tau,\sigma^u), {}^3M_{(b)}(\tau,\sigma'^u)\} &= \epsilon_{(a)(b)(c)}\, {}^3M_{(c)}(\tau,\sigma^u)\delta^3(\sigma,\sigma'^u),\\
\{{}^3M_{(a)}(\tau,\sigma^u), {}^3\Theta_r(\tau,\sigma'^u)\} &= {}^3M_{(a)}(\tau,\sigma'^u)\, \frac{\partial\delta^3(\sigma^u,\sigma'^u)}{\partial\sigma^r},
\end{aligned}
$$

$$\{{}^3\Theta_r(\tau,\sigma^u), {}^3\Theta_s(\tau,\sigma^{'u})\} = \Big[{}^3\Theta_r(\tau,\sigma^{'u})\frac{\partial}{\partial\sigma^s} + {}^3\Theta_s(\tau,\sigma^u)\frac{\partial}{\partial\sigma^r}\Big]\,\delta^3(\sigma^u,\sigma^{'u})$$
$$- \delta_{ru}\,\delta_{sv}\Big[{}^3e^u_{(a)}\,{}^3e^m_{(a)}\,{}^3e^v_{(b)}\,{}^3e^n_{(b)}\,F_{mn}\Big]$$
$$(\tau,\sigma^u)\,\Gamma(\tau,\sigma^u)\,\delta^3(\sigma^u,\sigma^{'u}),$$
$$\{\mathcal{H}(\tau,\sigma^u), {}^3\Theta_r(\tau,\sigma^{'u})\} = \mathcal{H}(\tau,\sigma^{'u})\,\frac{\partial\delta^3(\sigma,\sigma^{'u})}{\partial\sigma^r}$$
$$+\,\delta_{rs}\,\frac{\pi^s(\tau,\sigma^u)}{{}^3e(\tau,\sigma^u)}\Gamma(\tau,\sigma^u)\,\delta^3(\sigma^u,\sigma^{'u}),$$
$$\{\mathcal{H}(\tau,\sigma^u), \mathcal{H}(\tau,\sigma^{'u})\} = \Big[{}^3e^r_{(a)}(\tau,\sigma^u)\,\mathcal{H}_{(a)}(\tau,\sigma^u)$$
$$+\,{}^3e^r_{(a)}(\tau,\sigma^{'u})\,\mathcal{H}_{(a)}(\tau,\sigma^{'u})\Big]\frac{\partial\delta^3(\sigma^u,\sigma^{'u})}{\partial\sigma^r}$$
$$= \Big(\Big[{}^3e^r_{(a)}\,{}^3e^s_{(a)}\,[{}^3\Theta_s + {}^3\omega_{s(b)}\,{}^3M_{(b)}]\Big](\tau,\sigma^u)$$
$$+\Big[{}^3e^r_{(a)}\,{}^3e^s_{(a)}\,[{}^3\Theta_s + {}^3\omega_{s(b)}\,{}^3M_{(b)}]\Big](\tau,\sigma^{'u})\Big)\frac{\partial\delta^3(\sigma^u,\sigma^{'u})}{\partial\sigma^r}. \tag{6.35}$$

In equation 3.11 of Ref. [95] there is the explicit form of the energy–momentum tensor and of the Bianchi identity.

The non-inertial electric and magnetic fields have the form $E_r(\tau,\sigma^u) = \Big(\frac{\partial A_\mathcal{T}}{\partial\eta_i^r} - \frac{\partial A_r}{\partial\tau}\Big)(\tau,\sigma^u) - F_{\tau r}(\tau,\sigma^u)$ and $B_r(\tau,\sigma^u) = \frac{1}{2}\,\varepsilon_{ruv}\,F_{uv}(\tau,\sigma^u) = \epsilon_{ruv}\,\partial_u\,A_{\perp\,v}(\tau,\sigma^u) \Rightarrow F_{uv}(\tau,\sigma^u) = \varepsilon_{uvr}\,B_r(\tau,\sigma^u)$, so that the homogeneous Maxwell equations, allowing the introduction of the electromagnetic potentials, have the standard inertial form $\epsilon_{ruv}\,\partial_u\,B_v(\tau,\sigma^u) = 0$, $\epsilon_{ruv}\,\partial_u\,E_v(\tau,\sigma^u) + \frac{1}{c}\frac{\partial\,B_r(\tau,\sigma^u)}{\partial\tau} = 0$.

For the electromagnetic field, one can find an analogue of the canonical basis (Eq. 4.55).

If $\Delta = \sum_r \partial_r^2$ is the non-covariant flat Laplacian, associated with the asymptotic Minkowski metric and acting in the instantaneous non-Euclidean 3-space Σ_τ, its inverse defines the following non-covariant distribution:

$$\frac{1}{\Delta}\,\delta^3(\sigma^u,\sigma^{'u}) = c(\sigma^u,\sigma^{'u}) = -\frac{1}{4\pi}\,\frac{1}{\sqrt{\sum_u(\sigma^u-\sigma^{'u})^2}}, \tag{6.36}$$

with $\delta^3(\sigma^u,\sigma^{'u})$ being the delta function for Σ_τ.

Then we can define the following non-covariant decomposition of the vector potential and its conjugate momentum ($\Gamma(\tau,\sigma^u)\approx 0$ is Gauss's law of Eq. (6.33); $\Delta\,\eta_{em}(\tau,\sigma^u) = \delta^{rs}\,\partial_r\,A_s(\tau,\sigma^u)$; $\eta_{em}(\tau,\sigma^u)$ describes a Coulomb cloud of longitudinal photons as said after Eq. (4.42)):

$$A_r(\tau,\sigma^u) = A_{\perp\,r}(\tau,\sigma^u) - \partial_r\,\eta_{em}(\tau,\sigma^u),$$
$$\pi^r(\tau,\sigma^u) = \pi^r_\perp(\tau,\sigma^u) + \delta^{rs}\,\partial_s\int d^3\sigma'\,c(\sigma^u,\sigma^{'u})\Big(\Gamma(\tau,\sigma^{'u})$$
$$-\sum_i Q_i\,\eta_i\,\delta^3(\sigma^{'u},\eta_i^u(\tau))\Big),$$

$$\eta_{em}(\tau,\sigma^u) = -\int d^3\sigma'\, c(\sigma^u,\sigma'^u)\, \Big(\delta^{rs}\, \partial_r\, A_s(\tau,\sigma'^u)\Big),$$
$$\{\eta_{em}(\tau,\sigma^u), \Gamma(\tau,\sigma'^u)\} = \delta^3(\sigma^u,\sigma'^u),$$
$$A_{\perp r}(\tau,\sigma^u) = \delta_{ru}\, P_\perp^{us}(\sigma^u)\, A_s(\tau,\sigma^u), \qquad \pi^r_\perp(\tau,\sigma^u) = \sum_s P_\perp^{rs}(\sigma^u)\, \pi^s(\tau,\sigma^u), \tag{6.37}$$

where we introduced the projector $P_\perp^{rs}(\sigma^u) = \delta^{rs} - \delta^{ru}\, \delta^{sv}\, \frac{\partial_u\, \partial_v}{\Delta}$.

If we introduce the following new Coulomb-dressed momenta for the particles:

$$\check{\kappa}_{ir}(\tau) = \kappa_{ir}(\tau) + \frac{Q_i}{c}\, \frac{\partial}{\partial \eta_i^r}\, \eta_{em}(\tau, \vec{\eta}_i(\tau)),$$
$$\Rightarrow \quad \kappa_{ir}(\tau) - \frac{Q_i}{c}\, A_r(\tau, \eta_i^u(\tau)) = \check{\kappa}_{ir}(\tau) - \frac{Q_i}{c}\, A_{\perp\, r}(\tau, \eta_i^u(\tau)), \tag{6.38}$$

we arrive at the Shanmugadhasan canonical basis, like Eq. (4.55). The transverse potential satisfies $\{A_{\perp r}(\tau,\sigma^u), \pi^s_\perp(\tau,\sigma'^u)\} = c\, \delta_{ru}\, P_\perp^{us}(\sigma^u)\, \delta^3(\sigma^u,\sigma'^u)$.

The non-covariant radiation gauge is defined by adding the gauge fixing $\eta_{em}(\tau,\sigma^u) \approx 0$. As shown in Refs. [98, 99], the τ-constancy, $\frac{\partial \eta_{em}(\tau,\sigma^u)}{\partial \tau} = \{\eta_{em}(\tau,\sigma^u), H_D\} \approx 0$, of this gauge fixing implies the secondary gauge fixing for the primary constraint $\pi^\tau(\tau,\sigma^u) \approx 0$

$$A_\tau(\tau,\sigma^u) \approx -\int d^3\sigma'\, c(\sigma^u,\sigma'^u)\, \frac{\partial}{\partial \sigma'^r}\, \Big[[{}^3e^s_{(a)}\, n_{(a)}](\tau,\sigma'^u)\, F_{sr}(\tau,\sigma'^u)$$
$$+ \frac{\Big(1 + n(\tau,\sigma'^u)\Big)\, [{}^3e_{(a)r}\, {}^3e_{(a)s}](\tau,\sigma'^u)}{{}^3e(\tau,\sigma'^u)}\, \Big(\pi^s_\perp(\tau,\sigma'^u)$$
$$- \delta^{sn}\, \sum_j Q_j\, \eta_j\, \frac{\partial\, c(\sigma^u, \eta_j^u(\tau))}{\partial\, \sigma'^n}\Big)\Big). \tag{6.39}$$

If we eliminate the electromagnetic variables $A_\tau(\tau,\sigma^u)$, $\pi^\tau(\tau,\sigma^u)$, $\eta_{em}(\tau,\sigma^u)$, and $\Gamma(\tau,\sigma^u)$ by going to Dirac brackets (still denoted $\{.,.\}$ for simplicity), we remain with only the transverse fields $A_{\perp\, r}(\tau,\sigma^u)$ and $\pi^r_\perp(\tau,\sigma^u)$ ($F_{rs}(\tau,\sigma^u) = \partial_r\, A_{\perp s}(\tau,\sigma^u) - \partial_s\, A_{\perp r}(\tau,\sigma^u)$).

Let us remark that in the radiation gauge the non-inertial magnetic field (see after Eq. (6.35)) is transverse: $B_r(\tau,\sigma^u) = \epsilon_{ruv}\, \partial_u\, A_{\perp\, v}(\tau,\sigma^u)$. But the non-inertial electric field $E_r(\tau,\sigma^u) = -F_{\tau r}(\tau,\sigma^u) = -\partial_\tau\, A_{\perp\, r}(\tau,\sigma^u) + \partial_r\, A_\tau(\tau,\sigma^u)$ is not transverse: it has $E_{\perp\, r}(\tau,\sigma^u) = -\partial_\tau\, A_{\perp\, r}(\tau,\sigma^u)$ as a transverse component. Instead, the transverse quantity is $\pi^r_\perp(\tau,\sigma^u)$ (it coincides with $\delta^{rs}\, E_{\perp\, s}(\tau,\sigma^u)$ only in the inertial frames of Minkowski space-time), whose expression in terms of the electric and magnetic fields is $\pi^r_\perp(\tau,\sigma^u) = \Big[\frac{\sqrt{\gamma}}{1+n}\, {}^3e^r_{(a)}\, {}^3e^s_{(a)}\, (E_s - \epsilon_{suv}\, n^u\, B_v)\Big](\tau,\sigma^u) - \delta^{rs}\, \sum_i Q_i\, \eta_i\, \partial_s\, c(\sigma^u, \eta_i^u(\tau))$.

In Ref. [95] there is the evaluation of the Hamilton equations of motion for the tetrads, the particles, and the electromagnetic field. The first half give the expression of the velocities $\partial_\tau\, R_{\bar a}(\tau,\sigma^u)$, $\partial_\tau\, \eta^r_i(\tau)$, $\partial_\tau\, A_{\perp r}(\tau,\sigma^u)$ in terms of the canonical variables. These equations can be inverted to get the momenta $\Pi_{\bar a}(\tau,\sigma^u)$, $\check\kappa_{ir}(\tau)$, $\pi^\tau_\perp(\tau,\sigma^u)$ in terms of the canonical configuration variables and of their velocities. By putting these solutions inside the second half of the Hamilton equations (giving the τ-derivatives of the momenta), we get the second-order equations of motion for the configuration variables.

For the tidal momenta we get (see Eq. (6.19) for the definition of $\Gamma_a(\tau,\sigma^u)$)

$$\begin{aligned}\Pi_{\bar a}(\tau,\sigma^u) &= -\frac{c^3}{8\pi\,G}\Big[\tilde\phi\sum_a \gamma_{\bar a a}\,\sigma_{(a)(a)}\Big](\tau,\sigma^u)\\ &\circeq \frac{c^3}{8\pi\,G}\,\frac{\tilde\phi(\tau,\sigma^u)}{1+n(\tau,\sigma^u)}\Big[\partial_\tau\,R_{\bar a}+\tilde\phi^{-1/3}\sum_a Q_a^{-1}\\ &\quad\Big(\Big[\gamma_{\bar a a}\,(2\,\partial_a\,q+\partial_a\,\Gamma^{(1)}_a)-\partial_a\,R_{\bar a}\Big]\,\bar n_{(a)}-\gamma_{\bar a a}\,\partial_a\,\bar n_{(a)}\Big)\Big](\tau,\sigma^u).\end{aligned} \tag{6.40}$$

As a consequence, for the tidal variables $R_{\bar a}$ we get from the equation for $\partial_\tau\Pi_{\bar a a}(\tau,\sigma^u)$ the following second-order equation ($q(\tau,\sigma^u)=\frac{1}{6}\ln\tilde\phi(\tau,\sigma^u)$):

$$\begin{aligned}\partial^2_\tau\,R_{\bar a}(\tau,\sigma^u)\circeq\approx &\Big[\tilde\phi^{-1/3}\sum_a Q_a^{-1}\,\bar n_{(a)}\sum_{\bar b}(\gamma_{\bar a a}\,\gamma_{\bar b a}-\delta_{\bar a\bar b})\,\partial_a\,\partial_\tau\,R_{\bar b}\\ &+\tilde\phi^{-1/3}\sum_a Q_u^{-1}\Big[\Big(\gamma_{\bar a u}\,(2\,\partial_u q+\partial_a\,\Gamma_a)\quad\partial_a\,R_{\bar a}\Big)\,\bar n_{(a)}\\ &-\gamma_{\bar a a}\,\partial_a\,\bar n_{(a)}\Big]\,\partial_\tau\,\Gamma_a\\ &-\tilde\phi^{-1}\Big[\partial_\tau\,R_{\bar a}+\frac{2}{3}\,\tilde\phi^{-1/3}\sum_a Q_a^{-1}\Big(\Big[\gamma_{\bar a a}\,(-\partial_a\,q+\partial_a\,\Gamma_a)\\ &-\partial_a\,R_{\bar a}\Big]\,\bar n_{(a)}-\gamma_{\bar a a}\,\partial_a\,\bar n_{(a)}\Big)\Big]\,\partial_\tau\,\tilde\phi\\ &-\frac{1}{3}\,\tilde\phi^{-4/3}\sum_a\gamma_{\bar a a}\,Q_a^{-1}\,\bar n_{(a)}\,\partial_a\,\partial_\tau\,\tilde\phi\\ &+\Big[\partial_\tau\,R_{\bar a}+\tilde\phi^{-1/3}\sum_a Q_a^{-1}\Big(\Big[\gamma_{\bar a a}\,(2\,\partial_a\,q+\partial_a\,\Gamma_b)\\ &-\partial_a\,R_{\bar a}\Big]\,\bar n_{(a)}-\gamma_{\bar a a}\,\partial_a\,\bar n_{(a)}\Big)\Big]\,\frac{\partial_\tau\,n}{1+n}\\ &-\tilde\phi^{-1/3}\sum_a Q_a^{-1}\Big(\Big[\gamma_{\bar a a}\,(2\,\partial_a\,q+\partial_a\,\Gamma_a)-\partial_a\,R_{\bar a}\Big]\,\partial_\tau\,\bar n_{(a)}\\ &-\gamma_{\bar a a}\,\partial_a\,\partial_\tau\,\bar n_{(a)}\Big)\Big](\tau,\sigma^u)\end{aligned}$$

$$\begin{aligned}&+\frac{1}{2}\Big(\tilde\phi^{-1}\,(1+n)\Big)(\tau,\sigma^u)\int d^3\sigma_1\Big[(1+n)(\tau,\sigma_1^u)\,\frac{\delta\,\mathcal{S}(\tau,\sigma_1^u)}{\delta\,R_{\bar a}(\tau,\sigma^u)}|_{\theta^i=0}\\ &+n(\tau,\sigma_1^u)\,\frac{\delta\,\mathcal{T}(\tau,\sigma_1^u)}{\delta\,R_{\bar a}(\tau,\sigma^u)}|_{\theta^i=0}\Big]\end{aligned}$$

$$
\begin{aligned}
&+\Big(\tilde{\phi}^{-1}\,(1+n)\Big)(\tau,\sigma^u)\,\Big(\frac{\tilde{\phi}^{2/3}}{1+n}\,\sum_a\,Q_a^{-1}\,\Big[\Big(\partial_a\,\bar{n}_{(a)}\\
&+\bar{n}_{(a)}\,(4\,\partial_a\,q-\partial_a\,\Gamma_a-\frac{\partial_a\,n}{1+n})\Big)\,\Big(\partial_\tau\,R_{\bar{a}}\\
&+\tilde{\phi}^{-1/3}\,\sum_c\,Q_c^{-1}\,\Big[\Big(\gamma_{\bar{a}c}\,(2\,\partial_c\,q+\partial_c\,\Gamma_c)\,-\,\partial_c\,R_{\bar{a}}\Big)\,\bar{n}_{(c)}-\gamma_{\bar{a}c}\,\partial_c\,\bar{n}_{(c)}\Big]\Big)\\
&+\bar{n}_{(a)}\,\partial_a\,\Big(\partial_\tau\,R_{\bar{a}}+\tilde{\phi}^{-1/3}\,\sum_c\,Q_c^{-1}\,\Big[\Big(\gamma_{\bar{a}c}\,(2\,\partial_c\,q\\
&+\partial_c\,\Gamma_c)\,-\,\partial_c\,R_{\bar{a}}\Big)\,\bar{n}_{(c)}-\gamma_{\bar{a}c}\,\partial_c\,\bar{n}_{(c)}\Big]\Big)\Big]\\
&+\tilde{\phi}^{2/3}\,\sum_{ab,a\neq b}\,Q_b^{-1}\,(\gamma_{\bar{a}a}-\gamma_{\bar{a}b})\,\Big[\partial_b\,\bar{n}_{(a)}-(2\,\partial_b\,q+\partial_b\,\Gamma_a)\,\bar{n}_{(a)}\Big]\,\sigma_{(a)(b)}\Big)(\tau,\sigma^u)\\
&-\frac{8\pi\,G}{c^3}\,\Big(\tilde{\phi}^{-1}\,(1+n)\Big)(\tau,\sigma^u)\,\int d^3\sigma_1\,(1+n)(\tau,\sigma_1^u)\,\frac{\delta\,\mathcal{M}(\tau,\sigma_1^u)}{\delta\,R_{\bar{a}}(\tau,\sigma^u)}|_{\theta^i=0},\\
&\int d^3\sigma_1\,\left[1+n(\tau,\sigma_1^u)\right]\,\frac{\delta\,\mathcal{S}(\tau,\sigma_1^u)}{\delta\,R_{\bar{a}}(\tau,\sigma^u)}|_{\theta^i=0}\\
&=2\,\Big(\tilde{\phi}^{1/3}\,\sum_a\,Q_a^{-2}\,\Big[\partial_a\,n\,\Big(2\,\gamma_{\bar{a}a}\,\partial_a\,q-\sum_{\bar{b}}\,(2\,\gamma_{\bar{a}a}\,\gamma_{\bar{b}a}-\delta_{\bar{a}\bar{b}})\,\partial_a\,R_{\bar{b}}\Big)\\
&-(1+n)\,\Big(2\,\gamma_{\bar{a}a}\,\Big(-\,\partial_a^2\,q+2\,(\partial_a\,q)^2\Big)\\
&+\sum_{\bar{b}}\,(2\,\gamma_{\bar{a}a}\,\gamma_{\bar{b}a}-\delta_{\bar{a}\bar{b}})\,(\partial_a^2\,R_{\bar{b}}+2\,\partial_a\,q\,\partial_a\,R_{\bar{b}})\\
&+\sum_{\bar{b}\bar{c}}\,\Big(2\,\gamma_{\bar{b}a}\,(\delta_{\bar{a}\bar{c}}-\gamma_{\bar{a}a}\,\gamma_{\bar{c}a})-\gamma_{\bar{a}a}\,\delta_{\bar{b}\bar{c}}\Big)\,\partial_a\,R_{\bar{b}}\,\partial_a\,R_{\bar{c}}\Big)\Big]\Big)(\tau,\sigma^u),\\
&\int d^3\sigma_1\,n(\tau,\sigma_1^u)\,\frac{\delta\,\mathcal{T}(\tau,\sigma_1^u)}{\delta\,R_{\bar{a}}(\tau,\sigma^u)}|_{\theta^i=0}\\
&=2\,\Big[\tilde{\phi}^{1/3}\,\sum_a\,\gamma_{\bar{a}a}\,Q_a^{-2}\,\Big(\partial_a^2\,n-6\,\partial_a\,q\,\partial_a\,n\Big)\Big](\tau,\sigma^u),
\end{aligned}
$$

$$
\begin{aligned}
&\int d^3\sigma_1\,\Big(1+n(\tau,\sigma_1^u)\Big)\,\frac{\delta\,\mathcal{M}(\tau,\sigma_1^u)}{\delta\,R_{\bar{a}}(\tau,\sigma^u)}=-\sum_i\,\delta^3(\sigma^u,\eta_i^u(\tau))\,\eta_i\,\Bigg((1+n)\\
&\frac{\tilde{\phi}^{-2/3}\,\sum_a\,\gamma_{\bar{a}a}\,Q_a^{-2}\,\Big(\kappa_{ia}(\tau)-\frac{Q_i}{c}\,A_{\perp\,a}\Big)^2}{\sqrt{m_i^2\,c^2+\tilde{\phi}^{-2/3}\,\sum_a\,Q_a^{-2}\,\Big(\kappa_{ia}(\tau)-\frac{Q_i}{c}\,A_{\perp\,a}\Big)^2}}\Bigg)(\tau,\sigma^u)\\
&+\Big(1+n(\tau,\sigma^u)\Big)\,\Bigg(\tilde{\phi}^{-1/3}\,\Bigg[\frac{1}{c}\,\sum_{ars}\,\gamma_{\bar{a}a}\,Q_a^2\,\delta_{ra}\,\delta_{sa}\,\pi_\perp^r\,\pi_\perp^s\\
&-\frac{1}{2c}\,\sum_{ab}\,(\gamma_{\bar{a}a}+\gamma_{\bar{a}b})\,Q_a^{-2}\,Q_b^{-2}\,F_{ab}\,F_{ab}
\end{aligned}
$$

$$-\frac{1}{c}\sum_{arsn}\gamma_{\bar{a}a}\,Q_a^2\,\delta_{ra}\,\delta_{sa}\,\Big(2\,\pi_\perp^r-\sum_m\delta^{rm}\sum_i Q_i\,\eta_i\,\partial_m\,c(\sigma^v,\eta_i^v(\tau))\Big)$$

$$\delta^{sn}\sum_j Q_j\,\eta_j\,\partial_n\,c(\sigma^v,\eta_j^v(\tau))\Big]\Big]\Big)(\tau,\sigma^u). \qquad (6.41)$$

The expression of the last integral in Eq. (6.41) has been obtained by using the super-momentum constraints (Eq. 6.33).

To get the final form of the second-order equations for $R_{\bar{a}}$ we have to use the three groups (A), (B), and (C) of the Hamilton equations quoted after Eq. (6.25), which are given in Ref. [95].

By using equations 6.14 and 6.15 of Ref. [95], the first half of the Hamilton equations for the particles implies

$$\eta_i\,\dot{\eta}_i^r(\tau)\overset{\circ}{=}\eta_i\left(\frac{\tilde{\phi}^{-2/3}\,(1+n)\,Q_r^{-2}\,\Big(\kappa_{ir}(\tau)-\frac{Q_i}{c}\,A_{\perp\,r}\Big)}{\sqrt{m_i^2\,c^2+\tilde{\phi}^{-2/3}\sum_c Q_c^{-2}\,\Big(\kappa_{ic}(\tau)-\frac{Q_i}{c}\,A_{\perp\,c}\Big)^2}}\right.$$

$$\left.-\,\phi^{-2}\,Q_r^{-1}\,\bar{n}_{(r)}\right)(\tau,\eta_i^u(\tau)),$$

$$\Downarrow$$

$$\kappa_{ir}(\tau)=\frac{Q_i}{c}\,A_{\perp\,r}(\tau,\eta_i^u(\tau))+m_i\,c\,\Big[\tilde{\phi}^{2/3}\,Q_r^2\,\Big(\dot{\eta}_i^r(\tau)+\tilde{\phi}^{-1/3}\,Q_r^{-1}\,\bar{n}_{(r)}\Big)\,\Big(\Big(1+n\Big)^2$$

$$-\tilde{\phi}^{2/3}\sum_c Q_c^2\,\Big(\dot{\eta}_i^c(\tau)+\tilde{\phi}^{-1/3}\,Q_c^{-1}\,\bar{n}_{(c)}\Big)^2\Big)^{-1/2}\Big](\tau,\eta_i^u(\tau)). \qquad (6.42)$$

so that the second half of the Hamilton equations becomes

$$\eta_i\,\frac{d}{d\tau}\left(m_i\,c\left(\frac{\tilde{\phi}^{2/3}\,Q_r^2\,\Big(\dot{\eta}_i^r(\tau)+\tilde{\phi}^{-1/3}\,Q_r^{-1}\,\bar{n}_{(r)}\Big)}{\sqrt{\Big(1+n\Big)^2-\tilde{\phi}^{2/3}\sum_c Q_c^2\,\Big(\dot{\eta}_i^c(\tau)+\tilde{\phi}^{-1/3}\,Q_c^{-1}\,\bar{n}_{(c)}\Big)^2}}\right)(\tau,\eta_i^u(\tau))\right.$$

$$\overset{\circ}{=}\Big(-\frac{\partial}{\partial\,\eta_i^r}\,\mathcal{W}+\frac{\eta_i\,Q_i}{c}\,\Big(\dot{\eta}_i^s(\tau)\,\frac{\partial\,A_{\perp\,s}}{\partial\,\eta_i^r}-\frac{d\,A_{\perp\,r}}{d\tau}\Big)+\eta_i\,\check{F}_{ir}\Big)(\tau,\eta_i^u(\tau)),$$

$$\mathcal{W}=\int d^3\sigma\,\Big[(1+n)\,\mathcal{W}_{(n)}+\tilde{\phi}^{-1/3}\sum_a Q_a^{-1}\,\bar{n}_{(a)}\,\mathcal{W}_a\Big](\tau,\sigma^u),$$

$$\mathcal{W}_{(n)}(\tau,\sigma^u)=-\frac{1}{2c}\Big[\tilde{\phi}^{-1/3}\sum_a Q_a^2\,\Big(2\,\pi_\perp^a-\delta^{am}\sum_i Q_i\,\eta_i\,\frac{\partial\,c(\sigma^v,\eta_i^v(\tau))}{\partial\,\sigma^m}\Big)$$

$$\delta^{an}\sum_j Q_j\,\eta_j\,\frac{\partial\,c(\sigma^v,\eta_j^v(\tau))}{\partial\,\sigma^n}\Big](\tau,\sigma^u),$$

$$
\begin{aligned}
\mathcal{W}_r(\tau,\sigma^u) &= -\frac{1}{c}\sum_s F_{rs}(\tau,\sigma^u)\,\delta^{sn}\sum_i Q_i\,\eta_i\,\frac{\partial\, c(\sigma^v,\eta_i^v(\tau))}{\partial\,\sigma^n},\\
F_{ir}(\tau,\sigma^u) &= m_i\,c\Big[\Big(\Big(1+n\Big)^2-\tilde{\phi}^{2/3}\sum_c Q_c^2\,\Big(\dot{\eta}_i^c(\tau)+\tilde{\phi}^{-1/3}\,Q_c^{-1}\,\bar{n}_{(c)}\Big)^2\Big)(\tau,\sigma^u)\Big]^{-1/2}\\
&\quad\Big[-(1+n)\,\frac{\partial\,n}{\partial\,\eta_i^r}+\tilde{\phi}^{1/3}\sum_a Q_a\,\Big(\frac{\partial\,\bar{n}_{(a)}}{\partial\,\eta_i^r}\\
&\quad-(2\,\partial_r\,q+\sum_{\bar{a}}\gamma_{\bar{a}a}\,\partial_r\,R_{\bar{a}})\,\bar{n}_{(a)}\Big)\,\Big(\dot{\eta}_i^a(\tau)+\tilde{\phi}^{-1/3}\,Q_a^{-1}\,\bar{n}_{(a)}\Big)\\
&\quad+\tilde{\phi}^{2/3}\sum_a Q_a^2\,(2\,\partial_r\,q+\sum_{\bar{a}}\gamma_{\bar{a}a}\,\partial_r\,R_{\bar{a}})\,\Big(\dot{\eta}_i^a(\tau)\\
&\quad+\tilde{\phi}^{-1/3}\,Q_a^{-1}\,\bar{n}_{(a)}\Big)^2\ \Big](\tau,\sigma^u).
\end{aligned}
\tag{6.43}
$$

Here, $\mathcal{W}$ is the non-inertial Coulomb potential, $\check{F}_{ir}$ are inertial relativistic forces, and the other terms correspond to the non-inertial Lorentz force [98, 99].

Finally, from equation 6.16 of Ref. [95], the Hamilton equations for the transverse electromagnetic fields in the radiation gauge become ($P_{\perp}^{rs}(\sigma^n)=\delta^{rs}-\sum_{uv}\delta^{ru}\,\delta^{sv}\,\frac{\partial_u\,\partial_v}{\Delta}$)

$$
\begin{aligned}
\partial_\tau\,A_{\perp\,r}(\tau,\sigma^u) &\mathrel{\circ\!\!=} \sum_{nua}\delta_{rn}\,P_{\perp}^{nu}(\sigma^v)\,\Big[\tilde{\phi}^{-1/3}\,(1+n)\,Q_a^2\,\delta_{ua}\\
&\quad\Big(\pi_{\perp}^a-\sum_m\delta^{am}\sum_i Q_i\,\eta_i\,\frac{\partial\,c(\sigma^v,\eta_i^v(\tau))}{\partial\,\sigma^m}\Big)\\
&\quad+\tilde{\phi}^{-1/3}\,Q_a^{-1}\,\bar{n}_{(a)}\,F_{au}\Big](\tau,\sigma^u),\\
\partial_\tau\,\pi_{\perp}^r(\tau,\sigma^u) &\mathrel{\circ\!\!=} \sum_m P_{\perp}^{rm}(\sigma^u)\,\Big(\sum_a\delta_{ma}\sum_i\eta_i\,Q_i\,\delta^3(\sigma^u,\eta_i^u(\tau))\\
&\quad\Bigg[\frac{\tilde{\phi}^{-2/3}\,(1+n)\,Q_a^{-2}\,\kappa_{ia}(\tau)}{\sqrt{m_i^2\,c^2+\tilde{\phi}^{-2/3}\sum_b Q_b^{-2}\,\Big(\kappa_{ib}(\tau)-\frac{Q_i}{c}\,A_{\perp\,b}\Big)^2}}\\
&\quad-\tilde{\phi}^{-1/3}\,Q_a^{-1}\,\bar{n}_{(a)}\Bigg]\,(\tau,\eta_i^u(\tau))\\
&\quad-\Big[2\,\tilde{\phi}^{-1/3}\,(1+n)\sum_{ab}Q_a^{-2}\,Q_b^{-2}\,\delta_{ma}\,\Big(\partial_b\,F_{ab}\\
&\quad-\Big[2\,\partial_b\,q+2\,\partial_b\,(\Gamma_a+\Gamma_b)\Big]\,F_{ab}\Big)\\
&\quad+2\,\tilde{\phi}^{-1/3}\sum_{ab}Q_a^{-2}\,Q_b^{-2}\,\delta_{ma}\,\partial_b\,n\,F_{ab}\\
&\quad-\tilde{\phi}^{-1/3}\sum_a\bar{n}_{(a)}\,Q_a^{-1}\,\Big(\partial_a\,\pi_{\perp}^m-\Big[2\,\partial_a\,q+\partial_a\,\Gamma_a\Big]\,\pi_{\perp}^m
\end{aligned}
$$

$$
\begin{aligned}
&+ \delta_{ma} \sum_n \Big[2\,\partial_n\, q + \partial_n\, \Gamma_a \Big]\, \pi^n_\perp \\
&+ \sum_i \eta_i\, Q_i \Bigg[\Big(2\,\partial_a\, q + \partial_a\, \Gamma_a \Big)\, \frac{\partial\, c(\sigma^v, \eta^v_i(\tau)))}{\partial\, \sigma^m} - \frac{\partial^2\, c(\sigma^v, \eta^u_i(\tau)))}{\partial\, \sigma^m\, \partial\, \sigma^a} \\
&- \delta_{ma} \sum_n \Bigg(\Big[2\,\partial_n\, q + \partial_n\, \Gamma_a \Big]\, \frac{\partial\, c(\vec{\sigma}, \vec{\eta}_i(\tau)))}{\partial\, \sigma^n} - \frac{\partial^2\, c(\sigma^v, \eta^v_i(\tau)))}{\partial\, \sigma^n\, \partial\, \sigma^n} \Bigg) \Bigg] \Bigg) \\
&+ \tilde{\phi}^{-1/3} \sum_a Q_a^{-1} \sum_i \eta_i\, Q_i\, \Bigg(\partial_a\, \bar{n}_{(a)}\, \frac{\partial\, c(\sigma^v, \eta^v_i(\tau))}{\partial\, \sigma^m} \\
&- \delta_{ma} \sum_n \partial_n\, \bar{n}_{(a)}\, \frac{\partial\, c(\sigma^v, \eta^v_i(\tau))}{\partial\, \sigma^n} \Bigg) \Bigg] (\tau, \sigma^u) \Bigg). \qquad (6.44)
\end{aligned}
$$

While the weak ADM energy $\hat{P}^\tau_{ADM} = \frac{1}{c}\, \hat{E}_{ADM}$ is given in Eq. (6.22), the other weak ADM Poincaré generators have the following expressions in the 3-orthogonal Schwinger time gauges:

$$
\begin{aligned}
\hat{P}^r_{ADM} &= \int d^3\sigma\, \Big[{}^3g^{rs}\, \mathcal{M}_s - 2\, {}^3\Gamma^r_{su}(\tau, \vec{\sigma})\, {}^3\Pi^{su} \Big] (\tau, \sigma^u) \\
&= 2 \int d^3\sigma\, \Big\{ \tilde{\phi}^{-2/3} \sum_{\bar{b}} Q_r^{-2} \Big(2\, \gamma_{\bar{b}r}\, \partial_r\, q + \sum_{\bar{a}} (\gamma_{\bar{b}r} \gamma_{\bar{a}r} - \frac{1}{2}\, \delta_{\bar{a}\bar{b}})\, \partial_r R_{\bar{a}} \Big)\, \Pi_{\bar{b}} \\
&\quad \frac{c^3}{12\pi\, G}\, \tilde{\phi}^{1/3}\, Q_r^{-2} \Big(4\, \partial_r\, q + \partial_r\, \Gamma_r \Big)\, {}^3K \\
&\quad + \frac{c^3}{8\pi G} \tilde{\phi}^{1/3} \sum_d Q_r^{-1} Q_d^{-1} \Big(2\, \partial_d\, q + \partial_d\, \Gamma_r \Big)\, \sigma_{(r)(d)} + \frac{1}{2} \tilde{\phi}^{-2/3}\, Q_r^{-2}\, \mathcal{M}_r \Big\} \approx 0, \\
\hat{J}^{rs}_{ADM} &= \int d^3\sigma\, \Big[-2(\sigma^r\, {}^3\Gamma^s_{uv} - \sigma^s\, {}^3\Gamma^r_{uv})\, {}^3\Pi^{uv} + (\sigma^r\, {}^3g^{su} - \sigma^s\, {}^3g^{ru})\, \mathcal{M}_u \Big] (\tau, \sigma^u) \\
&= 2 \int d^3\sigma\, \Big\{ \sigma^r\, \Big[\tilde{\phi}^{-2/3} \sum_{\bar{b}} Q_s^{-2} \Big(2\, \gamma_{\bar{b}s}\, \partial_s\, q + \sum_{\bar{a}} (\gamma_{\bar{b}s} \gamma_{\bar{a}s} - (1/2) \delta_{\bar{a}\bar{b}})\, \partial_s R_{\bar{a}} \Big)\, \Pi_{\bar{b}} \\
&\quad - \frac{c^3}{12\pi\, G}\, \tilde{\phi}^{1/3}\, Q_s^{-2}\, \Big(4\, \partial_s\, q + \partial_s\, \Gamma_s \Big)\, {}^3K \\
&\quad + \frac{c^3}{8\pi G} \tilde{\phi}^{1/3} \sum_d Q_s^{-1} Q_d^{-1} \Big(2\, \partial_d\, q + \partial_d\, \Gamma_s \Big)\, \sigma_{(s)(d)} + \frac{1}{2} \tilde{\phi}^{-2/3}\, Q_s^{-2}\, \mathcal{M}_s \Big] \\
&\quad - \sigma^s\, \Big[\tilde{\phi}^{-2/3} \sum_{\bar{b}} Q_r^{-2} \Big(2\, \gamma_{\bar{b}r}\, \partial_r\, q + \sum_{\bar{a}} (\gamma_{\bar{b}r} \gamma_{\bar{a}r} - (1/2) \delta_{\bar{a}\bar{b}})\, \partial_r R_{\bar{a}} \Big)\, \Pi_{\bar{b}} \\
&\quad - \frac{c^3}{12\pi\, G}\, \tilde{\phi}^{1/3}\, Q_r^{-2}\, \Big(4\, \partial_r\, q + \partial_r\, \Gamma_r \Big)\, {}^3K \\
&\quad + \frac{c^3}{8\pi G} \tilde{\phi}^{1/3} \sum_d Q_r^{-1} Q_d^{-1} \Big(2\, \partial_d\, q + \partial_d\, \Gamma_r \Big)\, \sigma_{(r)(d)} + \frac{1}{2} \tilde{\phi}^{-2/3}\, Q_r^{-2}\, \mathcal{M}_r \Big] \Big\},
\end{aligned}
$$

$$
\begin{aligned}
\hat{K}^r_{ADM} &= \hat{J}^{\tau r}_{ADM} = -\hat{J}^{r\tau}_{ADM} \\
&= \int d^3\sigma \Big(\sigma^r \Big[\frac{c^3}{16\pi\, G}\, \sqrt{\gamma}\; {}^3g^{ns}\, ({}^3\Gamma^u_{nv}\, {}^3\Gamma^v_{su} - {}^3\Gamma^u_{ns}\, {}^3\Gamma^v_{vu}) \\
&\quad - \frac{8\pi\, G}{c^3\, \sqrt{\gamma}}\, {}^3G_{nsuv}\, {}^3\Pi^{ns}\, {}^3\Pi^{uv} - \mathcal{M} \Big] \\
&\quad + \frac{c^3}{16\pi\, G}\, \delta^r_u\, ({}^3g_{vs} - \delta_{vs})\, \partial_n \Big[\sqrt{\gamma}\, ({}^3g^{ns}\, {}^3g^{uv} - {}^3g^{nu}\, {}^3g^{sv}) \Big] \Big)(\tau, \sigma^u) \\
&= \int d^3\sigma \Big\{ \sigma^r \Big[\frac{c^3}{16\pi G}\, \mathcal{S} - \mathcal{M} - \frac{4\pi G}{c^3}\, \tilde{\phi}^{-1} \sum_{\bar{b}} \Pi^2_{\bar{b}} \\
&\quad - \tilde{\phi}\, \frac{c^3}{16\pi G} \Big(\sum_{a\neq b} \sigma^2_{(a)(b)} - \frac{2}{3}\, ({}^3K)^2 \Big) \Big] \\
&\quad - \frac{c^3}{16\pi G} \tilde{\phi}^{-1/3}\, Q_r^{-2} \sum_s (\tilde{\phi}^{2/3} - Q_s^{-2}) \Big[\delta_{rs}\, \partial_s\, (\Gamma_r + \Gamma_s + q) \\
&\quad - \partial_r\, (\Gamma_r + \Gamma_s + q) \Big] \Big\}(\tau, \sigma^u) \approx 0.
\end{aligned}
$$

$$
\begin{aligned}
\mathcal{S}|_{\theta^i=0}(\tau, \sigma^u) &= \Big[\phi^2 \sum_a Q_a^{-2} \Big(\sum_{\bar{b}} \Big[\sum_{\bar{c}} (2\, \gamma_{\bar{b}a}\, \gamma_{\bar{c}a} - \delta_{\bar{b}\bar{c}})\, \partial_a\, R_{\bar{b}}\, \partial_a\, R_{\bar{c}} - \\
&\quad - 4\, \gamma_{\bar{b}a}\, \phi^{-1}\, \partial_a\, \phi\, \partial_a\, R_{\bar{b}} \Big] + 8\, (\phi^{-1}\, \partial_a\, \phi)^2 \Big) \Big](\tau, \sigma^u). \qquad (6.45)
\end{aligned}
$$

The gauge fixing $\hat{K}^r_{ADM} \approx 0$ eliminates the internal center of mass of the isolated system gravitational field, plus particles, plus electromagnetic field in the non-inertial rest-frames. Since no one is able to solve these equations, in the next chapter we will look at possible approximations.

7

Post-Minkowskian and Post-Newtonian Approximations

In this chapter I will describe the Hamiltonian post-Minkowskian (HPM) approximation, followed by the Hamiltonian post-Newtonian (HPN) one.

7.1 The Post-Minkowskian Approximation in the 3-Orthogonal Gauges

In Ref. [96] it is shown that in the family of non-harmonic 3-orthogonal Schwinger gauges it is possible to define a consistent linearization of the ADM canonical tetrad gravity plus matter in the non-inertial rest-frames described in Chapter 6, in the weak field approximation, to obtain a formulation of HPM gravity with non-flat Riemannian 3-spaces and asymptotic Minkowski background.

In the standard linearization [4, 8, 10, 11, 355] one introduces a fixed Minkowski background space-time, introduces the decomposition ${}^4g_{\mu\nu}(x) = {}^4\eta_{\mu\nu} + {}^4h_{\mu\nu}(x)$ in an inertial frame, and studies the linearized equations of motion for the small Minkowskian fields ${}^4h_{\mu\nu}(x)$. The approximation, usually a post-Newtonian (PN) one, is assumed valid over a big enough characteristic length L interpretable as the reduced wavelength $\lambda/2\pi$ of the resulting gravitational waves (GWs) (only for higher distances of L the linearization breaks down and curved space-time effects become relevant). For the Solar System there is a PN approximation in harmonic gauges, which is adopted in the BCRS [32–35] and whose 3-spaces $t_B =$ const. have deviations of order c^{-2} from Euclidean 3-spaces.

See Refs. [11, 278, 279, 376–401] and appendix A of Ref. [96] for a review of all the results of the standard approach and of the existing points of view on the subject.

In the class of asymptotically Minkowskian space-times without super-translations the 4-metric tends to an asymptotic Minkowski metric at spatial infinity, ${}^4g_{AB} \rightarrow {}^4\eta_{AB}$, which can be used as an asymptotic background. The decomposition ${}^4g_{AB} = {}^4\eta_{AB} + {}^4h_{(1)AB}$, with a first-order perturbation ${}^4h_{(1)AB}$ vanishing at spatial infinity, is defined in a global non-inertial rest-frame of an asymptotically Minkowskian space-time deviating for first-order effects from a global inertial rest-frame of an abstract Minkowski space-time $M_{(\infty)}$. The non-Euclidean 3-spaces Σ_τ will deviate by first-order effects from the Euclidean 3-spaces $\Sigma_{\tau(\infty)}$ of the inertial rest-frame of $M_{(\infty)}$, coinciding with the limit of Σ_τ at spatial infinity. When needed, differential operators like the Laplacian in Σ_τ will be approximated with the flat Laplacian in $\Sigma_{\tau(\infty)}$.

If $\zeta \ll 1$ is a small a-dimensional parameter, a consistent Hamiltonian linearization implies the following restrictions on the variables of the York canonical basis in the family of 3-orthogonal gauges with ${}^3K(\tau, \sigma^u) = F(\tau, \sigma^u) =$ numerical function (in this section one uses the notation $\vec{\sigma}$ for the curvilinear 3-coordinates σ^r, which become Cartesian in the background space-time $\Sigma_{\tau(\infty)}$):

$$
\begin{aligned}
R_{\bar a}(\tau, \vec{\sigma}) &= R_{(1)\bar a}(\tau, \vec{\sigma}) = O(\zeta) \ll 1,\\
\Pi_{\bar a}(\tau, \vec{\sigma}) &= \Pi_{(1)\bar a}(\tau, \vec{\sigma}) = \frac{1}{LG}\, O(\zeta),
\end{aligned}
$$

$$
\begin{aligned}
\tilde{\phi}(\tau, \vec{\sigma}) &= \sqrt{\det {}^3g_{rs}(\tau, \vec{\sigma})} = 1 + 6\,\phi_{(1)}(\tau, \vec{\sigma}) + O(\zeta^2),\\
N(\tau, \vec{\sigma}) &= 1 + n(\tau, \vec{\sigma}) = 1 + n_{(1)}(\tau, \vec{\sigma}) + O(\zeta^2),\\
\epsilon\, {}^4g_{\tau\tau}(\tau, \vec{\sigma}) &= 1 + \epsilon\, {}^4h_{(1)\tau\tau}(\tau, \vec{\sigma}) = 1 + 2\, n_{(1)}(\tau, \vec{\sigma}) + O(\zeta^2),\\
\bar{n}_{(a)}(\tau, \vec{\sigma}) &= -\epsilon\, {}^4g_{\tau a}(\tau, \vec{\sigma}) = -\epsilon\, {}^4h_{(1)\tau r}(\tau, \vec{\sigma}) = \bar{n}_{(1)(a)}(\tau, \vec{\sigma}) + O(\zeta^2),\\
{}^3K(\tau, \vec{\sigma}) &= \frac{12\pi\, G}{c^3}\, \pi_{\tilde\phi}(\tau, \vec{\sigma}) = {}^3K_{(1)}(\tau, \vec{\sigma}) = \frac{12\pi\, G}{c^3}\, \pi_{(1)\tilde\phi}(\tau, \vec{\sigma}) = \frac{1}{L}\, O(\zeta),\\
\sigma_{(a)(b)}|_{a\neq b}(\tau, \vec{\sigma}) &= \sigma_{(1)(a)(b)}|_{a\neq b}(\tau, \vec{\sigma}) = \frac{1}{L}\, O(\zeta),
\end{aligned}
$$

$$
\begin{aligned}
{}^3g_{rs}(\tau, \vec{\sigma}) &= -\epsilon\, {}^4g_{rs}(\tau, \vec{\sigma}) = \delta_{rs} - \epsilon\, {}^4h_{(1)rs}(\tau, \vec{\sigma})\\
&= [1 + 2\,(\Gamma_r(\tau, \vec{\sigma}) + 2\,\phi_{(1)}(\tau, \vec{\sigma}))]\, \delta_{rs} + O(\zeta^2),\\
\Gamma_a(\tau, \vec{\sigma}) &= \sum_{\bar a r=1}^{2} \gamma_{\bar a a}\, R_{\bar a}(\tau, \vec{\sigma}), \qquad R_{\bar a}(\tau, \vec{\sigma}) = \sum_{a=1}^{3} \gamma_{\bar a a}\, \Gamma_a(\tau, \vec{\sigma}). \qquad (7.1)
\end{aligned}
$$

The tidal variables $R_{\bar a}(\tau, \vec{\sigma})$ are slowly varying over the length L and times L/c; one has $(\frac{L}{{}^4\mathcal{R}})^2 = O(\zeta)$, where ${}^4\mathcal{R}$ is the mean radius of curvature of space-time.

The consistency of the Hamiltonian linearization requires the introduction of an ultraviolet cutoff M for matter. For the particles, described by the canonical variables $\vec{\eta}_i(\tau)$ and $\vec{\kappa}_i(\tau)$, this implies the conditions $\frac{m_i}{M}, \frac{\vec{\kappa}_i}{M} = O(\zeta)$. With similar restrictions on the electromagnetic field, one gets that the energy–momentum tensor of matter is $T^{AB}(\tau, \vec{\sigma}) = T^{AB}_{(1)}(\tau, \vec{\sigma}) + O(\zeta^2)$. Therefore, also the mass and momentum densities have the behavior $\mathcal{M}(\tau, \vec{\sigma}) = \mathcal{M}_{(1)}(\tau, \vec{\sigma}) + O(\zeta^2)$, $\mathcal{M}_r(\tau, \vec{\sigma}) = \mathcal{M}_{(1)r}(\tau, \vec{\sigma}) + O(\zeta^2)$. This approximation is not reliable at distances from the point particles less than the gravitational radius $R_M = \frac{MG}{c^2} \approx 10^{-29}\, M$ determined by the cutoff mass. The weak ADM Poincaré generators become equal to the Poincaré generators of this matter in the inertial rest-frame of the Minkowski space-time $M_{(\infty)}$ plus terms of order $O(\zeta^2)$ containing GWs and matter. Finally, the GW described by this linearization must have wavelengths satisfying $\lambda/2\pi \approx L \gg R_M$. If all the particles are contained in a compact set of radius l_c (the source), one must have $l_c \gg R_M$ for particles with relativistic velocities and $l_c \geq R_M$ for slow particles (like in binaries). See Ref. [11] for more details.

With this Hamiltonian linearization, one can avoid making PN expansions: One gets fully relativistic expressions, i.e., a HPM formulation of gravity.

The effective Hamiltonian adapted to the 3-orthogonal gauges and replacing the weak ADM energy is $\frac{1}{c}\Big(\hat{E}_{ADM(1)} + \hat{E}_{ADM(2)}\Big) + \frac{c^3}{12\pi G}\int d^3\epsilon \Big(\partial_\tau\, {}^3K_{(1)}\Big[1 + \frac{6}{\triangle}\Big(\frac{1}{4}\sum_a \partial_a^2 \Gamma_a - \frac{2\pi G}{c^3}\mathcal{M}_{(1)}\Big)\Big]\Big)(\tau, \sigma^u) + O(\zeta^3)$ in the HPM linearized theory ($\Gamma_a(\tau,\vec{\sigma}) = \sum_{\bar{a}} \gamma_{\bar{a}a}\, R_{\bar{a}}(\tau,\vec{\sigma}) = \Gamma_{(1)a}(\tau,\vec{\sigma})$ from Eq. (7.1)).

In Ref. [96] one has found the solutions of the super-momentum and super-Hamiltonian constraints and of the equations for the lapse and shift functions with the Bianchi identities satisfied. Therefore, one knows the first-order quantities $\pi^{(\theta)}_{(1)i}(\tau,\vec{\sigma})$, $\tilde{\phi}(\tau,\vec{\sigma}) = 1 + 6\,\phi_{(1)}(\tau,\vec{\sigma})$, $1 + n_{(1)}(\tau,\vec{\sigma})$, $\bar{n}_{(1)(a)}(\tau,\vec{\sigma})$ (the quantities containing the action-at-a-distance part of the gravitational interaction in the 3-orthogonal gauges) with an explicit expression for the HPM Newton and gravito-magnetic potentials. In the absence of the electromagnetic field they are (the terms in $\Gamma_a(\tau,\vec{\sigma})$ describe the contribution of GWs; ${}^3\mathcal{K}_{(1)}(\tau,\vec{\sigma}) = \frac{1}{\triangle}\,{}^3K(\tau,\vec{\sigma}$ is a non-local York time):[1]

$$
\begin{aligned}
\tilde{\phi}(\tau,\vec{\sigma}) &= 1 + 6\,\phi_{(1)}(\tau,\vec{\sigma})\\
&= 1 + \frac{3G}{c^3}\sum_i \eta_i \frac{\sqrt{m_i^2 c^2 + \vec{\kappa}_i^2(\tau)}}{|\vec{\sigma} - \vec{\eta}_i(\tau)|}\\
&\quad - \frac{3}{8\pi}\int d^3\sigma_1 \frac{\sum_a \partial^2_{1a}\,\Gamma_a(\tau,\vec{\sigma}_1)}{|\vec{\sigma} - \vec{\sigma}_1|},
\end{aligned}
$$

$$
\begin{aligned}
\epsilon\, {}^4g_{\tau\tau}(\tau,\vec{\sigma}) &= 1 + 2\,n_{(1)}(\tau,\vec{\sigma}) = 1 - 2\,\partial_\tau\, {}^3\mathcal{K}_{(1)}(\tau,\vec{\sigma})\\
&\quad - \frac{2G}{c^3}\sum_i \eta_i \frac{\sqrt{m_i^2 c^2 + \vec{\kappa}_i^2(\tau)}}{|\vec{\sigma} - \vec{\eta}_i(\tau)|}\left(1 + \frac{\vec{\kappa}_i^2(\tau)}{m_i^2 c^2 + \vec{\kappa}_i^2(\tau)}\right),\\
-\epsilon\, {}^4g_{\tau a}(\tau,\vec{\sigma}) &= \bar{n}_{(1)(a)}(\tau,\vec{\sigma}) = \partial_a\, {}^3\mathcal{K}_{(1)}(\tau,\vec{\sigma})\\
&\quad - \frac{G}{c^3}\sum_i \frac{\eta_i}{|\vec{\sigma} - \vec{\eta}_i(\tau)|}\Big(\frac{7}{2}\kappa_{ia}(\tau)\\
&\quad - \frac{1}{2}\frac{(\sigma^a - \eta_i^a(\tau))\,\vec{\kappa}_i(\tau)\cdot(\vec{\sigma} - \vec{\eta}_i(\tau))}{|\vec{\sigma} - \vec{\eta}_i(\tau)|^2}\Big)
\end{aligned}
$$

[1] Quantities like $|\vec{\eta}_i(\tau) - \vec{\eta}_j(\tau)|$ are the Euclidean 3-distance between the two particles in the asymptotic 3-space $\Sigma_{\tau(\infty)}$, which differs by quantities of order $O(\zeta)$ from the real non-Euclidean 3-distance in Σ_τ, as shown in equation 3.3 of Ref. [97].

$$- \int \frac{d^3\sigma_1}{4\pi\, |\vec{\sigma} - \vec{\sigma}_1|}\, \partial_{1a}\, \partial_\tau\, \Big[2\, \Gamma_a(\tau, \vec{\sigma}_1)$$

$$- \int d^3\sigma_2\, \frac{\sum_c \partial^2_{2c}\, \Gamma_c(\tau, \vec{\sigma}_2)}{8\pi\, |\vec{\sigma}_1 - \vec{\sigma}_2|} \Big],$$

$$\sigma_{(1)(a)(b)}\big|_{a\neq b}(\tau, \vec{\sigma}) = \frac{1}{2}\, \Big(\partial_a\, \bar{n}_{(1)(b)} + \partial_b\, \bar{n}_{(1)(a)} \Big)\big|_{a\neq b}(\tau, \vec{\sigma}). \tag{7.2}$$

Instead, the linearization of the Hamilton equations for the tidal variables $R_{\bar{a}}(\tau, \vec{\sigma})$ implies that they satisfy the following wave equation ($\triangle$ and $\Box$ are the flat Laplacian and the flat D'Alambertian on $\Sigma_{\tau(\infty)}$):[2]

$$\partial^2_\tau\, R_{\bar{a}}(\tau, \vec{\sigma}) = \triangle\, R_{\bar{a}}(\tau, \vec{\sigma}) + \sum_a \gamma_{\bar{a}a}\, \Big[\partial_\tau\, \partial_a\, \bar{n}_{(1)(a)}$$

$$+ \partial^2_a\, n_{(1)} + 2\, \partial^2_a\, \phi_{(1)} - 2\, \partial^2_a\, \Gamma_a + \frac{8\pi\, G}{c^3}\, T^{aa}_{(1)} \Big](\tau, \vec{\sigma}). \tag{7.3}$$

By using Eq. (7.2), this wave equation becomes

$$\Box \sum_{\bar{b}} M_{\bar{a}\bar{b}}\, R_{\bar{b}}(\tau, \vec{\sigma}) = E_{\bar{a}}(\tau, \vec{\sigma}),$$

$$M_{\bar{a}\bar{b}} = \delta_{\bar{a}\bar{b}} - \sum_a \gamma_{\bar{a}a}\, \frac{\partial^2_a}{\triangle}\, \Big(2\, \gamma_{\bar{b}a} - \frac{1}{2} \sum_b \gamma_{\bar{b}b}\, \frac{\partial^2_b}{\triangle} \Big),$$

$$E_{\bar{a}}(\tau, \vec{\sigma}) = \frac{4\pi\, G}{c^3} \sum_a \gamma_{\bar{a}a}\, \Big[\frac{\partial_\tau\, \partial_a}{\triangle}\, \Big(4\, \mathcal{M}_{(1)a} - \frac{\partial_a}{\triangle} \sum_c \partial_c\, \mathcal{M}_{(1)c} \Big)$$

$$+ 2\, T^{aa}_{(1)} + \frac{\partial^2_a}{\triangle} \sum_b T^{bb}_{(1)} \Big](\tau, \vec{\sigma}),$$

$$\Downarrow$$

$$\Box \sum_b \tilde{M}_{ab}\, \Gamma_b(\tau, \vec{\sigma}) = \sum_{\bar{a}} \gamma_{\bar{a}a}\, E_{\bar{a}}(\tau, \vec{\sigma}),$$

$$\tilde{M}_{ab} = \sum_{\bar{a}\bar{b}} \gamma_{\bar{a}a}\, \gamma_{\bar{b}b}\, M_{\bar{a}\bar{b}} = \delta_{ab}\, \Big(1 - 2\, \frac{\partial^2_a}{\triangle} \Big) + \frac{1}{2}\, \Big(1 + \frac{\partial^2_a}{\triangle} \Big)\, \frac{\partial^2_b}{\triangle},$$

$$\sum_a \tilde{M}_{ab} = 0, \qquad M_{\bar{a}\bar{b}} = \sum_{ab} \gamma_{\bar{a}a}\, \gamma_{\bar{b}b}\, \tilde{M}_{ab}. \tag{7.4}$$

To understand the meaning of the spatial operators $M_{\bar{a}\bar{b}}$ and $\tilde{M}_{ab}$, one must consider the perturbation ${}^4h_{(1)rs}(\tau, \vec{\sigma}) = -2\, \epsilon\, \delta_{rs}\, (\Gamma_r + 2\, \phi_{(1)})(\tau, \vec{\sigma})$ of Eq. (7.1), and apply to it the following decomposition, given in Ref. [401]:

[2] For the tidal momenta one gets $\frac{8\pi\, G}{c^3}\, \Pi_{\bar{a}}(\tau, \vec{\sigma}) = [\partial_\tau\, R_{\bar{a}} - \sum_a \gamma_{\bar{a}a}\, \partial_a\, \bar{n}_{(1)(a)}](\tau, \vec{\sigma}) + O(\zeta^2)$, so that the diagonal elements of the shear are $\sigma_{(1)(a)(a)}(\tau, \vec{\sigma}) = [- \sum_{\bar{a}} \gamma_{\bar{a}a}\, \partial_\tau\, R_{\bar{a}} + \bar{n}_{(1)(a)} - \frac{1}{3} \sum_b \bar{n}_{(1)(b)}](\tau, \vec{\sigma}) + O(\zeta^2)$.

$$
{}^4h_{(1)rs}(\tau,\vec{\sigma}) = \Big({}^4h^{TT}_{(1)rs} + \frac{1}{3}\,\delta_{rs}\,H_{(1)} + \frac{1}{2}\,(\partial_r\,\epsilon_{(1)s} + \partial_s\,\epsilon_{(1)r})
+ (\partial_r\,\partial_s - \frac{1}{3}\,\delta_{rs}\,\triangle)\,\lambda_{(1)}\Big)(\tau,\vec{\sigma}), \tag{7.5}
$$

with $\sum_r \partial_r\,\epsilon_{(1)r} = 0$ and ${}^4h^{TT}_{(1)rs}$ traceless and transverse (TT), i.e., $\sum_r {}^4h^{TT}_{(1)rr} = 0$, $\sum_r \partial_r\,{}^4h^{TT}_{(1)rs} = 0$. Since one finds $H_{(1)}(\tau,\vec{\sigma}) = -12\,\epsilon\,\phi_{(1)}(\tau,\vec{\sigma})$, $\lambda_{(1)}(\tau,\vec{\sigma}) = -3\,\epsilon\,\sum_u \frac{\partial_u^2}{\triangle^2}\,\Gamma_u(\tau,\vec{\sigma})$ and $\epsilon_{(1)r}(\tau,\vec{\sigma}) = -4\,\epsilon\,\frac{\partial_r}{\triangle}\,\Big(\Gamma_r - \sum_u \frac{\partial_u^2}{\triangle}\,\Gamma_u\Big)(\tau,\vec{\sigma})$, it turns out that the TT part of the spatial metric is independent from $\phi_{(1)}$ and has the expression

$$
\begin{aligned}
{}^4h^{TT}_{(1)rs}(\tau,\vec{\sigma}) &= -\epsilon\,\Big[\Big(2\,\Gamma_r + \sum_u \frac{\partial_u^2}{\triangle}\,\Gamma_u\Big)\,\delta_{rs} \\
&\quad -2\,\frac{\partial_r\,\partial_s}{\triangle}\,(\Gamma_r + \Gamma_s) + \frac{\partial_r\,\partial_s}{\triangle}\,\sum_u \frac{\partial_u^2}{\triangle}\,\Gamma_u\Big](\tau,\vec{\sigma}), \\
\Rightarrow \quad {}^4h^{TT}_{(1)aa}(\tau,\vec{\sigma}) &= -2\,\epsilon\,\sum_b \tilde{M}_{ab}\,\Gamma_b(\tau,\vec{\sigma}).
\end{aligned} \tag{7.6}
$$

Therefore, the spatial operator $\tilde{M}_{ab}$ connects the tidal variables $R_{\bar{a}}(\tau,\vec{\sigma})$ of the York canonical basis to the TT components of the 3-metric. By applying the decomposition (Eq. 7.5) to the spatial part $T^{rs}_{(1)}(\tau,\vec{\sigma})$ of the energy–momentum, one verifies that like in the harmonic gauges [11] the TT part of the 3-metric satisfies the wave equation $\Box\,{}^4h^{TT}_{rs}(\tau,\vec{\sigma}) = -\epsilon\,\frac{16\pi G}{c^3}\,T^{(TT)}_{(1)rs}(\tau,\vec{\sigma})$.

The retarded solution of the wave equation, with a no-incoming radiation condition, gives the following expression for the tidal variables (the HPM-GW):

$$
\begin{aligned}
R_{\bar{a}}(\tau,\vec{\sigma}) &= -\sum_a \gamma_{\bar{a}a}\,\Gamma_a(\tau,\vec{\sigma}) \overset{\circ}{=} \sum_{ab} \gamma_{\bar{a}a}\,\tilde{M}^{-1}_{ab}(\tau,\vec{\sigma}) \\
&\quad \frac{2\,G}{c^3}\int d^3\sigma_1\,\frac{T^{(TT)bb}_{(1)}(\tau - |\vec{\sigma}-\vec{\sigma}_1|;\vec{\sigma}_1)}{|\vec{\sigma}-\vec{\sigma}_1|}, \\
\frac{8\pi\,G}{c^3}\,\Pi_{\bar{a}}(\tau,\vec{\sigma}) &= \Bigg(\sum_{\bar{b}} M_{\bar{a}\bar{b}}\,\partial_\tau\,R_{\bar{b}} - \sum_a \gamma_{\bar{a}a}\,\Bigg[\frac{4\pi\,G}{c^3}\,\frac{1}{\triangle}\,\Bigg(4\,\partial_a\,\mathcal{M}_{(1)a} \\
&\quad -\frac{\partial_a^2}{\triangle}\sum_c \partial_c\,\mathcal{M}_{(1)c}\Bigg) + \partial_a^2\,{}^3\mathcal{K}_{(1)}\Bigg]\Bigg)(\tau,\vec{\sigma}).
\end{aligned} \tag{7.7}
$$

The explicit form of the inverse operator is given in Ref. [96]. By using the multipolar expansion of the energy–momentum $T^{AB}_{(1)}$ of Ref. [152–154] in the HPM version adapted to the rest-frame instant form of dynamics of Ref. [100, 101], one gets

$$
R_{\bar{a}}(\tau,\vec{\sigma}) = -\frac{G}{c^3}\sum_{ab} \gamma_{\bar{a}a}\,\tilde{M}^{-1}_{ab}\,\frac{\partial_\tau^2\,q^{(TT)aa|\tau\tau}(\tau - |\vec{\sigma}|)}{|\vec{\sigma}|} + (\text{higher multipoles}), \tag{7.8}
$$

where $q^{(TT)aa|\tau\tau}(\tau)$ is the TT mass quadrupole with respect to the center of energy (put in the origin of the radar 4-coordinates). An analogous result holds for ${}^4h^{TT}_{rs}(\tau,\vec{\sigma})$, and this implies a HPM relativistic version of the standard mass quadrupole emission formula.

Moreover, notwithstanding there is no gravitational self-energy due to the Grassmann regularization, the energy, 3-momentum, and angular momentum balance equations in HPM-GW emission are verified by using the conservation of the asymptotic ADM Poincaré generators (the same happens with the asymptotic Larmor formula of the electromagnetic case with Grassmann regularization, as shown in the last paper of Ref. [149–151]). See Refs. [11, 390–393, 402–407] for the use of the self-energy in the standard derivation of this result by means of PN expansions.

Eqs. (7.2) and (7.7) show that the HPM linearization with no-incoming radiation condition and Grassmann regularization is a theory with only dynamical matter interacting through suitable action-at-a-distance and retarded effective potentials. Instead, in relativistic atomic physics in special relativity (SR), the no-incoming radiation condition and the Grassmann regularization kill also the retardation, leaving only the action-at-a-distance interparticle Coulomb plus Darwin potentials. See equation 7.22 of Ref. [96] for the expression of the weak ADM energy till order $O(\zeta^3)$.

Moreover, it can be shown that the coordinate transformation $\bar{\tau} = \tau$, $\bar{\sigma}^r = \sigma^r + \frac{1}{2}\frac{\partial_r}{\triangle}\Big(4\,\Gamma^{(1)}_r - \sum_c \frac{\partial_c^2}{\triangle}\Gamma^{(1)}_c\Big)(\tau,\vec{\sigma})$, introducing new τ-dependent radar 3-coordinates on the 3-space Σ_τ, allows one to make a transition from the 3-orthogonal gauge with the 4-metric given by Eqs. (7.1) and (7.2) to a generalized non-3-orthogonal TT gauge containing the TT 3-metric (Eq. 7.6):

$$
\begin{aligned}
{}^4g_{(1)AB} &= {}^4\eta_{AB} \\
&+ \epsilon \begin{pmatrix} -2\,\frac{\partial_\tau}{\triangle}\,{}^3K_{(1)} + \alpha(\text{matter}) & -\frac{\partial_r}{\triangle}\,{}^3K_{(1)} + A_r(\Gamma_a) + \beta_r(\text{matter}) \\ -\frac{\partial_s}{\triangle}\,{}^3K_{(1)} + A_s(\Gamma_a) + \beta_s(\text{matter}) & \Big[B_r(\Gamma_a) + \gamma(\text{matter})\Big]\,\delta_{rs} \end{pmatrix} \\
&+ O(\zeta^2),
\end{aligned}
$$

$$\Downarrow$$

$$
\begin{aligned}
{}^4\bar{g}_{AB} &= {}^4\eta_{AB} + \epsilon \begin{pmatrix} -2\,\frac{\partial_\tau}{\triangle}\,{}^3K_{(1)} + \alpha(\text{matter}) & -\frac{\partial_r}{\triangle}\,{}^3K_{(1)} + \beta_r(\text{matter}) \\ -\frac{\partial_s}{\triangle}\,{}^3K_{(1)} + \beta_s(\text{matter}) & \epsilon\,{}^4h^{TT}_{(1)rs} + \delta_{rs}\,\gamma(\text{matter}) \end{pmatrix} \\
&+ O(\zeta^2).
\end{aligned} \tag{7.9}
$$

The functions appearing in Eq. (7.9) are: $A_r(\Gamma_a) = -\frac{1}{2}\,\partial_\tau\,\frac{\partial_r}{\triangle}\Big(4\,\Gamma_r - \sum_c \frac{\partial_c^2}{\triangle}\Gamma_c\Big)$, $B_r(\Gamma_a) = -2\Big(\Gamma_r + \frac{1}{2}\sum_c \frac{\partial_c^2}{\triangle}\Gamma_c\Big)$, $\alpha(\text{matter}) = \frac{8\pi\,G}{c^3}\,\frac{1}{\triangle}\Big(\mathcal{M}_{(1)} + \sum_c T^{cc}_{(1)}\Big)$, $\beta_r(\text{matter}) = -\frac{4\pi\,G}{c^3}\,\frac{1}{\triangle}\Big(4\,\mathcal{M}_{(1)r} - \frac{\partial_r}{\triangle}\sum_c \partial_c\,\mathcal{M}_{(1)c}\Big)$, $\gamma(\text{matter}) = \frac{8\pi\,G}{c^3}\,\frac{1}{\triangle}\,\mathcal{M}_{(1)}$.

Also, in the absence of matter, this TT gauge differs from the usual harmonic ones for the non-spatial terms, depending upon the inertial gauge variable non-local York time:

$$ {}^3\mathcal{K}_{(1)}(\tau,\vec{\sigma}) = \frac{1}{\triangle}\, {}^3K_{(1)}(\tau,\vec{\sigma}), \tag{7.10} $$

describing the HPM form of the gauge freedom in clock synchronization.

If one uses the coordinate system of the generalized TT gauge, one can introduce the standard polarization pattern of GWs for ${}^4h^{TT}_{rs}$ (see Refs. [11, 378, 401]), and then the inverse transformation gives the polarization pattern of HPM-GW in the family of 3-orthogonal gauges.

If the matter sources have a compact support and if the matter terms $\frac{1}{\triangle}\mathcal{M}_{(1)}(\tau,\vec{\sigma})$ and $\frac{1}{\triangle}\mathcal{M}_{(1)r}(\tau,\vec{\sigma})$ are negligible in the radiation zone far away from the sources, then Eq. (7.9) gives a spatial TT gauge with still the explicit dependence on the inertial gauge variable ${}^3\mathcal{K}_{(1)}(\tau,\vec{\sigma})$ (non-existent in Newtonian gravity), which determines the non-Euclidean nature of the instantaneous 3-spaces. Then, one can study the far field of compact matter sources: The restriction to the Solar System of the resulting HPM 4-metric[3] is compatible with the harmonic PN 4-metric of BCRS [32–35] if the non-local York time is of order c^{-2}. The resulting shift function should be used for the HPM description of gravito-magnetism (see Refs. [10, 408–414] for the Lense–Thirring and other associated effects).

The TT gauge allows one to reproduce the various descriptions of the GW detectors and of the reference frames used in GW detection in terms of HPM-GW: This is done in subsection VIID of Ref. [96], where the effect of a HPM-GW on a test mass is given in terms of the proper distance between two nearby geodesics.

The HPM-GW propagate in non-Euclidean instantaneous 3-spaces Σ_τ, differing from the inertial asymptotic Euclidean 3-spaces $\Sigma_{\tau(\infty)}$ at the first order. In the family of 3-orthogonal gauges with York time ${}^3K_{(1)}(\tau,\vec{\sigma}) \approx F_{(1)}(\tau,\vec{\sigma}) =$ numerical function, the dynamically determined 3-spaces Σ_τ have an intrinsic 3-curvature ${}^3\hat{R}|_{\theta^i=0}(\tau,\vec{\sigma}) = 2\sum_a \partial_a^2\,\Gamma_a(\tau,\vec{\sigma})$ determined only by the HPM-GW (and therefore by the matter energy–momentum tensor in the past as shown by Eq. (7.7)). Their extrinsic curvature tensor as sub-manifolds of space-time is

$$ \begin{aligned} {}^3K_{(1)rs}(\tau,\vec{\sigma}) &\approx \sigma_{(1)(r)(s)}|_{r\neq s}(\tau,\vec{\sigma}) \\ &\quad + \delta_{rs}\left(\frac{1}{3}F_{(1)} - \partial_\tau\,\Gamma_r + \partial_r\,\bar{n}_{(1)(r)} - \sum_a \partial_a\,\bar{n}_{(1)(a)}\right)(\tau,\vec{\sigma}), \end{aligned} \tag{7.11} $$

[3] See equation 7.20 in Ref. [96] where ${}^4g_{\tau\tau}(\tau,\vec{\sigma})$ and ${}^4g_{\tau r}(\tau,\vec{\sigma})$ explicitly depend on the non-local York time.

with $\bar{n}_{(1)(r)}(\tau,\vec{\sigma})$, $\sigma_{(1)(r)(s)}|_{r\neq s}(\tau,\vec{\sigma})$ and $\Gamma_r(\tau,\vec{\sigma})$ given in Eqs. (7.2) and (7.7). The York time appears only in Eq. (7.11): all the other PM quantities depend on the non-local York time ${}^3\mathcal{K}_{(1)}(\tau,\vec{\sigma}) \approx \frac{1}{\triangle} F_{(1)}(\tau,\vec{\sigma})$.

In Ref. [97], where the matter is restricted only to the particles,[4] one evaluates all the properties of these HPM space-times:

1. the 3-volume element, the 3-distance, and the intrinsic and extrinsic 3-curvature tensors of the 3-spaces Σ_τ;
2. the proper time of a time-like observer;
3. the time-like and null 4-geodesics (they are relevant for the definition of the radial velocity of stars, as shown in the IAU conventions of Ref. [415] and in studies of time delays [410–414, 416–418]);
4. the red-shift and luminosity distance. In particular, one finds that the recessional velocity of a star is proportional to its luminosity distance from the Earth, at least for small distances. This is in accord with the Hubble old red-shift–distance relation, which is formalized in the Hubble law (velocity–distance relation) when the standard cosmological model is used (see, for instance, Ref. [419] on these topics). These results are dependent on the non-local York time, which could play a role in giving a different interpretation of the data from super novae, which are used as a support for dark energy [420, 421].

Finally, in subsection IIIB of Ref. [96] it is shown that this HPM linearization can be interpreted as the first term of a HPM expansion in powers of the Newton constant G in the family of 3-orthogonal gauges. This expansion has still to be studied. In particular, it will be useful to check whether in the HPM formulation there are phenomena (appearing at high orders in the standard PN expansions) like the hereditary tails starting from 1.5PN $[O((\frac{v}{c})^3)]$ and the non-linear (Christodoulou) memory starting from 3PN (see Ref. [422] for a review)[5]. This would allow one to make a comparison with all the results of the PN expansions, in which today there is control on the GW solution and on the matter equations of motion till order 3.5PN $[O((\frac{v}{c})^7)]$ (for binaries, see the review in chapter 4 of Ref. [11]) and well-established connections with numerical relativity (see the review in Ref. [423]), especially for the binary black hole problem (see the review in Ref. [424]).

The PM Hamilton equations and their PN limit in 3-orthogonal gauges for a system of N scalar particles of mass m_i and Grassmann-valued signs of energy η_i can be treated by using the results of Ref. [97]. See Refs. [376–378] for classical

[4] The properties of HPM transverse electro-magnetic fields remain to be explored.

[5] They imply that GW propagate not only on the flat light-cone but also inside it (i.e., with all possible speeds $0 \leq v \leq c$): There is an instantaneous wavefront followed by a tail traveling at lower speed (it arrives later and then fades away) and a persistent zero-frequency non-linear memory.

texts on the motion of particles in gravitational fields and Refs. [388, 389, 420, 421, 425, 426] for more recent developments[6].

The treatment in the 3-orthogonal gauges of the PM Hamilton equations for the electromagnetic field in the radiation gauge is given in Ref. [96], while the PM Hamilton equations for perfect fluids are given in Ref. [104].

With only particles the PM approximation with the ultraviolet cutoff M implies $\vec{\kappa}_i(\tau) = \frac{m_i c \dot{\vec{\eta}}_i(\tau)}{\sqrt{1-\dot{\vec{\eta}}_i^2(\tau)}} + Mc\,O(\zeta)$, $\mathcal{M}_{(1)}(\tau,\vec{\sigma}) = \sum_i \delta^3(\vec{\sigma},\vec{\eta}_i(\tau))\,\eta_i\,\frac{m_i c}{\sqrt{1-\dot{\vec{\eta}}_i^2(\tau)}} + O(\zeta^2)$, $\mathcal{M}_{(1)r}(\tau,\vec{\sigma}) = \sum_i \delta^3(\vec{\sigma},\vec{\eta}_i(\tau))\,\eta_i\,\frac{m_i c \dot{\eta}_{ir}(\tau)}{\sqrt{1-\dot{\vec{\eta}}_i^2(\tau)}} + O(\zeta^2)$. Moreover, one has $\ddot{\eta}_i^r(\tau) = O(\zeta)$. The notation $\dot{a}(\tau) = \frac{da(\tau)}{d\tau}$ is used.

One can make an equal time development of the retarded kernel in Eq. (7.7), as in Refs. [149–151] for the extraction of the Darwin potential from the Lienard–Wiechert solution (see equations 5.1–5.21 of Refs. [149–151] with $\sum_s P_\perp^{rs}(\vec{\sigma})\,\dot{\eta}_i^s(\tau) \mapsto \sum_{uv} \mathcal{P}_{bbuv}(\vec{\sigma})\,\frac{\dot{\eta}_i^u(\tau)\,\dot{\eta}_i^v(\tau)}{\sqrt{1-\dot{\eta}_i^2(\tau)}}$). In this way, one gets the following expression of the HPM-GW from point masses:

$$
\begin{aligned}
\Gamma_a(\tau,\vec{\sigma}) &\circeq -\frac{2G}{c^2} \sum_b \tilde{M}^{-1}_{ab}(\vec{\sigma}) \sum_i \eta_i\, m_i \sum_{uv} \mathcal{P}_{bbuv}(\vec{\sigma})\, \frac{\dot{\eta}_i^u(\tau)\,\dot{\eta}_i^v(\tau)}{\sqrt{1-\dot{\eta}_i^2(\tau)}} \\
&\quad \Big[|\vec{\sigma}-\vec{\eta}_i(\tau)|^{-1} + \sum_{m=1}^{\infty} \frac{1}{(2m)!} \left(\dot{\vec{\eta}}_i(\tau)\cdot \frac{\partial}{\partial \vec{\sigma}} \right)^{2m} |\vec{\sigma}-\vec{\eta}_i(\tau)|^{2m-1} \Big] \\
&\quad + O(\zeta^2), \\
\mathcal{P}_{rsuv}(\vec{\sigma}) &= \frac{1}{2}\,(\delta_{ru}\,\delta_{sv} + \delta_{rv}\,\delta_{su}) \\
&\quad -\frac{1}{2}\left(\delta_{rs} - \frac{\partial_r\,\partial_s}{\triangle}\right)\delta_{uv} + \frac{1}{2}\left(\delta_{rs} + \frac{\partial_r\,\partial_s}{\triangle}\right)\frac{\partial_u\,\partial_v}{\triangle} \\
&\quad -\frac{1}{2}\left[\frac{\partial_u}{\triangle}\,(\delta_{rv}\,\partial_s + \delta_{sv}\,\partial_r) + \frac{\partial_v}{\triangle}\,(\delta_{ru}\,\partial_s + \delta_{su}\,\partial_r)\right],
\end{aligned}
\tag{7.12}
$$

with the retardation effects pushed to order $O(\zeta^2)$.

If the lapse and shift functions are rewritten in the form $n_{(1)}(\tau,\vec{\sigma}) = \check{n}_{(1)}(\tau,\vec{\sigma}) - \partial_\tau\,{}^3\mathcal{K}_{(1)}(\tau,\vec{\sigma})$, $\bar{n}_{(1)(r)}(\tau,\vec{\sigma}) = \check{\bar{n}}_{(1)(r)}(\tau,\vec{\sigma}) + \partial_r\,{}^3\mathcal{K}_{(1)}(\tau,\vec{\sigma})$, to display their dependence on the inertial gauge variable non-local York time, it can be shown that the PM Hamilton equations for the particles imply the following form of the PM Grassmann regularized second-order equations of motion, showing explicitly the equality of the inertial and gravitational masses of the particles:

[6] In this approach, point particles are considered as independent matter degrees of freedom with a Grassmann regularization of the self-energies to get well-defined world-lines (see also Refs. [429, 430]): They are not considered as point-like singularities of solutions of Einstein's equations (the point of view of Ref. [431]). Solutions of this type have to be described with distributions and, as shown in Ref. [432], the most general class of such solutions under mathematical control includes singularities simulating matter shells, but not either strings or particles. See also Ref. [433].

$$m_i\,\eta_i\,\ddot{\eta}_i^r(\tau) \stackrel{\circ}{=} \eta_i\,\sqrt{1-\dot{\vec{\eta}}_i^2(\tau)}\,\Big(\mathcal{F}_i^r - \dot{\eta}_i^r(\tau)\,\dot{\vec{\eta}}_i(\tau)\cdot\vec{\mathcal{F}}_i\Big)(\tau|\vec{\eta}_i(\tau)|\vec{\eta}_{k\neq i}(\tau))$$
$$\stackrel{\text{def}}{=} \eta_i\,F_i^r(\tau|\vec{\eta}_i(\tau)|\vec{\eta}_{k\neq i}(\tau)),$$

$$\begin{aligned}
\eta_i\,\mathcal{F}_i^r(\tau|\vec{\eta}_i(\tau)|\vec{\eta}_{k\neq i}(\tau)) &= \frac{m_i\,\eta_i}{\sqrt{1-\dot{\vec{\eta}}_i^2(\tau)}}\Big(-\frac{\partial\,\check{n}_{(1)}(\tau,\vec{\eta}_i(\tau))}{\partial\,\eta_i^r}\\
&+\frac{\dot{\eta}_i^r(\tau)}{1-\dot{\vec{\eta}}_i^2(\tau)}\sum_u\Big[\dot{\eta}_i^u(\tau)\,\frac{\partial\,\check{n}_{(1)}}{\partial\,\eta_i^u}+\sum_{j\neq i}\dot{\eta}_j^u(\tau)\,\frac{\partial\,\check{n}_{(1)}}{\partial\,\eta_j^u}\Big](\tau,\vec{\eta}_i(\tau))\\
&+\Big(\sum_u\dot{\eta}_i^u(\tau)\Big[\frac{\partial\,\check{\bar{n}}_{(1)(u)}}{\partial\,\eta_i^r}-\frac{\partial\,\check{\bar{n}}_{(1)(r)}}{\partial\,\eta_i^u}\Big]-\sum_{j\neq i}\sum_u\dot{\eta}_j^u(\tau)\,\frac{\partial\,\check{\bar{n}}_{(1)(r)}}{\partial\,\eta_j^u}\\
&-\frac{\dot{\eta}_i^r(\tau)}{1-\dot{\vec{\eta}}_i^2(\tau)}\sum_u\dot{\eta}_i^u(\tau)\sum_s\Big[\dot{\eta}_i^s(\tau)\,\frac{\partial\,\check{\bar{n}}_{(1)(u)}}{\partial\,\eta_i^s}\\
&+\sum_{j\neq i}\dot{\eta}_j^s(\tau)\,\frac{\partial\,\check{\bar{n}}_{(1)(u)}}{\partial\,\eta_j^s}\Big]\Big)(\tau,\vec{\eta}_i(\tau))+\Big(\sum_u(\dot{\eta}_i^u(\tau))^2\,\frac{\partial\,(\Gamma_u+2\,\phi_{(1)})}{\partial\,\eta_i^r}\\
&-\dot{\eta}_i^r(\tau)\sum_u\Big[\dot{\eta}_i^u(\tau)\Big(2\,\frac{\partial\,(\Gamma_r+2\,\phi_{(1)})}{\partial\,\eta_i^u}+\sum_c\frac{(\dot{\eta}_i^c(\tau))^2}{1-\dot{\vec{\eta}}_i^2(\tau)}\,\frac{\partial\,(\Gamma_c+2\,\phi_{(1)})}{\partial\,\eta_i^u}\Big)\\
&+\sum_{j\neq i}\dot{\eta}_j^u(\tau)\Big(2\,\frac{\partial\,(\Gamma_r+2\,\phi_{(1)})}{\partial\,\eta_j^u}\\
&+\sum_c\frac{(\dot{\eta}_i^c(\tau))^2}{1-\dot{\vec{\eta}}_i^2(\tau)}\,\frac{\partial\,(\Gamma_c+2\,\phi_{(1)})}{\partial\,\eta_j^u}\Big)\Big]\Big)(\tau,\vec{\eta}_i(\tau))\\
&-\frac{\dot{\eta}_i^r(\tau)}{1-\dot{\vec{\eta}}_i^2(\tau)}\Big[\partial_\tau^2|_{\vec{\eta}_i}\,{}^3\mathcal{K}_{(1)}+2\sum_s\dot{\eta}_i^s(\tau)\,\frac{\partial\,\partial_\tau|_{\vec{\eta}_i}\,{}^3\mathcal{K}_{(1)}}{\partial\,\eta_i^s}\\
&+\sum_{su}\dot{\eta}_i^s(\tau)\,\dot{\eta}_i^u(\tau)\,\frac{\partial^2\,{}^3\mathcal{K}_{(1)}}{\partial\,\eta_i^u\,\partial\,\eta_i^s}\Big](\tau,\vec{\eta}_i(\tau))\Big)+O(\zeta^2).
\end{aligned}\tag{7.13}$$

The effective action-at-a-distance force $\vec{F}_i(\tau)$ contains

1. the contribution of the lapse function $\check{n}_{(1)}$, which generalizes the Newton force;
2. the contribution of the shift functions $\check{\bar{n}}_{(1)(r)}$, which gives the gravito-magnetic effects;
3. the retarded contribution of HPM-GW, described by the functions Γ_r of Eq. (7.12);
4. the contribution of the volume element $\phi_{(1)}$ ($\tilde{\phi} = 1 + 6\,\phi_{(1)} + O(\zeta^2)$), always summed to the HPM-GW, giving forces of Newton type; and
5. the contribution of the inertial gauge variable (the non-local York time) ${}^3\mathcal{K}_{(1)} = \frac{1}{\triangle}\,{}^3K_{(1)}$.

In the electromagnetic case in SR [100, 101], the regularized coupled second-order equations of motion of the particles obtained by using the Lienard Wiechert solutions for the electromagnetic field are independent of the type of Green's function (retarded, advanced, or symmetric) used. The electromagnetic retardation effects, killed by the Grassmann regularization, are connected with QED radiative corrections to the one-photon exchange diagram. This is not strictly true in the gravitational case. The effect of retardation is not killed by the Grassmann regularization but only pushed to $O(\zeta^2)$: At this order it should give extra contributions to the second-order equations of motion. This shows that our semi-classical approximation, obtained with our Grassmann regularization, of an unspecified quantum gravity theory does not take into account only a one-graviton exchange diagram: In the spin-2 case there is an extra retardation effect that shows up only at higher HPM orders.[7]

In this approach the 3-universe is described in a non-inertial rest-frame with non-Euclidean 3-spaces Σ_τ tending to Euclidean inertial ones $\Sigma_{\tau(\infty)}$ at spatial infinity. Both matter and gravitational degrees of freedom live inside Σ_τ and their internal 3-center of mass is eliminated by the rest-frame condition $\hat{P}^r_{ADM} \approx 0$ (implied by the absence of super-translations) if also the condition $\hat{K}^r_{ADM} = \hat{J}^{\tau r}_{ADM} \approx 0$ is added, as in SR. The 3-universe may be described as an external decoupled center of mass carrying a pole–dipole structure: $\hat{E}_{ADM}$ is the invariant mass and $\hat{J}^{rs}_{ADM}$ the rest spin. As in SR, the condition $\hat{K}^r_{ADM} \approx 0$ selects the Fokker–Pryce center of inertia as the natural time-like observer origin of the radar coordinates: It follows a non-geodetic straight world-line like the asymptotic inertial observers existing in these space-times.

This is a way out from the problem of the center of mass in GR and of its world-line, a still open problem in generic space-times as can be seen in Refs. [152–154, 394–400] (see Refs. [425–428] for the PN approach). Usually, by means of some supplementary condition, the center of mass is associated to the monopole of a multipolar expansion of the energy–momentum of a small body (see Refs. [47, 155] for the special relativistic case).

In SR, the elimination of the internal 3-center of mass leads to describing the dynamics inside Σ_τ only in terms of relative variables (see Sections 3.2 and 3.3 of in the case of particles). However, relative variables do not exist in the non-Euclidean 3-spaces of curved space-times, where flat objects like $\vec{r}_{ij}(\tau) = \vec{\eta}_i(\tau) - \vec{\eta}_j(\tau)$ have to be replaced with a quantity proportional to the tangent vector to the space-like 3-geodesics joining the two particles in the non-Euclidean 3-space Σ_τ (see Ref. [434] for an implementation of this idea). Quantities like $\vec{r}^{\,2}_{ij}(\tau)$ have to be replaced with the Synge world function [376, 416–418, 426].[8]

[7] In the electromagnetic case the Grassmann regularization implies $Q_i\, \dot{\eta}^r_i(\tau - |\vec{\sigma}|) = Q_i\, i\dot{\eta}^r_i(\tau)$ and equations of motion of the type $\ddot{\eta}^r_i(\tau) = Q_i \ldots$ with $Q_i^2 = 0$. In the gravitational case the equations of motion are of the type $\eta_i\, \ddot{\eta}^r_i(\tau) = \eta_i \ldots$ with $\eta_i^2 = 0$, but the Grassmann regularized retardation in Eq. (7.7) gives Eq. (7.13) only at the lowest order in ζ and has contributions of every order $O(\zeta^k)$.

[8] It is a bi-tensor, i.e., a scalar in both the points $\vec{\eta}_i(\tau)$ and $\vec{\eta}_j(\tau)$, defined in terms of the space-like geodesic connecting them in Σ_τ. See equation 3.13 of Ref. [97].

This problem is another reason why extended objects tend to be replaced with point-like multipoles, which, however, do not span a canonical basis of phase space (see Refs. [47, 155] for SR).

However, at the level of the HPM approximation, one can introduce relative variables for the particles, like the SR ones of Section 3.3, defined as 3-vectors in the asymptotic inertial rest-frame $\Sigma_{\tau(\infty)}$ by putting $\vec{\eta}_i(\tau) = \vec{\eta}_{(o)i}(\tau) + \vec{\eta}_{(1)i}(\tau)$ and $\vec{\kappa}_i(\tau) = \vec{\kappa}_{(o)i}(\tau) + \vec{\kappa}_{(1)i}(\tau)$, with $\vec{\eta}_{(o)i}(\tau)$, $\vec{\kappa}_{(o)i}(\tau) = O(\zeta)$. This allows one to define HPM collective and relative canonical variables for the particles, with the collective variables eliminated by the conditions $\hat{P}^r_{ADM} \approx 0$ and $\hat{K}^r_{ADM} \approx 0$ (at the lowest order they become the SR conditions).

In the case of two particles (with total and reduced masses $M = m_1 + m_2$ and $\mu = \frac{m_1 m_2}{M}$) one puts $\vec{\eta}_1(\tau) = \vec{\eta}_{12}(\tau) + \frac{m_2}{M}\,\vec{\rho}_{12}(\tau)$, $\vec{\eta}_2(\tau) = \vec{\eta}_{12}(\tau) - \frac{m_1}{M}\,\vec{\rho}_{12}(\tau)$, $\vec{\kappa}_1(\tau) = \frac{m_1}{M}\,\vec{\kappa}_{12}(\tau) + \vec{\pi}_{12}(\tau)$, $\vec{\kappa}_2(\tau) = \frac{m_2}{M}\,\vec{\kappa}_{12}(\tau) - \vec{\pi}_{12}(\tau)$, and goes to the new canonical basis $\vec{\eta}_{12}(\tau) = \frac{m_1\,\vec{\eta}_1(\tau) + m_2\,\vec{\eta}_2}{M}$, $\vec{\rho}_{12}(\tau) = \vec{\eta}_1(\tau) - \vec{\eta}_2(\tau)$, $\vec{\kappa}_{12}(\tau) = \vec{\kappa}_1(\tau) + \vec{\kappa}_2(\tau)$, $\vec{\pi}_{12}(\tau) = \frac{m_2\,\vec{\kappa}_1(\tau) - m_1\,\vec{\kappa}_2(\tau)}{M}$.

It can be shown that the conditions $\hat{P}^r_{ADM} \approx 0$ and $\hat{K}^r_{ADM} \approx 0$ imply

$$
\begin{aligned}
\vec{\eta}_1(\tau) &\approx \Big(\frac{m_2}{M} - A_{(o)}(\tau)\Big)\,\vec{\rho}_{(o)12}(\tau) + \frac{m_2}{M}\,\vec{\rho}_{(1)12}(\tau) + \vec{f}_{(1)}(\tau)[\text{rel.var., GW}],\\
\vec{\eta}_2(\tau) &\approx -\Big(\frac{m_1}{M} + A_{(o)}(\tau)\Big)\,\vec{\rho}_{(o)12}(\tau) - \frac{m_1}{M}\,\vec{\rho}_{(1)12}(\tau) + \vec{f}_{(1)}(\tau)[\text{rel.var., GW}],\\
A_{(o)}(\tau) &= \frac{\frac{m_2}{M}\sqrt{m_1^2\,c^2 + \vec{\pi}^2_{(1)12}(\tau)} - \frac{m_1}{M}\sqrt{m_2^2\,c^2 + \vec{\pi}^2_{(1)12}(\tau)}}{\sqrt{m_1^2\,c^2 + \vec{\pi}^2_{(1)12}(\tau)} + \sqrt{m_2^2\,c^2 + \vec{\pi}^2_{(1)12}(\tau)}},
\end{aligned}
\tag{7.14}
$$

for some function $\vec{f}_{(1)}[\text{rel.var., GW}](\tau) \approx \vec{\eta}_{(1)12}(\tau)$ depending on the relative variables and the HPM-GW of Eq. (7.12) in the absence of incoming radiation. Then, the equations of motion (7.13) imply

$$
\mu\,\ddot{\rho}^r_{(o)12}(\tau) \stackrel{\circ}{=} \frac{m_2}{M}\,F^r_1(\tau|\vec{\eta}_{(o)1}(\tau)|\vec{\eta}_{(o)2}(\tau)) - \frac{m_1}{M}\,F^r_2(\tau|\vec{\eta}_{(o)2}(\tau)|\vec{\eta}_{(o)1}(\tau))
\tag{7.15}
$$

for the relative configurational variable. The collective configurational variable has $\vec{\eta}_{(o)12}(\tau) \approx -A_{(o)}(\tau)\,\vec{\rho}_{(o)12}(\tau)$ at the lowest order, while at the first order there is an equation of motion equivalent to $\ddot{\eta}^r_{(1)12}(\tau) \approx \frac{d^2}{d\tau^2}\,\vec{f}_{(1)}[\text{rel.var., GW}]\,(\tau)$.

7.2 The Post-Newtonian Expansion of the Post-Minkowskian Linearization

If all the particles are contained in a compact set of radius l_c, one can add a slow-motion condition in the form $\sqrt{\epsilon} = \frac{v}{c} \approx \sqrt{\frac{R_{m_i}}{l_c}}$, $i = 1, \ldots, N$ ($R_{m_i} = \frac{2\,G\,m_i}{c^2}$

is the gravitational radius of particle i) with $l_c \geq R_M$ and $\lambda \gg l_c$. In this case one can do the PN expansion of Eq. (7.13).

After having put $\tau = ct$, one makes the following change of notation:

$$\vec{\eta}_i(\tau) = \vec{\tilde{\eta}}_i(t), \qquad \vec{v}_i(t) = \frac{d\vec{\tilde{\eta}}_i(t)}{dt}, \qquad \vec{a}_i(t) = \frac{d^2\vec{\tilde{\eta}}_i(t)}{dt^2},$$
$$\dot{\vec{\eta}}_i(\tau) = \frac{\vec{v}_i(t)}{c}, \qquad \ddot{\vec{\eta}}_i(\tau) = \frac{\vec{a}_i(t)}{c^2}. \tag{7.16}$$

For the non-local York time one uses the notation ${}^3\tilde{\mathcal{K}}_{(1)}(t,\vec{\sigma}) = {}^3\mathcal{K}_{(1)}(\tau,\vec{\sigma})$.

Then, one studies the PN expansion of the equations of motion (Eq. 7.13) with the result (kPN means of order $O(c^{-2k})$)

$$m_i \frac{d^2\,\tilde{\eta}_i^r(t)}{dt^2} = m_i \Big[-G\,\frac{\partial}{\partial\,\tilde{\eta}_i^r} \sum_{j\neq i} \frac{m_j}{|\vec{\tilde{\eta}}_i(t) - \vec{\tilde{\eta}}_j(t)|} - \frac{1}{c}\,\frac{d\tilde{\eta}_i^r(t)}{dt} \Big(\partial_t^2|_{\vec{\tilde{\eta}}_i}\, {}^3\tilde{\mathcal{K}}_{(1)}$$
$$+2 \sum_u v_i^u(t)\, \frac{\partial\,\partial_t|_{\vec{\tilde{\eta}}_i}\, {}^3\tilde{\mathcal{K}}_{(1)}}{\partial\,\tilde{\eta}_i^u} + \sum_{uv} v_i^u(t)\, v_i^v(t)\, \frac{\partial^2\, {}^3\tilde{\mathcal{K}}_{(1)}}{\partial\,\tilde{\eta}_i^u\,\partial\,\tilde{\eta}_i^v} \Big)(t, \vec{\tilde{\eta}}_i(t)) \Big]$$
$$+F^r_{i(1PN)}(t) + (\text{higher PN orders}). \tag{7.17}$$

At the lowest order, one finds the standard Newton gravitational force $\vec{F}_{i(Newton)}(t) = -m_i\, G\, \frac{\partial}{\partial\,\tilde{\eta}_i^r} \sum_{j\neq i} \frac{m_j}{|\vec{\tilde{\eta}}_i(t) - \vec{\tilde{\eta}}_j(t)|}$.

The unexpected result is a 0.5PN force term containing all the dependence upon the non-local York time. The (arbitrary in these gauges) double rate of change in time of the trace of the extrinsic curvature creates a 0.5PN damping (or anti-damping since the sign of the inertial gauge variable ${}^3\mathcal{K}_{(1)}$ of Eq. (7.10) is not fixed) effect on the motion of particles. This is an inertial effect (hidden in the lapse function) not existing in Newton theory where the Euclidean 3-space is absolute.

Then there are all the other kPN terms with $k = 1, 1.5, 2, \ldots$ Since these results have been obtained without introducing ad hoc Lagrangians for the particles, are not in the harmonic gauge, and do not contain terms of order $O(\zeta^2)$ and higher, it is not possible to make a comparison with the standard PN expansion (whose terms are known till the order 3.5PN [11]). Therefore, only the 1PN and 0.5PN terms will be considered in the next two sections.

Since in the next section the 0.5PN term depending on the non-local York time will be connected with dark matter at the level of galaxies, and clusters of galaxies and since there is no convincing evidence of dark matter in the Solar System and near the galactic plane of the Milky Way [435], it is reasonable to assume ${}^3\mathcal{K}_{(1)}(\tau,\vec{\sigma}) = \frac{1}{\triangle}\, F_{(1)}(\tau,\vec{\sigma}) \approx 0$ near a star with planets and near a binary.

In the description of the 1PN two-body problem, which is relevant for the treatment of binary systems,[9] as shown in chapter 4 of ref. [11] based on

[9] For binaries one assumes $\frac{v}{c} \approx \sqrt{\frac{R_m}{l_c}} \ll 1$, where $l_c \approx |\vec{r}|$, with $\vec{r}(t)$ being the relative separation after the decoupling of the center of mass. Often, one considers the case

Refs. [382–387, 390–393, 436, 437], it can be shown that the relative momentum in the rest-frame has the 1PN expression $\pi_{12}(\tau) = \pi_{(1)12}(\tau) \approx \mu\, \vec{v}_{(rel)(o)12}(t)\, \Big[1 + \frac{m_1^3+m_2^3}{2\,M^3}\, (\frac{\vec{v}_{(rel)(o)12}(t)}{c})^2\Big]$, where $\vec{v}_{(rel)(o)12}(t) = \frac{d\,\vec{\rho}_{(o)12}(t)}{dt}$ is the velocity of the lowest order $\vec{\rho}_{(o)12}(\tau)$ of the relative variable.

If one ignores the York time and considers only positive energy particles ($\eta_1, \eta_2 \mapsto +1$), the 1PN equations of motion for the relative variable of the binary implied by Eq. (7.15) and 1PN expression of the weak ADM energy $\hat{E}_{ADM}$ and of the rest spin $\hat{J}^{rs}_{ADM}$ (being determined by the boundary conditions they are constants of the motion implying planar motion in the plane orthogonal to the rest spin) can be shown to be

$$
\begin{aligned}
\frac{d\,\vec{v}_{(rel)(o)12}(t)}{dt} &= -G\,M\, \frac{\vec{\rho}_{(o)12}(t)}{|\vec{\rho}_{(o)12}(t)|^3} \Big[1 + \Big(1 + 3\frac{\mu}{M}\Big) \frac{v^2_{(rel)(o)12}(t)}{c^2} \\
&\quad - \frac{3\mu}{2M} \Big(\frac{\vec{v}_{(rel)(o)12}(t)\cdot\vec{\rho}_{(o)12}(t))}{|\vec{\rho}_{(o)12}(t)|}\Big)^2\Big] \\
&\quad - \frac{G\,M}{|\vec{\rho}_{(o)12}(t)|^3} \Big(4 - \frac{2\mu}{M}\Big) \vec{v}_{(rel)(o)12}(t)\, \frac{\vec{v}_{(rel)(o)12}(t)\cdot\vec{\rho}_{(o)12}(t))}{|\vec{\rho}_{(o)12}(t)|}.
\end{aligned}
$$

$$
\begin{aligned}
\hat{E}_{ADM(1PN)} &= \sum_i m_i\, c^2 + \mu\, \Big(\frac{1}{2}\, \vec{v}^2_{(rel)(o)12}(t)\, \Big[1 + \frac{m_1^3+m_2^3}{M^3}\, \Big(\frac{\vec{v}_{(rel)(o)12}(t)}{c}\Big)^2\Big] \\
&\quad - \frac{G\,M}{|\vec{\rho}_{(o)12}(t)|} \Big[1 + \frac{1}{2}\, \Big(\Big(3 + \frac{\mu}{M}\Big) \frac{\vec{v}^2_{(rel)(o)12}(t)}{c^2} \\
&\quad + \frac{\mu}{M} \Big(\frac{\vec{v}_{(rel)(o)12}(t)}{c} \cdot \frac{\vec{\rho}_{(o)12}(t)}{|\vec{\rho}_{(o)12}(t)|}\Big)^2\Big)\Big]\Big),
\end{aligned}
$$

$$
\begin{aligned}
\hat{J}^{rs}_{ADM(1PN)} &= \Big(\rho^r_{(o)12}(t)\, v^s_{(rel)(o)12}(t) - \rho^s_{(o)12}(t)\, v^r_{(rel)(o)12}(t)\Big) \\
&\quad \Big[1 + \frac{m_1^3+m_2^3}{2\,M^3}\, \Big(\frac{\vec{v}_{(rel)(o)12}(t)}{c}\Big)^2\Big]. \qquad (7.18)
\end{aligned}
$$

Our 1PN Eq. (7.18) in the 3-orthogonal gauges coincides with equations 2.5, 2.13, and 2.14 of Ref. [436] (without G^2 terms since they are $O(\zeta^2)$), which are obtained in the family of harmonic gauges starting from an ad hoc 1PN Lagrangian for the relative motion of two test particles (first derived by Infeld and Plebanski [431]).[10] These equations are the starting point for studying the post-Keplerian parameters of the binaries, which, together with the Roemer, Einstein, and Shapiro time delays (both near Earth and near the binary) in light propagation, allow one to fit the experimental data from the binaries (see Ref. [437] and Chapter 6 of Ref. [11]). Therefore these results are reproduced also in our 3-orthogonal gauge with ${}^3\mathcal{K}_{(1)}(\tau, \vec{\sigma}) = 0$.

$m_1 \approx m_2$. See chapter 4 of Ref. [11] for a review of the emission of GWs from circular and elliptic Keplerian orbits and of the induced inspiral phase.

10 This is also the starting point of the effective one-body description of the two-body problem of Refs. [388, 389].

7.3 Dark Matter as a Relativistic Inertial Effect and Relativistic Celestial Metrology

To study the effects induced by the 0.5PN velocity-dependent (friction or anti-friction) force term in Eq. (7.17), depending on the inertial gauge variable non-local York time ${}^3\tilde{\mathcal{K}}_{(1)}(t,\vec{\sigma}) = \frac{1}{\triangle}\,{}^3\tilde{K}_{(1)}(\tau,\vec{\sigma}) \approx \frac{1}{\triangle}\,F_{(1)}(\tau,\vec{\sigma})$, with $F_{(1)}(\tau,\vec{\sigma})$ arbitrary numerical function, it is convenient to rewrite such equations in the form

$$\frac{d}{dt}\Big[m_i\Big(1+\frac{1}{c}\frac{d}{dt}\,{}^3\tilde{\mathcal{K}}_{(1)}(t,\vec{\tilde{\eta}}_i(t))\Big)\frac{d\tilde{\eta}_i^r(t)}{dt}\Big] \stackrel{\circ}{=} -G\frac{\partial}{\partial\tilde{\eta}_i^r}\sum_{j\neq i}\eta_j\frac{m_i m_j}{|\vec{\tilde{\eta}}_i(t)-\vec{\tilde{\eta}}_j(t)|} +\mathcal{O}(\zeta^2), \tag{7.19}$$

because the damping or anti-damping factors in Eq. (7.17) are $\gamma_i(t,\vec{\tilde{\eta}}_i(t)) = \frac{d^2}{dt^2}\,{}^3\tilde{\mathcal{K}}_{(1)}(t,\vec{\tilde{\eta}}_i(t))$ and $\ddot{\vec{\tilde{\eta}}}_i(t) = O(\zeta)$.

As a consequence, the velocity-dependent force can be reinterpreted as the introduction of an effective (time-, velocity-, and position-dependent) inertial mass term for the kinetic energy of each particle:

$$m_i \mapsto m_i\Big(1+\frac{1}{c}\frac{d}{dt}\,{}^3\tilde{\mathcal{K}}_{(1)}(t,\vec{\tilde{\eta}}_i(t))\Big) = m_i + (\Delta\, m)_i(t,\vec{\tilde{\eta}}_i(t)), \tag{7.20}$$

in each instantaneous 3-space. Instead, in the Newton potential there are the gravitational masses of the particles, equal to the inertial ones in the 4-dimensional space-time due to the equivalence principle. Therefore, the effect is due to a modification of the effective inertial mass in each non-Euclidean 3-space depending on its shape as a 3-sub-manifold of space-time: It is the equality of the inertial and gravitational masses of Newtonian gravity to be violated! In Galilei space-time the Euclidean 3-space is an absolute time-independent notion like Newtonian time: The non-relativistic non-inertial frames live in this absolute 3-space differently from what happens in SR and GR, where they are (in general non-Euclidean) 3-sub-manifolds of the space-time.

Eqs. (7.17), (7.19), and (7.20) can be applied to the three main signatures of the existence of dark matter in the observed masses of galaxies and clusters of galaxies, where the 1PN forces are not important, namely the virial theorem [438–440], the weak gravitational lensing [438–442], and the rotation curves of spiral galaxies (see Refs. [443–445] for a review), to give a reinterpretation of dark matter as a relativistic inertial effect.

7.3.1 *Masses of Clusters of Galaxies from the Virial Theorem*

For a bound system of N particles of mass m (N equal mass galaxies) at equilibrium, the virial theorem relates the average kinetic energy $< E_{kin} >$ in the system to the average potential energy $< U_{pot} >$ in the system: $< E_{kin} >= -\frac{1}{2} < U_{pot} >$, assuming Newton gravity. For the average kinetic energy of a galaxy in the cluster, one takes $< E_{kin} >\approx \frac{1}{2} m < v^2 >$, where $< v^2 >$ is the average of the square of the radial velocity of single galaxies with respect to the center of the cluster (measured with Doppler shift methods; the velocity distribution is assumed isotropic). The average potential energy of the galaxy is assumed to be of the form $< U_{pot} >\approx -G \frac{m M}{\mathcal{R}}$, where $M = Nm$ is the total mass of the cluster and $\mathcal{R} = \alpha R$ is an "effective radius" depending on the cluster size R (the angular diameter of the cluster and its distance from Earth are needed to find R) and on the mass distribution on the cluster (usually $\alpha \approx 1/2$). Then, the virial theorem implies $M \approx \frac{\mathcal{R}}{G} < v^2 >$. It turns out that this mass M of the cluster is usually at least an order of magnitude bigger that the baryonic matter of the cluster $M_{bar} = Nm$ (spectroscopically determined). By applying Eq. (7.17) to the equilibrium condition for a self-gravitating system, i.e., $\frac{d^2}{dt^2} \sum_i m_i \mid \vec{\eta}_i(t) \mid^2 = 0$, with $m_i = m$, one gets $\sum_i m_i v_i^2(t) - G \sum_{i>j} \frac{m_i m_j}{|\vec{\eta}_i(t) - \vec{\eta}_j(t)|} - \frac{1}{c} \sum_i m_i \left(\vec{\eta}_i(t) \cdot \vec{v}_i(t)\right) \gamma_i(t, \vec{\eta}_i(t)) = 0$, with $m_i = m_j = m$. Therefore, one can write $\langle U_{pot} \rangle = -\frac{1}{N} \sum_{i>j} \frac{G m^2}{|\vec{\eta}_i(t) - \vec{\eta}_j(t)|} \approx G \frac{m M_{bar}}{\mathcal{R}}$ (with $\mathcal{R} = R/2$) and $\frac{1}{2} m \langle v^2 \rangle = -\frac{1}{2} \langle U_{pot} \rangle + \frac{m}{2c} \left\langle \left(\vec{\eta} \cdot \vec{v}\right) \gamma(t, \vec{\eta}) \right\rangle$ with the notation $\left\langle \left(\vec{\eta} \cdot \vec{v}\right) \gamma(t, \vec{\eta}) \right\rangle = \frac{1}{N} \sum_i \left(\vec{\eta}_i(t) \cdot \vec{v}_i(t)\right) \gamma_i(t, \vec{\eta}_i(t))$ (it contains the non-local York time). Therefore, for the measured mass M (the effective inertial mass in 3-space), one has

$$M = \frac{\mathcal{R}}{G} < v^2 >= M_{bar} + \frac{\mathcal{R}}{G c} \left\langle \left(\vec{\eta} \cdot \vec{v}\right) \gamma(t, \vec{\eta}) \right\rangle \stackrel{\text{def}}{=} M_{bar} + M_{DM}, \tag{7.21}$$

and one sees that the non-local York time can give rise to a dark matter contribution $M_{DM} = M - M_{bar}$.

7.3.2 Masses of Galaxies or Clusters of Galaxies from Weak Gravitational Lensing

Usually one considers a galaxy (or a cluster of galaxies) of big mass M behind which a distant, bright object (often a galaxy) is located. The light from the distant object is bent by the massive one (the lens) and arrives on the Earth deflected from the original propagation direction. As shown in Refs. [441, 442], one has to evaluate Einstein deflection of light, emitted by a source S at distance d_S from the observer O on the Earth, generated by the big mass at a distance d_D from the observer O. The mass M, at distance d_{DS} from the source S, is considered as a point-like mass generating a 4-metric of the Schwarzschild type (Schwarzschild lens). The ray of light is assumed to propagate in Minkowski space-time till near M, to be deflected by an angle α by the local gravitational field of M and then to propagate in Minkowski space-time till the observer O. The distances d_S, d_D, d_{DS} are evaluated by the observer O at some reference time in some nearly inertial Minkowski frame with nearly Euclidean 3-spaces (in the Euclidean case $d_{DS} = d_S - d_D$). If $\xi = \theta\, d_D$ is the impact parameter of the ray of light at M and if $\xi \gg R_s = \frac{2\,G\,M}{c^2}$ (the gravitational radius), Einstein's deflection angle is $\alpha = \frac{2\,R_s}{\xi} = \frac{4\,G\,M}{c^2\,\xi}$ and the so-called Einstein radius (or characteristic angle) is $\alpha_o = \sqrt{2\,R_s\,\frac{d_{DS}}{d_D\,d_S}} = \sqrt{\frac{4\,G\,M}{c^2}\,\frac{d_{DS}}{d_D\,d_S}}$. A measurement of the deflection angle and of the three distances allows getting a value for the mass M of the lens, which usually turns out to be much larger than its mass inferred from the luminosity of the lens. For the calculation of the deflection angle, one considers the propagation of the ray of light in a stationary 4-metric of the BCRS type and uses a version of the Fermat principle containing an effective index of refraction n. One has $n = {}^4g_{\tau\tau} = \epsilon\,[1 - \frac{2w}{c^2} - 2\,\partial_\tau\,{}^3\mathcal{K}]$ in the PM approximation. Since one has $\frac{2\,w}{c^2} = -\frac{G\,M_{bar}}{c^2\,|\vec{\sigma}|}$, the definition $2\,\partial_\tau\,{}^3\mathcal{K}_{(1)} \overset{\text{def}}{=} -\frac{G\,M_{DM}}{c^2\,|\vec{\sigma}|}$ leads to an Einstein deflection angle

$$\alpha = \frac{4\,G\,M}{c^2\,\xi} \qquad \text{with} \quad M \overset{\text{def}}{=} M_{bar} + M_{DM}. \tag{7.22}$$

Therefore, also in this case the measured mass M is the sum of a baryonic mass M_{bar} and of a dark matter mass M_{DM} induced by the non-local York time at the location of the lens.

7.3.3 Masses of Spiral Galaxies from Their Rotation Curves

In this case one considers a two-body problem (a point-like galaxy and a body circulating around it) described in terms of an internal center of mass $\tilde{\vec{\eta}}_{12}(t) \approx \tilde{\vec{\eta}}_{(1)12}(t)$ ($\tilde{\vec{\eta}}_{(o)12}(t) = 0$ is the origin of the 3-coordinates) and a relative variable $\tilde{\vec{\rho}}_{12}(t)$. Then, the sum and difference of Eq. (7.17) imply the equations of motion for $\tilde{\vec{\eta}}_{(1)12}(t)$ and $\tilde{\vec{\rho}}_{12}(t)$. While the first equation implies a small motion of the overall system, the second one has the form

$$\frac{d^2\,\tilde{\rho}^r_{(1)12}(t)}{dt^2} \circeq -G\,M\,\frac{\tilde{\rho}^r_{12}(t)}{|\tilde{\vec{\rho}}_{12}(t)|^3} - \frac{1}{c}\,\frac{d\tilde{\rho}^r_{12}(t)}{dt}\,\gamma_+(t, \tilde{\vec{\rho}}_{12}(t), \vec{v}(t)),$$
$$\gamma_+(t, \tilde{\vec{\rho}}_{12}(t), \vec{v}(t)) = \frac{m_1}{M}\,\gamma_1\left(t, \frac{m_2}{M}\,\tilde{\vec{\rho}}_{12}(t), \vec{v}(t)\right) + \frac{m_2}{M}\,\gamma_2\left(t, -\frac{m_1}{M}\,\tilde{\vec{\rho}}_{12}(t), \vec{v}(t)\right), \tag{7.23}$$

where γ_i are the damping or anti-damping factors defined after Eq. (7.19). Eq. (7.23) gives the two-body Kepler problem with an extra perturbative force. Without it a Keplerian solution with circular trajectory such that $|\,\tilde{\vec{\rho}}_{12}(t)\,| = R =$ const. implies that the Keplerian velocity $\vec{v}_o(t) = v_o\,\hat{n}(t)$ has the modulus vanishing at large distances, $v_o = \sqrt{\frac{G\,M}{R}} \rightarrow_{R\rightarrow\infty} 0$. Instead, the rotation curves of spiral galaxies imply that the relative 3-velocity goes to constant for large R, i.e., $v = \sqrt{\frac{G\,(M_{bar} + \Delta\,M(r))}{r}} \rightarrow_{R\rightarrow\infty}$ const. (M_{bar} is the spectroscopically determined baryon mass), so that the extra required term $\Delta\, M(r)$ is interpreted as the mass M_{DM} of a dark matter halo.

The presence of the extra force term implies that the velocity must be written as $\vec{v}(t) = \vec{v}_o(t) + \vec{v}_1(t)$, with $v_1(t)$ a first-order perturbative correction satisfying $\frac{dv_1^r(t)}{dt} = -\frac{v_o^r}{c}\, \hat{n}(t)\, \gamma_+(t, \vec{\tilde{\rho}}_{12}(t), \vec{v}_o^r(t))$. Therefore, at the first order in the perturbation one gets $v^2(t) = v_o^2 \left(1 - \frac{2}{c}\, \hat{n}(t) \cdot \int_o^t dt_1\, \hat{n}(t_1)\, \gamma_+(t_1, \vec{\tilde{\rho}}_{12}(t_1),\, \vec{v}_o(t_1)) \right)$. After having taken a mean value over a period T (the time dependence of the mass of a galaxy is not known) the effective mass of the two-body system is

$$\begin{aligned} M_{eff} &= \frac{\langle v^2 \rangle\, R}{G} = M \left(1 - \left\langle \frac{2}{c}\, \hat{n}(t) \cdot \int^t dt_1\, \hat{n}(t_1)\, \gamma_+(t_1, \vec{\tilde{\rho}}_{12}(t_1), \vec{v}_o(t_1)) \right\rangle \right) \\ &= M_{bar} + M_{DM}, \end{aligned} \tag{7.24}$$

with a $\Delta\, M(r) = M_{DM}$ function only of the mean value of the total time derivative of the non-local ${}^3\mathcal{K}_{(1)}$ to be fitted to the experimental data.

Therefore, the existence of the inertial gauge variable York time, a property of the non-Euclidean 3-spaces as 3-sub-manifolds of Einstein space-times (connected only to the general relativistic remnant of the gauge freedom in clock synchronization, independently from cosmological assumptions) implies the possibility of describing part (or maybe all) dark matter as a relativistic inertial effect in Einstein gravity without alternative explanations using:

1. the non-relativistic MOND approach [446] (where one modifies Newton equations);
2. modified gravity theories like the $f(R)$ ones (see, for instance, Refs. [447] – here one gets a modification of the Newton potential); and
3. the assumption of the existence of WIMP particles [448].

Let us also remark that the 0.5PN effect has its origin in the lapse function and not in the shift one, as in the gravito-magnetic elimination of dark matter proposed in Ref. [449].

In Ref. [38] there is a first attempt to fit some data of dark matter by using a Yukawa-like ansatz on the non-local York time of a galaxy. In each galaxy the Yukawa-like potential of $f(R)$ theories [447] is put equal to a contribution to the extra potential depending on the non-local York time present in the lapse function appearing in Eq. (7.2): In this way the good fits of the rotation curves of galaxies obtainable with $f(R)$ theories can be reproduced inside Einstein's GR as an inertial gauge effect.

The open problem with this explanation of dark matter is the determination of the non-local York time from the data on dark matter. From what is known about dark matter in the Solar System and inside the Milky Way near the galactic plane, it seems that ${}^3\mathcal{K}_{(1)}(\tau, \vec{\sigma})$ is negligible near the stars inside a galaxy. Instead, the non-local York time (or better a mean value in time of its total time derivative) should be relevant around the galaxies and the clusters of galaxies, where there are big concentrations of mass and well-defined signatures of dark matter. Instead, there is no indication on its value in the voids existing among the clusters of galaxies.

Therefore, the known data on dark matter do not allow one to get an experimental determination of the York time ${}^3K_{(1)}(\tau,\vec{\sigma}) = \triangle\, {}^3\mathcal{K}_{(1)}(\tau,\vec{\sigma})$, because to do so one needs to know the non-local York time on all the 3-universe at a given τ.

Since at the experimental level the description of matter is intrinsically coordinate-dependent, namely is connected with the conventions used by physicists, engineers, and astronomers for the modeling of space-time, one has to choose a gauge (i.e., a 4-coordinate system) in non-modified Einstein gravity which is in agreement with the observational conventions in astronomy. This way out from the gauge problem in GR requires a choice of 3-coordinates on the instantaneous 3-spaces identified by a choice of time and by a clock synchronization convention, i.e., a fixation of the York time ${}^3K_{(1)}(\tau,\vec{\sigma})$. The convention resulting by one set of such choices would give a PM extension of ICRS, with BCRS being its quasi-Minkowskian approximation for the Solar System. Since the existing ICRS [32–35, 450–456] has diagonal 3-metric, 3-orthogonal gauges are a convenient choice.

The real problem is the extraction of an indication of which kind of function of time and 3-coordinates to use for the York time ${}^3K_{(1)}(\tau,\vec{\sigma})$ from astrophysical data different from the ones giving information about dark matter. Once one had a phenomenological parametrization of the York time, then the data on dark matter would put restrictions on the induced phenomenological parametrization of the non-local York time ${}^3\mathcal{K}_{(1)}(\tau,\vec{\sigma}) = \frac{1}{\triangle}\, {}^3K_{(1)}(\tau,\vec{\sigma})$. As will be delineated in Chapter 10, to implement this program one has to look at the astrophysical data on dark energy after having successfully interpreted it as a relativistic inertial effect in suitable cosmological space-times in which one can induce the distinction between inertial and tidal degrees of freedom of the gravitational field from the previously discussed Hamiltonian framework.

Part III

Dirac–Bergmann Theory of Constraints

In this part I will present a review of the main properties of constrained systems based on my personal viewpoint on the subject at the classical level, with some comments on the weak points of the existing quantization approaches.

Besides Dirac's and Bergmann works [20, 21, 457, 458] and Refs. [25, 30], I recommend Refs. [26, 27] for an extended treatment of many aspects of the theory also at the quantum level (included the BRST approach). Other works on the subject are Refs. [5, 6, 19, 28, 29]. However, there is no good treatment of constrained systems in mathematical physics and differential geometry, there are only partial treatments for finite-dimensional systems like presymplectic geometry [459–461] (see Refs. [462–467] and their bibliographies for recent contributions) without any extension to infinite-dimensional systems like field theory [468].

8

Singular Lagrangians and Constraint Theory

In this chapter, after a review of regular Lagrangians and of the first Noether theorem, I cover the theory of singular Lagrangians and of Dirac–Bergmann constraints in their Hamiltonian formulation.

In the final section there is a long review of the big difficulties one encounters when trying to understand all the aspects of both Lagrangian and Hamiltonian gauge transformations both in field theory and in general relativity (GR). There is a list of the mathematical problems to be solved and of the needed boundary conditions at spatial infinity to be imposed for defining the 3+1 approach and for using constraint theory in field theory and in GR.

8.1 Regular Lagrangians and the First Noether Theorem

Let us consider a finite-dimensional system whose configuration space Q is n-dimensional (either $Q = R^n$ or Q is an n-dimensional manifold with or without boundary), spanned by the configurational coordinates q^i, $i = 1, \ldots, n$. We shall give a short review of standard classical mechanics for such systems [469–472].

8.1.1 The Second-Order Lagrangian Formalism

Let the system be described by a time-independent Lagrangian $L(q(t), \dot{q}(t))$, where $q^i(t)$ is a curve in Q with time as a parameter and $\dot{q}^i(t) = \frac{dq^i(t)}{dt}$ are the velocities, and by the Lagrangian action $S = \int_{t_i}^{t_f} dt\, L(q, \dot{q})$.

The stationarity of the action, $\delta S = \int dt\, \frac{\delta S}{\delta q^i(t)}\, \delta q^i(t) = 0$, under variations $\delta q^i(t)$ $[\delta \dot{q}^i = \frac{d}{dt} \delta q^i]$, which vanish at the end-points t_i, t_f, identifies the classical motions of the system as those trajectories $q^i(t)$ which satisfy the Euler–Lagrange (EL) equations:

$$L_i = \frac{\partial L}{\partial q^i} - \frac{d}{dt} \frac{\partial L}{\partial \dot{q}^i} = -(A_{ij} \ddot{q}^j - \alpha_i) \overset{\circ}{=} 0,$$

$$\alpha_i(q,\dot q) = \frac{\partial L}{\partial q^i} - \frac{\partial^2 L}{\dot q^i\,\partial q^j}\,\dot q^j = \left(1 - \dot q^k\,\frac{\partial}{\partial \dot q^k}\right)\frac{\partial L}{\partial q^i} - R_{ij}\,\dot q^j,$$

$$R_{ij}(q,\dot q) = -R_{ji} = \frac{\partial^2 L}{\partial \dot q^i\,\partial q^j} - \frac{\partial^2 L}{\partial \dot q^j\,\partial q^i}. \tag{8.1}$$

The Hessian matrix is $A_{ij}(q,\dot q) = A_{ji} = \frac{\partial^2 L}{\partial \dot q^i\,\partial \dot q^j}$ and the Lagrangian is said to be regular when $\det A \neq 0$. If we denote $B = A^{-1}$ the inverse Hessian matrix, it follows that the EL equations can be put in the following normal form: $\ddot q^i - \Lambda^i \overset{\circ}{=} 0$, $\Lambda^i = B^{ij}\,\alpha_j$.

8.1.2 The First-Order Hamiltonian Formalism

The canonical momenta are defined by $p_i = \frac{\partial L(q,\dot q)}{\partial \dot q^i} = \mathcal{P}_i(q,\dot q)$ and the regularity condition $\det A \neq 0$ implies that these equations can be inverted to express the velocities $\dot q^i$ in terms of q^k and p_k.

By means of the Legendre transformation we can reformulate the second-order Lagrangian formalism in the first-order Hamiltonian on the phase space T^*Q (the co-tangent bundle) over Q with coordinates q^i, p_i. The Hamiltonian of the system is $\bar H = p_i\,\dot q^i - L$ (we shall denote $\bar f = \bar f(q,p)$ the functions on phase space) and the phase space action is $\bar S = \int_{t_i}^{t_f} dt\,\bar L$, with $\bar L = p_i\,\dot q^i - \bar H$. By asking the stationarity, $\delta\bar S = 0$, of this action under variations δq^i which vanish at the end-points t_i, t_f, and under arbitrary variations δp_i, we get the first-order differential Hamilton equations of motion, $\bar L_{qi} = \dot q^i - \frac{\partial \bar H}{\partial p_i} \overset{\circ}{=} 0$, $\bar L_{pi} = \dot p_i + \frac{\partial \bar H}{\partial q^i} \overset{\circ}{=} 0$. The first half of the Hamilton equations, $\bar L_{qi} \overset{\circ}{=} 0$, have a purely kinematical content: they give the inversion of the equations $p_i = \mathcal{P}_i(q,\dot q)$, i.e., $\dot q^i = \bar g^i(q,p)$.

By introducing the Poisson brackets[1] $\{\bar A(q,p), \bar B(q,p)\} = \frac{\partial \bar A}{\partial q^i}\frac{\partial \bar B}{\partial p^i} - \frac{\partial \bar A}{\partial p^i}\frac{\partial \bar B}{\partial q^i}$, we can rewrite the Hamilton equations in the form $\dot q^i \overset{\circ}{=} \{q^i, \bar H\} = \bar X_{\bar H}\,q^i$, $\dot p_i \overset{\circ}{=} \{p_i, \bar H\} = \bar X_{\bar H}\,p_i$, where we introduced the evolution Hamiltonian vector field $\bar X_{\bar H} = \{.,\bar H\}$.

In the regular case every function $f(q,\dot q)$ is projectable to phase space: $f(q,\dot q) = \bar f(q,p)$ by using $\dot q^i = \bar g^i(q,p)$.

Let us remark that there are two intrinsic formulations of the Hamiltonian description (we consider only the case of an exact symplectic structure arising when there is a well-defined Lagrangian):

1. a time-independent one on the symplectic manifold T^*Q (the symplectic structure) based on the Cartan–Liouville one-form $\bar\theta = p_i\,dq^i$ and on the closed symplectic two-form $\bar\omega = d\bar\theta = dp_i \wedge dq^i$, $d\bar\omega = 0$, where we gave the coordinate expression in Darboux coordinates adapted to the symplectic structure; and

[1] They satisfy: (1) $\{\bar f, \bar g\} = -\{\bar g, \bar f\}$; (2) $\{\bar f, \bar g_1 \bar g_2\} = \{\bar f, \bar g_1\}\,\bar g_2 + \bar g_1\,\{\bar f, \bar g_2\}$ (Leibnitz rule for derivations); and (3) $\{\{\bar f, \bar g\}, \bar u\} + \{\{\bar g, \bar u\}, \bar f\} + \{\{\bar u, \bar f\}, \bar g\} = 0$ (Jacobi identity).

2. a time-dependent one on $R \times T^*Q$ (the contact structure; R is the time axis; this formulation allows one to treat also time-dependent Lagrangians, $L(t, q, \dot{q})$), based on the Poincaré–Cartan one-form $\bar{\bar{\theta}} = \bar{L}\, dt = \bar{\theta} - \bar{H}\, dt$ and on the closed-contact two-form $\bar{\bar{\omega}} = d\bar{\bar{\theta}} = \bar{\omega} - d\bar{H} \wedge dt$.

In the regular case all these descriptions are equivalent.

8.1.3 The First-Order Velocity Space Formalism

When the second-order differential equations of motion are in the normal form ($\ddot{q}^i \overset{\circ}{=} \Lambda^i$), they can be rewritten as a set of first-order differential equations on the velocity space TQ (the tangent bundle) over Q with coordinates q^i, v^i (we shall denote $\tilde{f} = \tilde{f}(q, v)$ the functions on the velocity space and we have $f(q, \dot{q}) = \tilde{f}(q, v)|_{v=\dot{q}}$) with the following position:

$$\dot{q}^i \overset{\circ}{=} v^i, \qquad \dot{v}^i \overset{\circ}{=} \tilde{\Lambda}^i(q, v). \tag{8.2}$$

By introducing $\tilde{L}(q, v) = L(q, \dot{q})|_{\dot{q}=v}$ and the energy function $\tilde{E} = \frac{\partial \tilde{L}}{\partial v^i} v^i - \tilde{L}$, we can define the TQ action $\tilde{S} = \int_{t_i}^{t_f} dt\, \tilde{L}_v$ and Lagrangian $\tilde{L}_v = \frac{\partial \tilde{L}}{\partial v^i} \dot{q}^i - \tilde{E}$, whose stationarity yields the velocity space first-order differential equations of motion ($\tilde{A}_{ij}$, $\tilde{B}^{ij}$, $\tilde{R}_{ij}$ are the TQ expressions of A_{ij}, B^{ij}, R_{ij} respectively):

$$\begin{aligned} \tilde{L}_{qi} &= -\tilde{A}_{ij}\left[\dot{v}^j + \tilde{B}^{jk}\left(\frac{\partial \tilde{E}}{\partial q^k} + R_{kh}\, B^{hr}\, \frac{\partial \tilde{E}}{\partial v^r}\right)\right] - \tilde{R}_{ij}\left(\dot{q}^j - v^j\right) \overset{\circ}{=} 0, \\ \tilde{L}_{vi} &= \tilde{A}_{ij}\left(\dot{q}^j - v^j\right) \overset{\circ}{=} 0. \end{aligned} \tag{8.3}$$

The normal form of these equations is shown in Eq. (8.2), which can also be written in the form $\dot{q}^i \overset{\circ}{=} v^i = \{q^i, \tilde{E}\}_L$, $\dot{v}^i \overset{\circ}{=} -\tilde{B}^{ij}\left(\frac{\partial \tilde{E}}{\partial q^j} + \tilde{R}_{jk}\, \tilde{B}^{kh}\, \frac{\partial \tilde{E}}{\partial v^h}\right) = \tilde{B}^{ij}\, \tilde{\alpha}_j = \tilde{\Lambda}^i = \{v^i, \tilde{E}\}_L$, where we have introduced the (Lagrangian-dependent) TQ Poisson brackets $\{\tilde{f}, \tilde{g}\}_L = \tilde{B}^{ij}\left(\frac{\partial \tilde{f}}{\partial q^i}\frac{\partial \tilde{g}}{\partial v^j} - \frac{\partial \tilde{f}}{\partial v^j}\frac{\partial \tilde{g}}{\partial q^i}\right) - \frac{\partial \tilde{f}}{\partial v^i}\, \tilde{B}^{ih} \tilde{R}_{hk}\, \tilde{B}^{kj}\, \frac{\partial \tilde{g}}{\partial v^j}$, $\{q^i, q^j\}_L = 0$, $\{q^i, v^j\}_L = \tilde{B}^{ij}$, $\{v^i, v^j\}_L = \tilde{B}^{ih} \tilde{R}_{hk} \tilde{B}^{kj}$.

Therefore, we have the same symplectic structure in TQ and T^*Q. In this context the Legendre transformation is defined as the fiber derivative FL of $\tilde{L}(q, v)$: It is a linear and fiber-preserving mapping from TQ to T^*Q defined by $FL : (q, v) \in T_q Q \mapsto p_i\, dq^i \in T^*_q Q$ ($p_i = \frac{\partial \tilde{L}}{\partial v^i}$). When FL is a global diffeomorphism of TQ, the Lagrangian $\tilde{L}(q, v)$ is said to be hyper-regular and one has a global Hamiltonian formalism (i.e., $\bar{H}(q, p)$, the Legendre transform of $\tilde{E}(q, v)$, exists globally on T^*Q). When FL is only a local diffeomorphism of TQ, $\tilde{L}(q, v)$ is said to be regular and $\bar{H}(q, p)$ exists only locally. In the regular case, FL is a symplectomorphism that connects the symplectic structures of T^*Q and TQ.

The definition $\dot{q}^i \overset{\circ}{=} v^i \equiv \tilde{\Gamma} \cdot q^i$ and Eq. (8.2) imply $\ddot{q}^i \overset{\circ}{=} \frac{dv^i}{dt} \overset{\circ}{=} \tilde{\Lambda}^i = \tilde{\Gamma} \cdot v^i$ in the regular case: This is called the second-order differential equation (SODE) condition, ensuring that $\tilde{\Gamma}$ is a second-order vector field. See Ref. [473] for the study of the phase space over the velocity space, i.e., $T^*(TQ)$.

8.1.4 Symmetries, the First Noether Theorem, and its Extensions

The two Noether theorems are a basic ingredient in the study of the consequences of the invariances of Lagrangian systems under continuous symmetry transformations. Ref. [474] gives a review of their applications in theoretical physics, while Ref. [475] contains a review of the intrinsic geometrical formulations and of the various extensions of the first Noether theorem and Refs. [476–478] survey the use of the second theorem. In this subsection we shall review the first theorem and its extensions.

For finite-dimensional systems described by a regular (maybe time-dependent) Lagrangian $L(t, q, \dot{q})$, the first Noether theorem states that if the action functional $S = \int dt\, L$ is quasi-invariant under an r parameter group G_r of continuous transformations of t and q^i, then r linear independent combinations of the EL equations L_i reduce identically to total time derivatives. The converse is also true under appropriate hypotheses.

This means that if under an infinitesimal set of invertible local variations $\delta_a t = \bar{t}_a - t = \delta_a t(t, q)$, $\delta_{oa} q^i = \bar{q}^i_a(t) - q^i(t) = \delta_{oa} q^i(t, q, \dot{q})$,[2] $a = 1, \dots, r$, the total variation of L is a total time derivative (the following equation is a Killing-type equation; $\equiv$ means identically):

$$\begin{aligned}
\delta_a L &= L\Big(\bar{t}_a, \bar{q}_a(\bar{t}_a), \frac{d\bar{q}_a(\bar{t}_a)}{d\bar{t}_a}\Big)\, \frac{d\bar{t}_a}{dt} - L(t, q, \dot{q}) \\
&= \frac{\partial L}{\partial q^i}\, \delta_{oa} q^i + \frac{\partial L}{\partial \dot{q}^i}\, \delta_{oa} \dot{q}^i + \frac{d}{dt}(L\, \delta_a t) \\
&= \frac{\partial L}{\partial t}\, \delta_a t + \frac{\partial L}{\partial q^i}\, \delta_a q^i + \frac{\partial L}{\partial \dot{q}^i}\, \delta_a \dot{q}^i + L\, \frac{d\delta_a t}{dt} \\
&= \delta_{oa} q^i\, L_i + \frac{d}{dt}\Big(\frac{\partial L}{\partial \dot{q}^i}\, \delta_{oa}\dot{q}^i + L\, \delta_a t\Big) \equiv \frac{dF_a(t, q, \dot{q})}{dt},
\end{aligned} \tag{8.4}$$

then one obtains the following r Noether identities (both G_a and F_a are in general functions of t, q^i, and $\dot{q}^i$):

$$\begin{aligned}
&\frac{dG_a}{dt} \equiv -\delta_{oa} q^i\, L_i \overset{\circ}{=} 0, \\
&G_a = \frac{\partial L}{\partial \dot{q}^i}\, \delta_{oa} q^i - F_a + L\delta_a t = \frac{\partial L}{\partial \dot{q}^i}\, \delta_a q^i - F_a - \Big(\dot{q}^i\, \frac{\partial L}{\partial \dot{q}^i} - L\Big)\delta_a t.
\end{aligned} \tag{8.5}$$

The r quantities $G_a(q, \dot{q})$ are constants of the motion (in field theory one would obtain r conservation laws $\partial_\mu J^\mu_a \overset{\circ}{=} 0$). For $F_a \neq 0$ we speak of

[2] The associated global variations (corresponding to Lie derivatives) are $\delta_a q^i = \bar{q}^i_a(\bar{t}_a) - q^i(t) = \delta_{oa} q^i + \dot{q}^i\, \delta_a t$. The corresponding variations of the velocities are $\delta_{oa} \dot{q}^i = \frac{d}{dt}\, \delta_{oa} q^i$, $\delta_a \dot{q}^i = \delta_{oa} \dot{q}^i - \dot{q}^i\, \frac{d\delta_a t}{dt}$.

quasi-invariance, while for $F_a = 0$ of invariance. When we have $\delta_{oa}q^i(t,q)$, we get $F_a(t,q)$.

It is always possible to define a new set of variations in which t is not varied ($\delta^{'}_a t = 0$, $\delta^{'}_{oa}q^i = \delta_{oa}q^i$) and which gives rise to the same constants of motion G_a: the only difference is that now $\delta^{'}_a L \equiv \frac{dF^{'}_a}{dt}$ with $F^{'}_a = F_a - L\delta_a t$. In general, there is an infinite family of Noether symmetry transformations $\delta_a t$, $\delta_{oa}q^i$ associated with the same set of constants of motion G_a (even a change of the functional form of the Lagrangian is allowed: $\delta_a L = L^{'}$(barred variables) $- L$). See Ref. [475] for a critical review and the proposal of a preferred geometrical approach. Moreover, inside every family of Noether symmetry transformations there are always dynamical symmetry transformations, i.e., symmetry transformations of the EL differential equations mapping the space of its solutions onto itself (the sets of Noether symmetry and dynamical symmetry transformations of a Lagrangian system do not coincide but have an overlap).

The concept of a family of Noether transformations associated with a given set of constants of motion has also been analyzed by Candotti et al. [479, 480]. They point out that each family contains transformations $\tilde{\delta}_a t$, $\tilde{\delta}_{oa}q^i$ such that Eq. (8.4) becomes $\tilde{\delta}_a L \equiv \frac{d\tilde{F}_a}{dt} + f_a, f_a(t,q,\dot{q},\ddot{q}) \overset{\circ}{=} 0$. That is, we have a weak quasi-invariance, because $\delta_a L$ only becomes a total time derivative by using the EL equations. The Noether identities (Eq. 8.5) become $\frac{d\tilde{G}_a}{dt} \equiv -\tilde{\delta}_{oa}q^i\, L_i + f_a \overset{\circ}{=} 0$ and give rise to the same constants of motion G_a, if $\tilde{\delta}_a t$, $\tilde{\delta}_{oa}q^i$, $\tilde{F}_a$ are such that $\tilde{G}_a = \frac{\partial L}{\partial \dot{q}^i}\, \tilde{\delta}_{oa}q^i - \tilde{F}_a + L\tilde{\delta}_a t \overset{\circ}{=} G_a$. In the regular case, these extensions can be considered irrelevant, but it is not so in the singular case.

The generator of the Noether transformation is the vector field $Y = \delta t\, \frac{\partial}{\partial t} + \delta q^i\, \frac{\partial}{\partial q^i} + \delta\dot{q}^i\, \frac{\partial}{\partial \dot{q}^i}$ and in terms of it we get $\delta t = Yt$, $\delta q^i = Yq^i = \mathcal{L}_Y\, q^i$, $YL \equiv \dot{F} - L\, \frac{d\delta t}{dt}$. The constant of motion $G = \frac{\partial L}{\partial \dot{q}^i}\, \delta_o q^i - F + L\,\delta t$ is an invariant of the generator Y: $Y\,G = 0$.

The natural setting for the definition and study of the generator Y (and of the dynamical symmetries of differential equations) is the infinite jet bundle [481], where Y is a Lie–Bäcklund vector field (Y gives its truncation to the first derivatives). As shown in Ref. [481], there are only two kinds of invariance transformations when the number of degrees of freedom is higher than one: (1) the Lie point transformations of $R \times Q$ extended to the higher derivatives; and (2) the Lie–Bäcklund transformations (or tangent transformations of infinite order preserving the tangency of infinite order of two curves). These latter have δt and/or $\delta_o q^i$, depending on the velocities and possibly on the higher accelerations. For instance, the non-point canonical transformations of T^*Q become Lie–Bäcklund transformations with $\delta t = 0$ when rephrased through the inverse Legendre transformation in the second-order Lagrangian formalism.

By expressing the velocities in terms of the coordinates and canonical momenta, we get $\delta q^i(q,\dot{q}) = \bar{\delta q}^i(q,p)$, $F(q,\dot{q}) = \bar{F}(q,p)$, $G(q,\dot{q}) = \bar{G}(q,p) = p_i\, \bar{\delta q}^i(q,p) - \bar{F}(q,p)$. Since the Hamiltonian is defined as $\bar{H} = p_i\, \dot{q}^i - L$, the phase space Lagrangian satisfies $\bar{L}(q,p,\dot{q}) = p_i\,\dot{q}^i - \bar{H}(q,p) = L(q,\dot{q})$ and therefore will have the same invariance properties. This means $\delta\bar{L} = \dot{q}^i\,\bar{\delta p}_i + p_i\,\frac{d\bar{\delta q}^i}{dt} - \frac{\partial\bar{H}}{\partial q^i}\,\bar{\delta q}^i - \frac{\partial\bar{H}}{\partial p_i}\,\bar{\delta p}_i = \bar{\delta p}_i\,\bar{L}_{qi} - \bar{\delta q}^i\,\bar{L}_{pi} + \frac{d}{dt}(p_i\,\bar{\delta q}^i) \equiv \frac{d\bar{F}}{dt}$. Therefore, we get $\frac{d}{dt}\,\bar{G} = \frac{d}{dt}\,(p_i\,\bar{\delta q}^i - \bar{F}) \equiv -\bar{\delta p}_i\,\bar{L}_{qi} + \bar{\delta q}^i\,\bar{L}_{pi} \overset{\circ}{=} 0, \quad \Rightarrow \{\bar{G},\bar{H}\} \overset{\circ}{=} 0$, and $\bar{\delta q}^i \equiv \frac{\partial\bar{G}}{\partial p_i} = \{q^i,\bar{G}\}$, $\bar{\delta p}_i \equiv -\frac{\partial\bar{G}}{\partial q^i} = \{p_i,\bar{G}\}$. This is the phase space projected Noether identity associated to the constant of motion $\bar{G}$.

In this way we have found that the Hamiltonian Noether symmetry transformation is generated by the constant of motion $\bar{G}(q,p)$ ($\{\bar{G},\bar{H}\} = 0$), considered as the generator of a symmetry canonical transformation, i.e., such that the functional form of the Hamiltonian does not change ($\delta_o\,H = \bar{H}'(q,p) - \bar{H}(q,p) = 0$).

The intrinsic formulation of the first Noether theorem and of the reduction of dynamical systems with symmetry, when there is a free and proper symplectic action of a (connected) Lie group on a (connected) symplectic manifold (phase space of an autonomous regular Hamiltonian system with symmetry), is the momentum map approach [482]. For a weakly regular value of the momentum map associated with this action the reduced phase space has a structure of symplectic manifold and inherits a Hamiltonian dynamics. For a singular value of the momentum map, the reduced phase space is a stratified symplectic manifold [483].

8.2 Singular Lagrangians and Dirac–Bergmann Theory of Hamiltonian First- and Second-Class Constraints

Let us consider a finite-dimensional system with an n-dimensional configuration space Q admitting a global coordinate system q^i, $i = 1,\dots,n$.[3] Let its dynamics be described by a time-independent singular Lagrangian $L(q(t),\dot{q}(t))$ (the extension to the time-dependent case does not introduce further complications), namely such that its Hessian matrix is singular: $\det \frac{\partial^2 L(q,\dot{q})}{\partial\dot{q}^i\partial\dot{q}^j} = 0$.

In this section we introduce the Hamiltonian formalism for singular systems and then we come back to study the second-order Lagrangian formalism after having looked at the notion of Dirac observables (DOs). We shall follow Refs. [26, 473, 484–487].

8.2.1 Primary Hamiltonian Constraints and the Hamilton–Dirac Equations

When the Hessian matrix is singular, the EL equations (Eq. 8.4) cannot be put in normal form. This means that the accelerations $\ddot{q}^i$ cannot be uniquely determined in terms of q^i, $\dot{q}^i$ and that the solutions of the EL equations may depend on arbitrary functions of time.

[3] Otherwise the following treatment will only hold locally in a chart of the coordinate atlas of Q.

Moreover, $\det A_{ij}(q,\dot q) = 0$ implies that the canonical momenta cannot be inverted to get the velocities $\dot q^i$ in terms of q^i, p_i. The n functions $p_i = \mathcal{P}_i(q,\dot q)$ are not functionally independent, namely there are as many identities $\phi_A(q,\mathcal{P}(q,\dot q)) \equiv 0, A = 1,\dots,m$, as null eigenvalues of the Hessian matrix. In phase space (T^*Q) these identities become the primary Hamiltonian constraints (their functional form is highly arbitrary):

$$\bar\phi_A(q,p) = 0, \quad A = 1,\dots,m, \tag{8.6}$$

which identify the region γ of T^*Q allowed to the configurations of the singular system. Points outside γ are not accessible, but we go on to work in T^*Q to utilize its symplectic structure (γ in general has no such structure), i.e., its Poisson brackets.

Eq. (8.6) is usually written with Dirac's weak equality sign $\approx$, i.e., $\bar\phi_A(q,p) \approx 0, A = 1,\dots,m$. An equation $\bar f(q,p) \approx 0$ means that the function $\bar f$ vanishes on γ, but can be different from zero outside γ so that it cannot be put equal to zero inside the T^*Q Poisson brackets even when they are restricted to γ. Instead, the strong equality symbol $\equiv$ (like for identical) is used for a function $\bar f(q,p)$ vanishing on γ, $\bar f \approx 0$, and such that also its differential vanishes on γ, $d\bar f \approx 0$; such a function (for instance $\bar f = \bar\phi_A^2$) can be put equal to zero inside Poisson brackets restricted to γ.

Let us assume that the rank of the Hessian matrix $A_{ij}(q,\dot q)$ is constant and equal to m for every value of $(q,\dot q)$. See Section 8.5 and Ref. [484] for what may happen when we relax this assumption.

Let us also assume that the singular Lagrangian is such that the region γ defined by the primary constraints is a $(2n-m)$-dimensional sub-manifold of T^*Q (γ is the primary constraint sub-manifold). While $p_k \approx 0$ is an acceptable constraint, neither $p_k^2 \approx 0$ nor $\sqrt{p_k} \approx 0$ are acceptable. When the rank of the Hessian matrix is not constant, constraints of the type $p_k^2 \approx 0$ may appear. Constraints of the type $(p_k)^2 + (q^h)^2 \approx 0$ must be put in the form $p_k \approx 0$ and $q^h \approx 0$.

While in the regular case the invertibility of the equations $p_i = \mathcal{P}(q,\dot q)$ to $\dot q^i = \bar g^i(q,p)$ implies that all the velocities $\dot q^i$ are projectable to T^*Q, now there will be m independent (but with a not-uniquely determined functional form) functions of the velocities $g^A(q,\dot q)$ (named non-projectable velocity functions) not projectable to T^*Q.

Let us also assume that the singular Lagrangian admits a well-defined Legendre transformation. Then, if we introduce the function $H_c(q,\dot q) = p_i\,\dot q^i - L(q,\dot q) = \mathcal{P}_i(q,\dot q)\,\dot q^i - L(q,\dot q)$, we get $\delta H_c = \delta p_i\,\dot q^i + p_i\,\delta\dot q^i - \frac{\partial L}{\partial q^i}\,\delta q^i - \frac{\partial L}{\partial \dot q^i}\,\delta\dot q^i = \dot q^i\,\delta p_i - \frac{\partial L}{\partial q^i}\,\delta q^i = \delta\bar H_c$ as in the regular case. This means that $H_c(q,\dot q)$ is projectable to a well-defined canonical Hamiltonian also in the singular case:

$$H_c(q,\dot q) = \bar H_c(q,p), \quad \Big(\frac{\partial \bar H_c}{\partial q^i} + \frac{\partial L}{\partial q^i}\Big)\,\delta q^i + \Big(\frac{\partial \bar H_c}{\partial p_i} - \dot q^i\Big)\,\delta p_i = 0. \tag{8.7}$$

But in the singular case, Eq. (8.7) is meaningful only if $(\delta q^i, \delta p_i)$ is a vector tangent to the primary constraint sub-manifold γ, so that one gets (see Ref. [26] for a demonstration)

$$\dot{q}^i = \frac{\partial \bar{H}_c}{\partial p_i} + u^A \frac{\partial \bar{\phi}_A}{\partial p_i} = \{q^i, \bar{H}_c\} + u^A \{q^i, \bar{\phi}_A\},$$
$$\dot{p}_i = \frac{\partial L}{\partial q^i}|_{\dot{q}} = -\frac{\partial \bar{H}_c}{\partial q^i} - u^A \frac{\partial \bar{\phi}_A}{\partial q^i} = \{p_i, \bar{H}_c\} + u^A \{p_i, \bar{\phi}_A\}, \tag{8.8}$$

where the EL equations have been used in the second line. Since the velocities are not projectable to T^*Q in the singular case, the functions u^A in the first line of Eq. (8.8) cannot be functions on T^*Q but must depend also on the velocities: $u^A = u^A(q, p, \dot{q}) = u^A(q, \mathcal{P}(q, \dot{q}), \dot{q}) = v^A(q, \dot{q})$. These multipliers identify a canonical functional form $g^A_{(u)} = v^A$ of the non-projectable velocity functions $g^A(q, \dot{q})$.

However, since Eq. (8.8) has the Hamilton equations for the singular case (the so-called Hamilton–Dirac equations), their right sides cannot depend explicitly on the velocities. Therefore, a consistent Hamiltonian formalism is obtained by replacing the functions $u^A = v^A(q, \dot{q})$ with arbitrary multipliers $\lambda^A(t)$ (the so-called Dirac multipliers) by introducing the Dirac Hamiltonian

$$\bar{H}_D(q, p, \lambda) = \bar{H}_c(q, p) + \lambda^A(t)\, \bar{\phi}_A(q, p), \tag{8.9}$$

and by introducing the T^*Q action $\bar{S} = \int_{t_i}^{t_f} dt\, (p_i\, \dot{q}^i - \bar{H}_c - \lambda^A(t)\, \bar{\phi}_A)$.

In it, q^i, p_i, and λ^A are considered as independent variables. The stationarity, $\delta\bar{S} = 0$, under variations δq^i, δp_i, $\delta\lambda^A$ with the only restriction $\delta q^i(t_f) = \delta q^i(t_i) = 0$ yields the Hamilton–Dirac equations supplemented by the definition of the primary constraint sub-manifold:

$$\begin{aligned} \dot{q}^i &\stackrel{\circ}{=} \{q^i, \bar{H}_D(q, p, \lambda)\}, \quad \text{or} \quad \bar{L}^i_{Dq} = \dot{q}^i - \{q^i, \bar{H}_D\} \stackrel{\circ}{=} 0, \\ \dot{p}_i &\stackrel{\circ}{=} \{p_i, \bar{H}_D(q, p, \lambda)\}, \quad \text{or} \quad \bar{L}_{Dpi} = \dot{p}_i - \{p_i, \bar{H}_D\} \stackrel{\circ}{=} 0, \\ \bar{\phi}_A(q, p) &\stackrel{\circ}{=} 0. \end{aligned} \tag{8.10}$$

The kinematical equations $\dot{q}^i \stackrel{\circ}{=} \{q^i, \bar{H}_D\}$, $\bar{\phi}_A \stackrel{\circ}{=} 0$, determine the canonical momenta and the Dirac multipliers in terms of the coordinates and momenta (if suitable regularity conditions hold); namely, we get: (1) $p_i = \mathcal{P}_i(q, \dot{q})$; (2) the canonical functional form $g^A_{(\lambda)}$ of the non-projectable velocity functions $g^A(q, \dot{q})$ associated with the chosen functional form of the primary constraints as the $(q, \dot{q})$ space expression of the Dirac multipliers $g^A_{(\lambda)}(q, \dot{q}) \stackrel{\circ}{=} \lambda^A(t)$. Then, we can make the inverse Legendre transformation and recover the original singular Lagrangian: $p_i\, \dot{q}^i - \bar{H}_D \stackrel{\circ}{=} \mathcal{P}_i(q, \dot{q})\, \dot{q}^i - \bar{H}_c(q, \mathcal{P}(q, \dot{q})) = L(q, \dot{q})$.

8.2.2 Dirac's Algorithm for the Determination of the Final Constraint Sub-manifold

The inspection of the functional form of the canonical momenta $p_i = \mathcal{P}_i(q, \dot{q})$ identifies the primary constraint sub-manifold $\gamma \subset T^*Q$. The Hamiltonian formalism will produce a consistent treatment of singular systems only if γ does not change with time, namely if the primary constraints $\bar{\phi}_A(q,p) \approx 0$ are constant of motion with respect to the evolution generated by the Dirac Hamiltonian,

$$\begin{aligned}\frac{d\bar{\phi}_A(q,p)}{dt} &\stackrel{\circ}{=} \{\bar{\phi}_A(q,p), \bar{H}_D\} \\ &= \{\bar{\phi}_A(q,p), \bar{H}_c(q,p)\} + \lambda^B(t)\, \{\bar{\phi}_A(q,p), \bar{\phi}_B(q,p)\} \\ &\approx 0 \quad \text{on } \gamma,\ A = 1, \dots, m. \end{aligned} \tag{8.11}$$

Some of these equations may be void ($0 = 0$). The non-void ones, restricted to γ, have to be separated in two disjoint groups:

1. a set of $m_1 \leq m$ equations independent from the Dirac multipliers,

$$\bar{\chi}^{(1)}_{a_1}(q,p) \approx 0 \quad on\, \gamma, \quad a_1 = 1, \dots, m_1; \tag{8.12}$$

2. a set of h_1 equations ($h_1 \leq m$, $h_1 + m_1 \leq m$) for the Dirac multipliers,

$$\bar{f}_{\tilde{A}_1 B}(q,p)\, \lambda^B(t) + \bar{g}_{\tilde{A}_1}(q,p) \approx 0 \quad \text{on}\, \gamma, \quad \tilde{A}_1 = 1, \dots, h_1, \tag{8.13}$$

with $\bar{f}_{\tilde{A}_1} = \bar{h}^A_{\tilde{A}_1} \{\bar{\phi}_A, \bar{\phi}_B\}$, $\bar{g}_{\tilde{A}_1} = \bar{h}^A_{\tilde{A}_1} \{\bar{\phi}_A, \bar{H}_c\}$ for some functions $\bar{h}^A_{\tilde{A}_1}$.

Let us remark that without certain regularity conditions on the singular Lagrangian this separation cannot be done in a unique way: (1) we can get different separations in different regions of γ; (2) also in the same point of γ we can have alternative inequivalent separations. A regularity condition that eliminates most (if not all) of these possibilities is that the antisymmetric matrix $\bar{M}_{AB}(q,p) = \Big(\{\bar{\phi}_A(q,p), \bar{\phi}_B(q,p)\}\Big)$ of the Poisson brackets of the primary constraints has constant rank on γ.

Let us assume that there is a unique separation given by Eqs. (8.12) and (8.13). Eqs. (8.12) is called a secondary constraint and defines a secondary constraint sub-manifold γ_1 of γ to which the description of the singular system has to be restricted for consistency. Differently from the primary constraints, the secondary constraint is defined using the equations of motion. Eq. (8.13) shows that on the constraint sub-manifold $\gamma_1 \subset \gamma \subset T^*Q$ there may be less arbitrariness than on γ, because $h_1 \leq m$ Dirac multipliers $\lambda^A(t)$ are determined by these equations and an equal number of velocity functions $g^A_{(\lambda)}(q, \dot{q})$, not projectable onto γ, become projectable onto $\gamma_1 \subset \gamma$.

If $\bar{U}^A(q,p)$ is a particular solution of the inhomogeneous Eq. (8.13) and $\bar{V}^A_{A_1}(q,p)$, $A_1 = 1, \dots, k_1 = m - h_1$, are independent solutions of the homogeneous equations $\bar{f}_{\tilde{A}_1 B}\, \lambda^B(t) = \bar{h}^A_{\tilde{A}_1}\, \{\bar{\phi}_A, \bar{\phi}_B\}\, \lambda^B(t) = 0$ (namely $\{\bar{\phi}_A, \bar{\phi}_B\}\, \bar{V}^B_{A_1} \equiv 0$), then the general solution of Eq. (8.13) is

$$\lambda^A(t) \approx \bar{U}^A(q,p) + \lambda^{(1)\,A_1}(t)\, \bar{V}^A_{A_1}(q,p) \quad \text{on}\, \gamma_1, \quad A_1 = 1, \dots, k_1 = m - h_1, \tag{8.14}$$

with the $\lambda^{(1)A_1}(t)$ being new $k_1 = m - h_1$ arbitrary Dirac multipliers.

The Dirac Hamiltonian on $\gamma_1 \subset \gamma$ is

$$\begin{aligned}
\bar{H}_D(q,p,\lambda) &= \bar{H}^{(1)}_c(q,p) + \lambda^{(1)A_1}(t)\, \bar{\phi}^{(1)}_{A_1}(q,p) \quad \text{on}\, \gamma_1, \\
\bar{\phi}^{(1)}_{A_1}(q,p) &= \bar{V}^A_{A_1}(q,p)\, \bar{\phi}_A(q,p), \quad A_1 = 1, \dots, k_1 = m - h_1, \\
\bar{H}^{(1)}_c(q,p) &= \bar{H}_c(q,p) + \bar{U}^A(q,p)\, \bar{\phi}_A(q,p) \approx \bar{H}_c(q,p) \quad \text{on}\, \gamma_1.
\end{aligned} \tag{8.15}$$

The remaining Dirac multipliers $\lambda^{(1)A_1}(t)$ are now multiplied by the linear combinations $\bar{\phi}^{(1)}_{A_1}(q,p)$ of the original primary constraints.

When there are secondary constraints $\bar{\chi}^{(1)}_{a_1}(q,p) \approx 0$, for consistency we must ask that the secondary constraint sub-manifold $\gamma_1 \subset \gamma$ does not change with time: The secondary constraints must be constants of motion on γ_1 with respect to the Dirac Hamiltonian

$$\begin{aligned}
\frac{d\bar{\chi}^{(1)}_{a_1}(q,p)}{dt} &\stackrel{\circ}{=} \{\bar{\chi}^{(1)}_{a_1}(q,p), \bar{H}_D(q,p,\lambda)\} \\
&\approx \{\bar{\chi}^{(1)}_{a_1}(q,p), \bar{H}^{(1)}_c(q,p)\} + \lambda^{(1)\,A_1}(t)\, \{\bar{\chi}^{(1)}_{a_1}(q,p), \bar{\phi}^{(1)}_{A_1}(q,p)\} \approx 0 \quad \text{on}\, \gamma_1, \\
& A_1 = 1, \dots, k_1 = m - h_1, \quad a_1 = 1, \dots, m_1.
\end{aligned} \tag{8.16}$$

By assuming the regularity condition that the rank of the matrix $\left(\{\bar{\chi}^{(1)}_{a_1}, \bar{\phi}^{(1)}_{A_1}\}\right)$ is constant on γ_1, the non-void Eq. (8.16) may be separated into two disjoint sets:

$$1. \quad \bar{\chi}^{(2)}_{a_2}(q,p) \approx 0 \quad \text{on}\, \gamma_1, \quad a_2 = 1, \dots, m_2 \le m_1; \tag{8.17}$$

$$2. \quad \bar{f}_{\tilde{A}_2 B_1}(q,p)\, \lambda^{(1)B_1}(t) + \bar{g}_{\tilde{A}_2}(q,p) \approx 0 \quad \text{on}\, \gamma_1, \quad \tilde{A}_2 = 1, \dots, h_2 \le k_1. \tag{8.18}$$

Eq. (8.17) is the tertiary constraint and define a new constraint sub-manifold $\gamma_2 \subset \gamma_1$. With the same procedure delineated above, we arrive at the conclusion that only on γ_2 can there be a consistent dynamics for the singular system with (in general) a reduction of the number of independent Dirac multipliers, which are replaced by the new ones $\lambda^{(2)A_2}(t)$ due to Eq. (8.18). The dynamics is described in terms of the following quantities:

$$\begin{aligned}
\lambda^{(1)A_1}(t) &\approx \bar{U}^{(1)A_1}(q,p) + \lambda^{(2)A_2}(t)\, \bar{V}^{(1)A_1}_{A_2}(q,p) \quad \text{on}\, \gamma_2, \\
& A_2 = 1, \dots, k_2 = m - h_1 - h_2,
\end{aligned}$$

$$\begin{aligned}
\bar{H}_D(q,p,\lambda) &= \bar{H}^{(2)}_c(q,p) + \lambda^{(2)A_2}(t)\, \bar{\phi}^{(2)}_{A_2}(q,p), \quad \text{on}\, \gamma_2, \\
\bar{\phi}^{(2)}_{A_2}(q,p) &= \bar{V}^{(1)A_1}_{A_2}(q,p)\, \bar{\phi}^{(1)}_{A_1}(q,p) = \bar{V}^{(1)A_1}_{A_2}(q,p)\, \bar{V}^A_{A_1}(q,p)\, \bar{\phi}_A(q,p),
\end{aligned}$$

$$\bar{H}_c^{(2)}(q,p) = \bar{H}_c^{(1)}(q,p) + \bar{U}^{(1)A_1}(q,p)\bar{\phi}_{A_1}^{(1)}(q,p)$$
$$= \bar{H}_c(q,p) + \left(\bar{U}^A(q,p) + \bar{U}^{(1)A_1}(q,p)\,\bar{V}_{A_1}^A(q,p)\right)\bar{\phi}_A(q,p). \quad (8.19)$$

This procedure is iterated till a final stage (f) in which the final constraint sub-manifold $\bar{\gamma} = \gamma_f \subset \ldots \subset \gamma_1 \subset \gamma \subset T^*Q$ is determined by $(f+1)$-ary constraints:

$$\bar{\chi}_{a_f}^{(f)}(q,p) \approx 0 \quad \text{on}\, \gamma_{f-1}, \quad a_f = 1,\ldots,m_f \le m_{f-1} \le \ldots \le m. \quad (8.20)$$

On $\bar{\gamma} = \gamma_f$ we have

$$\bar{H}_D(q,p,\lambda) = \bar{H}_c^{(f)}(q,p) + \lambda^{(f)A_f}(t)\,\bar{\phi}_{A_f}^{(f)}(q,p), \quad \text{on}\ \gamma_f,$$

$$\bar{\phi}_{A_f}^{(f)}(q,p) = \bar{V}^{(f-1)}{}_{A_f}^{A_{f-1}}(q,p)\ \ldots\ \bar{V}_{A_1}^A(q,p)\,\bar{\phi}_A(q,p),$$
$$A_f = 1,\ldots,k_f = m - h_1 - \ldots - h_f,$$
$$\bar{H}_c^{(f)}(q,p) = \bar{H}_c(q,p) + \Big(\bar{U}^A(q,p) + \bar{U}^{(1)A_1}(q,p)\,\bar{V}_{A_1}^A(q,p) + \ldots$$
$$+\bar{U}^{(f-1)A_{f-1}}(q,p)\,\bar{V}^{(f-2)}{}_{A_{f-1}}^{A_{f-2}}(q,p)\ \ldots\ \bar{V}_{A_1}^A(q,p)\Big)\,\bar{\phi}_A(q,p)$$
$$\approx \bar{H}_c(q,p),$$

$$\frac{d\bar{\chi}_{a_f}^{(f)}(q,p)}{dt} \mathrel{\overset{\circ}{=}} \{\bar{\chi}_{a_f}^{(f)}(q,p), \bar{H}_D(q,p,\lambda)\} \quad \text{identically satisfied,} \quad (8.21)$$

with only $k_f = m - h_1 - \ldots - h_f$ independent final Dirac multipliers $\lambda^{(f)A_f}(t)$. Only an equal number of velocity functions $g_{(\lambda)}^A(q,\dot{q})$ cannot be projected onto $\bar{\gamma} = \gamma_f \subset T^*Q$.

The solutions of the Hamilton–Dirac equations on $\bar{\gamma}$ will depend on the k_f arbitrary functions of time $\lambda^{(f)A_f}(t)$, which describe the non-deterministic aspects of the time evolution of the singular system.

8.2.3 First- and Second-Class Constraints

The final constraint manifold $\bar{\gamma} = \gamma_f \subset \ldots \subset \gamma \subset T^*Q$ is determined by the full set of primary, secondary, etc. constraints $\bar{\phi}_A(q,p) \approx 0$ $(A = 1,\ldots,m)$, $\bar{\chi}_{a_1}^{(1)}(q,p) \approx 0$ $(a_1 = 1,\ldots,m_1)$, ..., $\bar{\chi}_{a_f}^{(f)}(q,p) \approx 0$ $(a_f = 1,\ldots,m_f)$. Let us denote all the $M = m + m_1 + \ldots + m_f$ constraints with the collective notation

$$\bar{\zeta}_{\mathcal{A}}(q,p) \approx 0, \quad \mathcal{A} = 1,\ldots,M, \quad (8.22)$$

because the property of being a primary, secondary, etc. constraint is not important.

Let us also denote with $\lambda^{\bar{A}}(t)$ and $\bar{\phi}_{\bar{A}}(q,p)$, with $\bar{A} = 1,\ldots,\bar{k} = m - h_1 - \ldots - h_f$, the final arbitrary Dirac multipliers and the associated linear combinations of primary constraints respectively, so that the Dirac Hamiltonian on $\bar{\gamma}$ is

$$\bar{H}_D(q,p,\lambda) = \bar{H}_c^{(F)}(q,p) + \lambda^{\bar{A}}(t)\,\bar{\phi}_{\bar{A}}(q,p), \quad \text{on}\ \bar{\gamma},$$

$$\bar{\phi}_{\bar{A}}(q,p) = \bar{V}^{(f-1)A_{f-1}}_{\bar{A}}(q,p) \ldots \bar{V}^{A}_{A_1}(q,p)\,\bar{\phi}_A(q,p), \quad \bar{A} = 1,\ldots,\bar{k},$$
$$\bar{H}^{(F)}_c(q,p) = \bar{H}_c(q,p) + \bar{U}^{(F)A}(q,p)\,\phi_A(q,p) \approx \bar{H}_c(q,p),$$
$$\bar{U}^{(F)A}(q,p) = \bar{U}^A(q,p) + \bar{U}^{(1)A_1}(q,p)\,\bar{V}^A_{A_1}(q,p) + \ldots$$
$$+\bar{U}^{(f-1)A_{f-1}}(q,p)\,\bar{V}^{(f-2)A_{f-2}}_{A_{f-1}}(q,p) \ldots \bar{V}^A_{A_1}(q,p). \tag{8.23}$$

As a result of the previous construction, all the constraints are preserved in time on $\bar{\gamma}$,

$$\frac{d\bar{\zeta}_{\mathcal{A}}}{dt} \stackrel{\circ}{=} \{\bar{\zeta}_{\mathcal{A}}, \bar{H}_D\} \approx 0 \Rightarrow \quad \{\bar{\zeta}_{\mathcal{A}}, \bar{H}^{(F)}_c\} \approx 0, \quad \{\bar{\zeta}_{\mathcal{A}}, \bar{\phi}_{\bar{A}}\} \approx 0. \tag{8.24}$$

Let us call a first-class function a function $\bar{f}(q,p)$ on T^*Q whose Poisson brackets with every constraint is weakly zero:

$$\{\bar{f}(q,p), \bar{\zeta}_{\mathcal{A}}(q,p)\} = \bar{F}^{\mathcal{B}}_{\mathcal{A}}(q,p)\,\bar{\zeta}_{\mathcal{B}}(q,p) \approx 0 \quad \text{on } \bar{\gamma}. \tag{8.25}$$

If two functions $\bar{f}$, $\bar{g}$ are first class, then also their Poisson bracket is a first-class function due to the Jacobi identity: $\{\{\bar{f},\bar{g}\},\bar{\zeta}_{\mathcal{A}}\} = \{\bar{f},\{\bar{g},\bar{\zeta}_{\mathcal{A}}\}\} - \{\bar{g},\{\bar{f},\bar{\zeta}_{\mathcal{A}}\}\} = \{\bar{f},\bar{G}^{\mathcal{B}}_{\mathcal{A}}\,\bar{\zeta}_{\mathcal{B}}\} - \{\bar{g},\bar{F}^{\mathcal{B}}_{\mathcal{A}}\,\bar{\zeta}_{\mathcal{B}}\} = \bar{K}^{\mathcal{B}}_{\mathcal{A}}\,\bar{\zeta}_{\mathcal{B}} \approx 0$.

All the functions that are not first class are named second-class functions.

Eq. (8.24) shows that both the final canonical Hamiltonian $\bar{H}^{(F)}_c(q,p)$ and the final combinations $\bar{\phi}_{\bar{A}}(q,p)$, $\bar{A} = 1,\ldots,\bar{k}$, of the primary constraints are first-class functions.

It is of fundamental importance in constraint theory to separate the constraints $\bar{\zeta}_{\mathcal{A}} = \left(\bar{\phi}_A,\, \bar{\chi}^{(1)}_{a_1},\, \ldots,\, \bar{\chi}^{(f)}_{a_f}\right)$ into two groups: (1) the first-class constraints $\bar{\Phi}_{(1)\mathcal{A}_1} = \bar{k}^{\mathcal{A}}_{\mathcal{A}_1}\,\bar{\zeta}_{\mathcal{A}} \approx 0$, $\mathcal{A}_1 = 1,\ldots,r_1$, $\{\bar{\Phi}_{(1)\mathcal{A}_1}, \bar{\zeta}_{\mathcal{A}}\} \approx 0$; and (2) the second-class constraints $\bar{\Phi}_{(2)\mathcal{A}_2} \approx 0$, $\mathcal{A}_2 = 1,\ldots,r_2$ $(r_1 + r_2 = M)$ with $\{\bar{\Phi}_{(2)\mathcal{A}_2}, \bar{\zeta}_{\mathcal{A}}\} \neq 0$ for some $\mathcal{A}$. Evidently we have $\{\bar{\Phi}_{(1)\mathcal{A}_1}, \bar{\Phi}_{(1)\mathcal{B}_1}\} \approx 0$, $\{\bar{\Phi}_{(1)\mathcal{A}_1}, \bar{\Phi}_{(2)\mathcal{A}_2}\} \approx 0$, $\det\left(\{\bar{\Phi}_{(2)\mathcal{A}_2}, \bar{\Phi}_{(2)\mathcal{B}_2}\}\right) \neq 0$.

The $\bar{\phi}_{\bar{A}}$ constitute a complete set of first-class primary constraints.

If a set of first-class constraints has the form $\bar{\Phi}_{(1)a} = p_a - \bar{K}_a(q^a, q^r, p_r) \approx 0$ with $r \neq a$ (i.e., they are solved in a subset of the momenta), then $\{\bar{\Phi}_{(1)a}, \bar{\Phi}_{(1)b}\} = \frac{\partial \bar{K}_a}{\partial q^b} - \frac{\partial \bar{K}_b}{\partial q^a} + \{\bar{K}_a, \bar{K}_b\} \equiv 0$.

Geometrically the Hamiltonian vector fields $\bar{X}_{(1)\mathcal{A}_1} = \{.,\bar{\Phi}_{(1)\mathcal{A}_1}\}$ and $\bar{X}_{(2)\mathcal{A}_2} = \{.,\bar{\Phi}_{(2)\mathcal{A}_2}\}$ are tangent and skew respectively to the constraint sub-manifold $\bar{\gamma}$ (see Ref. [26]; in Ref. [488] there is a study of the conditions for putting all the first-class constraints in this Abelianized form).

8.2.4 Chains of Constraints: Diagonalization of the Dirac Algorithm

We quote three theorems [462, 489, 490] on equivalent sets of constraints $\bar{\zeta}_{\mathcal{A}}(q,p) \approx 0$, $\tilde{\bar{\zeta}}_{\mathcal{A}}(q,p) \approx 0$, both defining $\bar{\gamma}$, valid when suitable regularity conditions hold, without reproducing the long unilluminating demonstrations based on inductive procedures. These theorems allow us to perform a diagonalization of the Dirac algorithm and to separate the constraints in chains (one for each primary constraint, namely for each null eigenvalue of the Hessian matrix), such that the time constancy of a constraint in the chain implies the next constraint in the chain. The time constancy of the last constraint in the chain either is automatically satisfied or determines the Dirac multiplier associated to the chain. A 0-chain has only the primary constraint, a 1-chain has the primary and a secondary, and so on.

The first theorem shows the existence of diagonalized chains.

Theorem 1 [462]: By taking suitable combinations $\bar{\phi}_{\bar{A}} = \bar{V}^{(f-1)}{}^{A_{f-1}}_{\bar{A}} \dots \bar{V}^{A}_{A_1} \bar{\phi}_A$ $(\bar{A} = 1, \dots, \bar{k} = m - h_1 - \dots - h_f)$ and $\bar{\phi}_{A'} = \bar{u}^{A}_{A'} \bar{\phi}_A$ $(A' = 1, \dots, m - \bar{k})$ of the primary constraints $\bar{\phi}_A$ $(A = 1, \dots, m)$ defining γ, then suitable combinations of the secondary $\bar{\chi}^{(1)}_{a_1}$ and primary $\bar{\phi}_A$ constraints defining γ_1 and so on, the final pattern of the chains of constraints can be put in the following form:

1. Chains of constraints starting from primary constraints whose Dirac multiplier $\lambda^A(t)$ is determined by the Dirac algorithm on $\bar{\gamma}$. We use the following notation: $\bar{\phi}_{(h)A'_h} \approx 0$ is the primary constraint of an h-chain of $h+1$ constraints (A'_h labels the various h-chains), $\bar{\phi}^{(1)}_{(h)A'_h} \approx \frac{d\bar{\phi}_{(h)A'_h}}{dt} \approx \left\{\bar{\phi}_{(h)A'_h}, \bar{H}^{(1)}_c\right\} \approx 0$ is the secondary, $\bar{\phi}^{(2)}_{(h)A'_h} \approx \frac{d\bar{\phi}^{(1)}_{(h)A'_h}}{dt} \approx \left\{\bar{\phi}^{(1)}_{(h)A'_h}, H^{(2)}_c\right\} \approx 0$ is the tertiary, and so on till $\bar{\phi}^{(f_h)}_{(h)A'_h} \approx \frac{d\bar{\phi}^{(f_h-1)}_{(h)A'_h}}{dt} \approx \left\{\bar{\phi}^{(f_h-1)}_{(h)A'_h}, \bar{H}^{(f_h)}_c\right\} \approx 0$; then $\frac{d\bar{\phi}^{(f_h)}_{(h)A'_h}}{dt} \approx 0$ determines the Dirac multiplier. Here, $\bar{H}_D$, $\bar{H}_c$, $\bar{H}^{(1)}_c, \dots$ are the quantities already introduced in the previous section.

$$
\begin{array}{llll|l}
\text{0-chains} & \text{1-chains} & & f\text{-chains} & \\
\bar{\phi}_{(o)A'_o}\ (A'_o = 1,\dots,k'_o) & \bar{\phi}_{(1)A'_1}\ (A'_1 = 1,\dots,k'_1) & \dots & \bar{\phi}_{(f)A'_f}\ (A'_f = 1,\dots,k'_f) & \text{primary} \\
\lambda^{(o)A'_o}\ \text{determined} & \bar{\phi}^{(1)}_{(1)A'_1} & \dots & \bar{\phi}^{(1)}_{(f)A'_f} & \text{secondary} \\
 & \lambda^{(1)A'_1}\ \text{determined} & \dots & \bar{\phi}^{(2)}_{(f)A'_f} & \text{tertiary} \\
\dots & \dots & \dots & \dots & \\
 & & \dots & \bar{\phi}^{(f)}_{(f)A'_f} & (f+1)\text{-ary} \\
 & & \dots & \lambda^{(f)A'_f}\ \text{determined} &
\end{array}
\qquad (8.26)
$$

2. Chains of constraints starting from the primary constraints $\bar{\phi}_{\bar{A}}$ with associated arbitrary Dirac multipliers on $\bar{\gamma}$. The same notation as in (1) is used, but now $\frac{d\bar{\phi}^{(f_h)}_{(h)\bar{A}_h}}{dt} \approx 0$ is identically satisfied without determining the Dirac multiplier.

$$\begin{array}{lllll|l}
\text{0-chains} & \text{1-chains} & & \text{f-chains} & \\
\bar{\phi}_{(o)\bar{A}_o}\ (\bar{A}_o = 1,\dots,k_o') & \bar{\phi}_{(1)\bar{A}_1}\ (\bar{A}_1 = 1,\dots,k_1') & \dots & \bar{\phi}_{(f)\bar{A}_f}\ (\bar{A}_f - 1,\dots,k_f') & \text{primary} \\
 & \bar{\phi}^{(1)}_{(1)\bar{A}_1} & \dots & \bar{\phi}^{(1)}_{(f)\bar{A}_f} & \text{secondary} \\
 & & \dots & \bar{\phi}^{(2)}_{(f)\bar{A}_f} & \text{tertiary} \\
\dots & \dots & \dots & \dots & \\
 & & \dots & \bar{\phi}^{(f)}_{(f)\bar{A}_f} & (f+1)\text{-ary}
\end{array} . \tag{8.27}$$

The second theorem shows that the diagonalized chains of theorem 1 can be redefined so that each chain has all the constraints either first or second class. Theorem 2 [489]: By leaving the primary constraints in the form of theorem 1, we can take linear combinations of all the other constraints to obtain the following pattern:

1. chains of second-class constraints $\bar{\chi}^{(h)}_{(k)A_k'}$ with the associated Dirac multiplier determined $\left(\det\left(\left\{\bar{\chi}^{(h)}_{(k)A_k'}, \bar{\chi}^{(h_1)}_{(k_1)A_{k_1}'}\right\}\right) \neq 0\right)$.

$$\begin{array}{llll|l}
\text{0-chains} & \text{1-chains} & & \text{f-chains} & \\
\bar{\phi}_{(o)A_o'} = \bar{\chi}^{(o)}_{(o)A_o'} & \bar{\phi}_{(1)A_1'} = \bar{\chi}^{(o)}_{(1)A_1'} & \dots & \bar{\phi}_{(f)A_f'} = \bar{\chi}^{(o)}_{(f)A_f'} & \text{primary} \\
\lambda^{(o)A_o'} \text{ determined} & \bar{\chi}^{(1)}_{(1)A_1'} & \dots & \bar{\chi}^{(1)}_{(f)A_f'} & \text{secondary} \\
 & \lambda^{(1)A_1'} \text{ determined} & \dots & \bar{\chi}^{(2)}_{(f)A_f'} & \text{tertiary} \\
\dots & \dots & \dots & \dots & \\
 & & \dots & \bar{\chi}^{(f)}_{(f)A_f'} & (f+1)\text{-ary} \\
 & & \dots & \lambda^{(f)A_f'} \text{ determined} &
\end{array} . \tag{8.28}$$

2. chains of first-class constraints $\bar{\chi}^{(h)}_{(k)\bar{A}_k}$ with arbitrary Dirac multipliers and with the property $\bar{\chi}^{(h+1)}_{(k)\bar{A}_k} = \left\{\bar{\chi}^{(h)}_{(k)\bar{A}_k}, \bar{H}^{(F)}_c\right\}$[4]

$$\begin{array}{llll|l}
\text{0-chains} & \text{1-chains} & & \text{f-chains} & \\
\bar{\phi}_{(o)\bar{A}_o} = \bar{\chi}^{(o)}_{(o)\bar{A}_o} & \bar{\phi}_{(1)\bar{A}_1} = \bar{\chi}^{(o)}_{(1)\bar{A}_1} & \dots & \bar{\phi}_{(f)\bar{A}_f} = \bar{\chi}^{(o)}_{(f)\bar{A}_f} & \text{primary} \\
 & \bar{\chi}^{(1)}_{(1)\bar{A}_1} & \dots & \bar{\chi}^{(1)}_{(f)\bar{A}_f} & \text{secondary} \\
 & & \dots & \bar{\chi}^{(2)}_{(f)\bar{A}_f} & \text{tertiary} \\
\dots & \dots & \dots & \dots & \\
 & & \dots & \bar{\chi}^{(f)}_{(f)\bar{A}_f} & (f+1)\text{-ary}
\end{array} . \tag{8.29}$$

The third theorem [490] gives a simple canonical form for the chains of second-class constraints: either $(2k+1)$-chains with k pairs of constraints or pairs of

[4] For each chain the time preservation of the gauge fixing to the primary constraint generates the gauge fixing for the secondary one, and so on; the time preservation of the generated gauge fixing for the last constraint in the chain determines the Dirac multiplier of the chain.

$2k$-chains with $k+1$ pairs of constraints. This theorem shows that under suitable regularity conditions on the singular Lagrangian we cannot obtain either a chain in which a primary first-class constraints generates a secondary second-class constraint, nor two chains whose primary constraints are a second-class pair and which generate secondary first-class constraints.

8.2.5 Second-Class Constraints and Dirac Brackets

Second-class constraints describe non-essential pairs of canonical variables, which can be eliminated to reduce the number of degrees of freedom carrying the dynamics of the singular system (maybe with the price of a breaking of manifest covariance and/or of the introduction of non-linearities).

When we have a singular system with the constraint sub-manifold $\bar{\gamma} \subset T^*Q$ described by the set $\bar{\tilde{\zeta}}_{\mathcal{A}} = \left(\bar{\Phi}_{(1)\mathcal{A}_1}, \bar{\Phi}_{(2)\mathcal{A}_2}\right)$ of first $\left(\bar{\Phi}_{(1)\mathcal{A}_1}\right)$, $\mathcal{A}_1 = 1, \dots, m-2s_2$, and second $\left(\bar{\Phi}_{(2)\mathcal{A}_2}\right)$, $\mathcal{A}_2 = 1, \dots, 2s_2$, class constraints, we can (at least implicitly) eliminate s_2 pairs of canonical variables with the following procedure. Let us assume that the second-class constraints define a $2(n-s_2)$-dimensional constraint sub-manifold $\gamma_{(2)}$ of T^*Q containing the final sub-manifold $\bar{\gamma} \subset \gamma_{(2)} \subset T^*Q$ and that the regularity conditions are such that the antisymmetric matrix,

$$\bar{\tilde{C}}_{\mathcal{A}_2\mathcal{B}_2} = \left(\{\bar{\Phi}_{(2)\mathcal{A}_2}, \bar{\Phi}_{(2)\mathcal{B}_2}\}\right), \tag{8.30}$$

which is invertible on $\bar{\gamma}$ ($\det \bar{\tilde{C}}_{\mathcal{A}_2\mathcal{B}_2}|_{\bar{\gamma}} \neq 0$), is invertible also on $\gamma_{(2)}$ ($\det \bar{\tilde{C}}_{\mathcal{A}_2\mathcal{B}_2}|_{\gamma_{(2)}} \neq 0$) with inverse matrix $\bar{\tilde{C}}^{\mathcal{A}_2\mathcal{B}_2}$ on $\gamma_{(2)}$. As shown by Dirac [20, 21], the sub-manifold $\gamma_{(2)}$ (in general it is not $T^*Q_{(2)}$ for some configuration space $Q_{(2)}$) has an induced symplectic structure whose Poisson brackets, named Dirac brackets, are

$$\{\bar{f}, \bar{g}\}^* = \{\bar{f}, \bar{g}\} - \{\bar{f}, \bar{\Phi}_{(2)\mathcal{A}_2}\}\, \bar{\tilde{C}}^{\mathcal{A}_2\mathcal{B}_2} \{\bar{\Phi}_{(2)\mathcal{A}_2}, \bar{g}\}. \tag{8.31}$$

Besides the standard properties $\{\bar{f}, \bar{g}\}^* = -\{\bar{g}, \bar{f}\}^*$, $\{\bar{f}, \bar{g}_1\bar{g}_2\}^* = \{\bar{f}, \bar{g}_1\}^*\, \bar{g}_2 + \bar{g}_1\, \{\bar{f}, \bar{g}_2\}^*$, $\{\{\bar{f}, \bar{g}\}^*, \bar{u}\}^* + \{\{\bar{u}, \bar{f}\}^*, \bar{g}\}^* + \{\{\bar{g}, \bar{u}\}^*, \bar{f}\}^* = 0$, the Dirac brackets have the extra, easily verified, properties

$$\begin{aligned}
&\{\bar{\Phi}_{(2)\mathcal{A}_2}, \bar{f}\}^* = 0 \quad \text{for every } \bar{f} \text{ and } \mathcal{A}_2, \Rightarrow \quad \bar{\Phi}_{(2)\mathcal{A}_2} \equiv 0 \quad \text{on } \gamma_{(2)},\\
&\{\bar{f}, \bar{g}_{(1)}\}^* \approx \{\bar{f}, \bar{g}_{(1)}\} \quad \text{on } \bar{\gamma} \subset \gamma_{(2)}, \text{ for } \bar{g}_{(1)} \text{ first class, } \bar{f} \text{ arbitrary},\\
&\{\bar{f}, \{\bar{g}_{(1)}, \bar{u}_{(1)}\}^*\}^* \approx \{\bar{f}, \{\bar{g}_{(1)}, \bar{u}_{(1)}\}\} \quad \text{on } \bar{\gamma} \subset \gamma_{(2)},\\
&\qquad \text{for } \bar{g}_{(1)} \text{ and } \bar{u}_{(1)} \text{ first class, } \bar{f} \text{ arbitrary.}
\end{aligned} \tag{8.32}$$

When we use the Dirac brackets, the Dirac Hamiltonian $\bar{H}_D = \bar{H}_c^{(F)} + \lambda^{\bar{A}}(t)\, \bar{\phi}_{\bar{A}}$ becomes $\bar{H}'_D = \bar{H}_c^{(F)'} + \lambda^{\bar{A}}(t)\, \bar{\phi}_{\bar{A}}$ with $\bar{H}_c^{(F)'} = \bar{H}_c^{(F)}|_{\gamma_{(2)}}$. Since the Dirac

Hamiltonian is a first-class function, for the Hamilton–Dirac equations we get $\frac{d\bar{f}}{dt} \stackrel{\circ}{=} \{\bar{f}, \bar{H}_D\} \approx \{\bar{f}, \bar{H}'_D\}^* \quad \text{on } \bar{\gamma} \subset \gamma_{(2)}$.

The second-class constraints are not generators of canonical transformations interpretable as Hamiltonian gauge transformations like first-class constraints, but they are the generators of local Noether extended symmetry transformations under which the singular Lagrangian has a generalized type of quasi-invariance (instead the first-class constraints generate local Noether symmetry transformations under which the singular Lagrangian is quasi-invariant).

8.3 The Gauge Transformations in Field Theory and General Relativity

In ADM canonical gravity there are eight first-class constraints, which are generators of Hamiltonian gauge transformations. Some general properties of these transformations will now be analyzed both in field theory and in GR to clarify which boundary conditions are needed at spatial infinity so as to be able to use constraint theory in the 3+1 approach. The choice of the Christodoulou–Klainermann boundary conditions [8] made in Section 5.1 is a consequence of the following set of problems.

In the Hamiltonian formulation of every gauge field theory, one has to make a choice of the boundary conditions of the canonical variables and of the parameters of the gauge transformations[5] in such a way as to give a meaning to integrations by parts, to the functional derivatives (and therefore to Poisson brackets) and to the proper gauge transformations connected with the identity.[6] In particular, the boundary conditions must be such that the variation of the final Dirac Hamiltonian H_D must be linear in the variations of the canonical variables[7] and this may require a redefinition of H_D, namely H_D has to be replaced by $\tilde{H}_D = H_D + H_\infty$, where H_∞ is a suitable integral on the surface at spatial infinity. When this is accomplished, one has a good definition of functional derivatives and Poisson brackets. Then, one must consider the most general generator of gauge transformations of the theory (it includes H_D as a special case), in which there are arbitrary functions (parametrizing infinitesimal gauge transformations) in front of all the first-class constraints and not only in front of the primary ones.[8] Also, the variations of this generator must be linear in the variations of the canonical variables: This implies that all the surface terms

[5] The infinitesimal ones are generated by the first-class constraints of the theory.

[6] The improper ones, including the rigid or global or first kind gauge transformations related to the non-Abelian charges, have to be treated separately; when there are topological numbers like winding number, they label disjoint sectors of gauge transformations and one speaks of large gauge transformations.

[7] The coefficients are the Dirac–Hamilton equations of motion.

[8] These are the generalized Hamiltonian gauge transformations of the Dirac conjecture. They are not generated by the Dirac Hamiltonian. However, their pullback to configuration space generates local Noether transformations under which the ADM action is quasi-invariant, in accord with the general theory of singular Lagrangians.

coming from integration by parts must vanish with the given boundary conditions on the canonical variables or must be compensated by the variation of H_∞. In this way, one gets boundary conditions on the parameters of the infinitesimal gauge transformations identifying the proper ones, which transform the canonical variables among themselves without altering their boundary conditions. Let us remark that in this way one is defining Hamiltonian boundary conditions that are not manifestly covariant; however, in Minkowski space-time they can be made Wigner covariant with the 3+1 formulation of parametrized Minkowski theories.

The proper gauge transformations are those that are connected to the identity and generated by the first-class constraints at the infinitesimal level.

The improper gauge transformations are a priori of four types:

1. global or rigid or first kind ones (the gauge parameter fields tend to constant at spatial infinity) connected with the group G (isomorphic to the structure group of the Yang–Mills principal bundle) generated by the non-Abelian charges;
2. the global or rigid ones in the center of the gauge group G (triality when $G = SU(3)$);
3. gauge transformations with non-vanishing winding number $n \in Z$ (large gauge transformations not connected with the identity; zeroth homotopy group of the gauge group);
4. other improper non-rigid gauge transformations. Since this last type of gauge transformation does not play any role in Yang–Mills dynamics, it was assumed [169, 170] that the choice of the function space for the gauge parameter fields $\alpha_a(\tau, \vec{\sigma})$ (describing the component of the gauge group connected with the identity) be such that for $r = |\vec{\sigma}| \to \infty$ one has

$$\alpha_a(\tau, \vec{\sigma}) \to \alpha_a^{(rigid)} + \alpha_a^{(proper)}(\tau, \vec{\sigma}), \tag{8.33}$$

with constant $\alpha_a^{(rigid)}$ and with $\alpha_a^{(proper)}(\tau, \vec{\sigma})$ tending to zero in a direction-independent way.

However, in gauge theories, in the framework of local quantum field theory, one does not consider the Abelian and non-Abelian charges generators of gauge transformations of first kind, but speaks of super-selection sectors determined by the charges. This is valid both for the electric charge, which is a physical observable, and for the color charge in quantum chromodynamics (QCD), where the hypothesis of quark confinement requires the existence only of color singlets, namely: (1) physical observables must commute with the non-Abelian charges; and (2) the $SU(3)$ color charges of isolated systems have to vanish themselves.

We will follow the same scheme in the analysis of the Hamiltonian gauge transformations of metric gravity.

The definition of an isolated system in GR is a difficult problem (see Ref. [491] for a review), since there is neither a background flat metric ${}^4\eta$ nor a natural global inertial coordinate system allowing one to define a preferred radial coordinate r and a limit ${}^4g_{\mu\nu} \to {}^4\eta_{\mu\nu} + O(1/r)$ for $r \to \infty$ along either spatial or null directions. Usually, one considers an asymptotic Minkowski metric ${}^4\eta_{\mu\nu}$ in rectangular coordinates and tries to get asymptotic statements with various types of definitions of r. However, it is difficult to correctly specify the limits for $r \to \infty$ in a meaningful, coordinate-independent way.

This led to the introduction of coordinate independent definitions of asymptotic flatness of a space-time:

1. Penrose [492, 493][9] introduced the notions of asymptotic flatness at null infinity (i.e., along null geodesics) and of asymptotic simplicity with his conformal completion approach. A smooth (time- and -space orientable) space-time $(M^4, {}^4g)$ is asymptotically simple if there exists another smooth Lorentz manifold $(\hat{M}^4, {}^4\hat{g})$ such that: (1) M^4 is an open sub-manifold of $\hat{M}^4$ with smooth boundary $\partial M^4 = \mathcal{S}$ (smooth conformal boundary); (2) there exists a smooth scalar field $\Omega \geq 0$ on $\hat{M}^4$, such that ${}^4\hat{g} = \Omega^2\, {}^4g$ on M^4 and $\Omega = 0$, $d\Omega \neq 0$ on $\mathcal{S}$; and (3) every null geodesic in M^4 acquires a future and past end-point on $\mathcal{S}$. An asymptotically simple space-time is asymptotically flat if vacuum Einstein equations hold in a neighborhood of $\mathcal{S}$.[10]

2. Geroch [496] introduced a definition of asymptotic flatness at spatial infinity in terms of the large-distance behavior of initial data on a Cauchy surface.

3. In the projective approach a time-like hyperboloid is introduced as the space-like boundary of space-time.

4. The two definitions of asymptotic flatness at null and spatial infinity were unified in the SPI formalism of Ashtekar and Hanson [298–300]. Essentially, in the SPI approach, the spatial infinity of the space-time M^4 is compactified to a point i^o and fields on M^4 have direction-dependent limits at i^o (this implies a peculiar differential structure on Σ_τ and awkward differentiability conditions of the 4-metric).

5. In Ref. [301], a new kind of completion, neither conformal nor projective, is developed by Ashtekar and Romano: Now the boundary of M^4 is a unit time-like hyperboloid like in the projective approach, which, however, has a well-defined contra-variant normal in the completion;[11] now, there is no need for awkward

[9] See also Refs. [494, 495]) for definitions of asymptotically simple and weakly asymptotically simple space-times, intended to ensure that the asymptotic structure be globally the same as that of Minkowski space-time.

[10] In this case the conformal boundary $\mathcal{S}$ is a shear-free smooth null hyper-surface with two connected components $\mathcal{I}^{\pm}$ (scri-plus and -minus), each with topology $S^2 \times R$ and the conformal Weyl tensor vanishes on it. In the conformal completion of Minkowski space-time $\mathcal{S}$, is formed by the future $\mathcal{I}^+$ and past $\mathcal{I}^-$ null infinity, which join in a point i^o representing the compactified space-like infinity: $\mathcal{I}^+$ terminates in the future at a point i^+ (future time-like infinity), while $\mathcal{I}^-$ terminates in the past at a point i^- (past time-like infinity).

[11] There are different conformal rescalings of the 4-metric ${}^4g \mapsto {}^4\tilde{g} = \Omega^2\, {}^4g$ ($\Omega \geq 0$, $\Omega = 0$ is the boundary 3-surface of the unphysical space-time $\tilde{M}^4$) and of the normal $n^\mu \mapsto \tilde{n}^\mu = \Omega^{-4}\, n^\mu$.

differentiability conditions. While in the SPI framework each hyper-surface Σ_τ has the sphere at spatial infinity compactified at the same point i^o, which is the vertex for both future $\mathcal{I}^+$ (scri-plus) and past $\mathcal{I}^-$ (scri-minus) null infinity, these properties are lost in the new approach: Each Σ_τ has as a boundary at spatial infinity the sphere $S^2_{\tau,\infty}$ cut by Σ_τ in the time-like hyperboloid; there is no relation between the time-like hyperboloid at spatial infinity and $\mathcal{I}^\pm$. This new approach simplifies the analysis of Ref. [498, 499] of uniqueness (modulo the logarithmic translations of Bergmann [500]) of the completion at space-like infinity.

6. See Refs. [501–503] and the reviews [504, 505] for the status of Friedrich's conformal field equations, derived from Einstein's equations, which arise in the study of the compatibility of Penrose's conformal completion approach with Einstein's equations. In the final description space-like infinity is a cylinder since each space-like hyper-surface has its point i^o blown up to a 2-sphere I^o at spatial infinity.[12] The cylinder meets future null infinity $\mathcal{I}^+$ in a sphere I^+ and past null infinity $\mathcal{I}^-$ in a sphere I^-. It is an open question whether the concepts of asymptotic simplicity and conformal completion are too strong requirements. Other reviews of the problem of consistency, i.e., whether the geometric assumptions inherent in the existing definitions of asymptotic flatness are compatible with Einstein equations, are given in Refs. [324, 502, 503], while in Ref. [506] a review is given about space-times with gravitational radiation (nearly all the results on radiative space-times are at null infinity, where, for instance, the SPI requirement of vanishing of the pseudo-magnetic part of the Weyl tensor to avoid super-translations is too strong and destroys radiation).

There are also coordinate-dependent formalisms:

1. The one of Beig and Schmidt [507, 508][13] whose relation to the new completion is roughly the same as that between Penrose's coordinate-independent approach to null infinity [492, 493] and Bondi's approach [509–513] based on null coordinates. The class of space-times studied in Refs. [507, 508] (called radially smooth of order m at spatial infinity) have 4-metrics of the type

$$ds^2 = d\rho^2\left(1 + \frac{{}^1\sigma}{\rho} + \frac{{}^2\sigma}{\rho^2} + \dots\right)^2 + \rho^2\left({}^o h_{rs} + \frac{1}{\rho}\,{}^1 h_{rs} + \dots\right) d\phi^r d\phi^s, \qquad (8.34)$$

where ${}^o h_{rs}$ is the 3-metric on the unit time-like hyperboloid, which completes M^4 at spatial infinity, and ${}^n\sigma$, ${}^n h_{rs}$, are functions on it. There are coordinate charts x^σ in $(M^4, {}^4g)$ where the 4-metric becomes

[12] It is interpretable as the space of space-like directions at i^o, namely the set of the end-points of space-like geodesics. This allows defining a regular initial value problem at space-like infinity with Minkowski data on I^o.

[13] It was developed to avoid the awkward differentiability conditions of the SPI framework and using polar coordinates like the standard hyperbolic ones for Minkowski space-time and agreeing with them at first order in $1/r$.

$$
{}^4g_{\mu\nu} = {}^4\eta_{\mu\nu} + \sum_{n=1}^{m} \frac{1}{\rho^n}\, {}^nl_{\mu\nu}\left(\frac{x^\sigma}{\rho}\right) + O(\rho^{-(m+1)}). \tag{8.35}
$$

2. The Christodoulou and Klainermann [8] result on the non-linear gravitational stability of Minkowski space-time implies a peeling behavior of the conformal Weyl tensor near null infinity which is weaker than the peeling behavior implied by asymptotic simplicity (see Refs. [492, 493, 509–513]) and this could mean that asymptotic simplicity can be established only, if at all, with conditions stronger than those required by these authors. In Ref. [8], one studies the existence of global, smooth, non-trivial solutions to Einstein's equations without matter, which look, in the large, like the Minkowski space-time,[14] are close to Minkowski space-time in all directions in a precise manner (for developments of the initial data sets uniformly close to the trivial one), and admit gravitational radiation in the Bondi sense. These authors reformulate Einstein's equations with the ADM variables (there are four constraint equations plus the equations for $\partial_\tau\, {}^3g_{rs}$ and $\partial_\tau\, {}^3K_{rs}$), put the shift functions equal to zero,[15] and add the maximal slicing condition ${}^3K = 0$. Then, they assume the existence of a coordinate system $\vec{\sigma}$ near spatial infinity on the Cauchy surfaces Σ_τ and of smoothness properties for ${}^3g_{rs}$, ${}^3K_{rs}$, such that for $r = \sqrt{\vec{\sigma}^2} \to \infty$ the initial data set $(\Sigma_\tau, {}^3g_{rs}, {}^3K_{rs})$ is strongly asymptotically flat, namely:[16]

$$
\begin{aligned}
{}^3g_{rs} &= \left(1 + \frac{M}{r}\right)\delta_{rs} + o_4(r^{-3/2}),\\
{}^3K_{rs} &= o_3(r^{-5/2}),
\end{aligned} \tag{8.36}
$$

where the leading term in ${}^3g_{rs}$ is called the Schwarzschild part of the 3-metric, also in the absence of matter; this asymptotic behavior ensures the existence of the ADM energy and angular momentum and the vanishing of the ADM momentum (center-of-mass frame). The addition of a technical global smallness assumption on the strongly asymptotically flat initial data leads to a unique, globally hyperbolic, smooth and geodesically complete solution of Einstein's equations without matter, which is globally asymptotically flat in the sense that its Riemann curvature tensor approaches zero on any causal or space-like geodesic. It is also shown that the 2-dimensional space of the dynamical degrees of freedom of the gravitational field at a point (the reduced configuration space) is the space of trace-free symmetric 2-covariant tensors on a 2-plane. A serious technical difficulty[17] derives from the "mass term" in the asymptotic

[14] These space-times are without singularities: Since the requirements needed for the existence of a conformal completion are not satisfied, it is possible to evade the singularity theorems.

[15] The lapse function is assumed equal to 1 at spatial infinity, but not everywhere because, otherwise, one should have a finite time breakdown.

[16] $f(\vec{\sigma})$ is $o_m(r^{-k})$ if $\partial^l f = o(r^{-k-l})$ for $l = 0, 1, \ldots, m$ and $r \to \infty$.

[17] It requires the definition of an "optical function" and reflecting the presence of gravitational radiation in any non-trivial perturbation of Minkowski space-time.

Schwarzschild part of the 3-metric: It has the long-range effect of changing the asymptotic position of the null geodesic cone relative to the maximal ($^3K = 0$) foliation.[18]

Let us now consider the problem of asymptotic symmetries [4] and of the associated conserved asymptotic charges containing the ADM Poincaré charges.

In the same way that null infinity admits an infinite-dimensional group (the BMS group [509–513]) of asymptotic symmetries, the SPI formalism admits an even bigger group, the SPI group [298–300], of such symmetries. Both BMS and SPI algebras have an invariant 4-dimensional subalgebra of translations, but they also have invariant infinite-dimensional Abelian subalgebras (including the translation subalgebra) of so-called super-translations or angle- (or direction-) dependent translations. Therefore, there is an infinite number of copies of Poincaré subalgebras in both BMS and SPI algebras, whose Lorentz parts are conjugate through super-translations.[19] All this implies that there is no unique definition of Lorentz generators and that in GR one cannot define intrinsically angular momentum and the Poincaré spin Casimir, so important for the classification of particles in Minkowski space-time. In Refs. [337–341] it is shown that the only known Casimirs of the BMS group are p^2 and its generalization involving super-translations. While Poincaré asymptotic symmetries correspond to the ten Killing fields of the Minkowski space-time,[20] super-translations are angle-dependent translations, which come just as close to satisfying Killing's equations asymptotically as any Poincaré transformation [4]. The problem seems to be that all known function spaces, used for the 4-metric and for Klein–Gordon and electromagnetic fields, do not put any restriction on the asymptotic angular behavior of the fields, but only restrict their radial decrease. Due to the relevance of the Poincaré group for particle physics in Minkowski space-time, and also to having a good definition of angular momentum in GR [4, 298–300, 342, 343], one usually restricts the class of space-times with boundary conditions such that super-translations are not allowed to exist. In the SPI framework [298–300], one asks that the pseudo-magnetic part of the limit of the conformally rescaled Weyl tensor vanishes at i^o. In Ref. [300] a 3+1 decomposition is made of the SPI framework; after having reexpressed the conserved quantities at i^o in terms of canonical initial data, it is shown that to remove ambiguities connected with the super-translations one must use stronger boundary conditions, again implying the vanishing of the pseudo-magnetic part of the Weyl tensor.

A related approach to these problems is given by Anderson [296]. He proved a slice theorem for the action of space-time diffeomorphisms asymptotic to

[18] These cones are expected to diverge logarithmically from their positions in flat space-time and to have their asymptotic shear drastically different from that in Minkowski space-time.

[19] The quotient of BMS and SPI groups with respect to super-translations is isomorphic to a Lorentz group.

[20] An asymptotically flat space-time tends asymptotically to Minkowski space-time in some way which depends on the chosen definition of asymptotic flatness.

Poincaré transformations on the set of asymptotically flat solutions of Einstein's equations in the context of spatial infinity, maximal slicing, and asymptotic harmonic coordinates (as gauge conditions). There is a heuristic extension of the momentum map method of reduction of dynamical systems with symmetries to the diffeomorphism group.

For metric general relativity, the spatially compact case has been solved in Ref. [516], with the result that, in the absence of Killing vector fields, the reduced phase space turns out to be a stratified symplectic manifold.[21]

In the spatially asymptotically flat case, one considers the group of those diffeomorphisms which preserve the conditions for asymptotic flatness and the nature of this group depends strongly on the precise asymptotic conditions. Apart from the compactification schemes of Geroch [496] and of Ashtekar and Hansen [298–300], three main types of asymptotic conditions have been studied: (1) the finite energy condition of O'Murchadha [517]; (2) the York quasi-isotropic (QI) gauge conditions [335]; and (3) the conditions of the type introduced by Regge-Teitelboim [294] with the parity conditions introduced by Beig and O'Murchadha [295] plus the gauge conditions of maximal slices and 3-harmonic asymptotic coordinates (their existence was shown in Ref. [517]).

These three types of asymptotic conditions have quite different properties.

1. In the case of the finite energy conditions, one finds that the group that leaves the asymptotic conditions invariant is a semidirect product $S\,|\!\times L$, where L is the Lorentz group and S consists of diffeomorphisms η such that roughly $D^2\eta \in L^2$, i.e., it is square integrable; S contains space and time translations. Under these conditions, it does not appear to be possible to talk about Hamiltonian dynamics. For a general element of the Lie algebra of $S\,|\times L$, the corresponding momentum integral does not converge, although for the special case of space and time translations the ADM 4-momentum is well defined.

2. QI gauge conditions of Ref. [335] have the desirable feature that no super-translations are allowed, but a more detailed analysis reveals that without extra conditions, the transformations corresponding to boosts are not well behaved; in any case, the QI asymptotic conditions do not give a well-defined angular and boost momentum and therefore are suitable only for the study of diffeomorphisms asymptotic to space and time translations.

3. To get a well-defined momentum for rotations and boosts, Anderson defines asymptotic conditions that contain the parity conditions of Ref. [296], but he replaces the 3-harmonic coordinates used in this paper with York's QI conditions. The space of diffeomorphisms $\mathrm{Diff}_P\, M^4$, which leaves invariant the space of solutions of the Einstein equations satisfying the parity conditions, is a semidirect

[21] In this case the space of solutions of Einstein's equations is a fibered space, which is smooth at $({}^3g, {}^3\Pi)$ if and only if the initial data $({}^3g, {}^3\tilde{\Pi})$ corresponds to a solution ${}^3g$ with no Killing field.

product $\mathrm{Diff}_P\, M^4 = \mathrm{Diff}_S\, M^4 \,|\!\times P$, where P is the Poincaré group and $\mathrm{Diff}_S\, M^4$ denotes the space of diffeomorphisms that are asymptotic to super-translations, which in this case are O(1) with odd leading term. When the QI conditions are added, the $\mathrm{Diff}_S\, M^4$ part is restricted to $\mathrm{Diff}_I\, M^4$, the space of diffeomorphisms which tend to the identity at spatial infinity.[22]

In this way, one obtains a realization of Bergmann's ideas based on his criticism [308–311] of general covariance: The group of coordinate transformations is restricted to containing an invariant Poincaré subgroup plus asymptotically trivial diffeomorphisms, analogously to what happens with the gauge transformations of electromagnetism.

It can be shown that the use of the parity conditions implies that the lapse and shift functions corresponding to the group of super-translations S have zero momentum. Thus, assuming the QI conditions, the ADM momentum appears as the momentum map with respect to the Poincaré group. Note that the classical form of the ADM momentum is correct only using the restrictive assumption of parity conditions, which are non-trivial restrictions not only on the gauge freedom, but also on the asymptotic dynamical degrees of freedom of the gravitational field (this happens also with Ashtekar–Hansen asymptotic condition on the Weyl tensor).

By assuming the validity of the conjecture on the global existence of solutions of Einstein's equations and of maximal slicing [10, 328–335] and working with Sobolev spaces with radial smoothness, Anderson demonstrates a slice theorem,[23] according to which, assuming the parity and QI conditions (which exclude the logarithmic translations of Bergmann [500]), for every solution ${}^3g_o$ of Einstein's equations one has that: (1) the gauge orbit of ${}^3g_o$ is a closed C^1 embedded sub-manifold of the manifold of solutions; and (2) there exists a sub-manifold containing ${}^3g_o$ which is a slice for the action of $\mathrm{Diff}_I\, M^4$. York's QI conditions should be viewed as a slice condition that fixes part of the gauge freedom at spatial infinity: (1) the $O(1/r^2)$ part of the trace of ${}^3\tilde{\Pi}^{rs}$ must vanish; (2) if ${}^3g = {}^3f + {}^3h$ (${}^3f$ is a flat metric) and if ${}^3h = {}^3h_{TT} = {}^3h_T + L_f(W)$ is the York decomposition[328–335] of ${}^3h$ with respect to ${}^3f$, then the $O(1)$ part of the longitudinal quantity W must vanish. In this way, one selects a QI asymptotically flat metric ${}^3g_{QI}$ and a preferred frame at spatial infinity, like in Refs. [308–311], i.e., preferred space-like hyper-surfaces corresponding to the intersections of the unit time-like hyperboloid at spatial infinity by spatial hyper-planes in R^4.

Since there is no agreement among the various viewpoints on the coordinate-independent definition of asymptotic flatness at spatial infinity, since we are

22 This result cannot be obtained with the finite energy conditions [517] or from boost theorems [518].

23 See appendix B of Anderson's paper [296] for the definition of slice.

interested in the coupling of general relativity to the standard $SU(3) \times SU(2) \times U(1)$ model and since we wish to recover the theory in Minkowski space-time if the Newton constant is switched off, in this book we shall use a coordinate-dependent approach and we shall work in the framework of Refs. [294, 295].

The boundary conditions and gauge fixings to be chosen will imply an angle- (i.e., direction-) independent asymptotic limit of the canonical variables, just as it is needed in Yang–Mills theory to have well-defined covariant non-Abelian charges[93, 94, 519, 520].[24] This is an important point for a future unified description of GR and of the standard model.

In particular, following Ref. [523], we will assume that at spatial infinity there is a 3-surface S_∞,[25] which intersects orthogonally the Cauchy surfaces Σ_τ. The 3-surface S_∞ is foliated by a family of 2-surfaces $S^2_{\tau,\infty}$ coming from its intersection with the slices Σ_τ. The normals $l^\mu(\tau,\vec{\sigma})$ to Σ_τ at spatial infinity, $l^\mu_{(\infty)}$, are tangent to S_∞ and normal to the corresponding $S^2_{\tau,\infty}$. The vector $b^\mu_\tau = z^\mu_\tau = N l^\mu + N^r b^\mu_r$ is not in general asymptotically tangent to S_∞. We assume that, given a subset $U \subset M^4$ of space-time, ∂U consists of two slices, Σ_{τ_i} (the initial one) and Σ_{τ_f} (the final one) with outer normals $-l^\mu(\tau_i,\vec{\sigma})$ and $l^\mu(\tau_f,\vec{\sigma})$ respectively, and of the surface S_∞ near space infinity. Since we will identify special families of hyper-surfaces Σ_τ asymptotic to Minkowski hyper-planes at spatial infinity, these families can be mapped onto the space of cross sections of the unit time-like hyperboloid by using a lemma of Ref. [296].

Let us add some information on the existence of the ADM Lorentz boost generators:

1. In Ref. [518] on the boost problem in GR, Christodoulou and O'Murchadha show (using weighted Sobolev spaces) that a very large class of asymptotically flat initial data for Einstein's equations have a development that includes complete space-like surfaces boosted relative to the initial surface. Furthermore, the asymptotic fall-off[26] is preserved along these boosted surfaces and there exists a global system of harmonic coordinates on such a development. As noted in Ref. [301], the results of Ref. [518] suffice to establish the existence of a large class of space-times which are asymptotically flat at i^o (in the sense of Refs. [298–300]) in all space-like directions along a family of Cauchy surfaces related to one another by "finite" boosts (it is hoped that new results will allow placing

[24] As shown in Ref. [93, 94], one needs a set of Hamiltonian boundary conditions both for the fields and the gauge transformations in the Hamiltonian gauge group, implying angle-independent limits at spatial infinity; it is also suggested that the elimination of Gribov ambiguity requires the use of the following weighted Sobolev spaces[521, 522]: $\vec{A}_a, \vec{E}_a \in W^{p,s-1,\delta-1}$, $\vec{B}_a \in W^{p,s-2,\delta+2}$, $\vec{\mathcal{G}} \in W^{p,s,\delta}$, with $p > 3$, $s \geq 3$, $0 \leq \delta \leq 1 - \frac{3}{p}$.

[25] It is not necessarily a time-like hyperboloid, but with outer unit (space-like) normal $n^\mu(\tau,\vec{\sigma})$, asymptotically parallel to the space-like hyper-surfaces Σ_τ.

[26] ${}^3g - {}^3f \in W^{2,s,\delta+1/2}(\Sigma)$, ${}^3K \in W^{2,s-1,\delta+1/2}(\Sigma)$, $s \geq 4$, $\delta > -2$.

control also on "infinite" boosts). The situation is unsettled with regard to the existence of space-times admitting both i^o (in the sense of Refs. [298–300]) as well as smooth $\mathcal{I}^{\pm}$.

2. In Ref. [498], Chruściel says that for asymptotically flat metrics ${}^3g = {}^3f + O(r^{-\alpha})$, $\frac{1}{2} < \alpha \leq 1$, it is not proved that the asymptotic symmetry conjecture, given any two coordinate systems of the previous type, all twice-differentiable coordinate transformations preserving these boundary conditions are of type $y^\mu = \Lambda^\mu_\nu x^\nu + \zeta^\mu$ (a Lorentz transformation + a super-translation $\zeta = O(r^{1-\alpha})$): This would be needed for having the ADM 4-momentum Lorentz covariant. By defining P_μ in terms of Cauchy data on a 3-end N (a space-like 3-surface Σ minus a ball), on which ${}^3g = {}^3f + O(r^{-\alpha})$, one can evaluate the invariant mass $m(N) = \sqrt{\epsilon P^\mu P_\mu}$. Then, provided the hyper-surfaces N_1 : x^o = const., N_2 : y^o = const., lie within a finite boost of each other or if the metric is a no-radiation metric, one can show the validity of the invariant mass conjecture $m(N_1) = m(N_2)$ for metrics satisfying vacuum Einstein equations. The main limitation is the lack of knowledge of long-time behavior of Einstein's equations. Ashtekar–Hansen and Beig–O'Murchadha requirements are much stronger and more restrictive than is compatible with Einstein's equations.

The counterpart of the Yang–Mills non-Abelian charges and also of the Abelian electric charge are the asymptotic Poincaré charges [278–281, 294–296]. In a natural way they should be connected with gauge transformations of first kind (there is no counterpart of the center of the gauge group in metric gravity).

However, in Ref. [524] two alternative options are presented for the asymptotic Poincaré charges of asymptotically flat metric gravity:

1. There is the usual interpretation [306], admitting gauge transformations of first kind, in which some observer is assumed to sit at or just outside the boundary at spatial infinity but he/she is not explicitly included in the action functional; this observer merely supplies a coordinate chart on the boundaries (perhaps through their parametrization clock), which may be used to fix the gauge of the system at the boundary (the asymptotic lapse function). If one wishes, this external observer may construct the clock to yield zero Poincaré charges[27] so that every connection with particle physics is lost.
2. Instead, Marolf's proposal [524] is to consider the system in isolation without the utilization of any structure outside the boundary at spatial infinity and to consider, at the quantum level, super-selection rules for the asymptotic

[27] In this way, one recovers a Machian interpretation [525] also in non-compact universes with boundary; there is a strong similarity with the results of Einstein–Wheeler cosmology [10], based on a closed compact universe without boundaries, for which Poincaré charges are not defined.

Poincaré Casimirs, in particular for the ADM invariant mass. In this viewpoint, the Poincaré charges are not considered generators of first kind gauge transformations and the open problem of boosts loses part of its importance.

In Ref. [322], Giulini considers a matter of physical interpretation of whether all 3-diffeomorphisms of Σ_τ into itself must be considered as gauge transformations. In the asymptotically flat open case, he studies large diffeomorphisms, but not the gauge transformations generated by the super-Hamiltonian constraint. After a 1-point compactification $\bar{\Sigma}_\tau$ of Σ_τ, there is a study of the quotient space $\mathrm{Riem}\,\bar{\Sigma}_\tau/\mathrm{Diff}_F\,\bar{\Sigma}_\tau$, where $\mathrm{Diff}_F\,\bar{\Sigma}_\tau$ are those 3-diffeomorphisms whose pullback goes to the identity at spatial infinity (the point of compactification) where a privileged oriented frame is chosen. Then there is a study of the decomposition of $\bar{\Sigma}_\tau$ into its prime factors as a 3-manifold, of the induced decomposition of $\mathrm{Diff}_F\,\bar{\Sigma}_\tau$ and of the evaluation of the homotopy groups of $\mathrm{Diff}_f\,\bar{\Sigma}_\tau$. The conclusion is that the Poincaré charges are not considered as generators of gauge transformations.

We shall take the point of view that the asymptotic Poincaré charges are not generators of first kind gauge transformations like in Yang–Mills theory (the ADM energy will be shown to be the physical Hamiltonian for the evolution in τ), that there are super-selection sectors labeled by the asymptotic Poincaré Casimirs, and that the parameters of the gauge transformations of ADM metric gravity have a clean separation between a rigid part (differently from Yang–Mills theory, (Eq. 8.33), it has both a constant and a term linear in $\vec{\sigma}$) and a proper one, namely we assume the absence of improper non-rigid gauge transformations like in Yang–Mills theory.

Let us now define the proper gauge transformations of the ADM metric gravity. In Refs. [521, 522, 526–528] it is noted that, in asymptotically flat space-times, the surface integrals arising in the transition from the Hilbert action to the ADM action and, then, from this to the ADM phase space action are connected with the ADM energy–momentum of the gravitational field of the linearized theory of metric gravity [278, 279], if the lapse and shift functions have certain asymptotic behaviors at spatial infinity. Extra complications for the differentiability of the ADM canonical Hamiltonian come from the presence of the second spatial derivatives of the 3-metric inside the ${}^3R$ term of the super-Hamiltonian constraint. In Refs. [521, 522] it is also pointed out that the Hilbert action for non-compact space-times is in general divergent and must be regularized with a reference metric (static solution of Einstein's equations): For space-times asymptotically flat at spatial infinity, one chooses a flat reference Minkowski metric.

By using the original ADM results [278, 279], Regge and Teitelboim [294] wrote the expression of the ten conserved Poincaré charges, by allowing the functions $N(\tau,\vec{\sigma})$, $N_{\check{r}}(\tau,\vec{\sigma})$, to have a linear behavior in $\vec{\sigma}$ for $r=|\vec{\sigma}|\rightarrow\infty$. These charges are surface integrals at spatial infinity, which have to be added to the Dirac Hamiltonian so that it becomes differentiable. In Ref. [294] there is a set of boundary conditions for the ADM canonical variables ${}^3g_{rs}(\tau,\vec{\sigma})$, ${}^3\tilde{\Pi}^{rs}(\tau,\vec{\sigma})$, so that it is possible to define ten surface integrals associated with the conserved Poincaré charges of the space-time (the translation charges are the ADM energy–momentum) and to show that the functional derivatives and Poisson brackets are well defined in metric gravity. There is no statement about gauge transformations and super-translations in this paper, but it is pointed out that the lapse and shift functions have the following asymptotic behavior at spatial infinity:

$$\begin{aligned}
N(\tau,\vec{\sigma}) &\to N_{(as)}(\tau,\vec{\sigma}) = -\tilde{\lambda}_{(\mu)}(\tau)l^{(\mu)}_{(\infty)} - l^{(\mu)}_{(\infty)}\tilde{\lambda}_{(\mu)(\nu)}(\tau)b^{(\nu)}_{(\infty)\check{s}}(\tau)\sigma^{\check{s}}\\
&= -\tilde{\lambda}_{\tau}(\tau) - \frac{1}{2}\tilde{\lambda}_{\tau\check{s}}(\tau)\sigma^{\check{s}},\\
N_{\check{r}}(\tau,\vec{\sigma}) &\to N_{(as)\check{r}}(\tau,\vec{\sigma}) = -b^{(\mu)}_{(\infty)\check{r}}(\tau)\tilde{\lambda}_{(\mu)}(\tau) - b^{(\mu)}_{(\infty)\check{r}}(\tau)\tilde{\lambda}_{(\mu)(\nu)}(\tau)b^{(\nu)}_{(\infty)\check{s}}(\tau)\sigma^{\check{s}}\\
&= -\tilde{\lambda}_{\check{r}}(\tau) - \frac{1}{2}\tilde{\lambda}_{\check{r}\check{s}}(\tau)\sigma^{\check{s}},
\end{aligned}$$

$$\begin{aligned}
\tilde{\lambda}_A(\tau) &= \tilde{\lambda}_{(\mu)}(\tau)b^{(\mu)}_{(\infty)A}(\tau), \qquad \tilde{\lambda}_{(\mu)}(\tau) = b^A_{(\infty)(\mu)}(\tau)\tilde{\lambda}_A(\tau),\\
\tilde{\lambda}_{AB}(\tau) &= \tilde{\lambda}_{(\mu)(\nu)}(\tau)[b^{(\mu)}_{(\infty)A}b^{(\nu)}_{(\infty)B} - b^{(\nu)}_{(\infty)A}b^{(\mu)}_{(\infty)B}](\tau) = 2[\tilde{\lambda}_{(\mu)(\nu)}b^{(\mu)}_{(\infty)A}b^{(\nu)}_{(\infty)B}](\tau),\\
\tilde{\lambda}_{(\mu)(\nu)}(\tau) &= \frac{1}{4}[b^A_{(\infty)(\mu)}b^B_{(\infty)(\nu)} - b^B_{(\infty)(\mu)}b^A_{(\infty)(\nu)}](\tau)\tilde{\lambda}_{AB}(\tau)\\
&= \frac{1}{2}[b^A_{(\infty)(\mu)}b^B_{(\infty)(\nu)}\tilde{\lambda}_{AB}](\tau),
\end{aligned} \tag{8.37}$$

Let us remark that with this asymptotic behavior any 3+1 splitting of the space-time M^4 is in some sense ill-defined because the associated foliation with leaves Σ_τ has diverging proper time interval $N(\tau,\vec{\sigma})d\tau$ and shift functions at spatial infinity for each fixed τ. Only in those gauges where $\tilde{\lambda}_{AB}(\tau) = -\tilde{\lambda}_{BA}(\tau) = 0$ do these problems disappear. These problems are connected with the previously quoted boost problem. They also suggest that the asymptotic Lorentz algebra (and therefore also the Poincaré and SPI algebras) need not be interpreted as generators of improper gauge transformations.

A more complete analysis, including also a discussion of super-translations in the ADM canonical formalism, has been given by Beig and O'Murchadha [295] (extended to Ashtekar's formalism in Ref. [306]). They consider 3-manifolds Σ_τ diffeomorphic to R^3, so that there exist global coordinate systems. If $\{\sigma^{\check{r}}\}$ is one of these global coordinate systems on Σ_τ, the 3-metric ${}^3g_{\check{r}\check{s}}(\tau,\sigma^{\check{t}})$, evaluated in this coordinate system, is assumed asymptotically Euclidean in the following sense: if $r = \sqrt{\delta_{\check{r}\check{s}}\sigma^{\check{r}}\sigma^{\check{s}}}$,[28] then one assumes

$$\begin{aligned}
{}^3g_{\check{r}\check{s}}(\tau,\vec{\sigma}) &= \delta_{\check{r}\check{s}} + \frac{1}{r}\,{}^3s_{\check{r}\check{s}}\left(\tau,\frac{\sigma^{\check{n}}}{r}\right) + {}^3h_{\check{r}\check{s}}(\tau,\vec{\sigma}), \quad r\to\infty,\\
{}^3s_{\check{r}\check{s}}\left(\tau,\frac{\sigma^{\check{n}}}{r}\right) &= {}^3s_{\check{r}\check{s}}\left(\tau,-\frac{\sigma^{\check{n}}}{r}\right), \quad \text{even parity},
\end{aligned}$$

[28] One could put $r = \sqrt{{}^3g_{\check{r}\check{s}}\sigma^{\check{r}}\sigma^{\check{s}}}$ and get the same kind of decomposition.

$$
\begin{aligned}
{}^3h_{\check r\check s}(\tau,\vec\sigma) &= O((r^{-(1+\epsilon)}), \quad \epsilon>0,\, for\, r\to\infty,\\
\partial_{\check u}\,{}^3h_{\check r\check s}(\tau,\vec\sigma) &= O(r^{-(2+\epsilon)}).
\end{aligned} \tag{8.38}
$$

The functions ${}^3s_{\check r\check s}(\tau,\frac{\sigma^{\check n}}{r})$ are C^∞ on the sphere $S^2_{\tau,\infty}$ at spatial infinity of Σ_τ; if they were of odd parity, the ADM energy would vanish. The difference ${}^3g_{\check r\check s}(\tau,\vec\sigma)-\delta_{\check r\check s}$ cannot fall off faster that $1/r$, because otherwise the ADM energy would be zero and the positivity energy theorem [529–534] would imply that the only solution of the constraints is flat space-time.

For the ADM momentum, one assumes the following boundary conditions:

$$
\begin{aligned}
{}^3\tilde\Pi^{\check r\check s}(\tau,\vec\sigma) &= \frac{1}{r^2}\,{}^3t^{\check r\check s}\left(\tau,\frac{\sigma^{\check n}}{r}\right)+{}^3k^{\check r\check s}(\tau,\vec\sigma), \quad r\to\infty,\\
{}^3t^{\check r\check s}\left(\tau,\frac{\sigma^{\check n}}{r}\right) &= -{}^3t^{\check r\check s}\left(\tau,-\frac{\sigma^{\check n}}{r}\right), \quad \text{odd parity},\\
{}^3k^{\check r\check s}(\tau,\vec\sigma) &= O(r^{-(2+\epsilon)}), \quad \epsilon>0, \quad r\to\infty.
\end{aligned} \tag{8.39}
$$

If ${}^3\tilde\Pi^{\check r\check s}(\tau,\vec\sigma)$ were to fall off faster than $1/r^2$, the ADM linear momentum would vanish and we could not consider Lorentz transformations. In this way, the integral $\int_{\Sigma_\tau} d^3\sigma[{}^3\tilde\Pi^{\check r\check s}\delta\,{}^3g_{\check r\check s}](\tau,\vec\sigma)$ is well defined and finite: since the integrand is of order $O(r^{-3})$, a possible logarithmic divergence is avoided due to the odd parity of ${}^3t^{\check r\check s}$.

These boundary conditions imply that functional derivatives and Poisson brackets are well defined [295]. In a more rigorous treatment one should use appropriately weighted Sobolev spaces.

The super-momentum and super-Hamiltonian constraints – see Eqs. (5.17), ${}^3\tilde{\mathcal H}^{\check r}(\tau,\vec\sigma)\approx 0$, and $\tilde{\mathcal H}(\tau,\vec\sigma)\approx 0$ – are even functions of $\vec\sigma$ of order $O(r^{-3})$. Their smeared version with the lapse and shift functions, appearing in the canonical Hamiltonian (Eq. 5.18), will give a finite and differentiable $H_{(c)ADM}$ if we assume [306]

$$
\begin{aligned}
N(\tau,\vec\sigma) &= m(\tau,\vec\sigma)=s(\tau,\vec\sigma)+n(\tau,\vec\sigma)\to_{r\to\infty}\\
&\to_{r\to\infty} k\left(\tau,\frac{\sigma^{\check n}}{r}\right)+O(r^{-\epsilon}), \quad \epsilon>0,\\
N_{\check r}(\tau,\vec\sigma) &= m_{\check r}(\tau,\vec\sigma)=s_{\check r}(\tau,\vec\sigma)+n_{\check r}(\tau,\vec\sigma)\to_{r\to\infty}\\
&\to_{r\to\infty} k_{\check r}\left(\tau,\frac{\sigma^{\check n}}{r}\right)+O(r^{-\epsilon}),\\
s(\tau,\vec\sigma) &= k\left(\tau,\frac{\sigma^{\check n}}{r}\right)=-k\left(\tau,-\frac{\sigma^{\check n}}{r}\right), \quad \text{odd parity},\\
s_{\check r}(\tau,\vec\sigma) &= k_{\check r}\left(\tau,\frac{\sigma^{\check n}}{r}\right)=-k_{\check r}\left(\tau,-\frac{\sigma^{\check n}}{r}\right), \quad \text{odd parity},
\end{aligned} \tag{8.40}
$$

with $n(\tau,\vec\sigma)$, $n_{\check r}(\tau,\vec\sigma)$ going to zero for $r\to\infty$ like $O(r^{-\epsilon})$ in an angle-independent way.

With these boundary conditions, one gets differentiability, i.e., $\delta H_{(c)ADM}$ is linear in $\delta\,{}^3g_{\check r\check s}$ and $\delta\,{}^3\tilde\Pi^{\check r\check s}$, with the coefficients being the Dirac–Hamilton equations of metric gravity. Therefore, since N and $N_{\check r}$ are a special case of the parameter fields for the most general infinitesimal gauge transformations generated by the first-class constraints $\tilde{\mathcal{H}}$, ${}^3\tilde{\mathcal{H}}^{\check r}$, with generator $G = \int d^3\sigma[\alpha\tilde{\mathcal{H}} + \alpha_{\check r}\,{}^3\tilde{\mathcal{H}}^{\check r}](\tau,\vec\sigma)$, the proper gauge transformations preserving Eqs. (8.38) and (8.39) have the multiplier fields $\alpha(\tau,\vec\sigma)$ and $\alpha_{\check r}(\tau,\vec\sigma)$ with the same boundary conditions (Eq. 8.40) of $m(\tau,\vec\sigma)$ and $m_{\check r}(\tau,\vec\sigma)$. Then, the Hamilton equations imply that also the Dirac multipliers $\lambda_N(\tau,\vec\sigma)$ and $\lambda^{\vec N}_{\check r}(\tau,\vec\sigma)$ have these boundary conditions [$\lambda_N \overset{\circ}{=} \delta N$, $\lambda^{\vec N}_{\check r} \overset{\circ}{=} \delta N_{\check r}$]. Instead, the momenta $\tilde\pi^N(\tau,\vec\sigma)$ and $\tilde\pi^{\check r}_{\vec N}(\tau,\vec\sigma)$, conjugate to N and $N_{\check r}$, must be of $O(r^{-(3+\epsilon)})$ to have $H_{(D)ADM}$ finite.

The angle-dependent functions $s(\tau,\vec\sigma) = k(\tau,\frac{\sigma^{\check n}}{r})$ and $s_{\check r}(\tau,\vec\sigma) = k_{\check r}(\tau,\frac{\sigma^{\check n}}{r})$ on $S^2_{\tau,\infty}$ are called odd time and space super-translations. The piece $\int d^3\sigma[s\,\tilde{\mathcal{H}} + s_{\check r}\,{}^3\tilde{\mathcal{H}}^{\check r}](\tau,\vec\sigma) \approx 0$ of the Dirac Hamiltonian is the Hamiltonian generator of super-translations (the zero momentum of super-translations of Ref. [296]). Their contribution to gauge transformations is to alter the angle-dependent asymptotic terms ${}^3s_{\check r\check s}$ and ${}^3t^{\check r\check s}$ in ${}^3g_{\check r\check s}$ and ${}^3\tilde\Pi^{\check r\check s}$. While Sachs [337–341] gave an explicit form of the generators (including super-translations) of the algebra of the BMS group of asymptotic symmetries, no such form is explicitly known for the generators of the SPI group.

With $N = m$, $N_{\check r} = m_{\check r}$ one can verify the validity of the smeared form of the Dirac algebra (Eq. 5.19) of the super-Hamiltonian and super-momentum constraints:

$$\begin{aligned}&\{H_{(c)ADM}[m_1, m_1^{\check r}], H_{(c)ADM}[m_2, m_2^{\check r}]\}\\ &\quad = H_{(c)ADM}[m_2^{\check r}\,{}^3\nabla_{\check r}m_1 - m_1^{\check r}\,{}^3\nabla_{\check r}m_2,\ \mathcal{L}_{\vec m_2}\,m_1^{\check r} + m_2\,{}^3\nabla^{\check r}m_1 - m_1\,{}^3\nabla^{\check r}m_2],\end{aligned} \tag{8.41}$$

with $m^{\check r} = {}^3g^{\check r\check s}m_{\check s}$ and with $H_{(c)ADM}[m, m^{\check r}] = \int d^3\sigma[m\tilde{\mathcal{H}} + m^{\check r}\,{}^3\tilde{\mathcal{H}}_{\check r}](\tau,\vec\sigma) = \int d^3\sigma[m\tilde{\mathcal{H}} + m_{\check r}\,{}^3\tilde{\mathcal{H}}^{\check r}](\tau,\vec\sigma)$.

When the functions $N(\tau,\vec\sigma)$ and $N_{\check r}(\tau,\vec\sigma)$ (and also $\alpha(\tau,\vec\sigma)$, $\alpha_{\check r}(\tau,\vec\sigma)$) do not have the asymptotic behavior of $m(\tau,\vec\sigma)$ and $m_{\check r}(\tau,\vec\sigma)$ respectively, one speaks of improper gauge transformations, because $H_{(D)ADM}$ is not differentiable even at the constraint hyper-surface.

At this point one has identified the following:

1. Certain global coordinate systems $\{\sigma^{\check r}\}$ on the space-like 3-surface Σ_τ, which hopefully define a minimal atlas $\mathcal{C}_\tau$ for the space-like hyper-surfaces Σ_τ foliating the asymptotically flat space-time M^4. With the $\mathcal{C}_\tau$ and the parameter τ as Σ_τ-adapted coordinates of M^4, one should build an atlas $\mathcal{C}$ of allowed coordinate systems of M^4.
2. A set of boundary conditions on the fields on Σ_τ (i.e., a function space for them) ensuring that the 3-metric on Σ_τ is asymptotically Euclidean in this minimal atlas (modulo 3-diffeomorphisms, see the next point).

3. A set of proper gauge transformations generated infinitesimally by the first-class constraints, which leave the fields on Σ_τ in the chosen function space. Since the gauge transformations generated by the super-momentum constraints ${}^3\tilde{\mathcal{H}}^{\check{r}}(\tau,\vec{\sigma}) \approx 0$ are the lift to the space of the tensor fields on Σ_τ (which contains the phase space of metric gravity) of the 3-diffeomorphisms Diff Σ_τ of Σ_τ into itself, the restriction of $N(\tau,\vec{\sigma})$, $N_{\check{r}}(\tau,\vec{\sigma})$ to $m(\tau,\vec{\sigma})$, $m_{\check{r}}(\tau,\vec{\sigma})$, ensures that these 3-diffeomorphisms are restricted to be compatible with the chosen minimal atlas for Σ_τ (this is the problem of the coordinate transformations preserving Eq. (8.38)). Also, the parameter fields $\alpha(\tau,\vec{\sigma})$, $\alpha_{\check{r}}(\tau,\vec{\sigma})$, of arbitrary (also improper) gauge transformations should acquire this behavior.

Since the ADM Poincaré charges are not considered as extra improper gauge transformations (Poincaré transformations at infinity) but as numbers individuating super-selection sectors, they cannot alter the assumed asymptotic behavior of the fields.

Let us remark at this point that the addition of gauge fixing constraints to the super-Hamiltonian and super-momentum constraints must happen in the chosen function space for the fields on Σ_τ. Therefore, the time constancy of these gauge fixings will generate secondary gauge fixing constraints for the restricted lapse and shift functions $m(\tau,\vec{\sigma})$, $m_{\check{r}}(\tau,\vec{\sigma})$.

These results, in particular Eqs. (8.37) and (8.40), suggest assuming the following form for the lapse and shift functions:

$$\begin{aligned} N(\tau,\vec{\sigma}) &= N_{(as)}(\tau,\vec{\sigma}) + \tilde{N}(\tau,\vec{\sigma}) + m(\tau,\vec{\sigma}), \\ N_{\check{r}}(\tau,\vec{\sigma}) &= N_{(as)\check{r}}(\tau,\vec{\sigma}) + \tilde{N}_{\check{r}}(\tau,\vec{\sigma}) + m_{\check{r}}(\tau,\vec{\sigma}), \end{aligned} \tag{8.42}$$

with $\tilde{N}$, $\tilde{N}_{\check{r}}$ describing improper gauge transformations not of first kind. Since, like in Yang–Mills theory, they do not play any role in the dynamics of metric gravity, we shall assume that they must be absent, so that (see Eq. (8.33)) we can parametrize the lapse and shift functions in the following form:

$$\begin{aligned} N(\tau,\vec{\sigma}) &= N_{(as)}(\tau,\vec{\sigma}) + m(\tau,\vec{\sigma}), \\ N_{\check{r}}(\tau,\vec{\sigma}) &= N_{(as)\check{r}}(\tau,\vec{\sigma}) + m_{\check{r}}(\tau,\vec{\sigma}). \end{aligned} \tag{8.43}$$

The improper parts $N_{(as)}$, $N_{(as)r}$, given in Eq. (8.37), behave as the lapse and shift functions associated with space-like hyper-planes in Minkowski space-time in parametrized Minkowski theories.

Let us add a comment on Dirac's original ideas. In Ref. [344] and in Ref. [20] (see also Ref. [294]), Dirac introduced asymptotic Minkowski Cartesian coordinates

$$z^{(\mu)}_{(\infty)}(\tau,\vec{\sigma}) = x^{(\mu)}_{(\infty)}(\tau) + b^{(\mu)}_{(\infty)\,\check{r}}(\tau)\sigma^{\check{r}} \tag{8.44}$$

in M^4 at spatial infinity $S_\infty = \cup_\tau S^2_{\tau,\infty}$.[29] For each value of τ, the coordinates $x^{(\mu)}_{(\infty)}(\tau)$ labels an arbitrary point, near spatial infinity chosen as the origin. On it there is a flat tetrad $b^{(\mu)}_{(\infty)\,A}(\tau) = (\, l^{(\mu)}_{(\infty)} = b^{(\mu)}_{(\infty)\,\tau} = \epsilon^{(\mu)}_{(\alpha)(\beta)(\gamma)} b^{(\alpha)}_{(\infty)\,\check{1}}(\tau) b^{(\beta)}_{(\infty)\,\check{2}}(\tau) b^{(\gamma)}_{(\infty)\,\check{3}}(\tau);\, b^{(\mu)}_{(\infty)\,\check{r}}(\tau)\,)$, with $l^{(\mu)}_{(\infty)}$ τ-independent, satisfying $b^{(\mu)}_{(\infty)\,A}\, {}^4\eta_{(\mu)(\nu)}\, b^{(\nu)}_{(\infty)\,B} = {}^4\eta_{AB}$ for every τ. There will be transformation coefficients $b^\mu_A(\tau,\vec{\sigma})$ from the adapted coordinates $\sigma^A = (\tau, \sigma^{\check{r}})$ to coordinates $x^\mu = z^\mu(\sigma^A)$ in an atlas of M^4, such that in a chart at spatial infinity one has $z^\mu(\tau,\vec{\sigma}) \to \delta^\mu_{(\mu)} z^{(\mu)}(\tau,\vec{\sigma})$ and $b^\mu_A(\tau,\vec{\sigma}) \to \delta^\mu_{(\mu)} b^{(\mu)}_{(\infty)A}(\tau)$.[30] The atlas $\mathcal{C}$ of the allowed coordinate systems of M^4 is assumed to have this property.

Dirac [344] and, then, Regge and Teitelboim [294] proposed that the asymptotic Minkowski Cartesian coordinates $z^{(\mu)}_{(\infty)}(\tau,\vec{\sigma}) = x^{(\mu)}_{(\infty)}(\tau) + b^{(\mu)}_{(\infty)\check{r}}(\tau)\sigma^{\check{r}}$ should define ten new independent degrees of freedom at the spatial boundary S_∞ (with ten associated conjugate momenta), as happens for Minkowski parametrized theories [8, 12, 13, 176, 178] when the extra configurational variables $z^{(\mu)}(\tau,\vec{\sigma})$ are reduced to ten degrees of freedom by the restriction to space-like hyper-planes, defined by $z^{(\mu)}(\tau,\vec{\sigma}) \approx x^{(\mu)}_s(\tau) + b^{(\mu)}_{\check{r}}(\tau)\sigma^{\check{r}}$.

In Dirac's approach to metric gravity the 20 extra variables of the Dirac proposal can be chosen as the set: $x^{(\mu)}_{(\infty)}(\tau)$, $p^{(\mu)}_{(\infty)}$, $b^{(\mu)}_{(\infty)A}(\tau)$,[31] $S^{(\mu)(\nu)}_{(\infty)}$, with the Dirac brackets $\{b^{(\rho)}_A, S^{(\mu)(\nu)}_s\} = {}^4\eta^{(\rho)(\mu)} b^{(\nu)}_A - {}^4\eta^{(\rho)(\nu)} b^{(\mu)}_A$, $\{S^{(\mu)(\nu)}_s, S^{(\alpha)(\beta)}_s\} = C^{(\mu)(\nu)(\alpha)(\beta)}_{(\gamma)(\delta)} S^{(\gamma)(\delta)}_s$, implying the orthonormality constraints $b^{(\mu)}_{(\infty)A}\, {}^4\eta_{(\mu)(\nu)} b^{(\nu)}_{(\infty)B} = {}^4\eta_{AB}$. Moreover, $p^{(\mu)}_{(\infty)}$ and $J^{(\mu)(\nu)}_{(\infty)} = x^{(\mu)}_{(\infty)} p^{(\nu)}_{(\infty)} - x^{(\nu)}_{(\infty)} p^{(\mu)}_{(\infty)} + S^{(\mu)(\nu)}_{(\infty)}$ satisfy a Poincaré algebra. In analogy with Minkowski parametrized theories restricted to space-like hyper-planes, one expects to have ten extra first-class constraints of the type

$$p^{(\mu)}_{(\infty)} - P^{(\mu)}_{ADM} \approx 0, \qquad S^{(\mu)(\nu)}_{(\infty)} - S^{(\mu)(\nu)}_{ADM} \approx 0, \tag{8.45}$$

with $P^{(\mu)}_{ADM}$, $S^{(\mu)(\nu)}_{ADM}$ related to the ADM Poincaré charges P^A_{ADM}, J^{AB}_{ADM} in place of $P^{(\mu)}_{sys}$, $S^{(\mu)(\nu)}_{sys}$ and ten extra Dirac multipliers $\tilde{\lambda}_{(\mu)}(\tau)$, $\tilde{\lambda}_{(\mu)(\nu)}(\tau)$, in front of them in the Dirac Hamiltonian. The origin $x^{(\mu)}_{(\infty)}(\tau)$ is going to play the role of an external decoupled observer with his parametrized clock. The main problem with respect to Minkowski parametrized theory on space-like hyper-planes is that it is not known which could be the ADM spin part $S^{(\mu)(\nu)}_{ADM}$ of the ADM Lorentz charge $J^{(\mu)(\nu)}_{ADM}$.

The way out from these problems is based on the following observation. If we replace $p^{(\mu)}_{(\infty)}$ and $S^{(\mu)(\nu)}_{(\infty)}$, whose Poisson algebra is the direct sum of an Abelian

[29] Here, $\{\sigma^{\check{r}}\}$ are the previous global coordinate charts of the atlas $\mathcal{C}_\tau$ of Σ_τ, not matching the spatial coordinates $z^{(i)}_{(\infty)}(\tau,\vec{\sigma})$.

[30] For $r \to \infty$ one has ${}^4g_{\mu\nu} \to \delta^{(\mu)}_\mu \delta^{(\nu)}_\nu {}^4\eta_{(\mu)(\nu)}$ and ${}^4g_{AB} = b^\mu_A\, {}^4g_{\mu\nu} b^\nu_B \to b^{(\mu)}_{(\infty)A}\, {}^4\eta_{(\mu)(\nu)} b^{(\nu)}_{(\infty)B} = {}^4\eta_{AB}$.

[31] With $b^{(\mu)}_{(\infty)\tau} = l^{(\mu)}_{(\infty)}$ τ-independent and coinciding with the asymptotic normal to Σ_τ, tangent to S_∞.

algebra of translations and of a Lorentz algebra, with the new variables (with indices adapted to Σ_τ),

$$p^A_{(\infty)} = b^A_{(\infty)(\mu)} p^{(\mu)}_{(\infty)}, \qquad J^{AB}_{(\infty)} \stackrel{\text{def}}{=} b^A_{(\infty)(\mu)} b^B_{(\infty)(\nu)} S^{(\mu)(\nu)}_{(\infty)} [\neq b^A_{(\infty)(\mu)} b^B_{(\infty)(\nu)} J^{(\mu)(\nu)}_{(\infty)}], \tag{8.46}$$

the Poisson brackets for $p^{(\mu)}_{(\infty)}$, $b^{(\mu)}_{(\infty)A}$, $S^{(\mu)(\nu)}_{(\infty)}$,[32] imply

$$\begin{aligned}
\{p^A_{(\infty)}, p^B_{(\infty)}\} &= 0,\\
\{p^A_{(\infty)}, J^{BC}_{(\infty)}\} &= {}^4g^{AC}_{(\infty)} p^B_{(\infty)} - {}^4g^{AB}_{(\infty)} p^C_{(\infty)},\\
\{J^{AB}_{(\infty)}, J^{CD}_{(\infty)}\} &= -(\delta^B_E \delta^C_F\, {}^4g^{AD}_{(\infty)} + \delta^A_E \delta^D_F\, {}^4g^{BC}_{(\infty)} - \delta^B_E \delta^D_F\, {}^4g^{AC}_{(\infty)} - \delta^A_E \delta^C_F\, {}^4g^{BD}_{(\infty)}) J^{EF}_{(\infty)}\\
&= -C^{ABCD}_{EF} J^{EF}_{(\infty)},
\end{aligned} \tag{8.47}$$

where ${}^4g^{AB}_{(\infty)} = b^A_{(\infty)(\mu)}\, {}^4\eta^{(\mu)(\nu)} b^B_{(\infty)(\nu)} = {}^4\eta^{AB}$ since the $b^{(\mu)}_{(\infty)A}$ are flat tetrad in both kinds of indices. Therefore, we get the algebra of a realization of the Poincaré group (this explains the notation $J^{AB}_{(\infty)}$) with all the structure constants inverted in the sign (transition from a left to a right action).

This implies that, after the transition to the asymptotic Dirac Cartesian coordinates, the Poincaré generators P^A_{ADM}, J^{AB}_{ADM} in Σ_τ-adapted coordinates should become a momentum $P^{(\mu)}_{ADM} = b^{(\mu)}_A P^A_{ADM}$ and only an ADM spin tensor $S^{(\mu)(\nu)}_{ADM}$.[33]

As a consequence of the previous results we shall assume the existence of a global coordinate system $\{\sigma^{\check r}\}$ on Σ_τ, in which we have

$$\begin{aligned}
N(\tau,\vec\sigma) &= N_{(as)}(\tau,\vec\sigma) + m(\tau,\vec\sigma),\\
N_{\check r}(\tau,\vec\sigma) &= N_{(as)\check r}(\tau,\vec\sigma) + m_{\check r}(\tau,\vec\sigma),
\end{aligned}$$

$$\begin{aligned}
N_{(as)}(\tau,\vec\sigma) &= -\tilde\lambda_{(\mu)}(\tau) l^{(\mu)}_{(\infty)} - l^{(\mu)}_{(\infty)} \tilde\lambda_{(\mu)(\nu)}(\tau) b^{(\nu)}_{(\infty)\check s}(\tau)\sigma^{\check s}\\
&= -\tilde\lambda_\tau(\tau) - \frac{1}{2}\tilde\lambda_{\tau\check s}(\tau)\sigma^{\check s},\\
N_{(as)\check r}(\tau,\vec\sigma) &= -b^{(\mu)}_{(\infty)\check r}(\tau)\tilde\lambda_{(\mu)}(\tau) - b^{(\mu)}_{(\infty)\check r}(\tau)\tilde\lambda_{(\mu)(\nu)}(\tau) b^{(\nu)}_{(\infty)\check s}(\tau)\sigma^{\check s}\\
&= -\tilde\lambda_{\check r}(\tau) - \frac{1}{2}\tilde\lambda_{\check r\check s}(\tau)\sigma^{\check s},
\end{aligned} \tag{8.48}$$

with $m(\tau,\vec\sigma)$, $m_{\check r}(\tau,\vec\sigma)$, given by Eq. (8.40): they still contain odd supertranslations.

This very strong assumption implies that one is selecting asymptotically at spatial infinity only coordinate systems in which the lapse and shift functions have behaviors similar to those of Minkowski space-like hyper-planes, so that the

[32] One has $\{b^A_{(\infty)(\gamma)}, S^{(\nu)(\rho)}_{(\infty)}\} = \eta^{(\nu)}_{(\gamma)} b^{A(\rho)}_{(\infty)} - \eta^{(\rho)}_{(\gamma)} b^{A(\nu)}_{(\infty)}$.

[33] To define an angular momentum tensor $J^{(\mu)(\nu)}_{ADM}$ one should find an "external center of mass of the gravitational field" $X^{(\mu)}_{ADM}[{}^3g, {}^3\tilde\Pi]$ (see Refs. [187, 188] for the Klein–Gordon case) conjugate to $P^{(\mu)}_{ADM}$, so that $J^{(\mu)(\nu)}_{ADM} = X^{(\mu)}_{ADM} P^{(\nu)}_{ADM} - X^{(\nu)}_{ADM} P^{(\mu)}_{ADM} + S^{(\mu)(\nu)}_{ADM}$.

allowed foliations of the 3+1 splittings of the space-time M^4 are restricted to have the leaves Σ_τ approaching these Minkowski hyper-planes at spatial infinity in a way independent from the direction. But this is coherent with Dirac's choice of asymptotic Cartesian coordinates (modulo 3-diffeomorphisms not changing the nature of the coordinates, namely tending to the identity at spatial infinity like in Ref. [296]) and with the assumptions used to define the asymptotic Poincaré charges. It is also needed to eliminate coordinate transformations not becoming the identity at spatial infinity, which are not associated with the gravitational fields of isolated systems [535].

By replacing the ADM configuration variables $N(\tau,\vec{\sigma})$ and $N_{\check{r}}(\tau,\vec{\sigma})$ with the new ones $\tilde{\lambda}_A(\tau) = \{\tilde{\lambda}_\tau(\tau); \tilde{\lambda}_{\check{r}}(\tau)\}$, $\tilde{\lambda}_{AB}(\tau) = -\tilde{\lambda}_{BA}(\tau)$, $n(\tau,\vec{\sigma})$, $n_{\check{r}}(\tau,\vec{\sigma})$[34] inside the ADM Lagrangian, one only gets the replacement of the primary first-class constraints of ADM metric gravity

$$\tilde{\pi}^N(\tau,\vec{\sigma}) \approx 0, \qquad \tilde{\pi}^{\check{r}}_{\vec{N}}(\tau,\vec{\sigma}) \approx 0, \tag{8.49}$$

with the new first-class constraints

$$\tilde{\pi}^n(\tau,\vec{\sigma}) \approx 0, \quad \tilde{\pi}^{\check{r}}_{\vec{n}}(\tau,\vec{\sigma}) \approx 0, \quad \tilde{\pi}^A(\tau) \approx 0, \quad \tilde{\pi}^{AB}(\tau) = -\tilde{\pi}^{BA}(\tau) \approx 0, \tag{8.50}$$

corresponding to the vanishing of the canonical momenta $\tilde{\pi}^A$, $\tilde{\pi}^{AB}$ conjugate to the new configuration variables.[35] The only change in the Dirac Hamiltonian of metric gravity $H_{(D)ADM} = H_{(c)ADM} + \int d^3\sigma[\lambda_N \tilde{\pi}^N + \lambda^{\vec{N}}_{\check{r}} \tilde{\pi}^{\check{R}}_{\vec{N}}](\tau,\vec{\sigma})$, $H_{(c)ADM} = \int d^3\sigma[N\tilde{\mathcal{H}} + N_{\check{r}}\tilde{\mathcal{H}}^{\check{r}}](\tau,\vec{\sigma})$ of Eq. (5.16) is

$$\int d^3\sigma[\lambda_N \tilde{\pi}^N + \lambda^{\vec{N}}_{\check{r}} \tilde{\pi}^{\check{r}}_{\vec{N}}](\tau,\vec{\sigma}) \mapsto \zeta_A(\tau)\tilde{\pi}^A(\tau) + \zeta_{AB}(\tau)\tilde{\pi}^{AB}(\tau)$$
$$+ \int d^3\sigma[\lambda_n \tilde{\pi}^n + \lambda^{\vec{n}}_{\check{r}} \tilde{\pi}^{\check{r}}_{\vec{n}}](\tau,\vec{\sigma}), \tag{8.51}$$

with $\zeta_A(\tau)$, $\zeta_{AB}(\tau)$ Dirac multipliers.

Finally let us remark that the presence of the terms $N_{(as)}$, $N_{(as)\check{r}}$ in Eq. (8.48) makes H_D not differentiable.

In Refs. [294, 295], following Refs. [526–528], it is shown that the differentiability of the ADM canonical Hamiltonian $[H_{(c)ADM} \to \hat{H}_{(c)ADM} + H_\infty]$ requires the introduction of the following surface term:

$$\begin{aligned} H_\infty = &- \int_{S^2_{\tau,\infty}} d^2\Sigma_u \{\epsilon k \sqrt{\gamma}\, {}^3g^{uv}\, {}^3g^{rs}[N(\partial_r\, {}^3g_{vs} - \partial_v\, {}^3g_{rs}) \\ &+ \partial_u N({}^3g_{rs} - \delta_{rs}) - \partial_r N({}^3g_{sv} - \delta_{sv})] - 2N_r\, {}^3\tilde{\Pi}^{ru}\}(\tau,\vec{\sigma}) \\ = &- \int_{S^2_{\tau,\infty}} d^2\Sigma_u \{\epsilon k \sqrt{\gamma}\, {}^3g^{uv}\, {}^3g^{rs}[N_{(as)}(\partial_r\, {}^3g_{vs} - \partial_v\, {}^3g_{rs}) \end{aligned}$$

[34] We use the notation n, $n_{\check{r}}$ for m, $m_{\check{r}}$; the reason for this will become clear in what follows.
[35] We assume the Poisson brackets $\{\tilde{\lambda}_A(\tau), \tilde{\pi}^B(\tau)\} = \delta^B_A$, $\{\tilde{\lambda}_{AB}(\tau), \tilde{\pi}^{CD}(\tau)\} = \delta^C_A\delta^D_B - \delta^D_A\delta^C_B$.

$$
\begin{aligned}
&+\partial_u N_{(as)}({}^3g_{rs} - \delta_{rs}) - \partial_r N_{(as)}({}^3g_{sv} - \delta_{sv})] \\
&-2N_{(as)r}\, {}^3\tilde{\Pi}^{ru}\}(\tau, \vec{\sigma}) \\
&= \tilde{\lambda}_A(\tau) P^A_{ADM} + \frac{1}{2}\tilde{\lambda}_{AB}(\tau) J^{AB}_{ADM}. \qquad (8.52)
\end{aligned}
$$

Indeed, by putting $N = N_{(as)}$, $N_{\check{r}} = N_{(as)\check{r}}$ in the surface integrals, the added term H_∞ becomes the given linear combination of the strong ADM Poincaré charges P^A_{ADM}, J^{AB}_{ADM} [294, 295] first identified in the linearized theory [278, 279]:

$$
\begin{aligned}
P^\tau_{ADM} &= \epsilon k \int_{S^2_{\tau,\infty}} d^2\Sigma_{\check{u}} [\sqrt{\gamma}\, {}^3g^{\check{u}\check{v}}\, {}^3g^{\check{r}\check{s}} (\partial_{\check{r}}\, {}^3g_{\check{v}\check{s}} - \partial_{\check{v}}\, {}^3g_{\check{r}\check{s}})](\tau, \vec{\sigma}), \\
P^{\check{r}}_{ADM} &= -2 \int_{S^2_{\tau,\infty}} d^2\Sigma_{\check{u}}\, {}^3\tilde{\Pi}^{\check{r}\check{u}}(\tau, \vec{\sigma}), \\
J^{\tau\check{r}}_{ADM} &= \epsilon k \int_{S^2_{\tau,\infty}} d^2\Sigma_{\check{u}} \sqrt{\gamma}\, {}^3g^{\check{u}\check{v}}\, {}^3g^{\check{n}\check{s}} \\
&\quad \cdot [\sigma^{\check{r}} (\partial_{\check{n}}\, {}^3g_{\check{v}\check{s}} - \partial_{\check{v}}\, {}^3g_{\check{n}\check{s}}) + \delta^{\check{r}}_{\check{v}} ({}^3g_{\check{n}\check{s}} - \delta_{\check{n}\check{s}}) - \delta^{\check{r}}_{\check{n}} ({}^3g_{\check{s}\check{v}} - \delta_{\check{s}\check{v}})](\tau, \vec{\sigma}), \\
J^{\check{r}\check{s}}_{ADM} &= \int_{S^2_{\tau,\infty}} d^2\Sigma_{\check{u}} [\sigma^{\check{r}}\, {}^3\tilde{\Pi}^{\check{s}\check{u}} - \sigma^{\check{s}}\, {}^3\tilde{\Pi}^{\check{r}\check{u}}](\tau, \vec{\sigma}),
\end{aligned}
$$

$$
\begin{aligned}
P^{(\mu)}_{ADM} &= l^{(\mu)} P^\tau_{ADM} + b^{(\mu)}_{(\infty)\check{r}}(\tau) P^{\check{r}}_{ADM} = b^{(\mu)}_{(\infty)A}(\tau) P^A_{ADM}, \\
S^{(\mu)(\nu)}_{ADM} &= [l^{(\mu)}_{(\infty)} b^{(\nu)}_{(\infty)\check{r}}(\tau) - l^{(\nu)}_{(\infty)} b^{(\mu)}_{(\infty)\check{r}}(\tau)] J^{\tau\check{r}}_{ADM} \\
&\quad + [b^{(\mu)}_{(\infty)\check{r}}(\tau) b^{(\nu)}_{(\infty)\check{s}}(\tau) - b^{(\nu)}_{(\infty)\check{r}}(\tau) b^{(\mu)}_{(\infty)\check{s}}(\tau)] J^{\check{r}\check{s}}_{ADM} \\
&= [b^{(\mu)}_{(\infty)A}(\tau) b^{(\nu)}_{(\infty)B}(\tau) - b^{(\nu)}_{(\infty)A}(\tau) b^{(\mu)}_{(\infty)B}(\tau)] J^{AB}_{ADM}. \qquad (8.53)
\end{aligned}
$$

Here, $J^{\tau\check{r}}_{ADM} = -J^{\check{r}\tau}_{ADM}$ by definition and the inverse asymptotic tetrads are defined by $b^A_{(\infty)(\mu)} b^{(\nu)}_{(\infty)B} = \delta^A_B$, $b^A_{(\infty)(\mu)} b^{(\nu)}_{(\infty)A} = \delta^{(\nu)}_{(\mu)}$.

This discussion of the problems behind gauge transformations is the origin of the choice of the boundary conditions of Yang–Mills fields given in Section 4.5 and of those for ADM metric gravity given in Section 5.1.

9

Dirac Observables Invariant under the Hamiltonian Gauge Transformations Generated by First-Class Constraints

On the final constraint sub-manifold $\bar{\gamma} \subset T^*Q$ the Dirac Hamiltonian depends on as many arbitrary Dirac multipliers $\lambda^{\bar{A}}(t)$ as primary first-class constraints $\bar{\phi}_{\bar{A}}(q,p) \approx 0$. As a consequence, the solutions $q^i(t)$, $p_i(t)$ of the Dirac–Hamilton equations equations are functionals of these arbitrary functions of time and cannot correspond to measurable observables, which must have a deterministic dependence on time starting from a given set of Cauchy initial data.

Let us give the canonical coordinates $q^i_o = q^i(t_o)$, $p_{oi} = p_i(t_o)$ at time t_o: This is interpreted as giving a physical initial state for the system. Let us consider the time evolution of a function $\bar{f}(q,p)$ from t_o to $t_o + \delta t$ generated by the Dirac Hamiltonian: $\bar{f}(q,p)|_{t_o+\delta t} \stackrel{\circ}{=} \bar{f}(q,p)|_{t_o} + \delta t\, \{\bar{f}(q,p), \bar{H}_D(q,p,\lambda)\} = \bar{f}(q,p)|_{t_o} + \delta t\, \{\bar{f}(q,p), \bar{H}^{(F)}_c(q,p)\} + \delta t\, \lambda^{\bar{A}}(t)\, \{\bar{f}(q,p), \bar{\phi}_{\bar{A}}(q,p)\}$. If we consider two sets of Dirac multipliers $\lambda^{\bar{A}}_1(t)$ and $\lambda^{\bar{A}}_2(t)$ coinciding at t_o $[\lambda^{\bar{A}}_1(t_o) = \lambda^{\bar{A}}_2(t_o)]$, we obtain the result that at time $t_o + \delta t$ there is no uniquely determined value for $q^i(t_o + \delta t)$, $p_i(t_o + \delta t)$, because for every function we get the following difference between the two time evolutions $\triangle_{12}\, \bar{f}(q,p)|_{t_o+\delta t} = \delta t\, [\lambda^{\bar{A}}_1(t_o) - \lambda^{\bar{A}}_2(t_o)]\, \{\bar{f}(q,p), \bar{\phi}_{\bar{A}}(q,p)\}|_{t_o}$.

The only way to recover a deterministic description of the physical states of the system as in the regular case is to abandon the assumption that a physical state is uniquely identified by one and only one set of values of the canonical coordinates at a given time. In the singular case at each instant of time many sets of canonical coordinates describe the same physical state. Two sets of canonical coordinates whose difference is $\triangle_{12}\, q^i$, $\triangle_{12}\, p_i$ are said to be gauge equivalent and the term $\lambda^{\bar{A}}(t)\, \bar{\phi}_{\bar{A}}(q,p)$ of the Dirac Hamiltonian is interpreted as the generator of a Hamiltonian gauge transformation. Therefore, the m primary first-class constraints $\bar{\phi}_{\bar{A}}$ are the generators for the Hamiltonian gauge transformations responsible of the non-deterministic time evolution. This means that in the singular case we must do the following:

1. Find which is the maximal set of Hamiltonian gauge transformations existing for each given singular system besides those appearing in the Dirac

Hamiltonian (they are the only ones allowed in the description of the time evolution).

2. Separate the canonical variables into three disjoint sets:

   a. the Hamiltonian gauge-invariant variables (the so-called Dirac observables, DOs), which have deterministic time evolution so that a complete set of them at one instant identifies the physical state of the system at that instant;
   b. the non-inessential pairs of canonical variables eliminable by means of the second-class constraints with the Dirac brackets; and
   c. the Hamiltonian gauge variables that are irrelevant for the identification of a physical state, because they have an arbitrary time evolution. The number of gauge variables will coincide with the number of functionally independent generators of infinitesimal Hamiltonian gauge transformations, which allow the reconstruction of their maximal set.

To find the maximal set of infinitesimal Hamiltonian gauge transformations, we shall assume that their generators $\bar{G}_a(q,p)$, $a = 1,\dots,g$ have the structure of a local Hamiltonian gauge algebra $\tilde{g}$ under the Poisson brackets, namely $\{\bar{G}_a(q,p), \bar{G}_b(q,p)\} = \bar{C}_{ab}{}^c(q,p)\,\bar{G}_c(q,p)$ with some set of structure functions $\bar{C}_{ab}{}^c(q,p)$. When the structure functions are constant on $\bar{\gamma}$, $\bar{C}_{ab}{}^c = C_{ab}{}^c = \text{const.}$, we speak of a Lie gauge algebra with structure constants $C_{ab}{}^c$.

Once the generators $\bar{G}_a$ are known, the next problem is to define a local Hamiltonian gauge group $\mathcal{G}$, i.e., the finite Hamiltonian gauge transformations that can be built with sequences of infinitesimal Hamiltonian gauge transformations, and then to see whether there exist Hamiltonian gauge transformations not connected to the identity (large gauge transformations) due to the topological properties of the system. The Hamiltonian gauge group is said to be local, because the space of its gauge parameters (i.e., the coordinates of its group manifold) is coordinatized by arbitrary functions of time $\epsilon^a(t)$, $a = 1,\dots,g$, and not by numerical constants $\epsilon^a = \text{const.}$

Let us remark that the Hamiltonian gauge transformations are defined off-shell, namely without using the equations of motion, and that in general they are of the standard Lagrangian gauge transformations. Only on-shell, namely on the space of the solutions of the equations of motion, do the two notions of gauge transformations coincide, and both of them are then also gauge dynamical symmetries of the equations of motion.

The primary first-class constraints $\bar{\phi}_{\bar{A}}(q,p)$ are in general only a subset of the $\bar{G}_a$. Their associated gauge parameters $\epsilon^{a=\bar{A}}(t)$ are the Dirac multipliers $\lambda^{\bar{A}}(t) \overset{\circ}{=} g^{\bar{A}}_{(\lambda)}(q,\dot{q})$, which identify the non-determined non-projectable velocity functions of the singular system associated with the non-invertibility of the equations $p_i = \mathcal{P}_i(q,\dot{q})$. Therefore, there will be an equal number of primary Hamiltonian gauge variables $\bar{Q}^{\bar{A}}(q,p)$, which are transformed among themselves by the Hamiltonian gauge transformations generated by the $\bar{\phi}_{\bar{A}}$, satisfy $\frac{d\bar{Q}^{\bar{A}}}{dt} \stackrel{\circ}{=} \{\bar{Q}^{\bar{A}}, \bar{H}_D\} = \lambda^{\bar{A}}(t) \stackrel{\circ}{=} g^{\bar{A}}_{(\lambda)}(q, \dot{q})$ and do not contribute to the identification of the physical state of the system.

However, the study of the Dirac–Hamilton equations (or of the Euler–Lagrange [EL] equations) will show that in general there exist secondary Hamiltonian (or Lagrangian) gauge variables $\bar{\mathcal{T}}^{\alpha}(q,p)$, which inherit the arbitrariness of the Dirac multipliers, being functionals of them on the solutions of the equations of motion. This implies that the gauge parameters $\epsilon^a(t)$, $a = 1, \ldots, g$, of the off-shell Hamiltonian gauge transformations have to be restricted to $\epsilon^{a=\bar{A}}(t) = \lambda^{\bar{A}}(t)$, $\epsilon^{a\neq\bar{A}}(t) = F^{a\neq\bar{A}}[\lambda^{\bar{B}}(t)]$ to be interpretable as the gauge parameters of the on-shell Hamiltonian gauge group.

The gauge algebra assumption implies that the Poisson bracket of two infinitesimal gauge transformations must be a gauge transformation. Therefore we must have $\{\bar{\phi}_{\bar{A}}(q,p), \bar{\phi}_{\bar{B}}(q,p)\} = \bar{C}_{\bar{A}\bar{B}}{}^{a}(q,p)\, \bar{G}_a(q,p)$.

Moreover, the gauge algebra must not change in time. This implies that the time derivative $\frac{d\bar{G}_a}{dt} \stackrel{\circ}{=} \{\bar{G}_a, \bar{H}_D\}$ of a Hamiltonian gauge transformation must be again a gauge transformation. Since we only know that $\{\bar{\phi}_{\bar{A}}, \bar{\phi}_{\bar{B}}\} = \bar{C}_{\bar{A}\bar{B}}{}^{a}\, \bar{G}_a$, we get the following requirement on the $\bar{\phi}_{\bar{A}}$: $\{\bar{\phi}_{\bar{A}}(q,p), \bar{H}^{(F)}_c(q,p)\} = \bar{V}^{a}_{\bar{A}}(q,p)\, \bar{G}_a(q,p)$, where $\bar{H}^{(F)}_c$ is the final canonical Hamiltonian (a first-class quantity).

Let us assume that the regularity conditions on the singular Lagrangian be such that theorem 2 on the diagonalization of chains of constraints holds. This means that we can find linear first-class combinations $\bar{\chi}^{(o)}_{(k)\bar{A}_k}$ of the primary constraints such that $\{\bar{\chi}^{(o)}_{(k)\bar{A}_k}, \bar{H}^{(F)}_c\} = \bar{\chi}^{(1)}_{(k)\bar{A}_k}$. Therefore, in this form the secondary first-class constraints are generators of Hamiltonian gauge transformations. The general result $\{\bar{\chi}^{(h)}_{(k)\bar{A}_k}, \bar{H}^{(F)}_c\} = \bar{\chi}^{(h+1)}_{(k)\bar{A}_k}$ shows that all the secondary, tertiary, etc. first-class constraints are generators of Hamiltonian gauge transformations, namely that $\bar{G}_{\mathcal{A}} = \bar{\Phi}_{(1)\mathcal{A}}$, $\mathcal{A} = 1, \ldots, M$. Since in general $\{\bar{\Phi}_{(1)\mathcal{A}}, \bar{\Phi}_{(1)\mathcal{B}}\} = \bar{C}_{\mathcal{A}\mathcal{B}}{}^{\mathcal{C}}\, \bar{\Phi}_{(1)\mathcal{C}} + \bar{C}'_{\mathcal{A}\mathcal{B}}{}^{\mathcal{C}'}\, \bar{\Phi}_{(2)\mathcal{C}'}$, we see that a true gauge algebra is obtained only near the second-class sub-manifold $\gamma_{(2)}$, which contains the final constraint sub-manifold $\bar{\gamma}$.

Therefore, under suitable regularity conditions on the singular Lagrangian, Dirac's conjecture [20, 21] that all the first-class constraints are generators of Hamiltonian gauge transformations is true. Dirac also proposed replacing the final Dirac Hamiltonian $\bar{H}_D = \bar{H}^{(F)}_c + \lambda^{\bar{A}}(t)\, \bar{\phi}_{\bar{A}}$ with the extended Hamiltonian,

$$\bar{H}_E = \bar{H}^{(F)}_c + \epsilon^{\mathcal{A}}(t)\, \bar{\Phi}_{(1)\mathcal{A}}, \tag{9.1}$$

including all the first-class constraints, each with an arbitrary multiplier. In this way the time evolution is split in a deterministic part governed by the final canonical Hamiltonian $\bar{H}^{(F)}_c$ (it generates a mapping from a gauge orbit to another one) and in a gauge part, which is the generator of the most general off-shell Hamiltonian gauge transformation.

Even if this extension does not change the on-shell dynamics (so that, as shown in Refs. [26, 28, 29], it is taken as the starting point of the BRST quantization program), it has the drawback that its inverse Legendre transformation does not reproduce the original singular Lagrangian, because the secondary gauge variables have an arbitrary gauge freedom instead of the reduced one ($\epsilon^{\mathcal{A}=\bar{A}}(t) = \lambda^{\bar{A}}(t)$, $\epsilon^{\mathcal{A}\neq\bar{A}}(t) = F^{\mathcal{A}\neq\bar{A}}[\lambda^{\bar{B}}(t)]$) associated with the on-shell Hamiltonian gauge transformations. Even if they have a reduced gauge freedom, this is a consequence of the fact that the secondary gauge variables have non-vanishing Poisson brackets with the generators $\bar{G}_{\mathcal{A}}$. The results: (1) $\{\bar{\chi}^{(h)}_{(k)\bar{A}_k}, \bar{H}^{(F)}_c\} = \bar{\chi}^{(h+1)}_{(k)\bar{A}_k}$ of theorem 2; and (2) $\bar{H}^{(F)}_c = \bar{H}_c +$ (combinations of primary both first- and second-class constraints), imply that all the secondary first-class constraints $\bar{\chi}^{(h)}_{(k)\bar{A}_k}$, $h \neq 0$, must already be present in the original canonical Hamiltonian $\bar{H}_c$ with some form of the primary $\bar{Q}^{\bar{A}}_{(k)} = \bar{\mathcal{T}}^{(o)\bar{A}_k}_{(k)}$ and secondary $\bar{\mathcal{T}}^{(h)\bar{A}_k}_{(k)}$, $h > 0$, gauge variables as coefficients (only the $\bar{\mathcal{T}}^{(f_k)\bar{A}_k}_{(k)}$ are not present in $\bar{H}_c$):

$$\bar{H}_c(q,p) = \bar{H}'_c(q,p) + \sum_k \bar{Q}^{\bar{A}=\bar{A}_k}_{(k)}(q,p)\,\bar{\chi}^{(1)}_{(k)\bar{A}_k}(q,p) + \sum_k \sum_{h=1}^{f_k-1} \bar{\mathcal{T}}^{(h)\bar{A}_k}_{(k)}(q,p)\,\bar{\chi}^{(h+1)}_{(k)\bar{A}_k}(q,p). \tag{9.2}$$

For instance, this is what happens in field theories like electromagnetism, Yang–Mills theory, and metric gravity, which have the secondary first-class constraints already present in the canonical Hamiltonian density, with the primary gauge variables in front.

Therefore, Dirac's proposal (Eq. 9.1) is already fulfilled for this class of singular Lagrangians, but with the gauge parameters of the on-shell Hamiltonian gauge group replacing those of the off-shell group present in the extended Hamiltonian.

In Refs. [485–487] it is shown that also the secondary, tertiary, etc. second-class constraints of this class of singular Lagrangians are present in the canonical Hamiltonian $\bar{H}_c$ in the form of quadratic combinations. This is clear if we use the form $\bar{\chi}^{(h)'}_{(k)A'_k}$ of the second-class constraints given in theorem 3. Since these constraints are generated by the equations $\{\bar{\chi}^{(h)'}_{(k)A'_k}, \bar{H}^{(F)}_D\} = \bar{\chi}^{(h+1)'}_{(k)A'_k}$, the final form of $\bar{H}_c$ implying these results will be

$$\begin{aligned}\bar{H}_c(q,p) = \bar{H}_d(q,p) &+ \sum_k \Big(\bar{Q}^{\bar{A}=\bar{A}_k}_{(k)}(q,p)\,\bar{\chi}^{(1)}_{(k)\bar{A}_k}(q,p) \\ &+ \sum_{h=1}^{f_k-1} \bar{\mathcal{T}}^{(h)\bar{A}_k}_{(k)}(q,p)\,\bar{\chi}^{(h+1)}_{(k)\bar{A}_k}(q,p)\Big) \\ &+ \sum_{k,h_1,h_2} \bar{S}^{(h_1h_2)A'_kB'_k}_{(k)}(q,p)\,\bar{\chi}^{(h_1)'}_{(k)A'_k}(q,p)\,\bar{\chi}^{(h_2)'}_{(k)B'_k}(q,p),\end{aligned} \tag{9.3}$$

with suitable functions $\bar{S}^{(h_1h_2)A'_kB'_k}_{(k)}$ consistent with the pattern of Poisson brackets of the second-class constraints given in theorem 3. In Eq. (9.3) $\bar{H}_d$ is the real first-class deterministic Hamiltonian generating a mapping among the gauge orbits.

See the bibliographies of Refs. [485–487, 536, 537] for the attempts to prove Dirac's conjecture and the use of the extended Hamiltonian. However, in many of these papers one uses singular Lagrangians with a singular Hessian matrix with non-constant rank.

We have found that the generators $\bar{G}_{\mathcal{A}}$ of the maximal set of Hamiltonian gauge transformations are all the first-class constraints $\bar{\Phi}_{(1)\mathcal{A}}$, $\mathcal{A} = 1, \ldots, M$. Therefore, as already said, from the $2n$ original canonical variables q^i, p_i we can form three groups of functions:

1. $2s_2$ functions which represent the non-essential degrees of freedom eliminable by going to Dirac brackets with respect to the second-class constraints $\bar{\Phi}_{(2)\mathcal{A}'} \approx 0$, $\mathcal{A}' = 1, \ldots, 2s_2$. They do not determine the physical state of the system, but only restrict the allowed region of T^*Q to the second-class $2(n - s_2)$-dimensional sub-manifold $\gamma_{(2)}$, on which the symplectic structure is given by the Dirac brackets, by eliminating degrees of freedom with trivial first-order dynamics. The Dirac Hamiltonian on $\gamma_{(2)}$ is $\bar{H}_D|_{\gamma(2)} = \bar{H}_d +$ (terms in the first-class constraints).
2. The non-deterministic M primary and secondary gauge variables, which also do not determine the physical state. Together with the first-class constraints $\bar{\Phi}_{(1)\mathcal{A}} \approx 0$, $\mathcal{A} = 1, \ldots, M$, they form a set of $2M$ functions not carrying dynamical information (except maybe a topological one). These constraints determine the final $[2(n - s_2) - M]$-dimensional sub-manifold $\bar{\gamma} \subset \gamma_{(2)} \subset T^*Q$. This sub-manifold, which can be odd-dimensional, does not admit a symplectic structure (no uniquely defined Poisson brackets exist for the functions on $\bar{\gamma}$), and is called a presymplectic (or co-isotropic) manifold co-isotropically embedded in T^*Q [459–461]. When suitable mathematical requirements are satisfied, it can be shown that the sub-manifold $\bar{\gamma}$ is foliated by M-dimensional diffeomorphic leaves, the Hamiltonian gauge orbits.
3. $2(n - M - s_2)$ independent functions $\bar{F}_\alpha(q, p)$ with (in general weakly) zero Poisson brackets with all the constraints, $\{\bar{F}_\alpha, \bar{\Phi}_{(1)\mathcal{A}}\} \approx 0$, $\{\bar{F}_\alpha, \bar{\Phi}_{(2)\mathcal{A}'}\} \approx 0$. They are the gauge-invariant classical DOs that parametrize the physical states of the system. The DOs are those functions on $\bar{\gamma}$ that are constant on the gauge orbits. One DO is the deterministic part $\bar{H}_d$ of the canonical Hamiltonian $\bar{H}_c$. If $\bar{F}$, $\bar{G}$ are DOs, then the Jacobi identity implies that also $\{\bar{F}, \bar{G}\}$ is a DO. Usually one eliminates the second-class constraints by introducing the associated Dirac brackets $\{.,.\}^*$ and considering $\bar{\gamma}$ a sub-manifold of the second-class sub-manifold $\gamma_{(2)}$.

In the case that the constraint sub-manifold $\bar{\gamma}$ is foliated by M-dimensional diffeomorphic Hamiltonian gauge orbits (nice foliation), we can go to the quotient with respect to the foliation and define the reduced phase space $\bar{\gamma}_R$, which will be a manifold if the projection $\pi \colon \bar{\gamma} \mapsto \bar{\gamma}_R$ is a submersion, but in general not a co-tangent bundle T^*Q_R over some reduced configuration space Q_R. In the nice case the reduced phase space is a symplectic manifold with a closed symplectic two-form and the Hamiltonian in $\bar{\gamma}_R$ is the deterministic part $\bar{H}_d$ of the canonical Hamiltonian and the Hamilton equations for the abstract DOs are $\frac{d\bar{F}_R}{dt} \overset{\circ}{=} \{\bar{F}_R, \bar{H}_{dR}\}_R = \{\bar{F}, \bar{H}_d\}^*$.

In general, things may be much more complicated: There can be non-diffeomorphic gauge orbits, there can be singular points in $\bar{\gamma}$, $\pi \colon \bar{\gamma} \mapsto \bar{\gamma}_R$ may not be a submersion. For more details, see Ref. [26].

To avoid technical problems with the definition of the reduced phase space $\bar{\gamma}_R$, usually we try to build a copy of it by adding M gauge fixing constraints $\bar{\rho}_{\mathcal{A}}(q,p) \approx 0$, $\mathcal{A} = 1, \ldots, M$, to eliminate the gauge freedom by choosing a definite gauge. The constraints $\bar{\rho}_{\mathcal{A}}(q,p) \approx 0$, $\bar{\Phi}_{(1)\mathcal{A}}(q,p) \approx 0$ must form a second-class set and we can define their Dirac brackets. Locally, the hyper-surface $\bar{\rho}_{\mathcal{A}}(q,p) \approx 0$ in T^*Q should intersect each Hamiltonian gauge orbit in $\bar{\gamma}$ in one and only one point (modulo global problems like the Gribov ambiguity in Yang–Mills theory [26]). The procedure for introducing the gauge fixing constraints has been delineated in Ref. [538].

Let us remark that due to the difficulties in trying to quantize the Dirac brackets after the elimination of an arbitrary set of second-class constraints, there have been some attempts to redefine the theory in such a way that only first-class constraints are present. The gauge fixings to the gauge freedom associated with these new constraints reproduce the original theory with its second-class constraints. The method [539–542] (see also exercise 1.22 of Ref. [26]) requires an enlarged phase space with as many new pairs of canonical variables as pairs of second-class constraints. The second-class constraints are transformed into first-class ones by inserting a suitable dependence on the new canonical variables. This method has a great degree of arbitrariness, modifies the theory off-shell and, having new gauge invariances, has to redefine the canonical Hamiltonian and the observables.

9.1 Shanmugadhasan Canonical Bases Adapted to the Constraints and the Dirac Observables

We have defined a DO as a first-class function on phase space restricted to the constraint sub-manifold $\bar{\gamma}$: This means that it must have weakly zero Poisson bracket with all the first- and second-class constraints and that, as a consequence, it is constant on the gauge orbits, namely that it is associated with a function on the reduced phase space. Since DOs describe the dynamical content of a singular dynamical system, it is important to find an algorithm for the determination of a canonical basis of them to be able to visualize such a content. This would allow determination of all possible DOs of the singular system and would open the path to the attempt to quantize only the dynamical degrees of freedom of the system as an alternative to Dirac quantization with subsequent reduction at the quantum level (see, for instance, the BRST observables of the BRST quantization [26]).

This strategy is possible due to the class of canonical transformations discovered by Shanmugadhasan [543], studying the reduction to normal form of a canonical differential system like the EL equations of singular Lagrangians. By using the Lie theory of function groups [544], Shanmugadhasan showed that in each neighborhood in T^*Q of a point of the constraint sub-manifold $\bar{\gamma}$ there exists local Darboux bases, whose restriction to $\bar{\gamma}$ allows one to separate the gauge variables from a local Darboux basis of DOs. These canonical transformations are implicitly used by Faddeev and Popov to define the measure of the phase space path integral and produce a trivialization of the BRST approach. See also Refs. [545, 546].

Given a $2n$-dimensional phase space T^*Q, the set G of all the functions $\phi(\bar{F}_a)$ of r independent functions $\bar{F}_1(q,p), \ldots, \bar{F}_r(q,p)$ (the basis of G) such that $\{\bar{F}_a, \bar{F}_b\} = \phi(\bar{F}_c)$ $(a, b, c = 1, \ldots, r)$ is said to be a function group of rank r. If $\phi_1, \phi_2 \in G$, then $\{\phi_1, \phi_2\} \in G$. When $\{\bar{F}_a, \bar{F}_b\} = 0$ for the values of a and b, the function group G is said to be commutative. A subset of G which forms a function group is a subgroup of G. If two function groups G_1, G_2 of rank r have p independent functions in common, they are the basis of a subgroup of both G_1 and G_2. A function $\phi \in G$ is said to be singular if it has zero Poisson bracket with all the functions of G; the independent singular functions of G form a subgroup. Given a function group G of rank r, it can be shown that the system of partial differential equations $\{\bar{g}, \bar{F}_a\} = 0$, $a = 1, \ldots, r$, admits $2n - r$ independent functions g_k, $k = 1, \ldots, 2n - r$, as solutions and they define a reciprocal function group G^r of rank $2n - r$. The basis functions of G and G^r are in involution under Poisson brackets.

The following two theorems on function groups and involutory systems [544] are the basis of Shanmugadhasan theory:

1. For a non-commutative function group G of rank r there exists a canonical basis $\bar{\phi}_1, \ldots, \bar{\phi}_{m+q}, \bar{\psi}_1, \ldots, \bar{\psi}_m$ with $2m + q = r$ such that

$$\{\bar{\phi}_\lambda, \bar{\phi}_\mu\} = \{\bar{\psi}_\alpha, \bar{\psi}_\beta\} = 0, \qquad \{\bar{\phi}_\alpha, \bar{\psi}_\lambda\} = \delta_{\alpha\lambda},$$
$$\alpha, \beta = 1, \ldots, m, \ \lambda, \mu = 1, \ldots, m + q. \tag{9.4}$$

As a corollary, a non-commutative function group G of rank r is a subgroup of a function group of rank $2n$, whose basis $\bar{\phi}_1, \ldots, \bar{\phi}_n, \bar{\psi}_1, \ldots, \bar{\psi}_n$ can be chosen so that

$$\{\bar{\phi}_i, \bar{\phi}_j\} = \{\bar{\psi}_i, \bar{\psi}_j\} = 0, \qquad \{\bar{\phi}_i, \bar{\psi}_j\} = \delta_{ij}, \quad i, j = 1, \ldots, n. \tag{9.5}$$

2. A system of $2m + q$ independent equations (defining a surface $\bar{\gamma}$ of dimension $2(n - m) - q$ in T^*Q),

$$\bar{\Omega}_a(q,p) = 0, \qquad a = 1, \ldots, 2m + q, \tag{9.6}$$

such that $rank\,\{\bar\Omega_a, \bar\Omega_b\} = 2m$, can be substituted by a locally equivalent system,

$$\bar\phi_\lambda(q,p) = 0, \qquad \lambda = 1,\ldots,m+q, \qquad\qquad \bar\psi_\alpha(q,p) = 0, \qquad \alpha = 1,\ldots,m, \tag{9.7}$$

for which the relations

$$\{\bar\phi_\lambda, \bar\phi_\mu\} = \{\bar\psi_\alpha, \bar\psi_\beta\} = 0, \qquad \{\bar\psi_\alpha, \bar\phi_\lambda\} = \delta_{\alpha\lambda}, \tag{9.8}$$

hold locally in T^*Q. Therefore, Eq. (9.6) is equivalent to the vanishing of the canonical basis of a non-commutative function group of rank $2m+q$.

Let us consider a dynamical system with an n-dimensional configuration space Q described by a singular Lagrangian, whose associated Hamiltonian description contains: (1) a set of first-class constraints $\bar\Phi_{(1)\mathcal{A}}(q,p) \approx 0$, $\mathcal{A} = 1,\ldots,M$, among which the primary ones are $\bar\phi_{(o)\bar A}(q,p) \approx 0$, $\bar A = 1,\ldots,m$; (2) a set of second-class constraints $\bar\Phi_{(2)\mathcal{A}'}(q,p) \approx 0$, $\mathcal{A}' = 1,\ldots,2s_2$; and (3) a final Dirac Hamiltonian $\bar H_D^{(F)} = \bar H_c^{(F)}(q,p) + \sum_{\bar a} \lambda^{(o)\bar A}(t)\, \bar\phi_{\bar A}(q,p)$. In the $2n$-dimensional phase space T^*Q the dynamics is restricted to the final $[2(n-s_2)-M]$-dimensional constraint sub-manifold $\bar\gamma \subset \ldots \subset \gamma \subset T^*Q$ (γ is the primary sub-manifold; if there are only first-class constraints $\bar\gamma$ is a presymplectic manifold; however, the term presymplectic manifold is often used to denote a generic $\bar\gamma$), whose closed degenerate 2-form $\bar\omega_{\bar\gamma}$ has $dimension\, ker\, \bar\omega_{\bar\gamma} = M$. We have $\bar\gamma \subset \gamma_{(2)} \subset T^*Q$, where $\gamma_{(2)}$ is the $2(n-s_2)$-dimensional sub-manifold defined by the second-class constraints with the symplectic 2-form $\bar\omega_{(2)}$ giving rise to the Dirac brackets.

Let the constraints form a function group of rank $2s_2+m$.

Theorem 2 ensures that in every neighborhood in T^*Q of a point of $\bar\gamma$ there exists a passive canonical transformation $(q^i, p_i) \mapsto (Q^i, P_i)$ in T^*Q such that in the new canonical basis the neighborhood is identified by the new constraints:

$$P_{\mathcal{A}} \approx 0, \qquad \mathcal{A} = 1,\ldots,m, \qquad\qquad Q^{a'} \approx 0, \quad P_{a'} \approx 0, \qquad a' = 1,\ldots,s_2. \tag{9.9}$$

Therefore, locally we obtain an Abelianization of first-class constraints and a canonical form of the second-class constraints $[Q^{a'} = \bar b^{a'}_{\mathcal{A}'}\, \bar\Phi_{(2)\mathcal{A}'},\ P_{a'} = \bar c_{a'\mathcal{A}'}\, \bar\Phi_{(2)\mathcal{A}'}]$ associated with this Abelianization. Eq. (9.9) gives the canonical form of a function group of rank $2s_2+m$. Due to theorem 1 the reciprocal function group of rank $2(n-s_2)-m$ has a basis formed by (1) m Abelianized gauge variables $Q^{\mathcal{A}}$ parametrizing the m-dimensional gauge orbits in $\bar\gamma$; and (2) a canonical basis of DOs associated with the Abelianization described by the $n-m-s_2$ pairs of canonical variables Q^a, P_a, $a = 1,\ldots,n-m-s_2$, which have zero Poisson brackets with the constraints in the form (Eq. 9.9) by construction. As a consequence, they have weakly zero Poisson brackets with all the original constraints, so that they are gauge-invariant. This is a local Darboux basis for the presymplectic sub-manifold $\bar\gamma$.

Let us remark that the T^*Q Poisson bracket $\{.,.\}_{Q,P}$ coincides with the Dirac bracket $\{.,.\}^*_{\gamma(2)}$ when restricted to the second-class sub-manifold $\gamma_{(2)}$. Therefore, it can be shown [545, 546] that the new Dirac Hamiltonian $\bar{H}_D^{(F)'}$ $[p_i\, dq^i - \bar{H}_D^{(F)}\, dt = P_i\, dQ^i - \bar{H}_D^{(F)'}\, dt - dF]$ is the first-class function

$$
\begin{aligned}
\bar{H}_D^{(F)'} &= \bar{H}_c^{(F)'}(Q,P) + \sum_{\bar{a}} \lambda^{(o)\bar{A}}(t)\, \bar{d}_{\bar{A}\mathcal{B}}\, P_{\mathcal{B}},\\
\bar{H}_c^{(F)'}(Q,P) &= \bar{K}_c^{(F)}(Q,P) - \tilde{\bar{\Phi}}_{(2)\mathcal{A}'}(Q,P)\, \tilde{\bar{c}}_{\mathcal{A}'\mathcal{B}'}(Q,P)\, \{\tilde{\bar{\Phi}}_{(2)\mathcal{B}'}(Q,P), \bar{K}_c^{(F)}(Q,P)\},\\
&\tilde{\bar{\Phi}}_{(2)\mathcal{A}'}(Q,P) = \bar{\Phi}_{(2)\mathcal{A}'}(q(Q,P), p(Q,P)),\\
&\tilde{\bar{c}}_{\mathcal{A}'\mathcal{C}'}\, \{\tilde{\bar{\Phi}}_{(2)\mathcal{C}'}, \tilde{\bar{\Phi}}_{(2)\mathcal{B}'}\} = \delta_{\mathcal{A}'\mathcal{B}'},\\
&\{\bar{K}_c^{(F)}, P_{\mathcal{A}}\} = \{\bar{K}_c^{(F)}, Q^{a'}\} = \{\bar{K}_c^{(F)}, P_{a'}\} = 0,
\end{aligned}
\tag{9.10}
$$

since we have $\bar{\Phi}_{(1)\mathcal{A}} = \bar{d}_{\mathcal{AB}}\, P_{\mathcal{B}}+$ (terms quadratic in the second-class constraints) $\equiv \bar{d}_{\mathcal{AB}}\, P_{\mathcal{B}}$ near $\bar{\gamma}$.

When we are able to solve the first-class constraints $\bar{\Phi}_{(1)\mathcal{A}}(q,p) \approx 0$ in a subset $p_{\mathcal{A}}$ of the momenta, as already said, a possible Abelianized form of the first-class constraints is

$$
P_{\mathcal{A}} = p_{\mathcal{A}} - \bar{\psi}_{\mathcal{A}}(q^i, p_{i\neq\mathcal{B}}) \approx 0. \tag{9.11}
$$

Therefore, with the Shanmugadhasan canonical transformations we are able to separate the gauge degrees of freedom (either non-essential variables or, in reparametrization-invariant theories, variables describing the generalized inertial effects [143–146]) from the physical ones at least locally in suitable open sets of T^*Q intersecting the constraint sub-manifold $\bar{\gamma}$.

Moreover, we see which kind of freedom we have in the choice of the functional form of the primary constraints: at least locally we can always make a choice ensuring the complete diagonalization of chains previously discussed. In a (in general local) Shanmugadhasan basis we have

$$
\begin{aligned}
&\{P_{\mathcal{A}}, P_{\mathcal{B}}\} = \{P_{\mathcal{A}}, Q^{a'}\} = \{P_{\mathcal{A}}, P_{a'}\} = 0,\\
&\{Q^{a'}, P_{b'}\} = \delta^{a'}_{b'}, \qquad \{Q^{a'}, Q^{b'}\} = \{P_{a'}, P_{b'}\} = 0,\\
&\{\bar{H}_d, P_{\mathcal{A}}\} = \{\bar{H}_d, Q^{a'}\} = \{\bar{H}_d, P_{a'}\} = 0.
\end{aligned}
\tag{9.12}
$$

An open fundamental problem is the determination of those singular systems that admit a subgroup of Shanmugadhasan canonical transformations globally defined in a neighborhood of the whole constraint sub-manifold $\bar{\gamma}$. When this class of transformations exist, we have a family of privileged canonical bases in which the constraint sub-manifold becomes the direct product of the reduced phase space $\bar{\gamma}_R$ (in the simplest case $\bar{\gamma}_R = T^*Q_R$ for some reduced configuration space Q_R) by a manifold Γ diffeomorphic to the gauge orbits, $\bar{\gamma} = \bar{\gamma}_R \times \Gamma$. When $\bar{\gamma}$ is a stratified sub-manifold, namely it is the disjoint union of different strata $\bar{\gamma}_a$, each one with different standard gauge orbit Γ_a (this may happen if the Hessian matrix has variable rank), the same result may be valid for each stratum, i.e.,

$\bar{\gamma}_a = \bar{\gamma}_{Ra} \times \Gamma_a$. The existence of privileged canonical bases is a phenomenon induced by the direct product structure and is similar to the existence of special coordinate systems for the separation of variables admitted by special partial differential equations.

In general, the topological structure of the original configuration space Q and/or of the constraint sub-manifold $\bar{\gamma} \subset T^*Q$ will not allow the existence of this privileged class of canonical transformations. For instance, this usually happens when the original configuration space Q is a compact manifold. In these cases the only way to study the constraint sub-manifold is to use the classical BRST cohomological method [26]. However, even in this case it is interesting to extrapolate and define new singular dynamical systems with this direct product structure from the local results for the original systems. The study of these new models can give an idea of the non-topological part of the dynamics of the original systems.

Moreover, special relativity (SR) induces a stratification of the constraint sub-manifold of relativistic singular systems (all having the Poincaré group as the kinematical global Noether symmetry group) according to the types of Poincaré orbit existing for the allowed configurations of the singular isolated system. Again, each Poincaré stratum has to be studied separately to see whether it admits the direct product structure. When such a structure is present, the privileged canonical bases have to be further restricted by selecting the ones whose coordinates are also adapted to the Poincaré group.

9.2 The Null Eigenvalues of the Hessian Matrix of Singular Lagrangians and Degenerate Cases

Having described the first-order Hamiltonian formalism for singular systems, let us come back to the second-order formalism based on the singular Lagrangian $L(q, \dot{q})$ and its EL equations. The singular nature of the system is associated with the $m \leq n$ null eigenvalues of the $n \times n$ Hessian matrix $A_{ij}(q, \dot{q})$, $\det\Big(A_{ij}(q, \dot{q})\Big) = 0$. This is the source of the m primary constraints $\bar{\phi}_A(q, p) \approx 0$, $A = 1, \dots, m$, when $rank\Big(A_{ij}(q, \dot{q})\Big) = n - m =$ const. everywhere in the $(q, \dot{q})$ space.

Since we have $\phi_A(q, \mathcal{P}(q, \dot{q})) = \bar{\phi}_A(q, p)|_{p=\mathcal{P}(q,\dot{q})} \equiv 0$, we get the identity

$$0 \equiv \frac{\partial}{\partial \dot{q}^i}\, \phi_A(q, \mathcal{P}(q, \dot{q})) = A_{ij}(q, \dot{q})\, \frac{\partial \bar{\phi}_A}{\partial p_j}|_{p=\mathcal{P}(q,\dot{q})}. \tag{9.13}$$

This means that $\frac{\partial \bar{\phi}_A}{\partial p_i}|_{p=\mathcal{P}(q,\dot{q})}$ is a (non-normalized) null eigenvector of the Hessian matrix and that, when the primary constraints are irreducible, each choice of their functional form generates a different basis of m (non-normalized) null eigenvectors for the m-dimensional null eigenspace of $A_{ij}(q, \dot{q})$.

If we saturate the EL equations with the null eigenvectors $\frac{\partial\bar{\phi}_A}{\partial p_i}|_{p=\mathcal{P}(q,\dot{q})}$ we get (some of these equations may be void, $0=0$)

$$\bar{\chi}_A(q,\dot{q}) = \frac{\partial\bar{\phi}_A}{\partial p_i}|_{p=\mathcal{P}(q,\dot{q})}\, L_i(q,\dot{q}) \equiv \frac{\partial\bar{\phi}_A}{\partial p_i}|_{p=\mathcal{P}(q,\dot{q})}\, \alpha_i(q,\dot{q}) \overset{\circ}{=} 0. \tag{9.14}$$

In the singular case the EL equations are an autonomous system of ordinary differential equations, which cannot be put in normal form and which, in general, contain equations of the second, first, and zeroth order, as shown by non-void equations (Eq. 9.14). The first-order EL equations are then divided into two groups according to whether they either are or are not projectable to phase space:

1. The zeroth-order EL equations are those non-void equations (Eq. 9.14) that depend only on the configuration coordinates q^i. They are holonomic Lagrangian constraints. Since we always include among the configuration variables q^i eventual (linear or non-linear) Lagrange multipliers, these Lagrangian constraints will appear as secondary Hamiltonian constraints $\bar{\chi}^{(1)}(q) \approx 0$ in T^*Q (the primary Hamiltonian constraint being given by the vanishing of the canonical momentum of the Lagrange multiplier).
2. The non-projectable first-order EL equations contained in Eq. (9.14) are the genuine first-order equations of motion. They are also called the primary SODE conditions. By using the extended Legendre transformation ($g^A_{(\lambda)}(q,\dot{q}) \mapsto \lambda^A(t)$ for the canonical form of the velocity functions) we get that their Hamiltonian version depends on the Dirac multipliers: In T^*Q these equations are recovered from the kinematical half of the Hamilton–Dirac equations on the final constraint sub-manifold $\bar{\gamma}$. These genuine first-order equations of motion determine the non-projectable primary (and by induction also the non-primary) velocity functions associated with the Hamiltonian second-class constraints (see the extended second Noether theorem in the next section). Actually, they are the counterpart in the second-order formalism of those Hamiltonian equations, like Eq. (8.13), which determine the Dirac multipliers associated with the primary second-class constraints (and therefore they determine the velocity functions in canonical form). From theorems 2 and 3 we deduce that the primary SODE conditions correspond to the determination of the Dirac multipliers of pairs of second-class 0-chains, while the higher SODE conditions correspond to the determination of the Dirac multipliers for all the second-class 1-, 2-, etc. chains.
3. The projectable first-order EL equations among Eq. (9.14) are those non-holonomic (also said to be an-holonomic or integrable) Lagrangian constraints that are projected to the secondary Hamiltonian constraints $\bar{\chi}^{(1)}_{a_1}(q,p)$ in T^*Q.
4. The remaining combinations of the EL equations, which depend on the accelerations $\ddot{q}^i$ in an essential way, are the genuine second order equations of motion.

As shown in Ref. [484] (where all the examples quoted in the reported bibliography are analyzed and clarified), when the Hessian matrix does not have a constant rank, many types of pathologies may appear. To control them the basic point is to look for a Hamiltonian formulation of these systems implying that the EL equations and the Hamilton equations have the same solutions.

The main pathologies are as follows:

1. *Third and fourth class-constraints.* These new types of non-primary constraints are at the basis of the failure of Dirac's conjecture for many singular Lagrangians with a Hessian matrix of variable rank. This happens because these constraints $\bar{\chi}(q,p) \approx 0$ look like first-class constraints. Their associated Hamiltonian vector fields $\bar{X}_{\bar{\chi}} = \{., \bar{\chi}\}$ are either first class, namely tangent to the constraint sub-manifold $\bar{\gamma}$, or vanishing on $\bar{\gamma}$ (but not near $\bar{\gamma}$). However, they are not generators of Hamiltonian gauge transformations. Instead, in general they generate spurious solutions of the Jacobi equations, which are not deviations between two neighboring solutions of the EL equations, due to the linearization instability present in these singular systems. As an example, consider a non-primary constraint $p_1 \approx 0$: (1) if it is first class, its conjugate variable q^1 is a gauge variable; (2) if it is second class there is another constraint determining q^1, so that the pair q^1, p_1 can be eliminated; (3) if it is third class, the conjugate variable q^1 is determined by one combination of the final Hamilton–Dirac equations and depends on the initial data.

   Instead, a fourth class or ineffective constraint is a non-primary constraint $\bar{\chi}(q,p) \approx 0$ generated inside a chain by the Dirac algorithm such that $d\bar{\chi}|_{\bar{\chi}=0} = 0$ even if all the other constraints in the chain have a non-vanishing differential. These constraints have weakly vanishing Poisson brackets with every function on T^*Q, so that they are first-class quantities. For the sake of simplicity, let us consider $p_1^2 \approx 0$ as a non-primary constraint of this type. For the determination of the constraint sub-manifold $\bar{\gamma}$ we must use its linearized form $p_1 \approx 0$. But we cannot use this linearized form in the final Dirac Hamiltonian generating the final Hamilton–Dirac equations (as instead is usually done), because otherwise the solutions of the Hamilton–Dirac and EL equations do not coincide.
2. *Proliferation of constraints and ramification of chains of constraints.* Let us consider the chains of constraints. If in a chain one gets a constraint like $q^1 q^2 \approx 0$ (this is possible only if the Hessian rank is not constant), then the chain gives rise to three distinct chains (ramification of chains) because the constraint gives rise to the following three sectors: (1) $q^1 \approx 0$, $q^2 \approx 0$ (proliferation of constraints); (2) $q^1 \approx 0$, $q^2 \neq 0$; (3) $q^2 \approx 0$, $q^1 \neq 0$.
3. *Joining of chains of constraints.* In certain examples after some steps after a ramification of chains there could be a joining of two of the new chains.

Look at Ref. [484] for all the examples of these pathologies and for what is known in mathematical physics on singular systems. Even if we discard all the pathological cases with Hessians of constant rank, there is not a consistent formulation of singular systems covering the second-order formalism, the tangent space one, and the co-tangent Hamiltonian one.

9.3 The Second Noether Theorem

The second Noether theorem states that if the action functional $S = \int dtL$ is quasi-invariant (i.e., its variation is a total time derivative) with respect to an infinite continuous group $G_{\infty r}$, involving up to order k derivatives (i.e., a group whose general transformations depend upon r essential arbitrary functions $\epsilon^a(t)$ and their first k time derivatives), then r identities exist among the EL equations L_i and their time derivatives up to order k. Under appropriate hypotheses the converse is also true.

This version of the theorem is oriented to the description of gauge theories and general relativity (GR), in which there is a singular Lagrangian invariant under local gauge transformations and/or space-time diffeomorphisms and giving rise only to first-class constraints at the Hamiltonian level (see, for instance, Ref. [474]). This is unsatisfactory, because at the Lagrangian level the fundamental property of singular Lagrangians is the number of null eigenvalues of the Hessian matrix and some of them may be associated with Hamiltonian second-class constraints (when present). As a consequence, in the literature there is no clear statement about the connection between the second Noether theorem and the canonical transformations generated by the constraints when some of them are second class [27]. An extension of the second Noether theorem is needed to include these cases. This was done in Ref. [547] along the lines of the Candotti–Palmieri–Vitale extension [479] of the first Noether theorem using the concept of weak quasi-invariance in the case of a Hessian matrix of constant rank.

The extended second Noether theorem may be expressed by saying that the action functional associated with a singular Lagrangian is weakly quasi-invariant (i.e., quasi-invariant only after having used combinations of the EL equations and of their time derivatives which are independent of the accelerations) under as many sets of local infinitesimal Noether transformations as is the number of null eigenvalues of the Hessian matrix. Each set of such transformations $\delta_A q^i$ depends on an arbitrary function $\epsilon^A(t)$ and its time derivatives up to order J_A and produces an identity that can be resolved in $J_A + 2$ Noether identities, each one being the time derivative of the previous one.

While the $\delta_A q^i$ associated with chains of first-class constraints will turn out to be the pull-back by means of the inverse Legendre transformation of the infinitesimal Hamiltonian gauge canonical transformations, the $\delta_A q^i$ associated with chains of second-class constraints will turn out to be the pull-back of the infinitesimal canonical transformations generated by the second-class constraints (they could be named pseudo-gauge transformations).

This local formulation of the extended theorem, which is based on a form of the infinitesimal Noether transformations $\delta_A q^i$, is valid if we use the orthonormal eigenvectors ${}_A\hat{\xi}^i_o(q,\dot{q}) = \tau^i_A(q,\dot{q}) = \frac{\partial \hat{\tilde{\phi}}_A(q,p)}{\partial p_i}|_{p=\mathcal{P}(q,\dot{q})}$ of the Hessian matrix. Their use corresponds to the diagonalization of the chains of constraints and allows us to show that the $J_A + 2$ Noether identities of the form of Eq. (8.5) implied by the generalized weak quasi-invariance are projectable to phase space, where they rebuild the whole Dirac algorithm (each chain of identities is connected with a chain in theorems 2 and 3).

9.4 Constraints in Field Theory

Let us now consider some aspects of the constraints appearing in classical field theory.

The naive extension of the previous results to classical field theory does not present conceptual problems [20, 21, 25, 27, 548, 549]. Instead, a more rigorous treatment would require much more sophisticated techniques – see, for instance, Refs. [463–467] for an introduction to infinite dimensional Hamiltonian systems. The new real phenomenon of field theory with constraints is the possible appearance of the zero modes of the elliptic operators associated with some constraints (due to the spatial gradients of the fields and/or the canonical momenta). It depends on the choice of the function space for the fields, an argument on which there is no general consensus, and creates obstructions to the existence of global gauge fixing constraints like the Gribov ambiguity in Yang–Mills theory (its existence depends upon the choice of the function space [550]) and the problem of the Lagrangian gauge fixings relevant for the BRS approach [551].

Let us suppose that we have a singular Lagrangian density $\mathcal{L}(\varphi^r(x), \varphi^r_{,\mu}(x))$ depending on a set of fields $\varphi^r(x)$, $r = 1, \ldots, n$ and their first derivatives $\varphi^r_{,\mu}(x) = \partial_\mu \varphi^r(x)$. The space-time manifold M of dimension $m+1$ (usually the 4-dimensional Minkowski space-time) has local Cartesian coordinates $x^\mu = (x^o, x^i)$ and Lorentzian metric $\eta^{\mu\nu} = \epsilon\,(1; -1, \ldots, -1)$. The action is the local functional $S = \int d^{m+1}x\, \mathcal{L}$ and a certain class of boundary conditions at $|\vec{x}| \to \infty$ for the fields has been chosen in some function space dictated by physical considerations.

With the usual non-covariant choice of x^o as time variable we have the following definition of Hessian matrix:

$$A_{rs}(x) \stackrel{def}{=} A^{oo}_{rs}(\varphi^s(x), \varphi^s_{,\mu}(x)) = \frac{\partial^2 \mathcal{L}}{\partial \varphi^r_{,o}(x)\, \partial \varphi^s_{,o}(x)}, \qquad A^{\mu\nu}_{rs} = \frac{\partial^2 \mathcal{L}}{\partial \varphi^r_{,\mu}\, \partial \varphi^s_{,\nu}}. \tag{9.15}$$

The EL equations implied by Hamilton's principle $\delta S = \delta \int_\Omega d^{m+1}x\, \mathcal{L} = 0$ for arbitrary variations $\delta \varphi^r(x)$ vanishing on the boundary $\partial\Omega$ of the compact region Ω are

$$L_r(x) = \frac{\partial \mathcal{L}}{\partial \varphi^r(x)} - \partial_\mu \frac{\partial \mathcal{L}}{\partial \varphi^r_{,\mu}(x)} = -(A^{\mu\nu}_{rs}(x)\, \varphi^s_{,\nu}(x) - \alpha_r(x))$$
$$= -(A_{rs}(x)\, \varphi^s_{,oo}(x) - \hat{\alpha}_r(x)),$$
$$\alpha_r - \frac{\partial \mathcal{L}}{\partial \varphi^r} - \frac{\partial^2 \mathcal{L}}{\partial \varphi^r_{,\mu}\, \partial \varphi^s} \varphi^s_{,\mu}, \qquad \hat{\alpha}_r = \alpha_r - 2\, A^{oi}_{rs}\, \varphi^s_{,oi} - A^{ij}_{rs}\, \varphi^s_{,ij}. \tag{9.16}$$

The canonical momenta and the standard Poisson brackets (after a suitable definition of functional derivative) are

$$\pi_r(x) = \Pi^o_r(x), \qquad \Pi^\mu_r(x) = \frac{\partial \mathcal{L}}{\partial \varphi^r_{,\mu}(x)}, \qquad \{\varphi^r(x^o, \vec{x}), \pi_s(x^o, \vec{y})\} = \delta^r_s\, \delta^m(\vec{x} - \vec{y}). \tag{9.17}$$

Since the Poisson brackets are local (i.e., they do not depend on primitives of the delta function, a property named local commutativity in Ref. [26]), the Poisson bracket of two local functionals is also a local functional, so that non-local terms cannot be generated through the operation of taking the bracket. For two functions $\bar{F}_a(\varphi^r(x^o, \vec{x}), \varphi^r_{,i}(x^o, \vec{x}), \pi_r(x^o, \vec{x}), \pi_{r\,,i}(x^o, \vec{x}))$, $a = 1, 2$, we have

$$\{\bar{F}_1(x^o, \vec{x}), \bar{F}_2(x^o, \vec{y})\} = \int d^{m+1}z \left[\frac{\delta \bar{F}_1(x^o, \vec{x})}{\delta \varphi^r(x^o, \vec{z})} \frac{\delta \bar{F}_2(x^o, \vec{x})}{\delta \pi_r(x^o, \vec{z})} - \frac{\delta \bar{F}_1(x^o, \vec{x})}{\delta \pi_r(x^o, \vec{z})} \frac{\delta \bar{F}_2(x^o, \vec{x})}{\delta \varphi^r(x^o, \vec{z})}\right]. \tag{9.18}$$

Since $\det A_{rs}(x) = 0$, there will be a certain number of null eigenvalues of the Hessian matrix with associated local orthonormal null eigenvectors $\tau^r_A(\varphi^s(x), \varphi^s_{,\nu}(x))$, $A = 1, \dots, n_1$. For the sake of simplicity we shall assume regularity conditions such that the Hessian matrix has a constant rank everywhere.

As in the finite-dimensional case, there are n_1 arbitrary velocity functions non-projectable to phase space and n_1 primary constraints $\bar{\phi}_A(\varphi^r, \varphi^r_{,i}, \pi_r, \pi_{r,i}) \approx 0$ such that $\bar{\phi}_A(\varphi^r, \varphi^r_{,i}, \pi_r, \pi_{r,i})|_{\pi_r = P_r(\varphi, \varphi_{,\mu})} \equiv 0$. Again, we have that $\frac{\partial \bar{\phi}_A}{\partial \pi_r}$ are null eigenvalues of the Hessian matrix. For the sake of simplicity we assume that there is a global functional form of the constraints producing the orthonormal eigenvectors ${}_A\xi^i_o$.

If the Lagrangian density is sufficiently regular that the Legendre transformation is well defined, the canonical Hamiltonian density is $\bar{\mathcal{H}}_c(\varphi^r(x), \varphi^r_{,i}(x), \pi_r(x), \pi_{r,i}(x)) = \varphi^r_{,o}(x)\, \pi_r(x) - \mathcal{L}(\varphi(x), \varphi_{,\mu}(x))$, while the Dirac Hamiltonian is

$$\bar{H}_D = \int d^m x \left(\bar{\mathcal{H}}_c(x^o, \vec{x}) + \sum_A \lambda^A(x^o, \vec{x})\, \bar{\phi}_A(x^o, \vec{x})\right) = \bar{H}_c + \sum_A \bar{H}_A. \tag{9.19}$$

Only by choosing consistent boundary conditions for the fields and the Dirac multipliers can we interpret the $\bar{H}_A$ as generators of local Noether transformations. The Hamilton equations are

$$\bar{L}^r_{D\varphi} = \varphi^r_{,o}(x^o,\vec{x}) - \{\varphi^r(x^o,\vec{x}), \bar{H}_D\} \stackrel{\circ}{=} 0,$$
$$\bar{L}_{Dpr} = p_{r,o}(x^o,\vec{x}) - \{p_r(x^o,\vec{x}), \bar{H}_D\} \stackrel{\circ}{=} 0. \tag{9.20}$$

Having the primary constraints $\bar{\phi}_A(x^o,\vec{x}) \approx 0$ and the Dirac Hamiltonian, Dirac's algorithm is plainly extended to field theory starting with the study of the time constancy of the primary constraints. One arrives at a final constraint sub-manifold $\bar{\gamma}$, divides the final set of constraints in first- and second-class ones, and determines the Dirac Hamiltonian of $\bar{\gamma}$ as $\bar{H}_D = \bar{H}_c + \int d^m x \sum_{\bar{A}} \lambda^{\bar{A}}(x^o,\vec{x})\, \bar{\phi}_{\bar{A}}(x^o,\vec{x})$, where $\bar{\phi}_{\bar{A}}(x^o,\vec{x}) \approx 0$ are the first-class primary constraints and $\lambda^{\bar{A}}(x^o,\vec{x})$ the Dirac multipliers, equal to the primary arbitrary velocity functions $g^{\bar{A}}_{(\lambda)}(\varphi^r(x^o,\vec{x}), \varphi^r_{,\mu}(x^o,\vec{x}))$ through the first half of the Hamilton–Dirac equations.

However, besides regularity conditions on the singular Lagrangian density $\mathcal{L}$ so to avoid the (non-explored) field theory counterparts of the pathologies associated to Hessians with variable rank, one has to consider extra requirements peculiar to field theory:

1. The constraints must define a sub-manifold of the infinite-dimensional phase space, whose properties depend on the choice of the boundary conditions and the function space for the fields and their canonical momenta. This function space must include all physically interesting solutions of the Hamilton–Dirac equations. The constraints must not only be local functionals of the fields but must also be locally complete [26]. This means that every phase space function vanishing on $\bar{\gamma}$ is zero by virtue of the constraints defining $\bar{\gamma}$ and their spatial derivatives of any order only, without having to invoke the boundary conditions. To put mathematical control on BRST cohomology (theorem 12.4 in Ref. [26]), one needs strong regularity conditions implying that every function vanishing on $\bar{\gamma}$ can be written as combination of the a constraints and a finite arbitrary number of their spatial derivatives.
2. In Ref. [552] it is pointed out that in field theory each constraint $\bar{\phi}(x^o,\vec{x}) \approx 0$ represents a continuous and infinite number of constraints characterized by the space label $\vec{x}$, so that problems may arise with the theory of distributions. Subtle difficulties may appear in the division of the constraints in the first- and second-class groups and in the mathematical definition of Dirac brackets where the inverse of continuous matrices $C(x^o,\vec{x},\vec{y})$ are needed.

A related problem is with the gauge transformations generated by first-class constraints $\bar{\phi}(x^o,\vec{x}) \approx 0$. If we consider the most general generator $\bar{G} = \int d^m x\, \alpha(x^o,\vec{x})\, \bar{\phi}(x^o,\vec{x})$, is $\bar{G}$ a generator of gauge transformations for every parameter function $\alpha(x^o,\vec{x})$? In Ref. [553] the following distinction between proper and improper gauge transformations was given:

1. Proper gauge transformations represent true gauge symmetries of the theory and do not change the physical state of the system. They can be eliminated by fixing the gauge.

2. Improper gauge transformations (they do not exist for finite-dimensional systems) do change the physical state of the system, mapping (on-shell) one physical solution onto a different physical solution. They cannot be eliminated by fixing the gauge but only by means of super-selection rules selecting a particular set of solutions.

   Given the function space $\mathcal{F}$ for the fields, the problem is the determination of the allowed function space of the parameter functions $\alpha(x^o, \vec{x})$, so that $\bar{G}$ is the generator of a proper gauge transformation. In order to not over-count the constraints, the space $\mathcal{F}_d$ of the allowed $\alpha(x^o, \vec{x})$ (the so-called dual space) must be such that, when α varies in this space, $\bar{G} \approx 0$ has information equivalent to the original constraints $\bar{\phi}(x^o, \vec{x}) \approx 0$. If $\alpha(x^o, \vec{x})$ does not belong to the dual space, then $\bar{G}$ is the generator of an improper gauge transformation. Under both proper and improper gauge transformations a field belonging to $\mathcal{F}$ must be transformed in a field still in $\mathcal{F}$. Therefore, if $\varphi(x^o, \vec{x}) \in \mathcal{F}$ and $\pi(x^o, \vec{x}) \in \mathcal{F}$ are the fields, then $\delta\,\varphi(x^o, \vec{x}) = \{\varphi(x^o, \vec{x}), \bar{G}\} = \frac{\delta \bar{G}}{\delta\pi(x^o, \vec{x})}$ and $\delta\,\pi(x^o, \vec{x}) = \{\pi(x^o, \vec{x}), \bar{G}\} = \frac{\delta \bar{G}}{\delta\varphi(x^o, \vec{x})}$ must be in $\mathcal{F}$. Moreover, the functional derivatives must be well defined; namely, we must have $\delta\,\bar{G} = \int d^m x \left(\frac{\delta \bar{G}}{\delta\varphi(x^o, \vec{x})}\, \delta\,\varphi(x^o, \vec{x}) + \frac{\delta \bar{G}}{\delta\pi(x^o, \vec{x})}\, \delta\,\pi(x^o, \vec{x}) \right)$. In general, to get this result one has to do a number of integrations by parts and to check whether the resulting surface terms vanish: If they vanish we have a proper gauge transformation with $\alpha(x^o, \vec{x}) \in \mathcal{F}_d$. If the $\alpha(x^o, \vec{x})$ are such that the surface terms do not vanish, we have to modify the generator $\bar{G}$ by adding a surface term, $\bar{G} \mapsto \bar{G}' = \bar{G} + \bar{G}_{ST}$, whose variation $\delta\,\bar{G}_{ST}$ cancels the unwanted surface terms. In this case $\bar{G}'$ is the generator of an improper gauge transformation and $\bar{G}' \neq 0$ on $\bar{\gamma}$, where it becomes the constant surface term $\bar{G}_{ST}$ commuting with the Hamiltonian. This constant surface term is a non-trivial constant of the motion which can be fixed only with a super-selection rule.
3. In Ref. [554] it is pointed out that the study of the formal integrability of the partial differential Hamilton equations requires the use of prolongation methods in the infinite jet bundle (namely, we have to consider derivatives of the original equations till the needed order), and in particular the determination of a system of equations in involution. While Dirac's algorithm considers all possible consequences of taking the time derivatives of the Hamilton equations, it says nothing about their spatial derivatives. Therefore, in field theory one has to check whether extra integrability conditions appear by considering these spatial derivatives.

In the regular case the first Noether theorem implies the existence of conservation laws $\partial_\mu\, G^\mu(x) \overset{\circ}{=} 0$, so that with suitable boundary conditions on the fields, conserved charges $Q = \int d^m x\, G^o(x^o, \vec{x})$, $\frac{dQ}{dx^o} \overset{\circ}{=} 0$ are obtained.

In the singular case, by using the orthonormal eigenvectors of the Hessian matrix the extended second Noether theorem states that each null eigenvalue of this matrix is associated with a local Noether transformation $\delta_A\, x^\mu = 0$,

$\delta_A\,\varphi^r(x) = \epsilon^A(x)\ {}_A\xi^r_{J_A}(x) + \sum_{j=1}^{J_A} \epsilon^A_{,\mu_1\dots\mu_j}(x)\ {}_A\xi^{r\,(\mu_1\dots\mu_j)}_{J_A-j}(x)$ ($(\mu_1\dots\mu_j)$ means symmetrization in the indices), under which we get the following weak quasi-invariance:

$$\delta_A\,\mathcal{L} = \delta_A\,\varphi^r\,L_r + \partial_\mu\left(\frac{\partial\mathcal{L}}{\partial\varphi^r_{,\mu}}\,\delta_A\,\varphi^r\right) \equiv \partial_\mu\,F^\mu_A + \epsilon^A(x)\,D_A \stackrel{\circ}{=} \partial_\mu\,F^\mu_A. \tag{9.21}$$

Here, $F^\mu_A = F^\mu_A(\varphi, \varphi_{,\nu}, \epsilon^A)$ and $D_A(\varphi,\varphi_{,\nu})\stackrel{\circ}{=}0$ by using the acceleration-independent consequences of the EL equations. By posing

$$\begin{aligned} F^\mu_A(\varphi,\varphi_{,\nu},\epsilon^A) &= \epsilon^A(x)\ {}_AF^\mu_{J_A}(\varphi,\varphi_{,\nu}) + \sum_{j=1}^{J_A} \epsilon^A_{,\mu_!\dots\mu_j}(x)\ {}_AF^{\mu\,(\mu_1\dots\mu_j)}_{J_A-j}(\varphi,\varphi_{,\nu}),\\ G^\mu_A(\varphi,\varphi_{,\nu},\epsilon^A) &= \frac{\partial\mathcal{L}}{\partial\varphi^r_{,\mu}}\,\delta_A\,\varphi^r - F^\mu_A = \epsilon^A(x)\ {}_AG^\mu_{J_A}\\ &\quad + \sum_{j=1}^{J_A} \epsilon^A_{,\mu_!\dots\mu_j}(x)\ {}_AG^{\mu\,(\mu_1\dots\mu_j)}_{J_A-j}(\varphi,\varphi_{,\nu}),\\ {}_AG^{\mu\,(\mu_1\dots\mu_j)}_{J_A-j} &= \frac{\partial\mathcal{L}}{\partial\varphi^r_{,\mu}}\ {}_A\xi^{r\,(\mu_1\dots\mu_j)}_{J_A-j} - {}_AF^{\mu\,(\mu_1\dots\mu_j)}_{J_A-j}, \end{aligned}\tag{9.22}$$

we get the following Noether identities:

$$\partial_\mu\,G^\mu_A \equiv \epsilon^A(x)\,D_A - \delta_A\,\varphi^r\,L_r \stackrel{\circ}{=} 0, \tag{9.23}$$

which imply

$$\begin{aligned} {}_AG_o^{(\mu_1\,(\mu_2\dots\mu_{J_A+1}))} &\equiv 0,\\ \partial_\mu\ {}_AG_o^{\mu\,(\mu_1\dots\mu_{J_A})} &\equiv -{}_AG_1^{(\mu_1\,(\mu_2\dots\mu_{J_A}))} - {}_A\xi_o^{r\,(\mu_1\dots\mu_{J_A})}\ L_r,\\ &\dots\\ \partial_\mu\ {}_AG^{\mu\,(\mu_1\dots\mu_j)}_{J_A-j} &\equiv -{}_AG^{(\mu_1\,(\mu_2\dots\mu_j))}_{J_A-j+1} - {}_A\xi^{r\,(\mu_1\dots\mu_j)}_{J_A-j}\ L_r, \quad j=1,\dots,J_A-1,\\ &\dots\\ \partial_\mu\ {}_AG^\mu_{J_A} &\equiv D_A - {}_A\xi^r_{J_A}\ L_r. \end{aligned}\tag{9.24}$$

These equations imply the following form of the Noether identities:

$$\begin{aligned} \partial_{\mu_1}\dots\partial_{\mu_{J_A+1}}\ {}_AG_o^{\mu_1\,(\mu_2\dots\mu_{J_A+1}))} &\equiv 0,\\ &\dots\\ \partial_{\mu_1}\dots\partial_{\mu_{j+1}}\ {}_AG^{\mu_1\,(\mu_2\dots\mu_{j+1}))}_{J_A-j} &\equiv \sum_{h=0}^{J_A-j-1} (-)^{J_A-j-h}\,\partial_{\mu_1}\\ &\quad \dots\partial_{\mu_{J_A-h}}\left({}_A\xi_h^{r\,(\mu_1\dots\mu_{J_A-h})}\ L_r\right)\stackrel{\circ}{=}0,\\ &\quad \dots\dots \quad j=1,\dots,J_A, \end{aligned}$$

$$D_A \equiv \sum_{h=0}^{J_A} (-)^{J_A - h}\, \partial_{\mu 1} \ldots \partial_{\mu_{J_A - h}} \left({}_A\xi^{r\,(\mu_1 \ldots \mu_{J_A - h})}\, L_r \right) \stackrel{\circ}{=} 0. \tag{9.25}$$

When we have $D_A \equiv 0$ (quasi-invariance, first-class constraints), we get the contracted Bianchi identities:

$$\sum_{h=0}^{J_A} (-)^{J_A - h}\, \partial_{\mu_1} \ldots \partial_{\mu_{J_A - h}} \left({}_A\xi^{r\,(\mu_1 \ldots \mu_{J_A - h})}\, L_r \right) \equiv 0. \tag{9.26}$$

We have given a formulation of the theorem based only on the variation of the Lagrangian density. Usually, in the absence of second-class constraints, namely with $D_A(x) \equiv 0$, and with $\delta_A\, \varphi^r$ depending only on $\epsilon^A(x)$ and $\partial_\mu\, \epsilon^A(x)$, one considers the variation of the action evaluated on a compact region Ω of the m-dimensional space bounded by two hyper-planes (Σ_f at x^o_f and Σ_i at x^o_i; the variations are assumed to vanish on the spatial boundary) and asks for $\delta\, S = 0$. Then, the identity (Eq. 9.23) becomes $\int_\Omega d^{m+1}x\, \delta_A\, \varphi^r\, L_r \equiv \int_{\Sigma_i} d^m\sigma_\mu\, G^\mu_A(x^o_i, \vec{x}) - \int_{\Sigma_f} d^m\sigma_\mu\, G^\mu_A(x^o_f, \vec{x})$ ($d^m\sigma_\mu = d^m x\, n_\mu$ with n_μ outer normal to the hyper-plane); namely, it takes the form of a term in the interior of Ω equated to boundary terms on the hyper-planes. If we ask that the arbitrary functions $\epsilon^A(x)$ and their derivatives vanish on the boundary, $\delta_A\, S = 0$ (the original Noether statement) implies the vanishing of the interior term, $\delta_A\, \varphi^r\, L_r \equiv 0$, and this gives the contracted Bianchi identities. By combining this result with the identities, one can recover Utiyama [555, 556] and Trautman [557, 558] results, i.e., their form of the identities. For a detailed discussion of this point and of the connected interpretative ambiguities, see Refs. [559, 560].

If in the Noether identity (Eq. 9.23) we put $\epsilon^A(x) = $ const., this global subgroup of gauge transformations gives rise to the first Noether theorem associated with the Noether transformation $\delta_A\, \varphi^r = \epsilon^A\ {}_A\xi^r_{J_A}$ as a sub-case of the second theorem and to the Noether identity,

$$\partial_\mu\ {}_AG^\mu_{J_A} \equiv D_A - {}_A\xi^r_{J_A}\ L_r \stackrel{\circ}{=} 0. \tag{9.27}$$

Eq. (9.27) comprises the weak conservation laws, and the weak (so-called) improper conserved (Noether) current is ${}_AG^\mu_{J_A}$. Instead, it can be checked that the strong conservation laws $\partial_\mu\, V^\mu_A \equiv 0$ hold independently from the EL equations for the following strong improper conserved current (it is not a Noether current):

$$\begin{aligned} V^\mu_A &= {}_AG^\mu_{J_A} - \sum_{h=0}^{J_A - 1} (-)^{J_A - h}\, \partial_{\mu_1} \ldots \partial_{\mu_{J_A - h - 1}} \left({}_A\xi_h^{r\,(\mu\mu_1 \ldots \mu_{J_A - h - 1})}\, L_r \right) \\ &= \partial_\nu\, U^{[\mu\nu]}_A \stackrel{\circ}{=} {}_AG^\mu_{J_A}, \end{aligned} \tag{9.28}$$

where $U_A^{[\mu\nu]}$ ($[\mu\nu]$ means antisymmetrization) is the following super-potential:

$$U_A^{[\mu\nu]} = \sum_{h=1}^{J_A-1} (-)^{J_A-h}\, \partial_{\mu_1} \dots \partial_{\mu_{J_A-h-1}} \Big({}_A G_h^{(\nu\,(\mu\mu_1\dots\mu_{J_A-h-1}))} - {}_A G_h^{(\mu\,(\nu\mu_1\dots\mu_{J_A-h-1}))} \Big). \tag{9.29}$$

The improper strong $Q_A^{(S)}$ and weak $Q_A^{(W)}$ conserved charges ($\frac{dQ_A^{(W)}}{dx^o} \overset{\circ}{=} 0$, $\frac{dQ_A^{(S)}}{dx^o} \equiv 0$ for suitable boundary conditions) coincide on the solutions of the acceleration-independent EL equations (Ω is a spatial volume with boundary $\partial\,\Omega$):

$$\begin{aligned} Q_A^{(S)} &= \int_\Omega d^m x\, V_A^o(x^o, \vec{x}) = \int_{\partial\Omega} d^{m-1}\Sigma_k\, U_A^{[ok]}(x^o, \vec{x}) \overset{\circ}{=} Q_A^{(W)} \\ &= \int_\Omega d^m x\; {}_A G_{J_A}^o(x^o, \vec{x}). \end{aligned} \tag{9.30}$$

The source of the ambiguities [559–561] is this doubling of the conserved currents and charges, which does not exist with the first Noether theorem applied to global symmetries.

Finally, the generalized Trautman strong conservation laws are (differently from Eq. (9.23), they hold independent of the EL equations):

$$\partial_\mu \Big[G_A^\mu - \sum_{h=0}^{J_A-1} \epsilon^A_{,\mu_1\dots\mu_h}(x) \sum_{j=h}^{J_A} (-)^{j-h}\, \partial_{\mu_{h+1}} \dots \partial_{\mu_j} \Big({}_A\xi_{J_A-j}^{r\,(\mu\mu_1\dots\mu_j)}\, L_r \Big) \Big] \equiv 0, \tag{9.31}$$

where for $j = h$ the derivatives acting on the round bracket are absent.

Let us remark (see Ref. [518]) that when a field theory has global symmetry quasi-invariances, the first Noether theorem implies the existence of a current that is conserved by using the EL equations; it is the analogue of the weak current, while there is no analogue of the strong current, which exists only with gauge symmetries. This gives rise to a conserved charge (the analogue of the weak charge; there is no strong charge in the form of the flux through the surface at infinity of some vector field) and the possibility of a symmetry reduction of the order of the system of equations of motion. Instead, in the case of local gauge symmetries we get super-selection rules, not symmetry reduction.

The Noether identities (Eqs. (9.24) and (9.25)) have not yet been studied in detail, such as in the finite-dimensional case, because they contain a lot of information that is not really needed for the Hamiltonian treatment.

10

Concluding Remarks and Open Problems

The 3+1 approach described in this book allows describing non-inertial frames parametrized with radar 4-coordinates in special relativity (SR) and in a family of globally hyperbolic, asymptotically Minkowskian at spatial infinity, without super-translations Einstein space-times (like those in Ref. [8]) admitting a Hamiltonian description and with the existence of the asymptotic ADM Poincaré generators at spatial infinity (so elementary particles can be assumed to be in irreducible representations of the Poincaré group).

In SR all the matter and field systems admitting a Lagrangian description can be reformulated as parametrized Minkowski theories, whose gauge variables are the embeddings describing the nice foliations of Minkowski space-time with instantaneous 3-spaces. The transition from either inertial or non-inertial frame to another is a gauge transformation, not changing the physics but only modifying the inertial forces.

Among the inertial frames there is the family of inertial rest-frames, whose 3-spaces are orthogonal to the 4-momentum of the isolated system.

This allows us to give a solution to the endless problem of the definition of the relativistic center of mass, to decouple it (being a non-local, non-measurable quantity) and to reformulate the measurable physics in terms of Wigner-covariant relative variables. As a consequence, it is possible to define a consistent relativistic quantum mechanics of particles.

All particle systems, classical fields, fluids, and statistical mechanics can be reformulated in this framework.

Moreover, there is the family of non-inertial rest-frames with non-Euclidean 3-spaces, which tend to flat space-like hyper-planes orthogonal to the total 4-momentum of the isolated system at spatial infinity.

Every isolated system can be described as a decoupled collective non-covariant, non-local, non-measurable external center of mass (described by frozen Jacobi data) carrying a pole–dipole structure (the mass and the rest spin of the isolated system). After the choice of the covariant non-canonical

Fokker–Pryce center of inertia as the observer origin of the 3-coordinates in the 3-spaces and the elimination of the gauge variable describing the internal 3-center-of-mass inside the 3-spaces, the dynamics is described by the Wigner-covariant relative variables and there is an internal non-faithful realization of the Poincaré algebra. In this way, the endless problem of the elimination of relative times in both classical and quantum relativistic bound states is completely solved.

Since in all these models there are Dirac–Bergmann constraints, in the third part of the book there are complete descriptions of this theory, of the physical Dirac observables (DOs), and of the two Noether theorems.

In Einstein GR in the described family of space-times, the absence of super-translations implies that the allowed foliations are non-inertial rest-frames, whose 3-spaces are asytmptotically orthogonal to the ADM 4-momentum.

This allows us to give a parametrization of metric and tetrad gravity and to choose the embeddings of the non-inertial rest-frames in which the system gravitational field plus matter is visualized as a decoupled external center of mass and an internal 3-space containing only (not yet known) relative variables, as in SR.

In these space-times, standard distributions like the Dirac delta function are used [433]. However, the validity of this point of view has to be better verified.

While in SR Minkowski space-time is an absolute notion, in Einstein general relativity (GR) also the space-time is a dynamical object since its 4-metric is the dynamical field. In Refs. [271–274] there is an analysis of all the foundational problems of GR. It is shown that after the Shanmugadhasan canonical transformation identifying the DOs, the active diffeomorphisms of the space-time under which the Lagrangian of GR is invariant (this is the origin of statements like the non-physical objectivity of the points of the space-time) correspond to the pull-back of passive Hamiltonian on-shell (i.e., on the solutions of Einstein equations) gauge transformations. As a consequence, there is a physical individuation of the point-events of the space-time: (1) the fixation of the inertial gauge variables is a metrological statement about clocks and measurements producing a phenomenological identification of space-time; and (2) the dynamics (the tidal effects) are specified by the evolution of the DOs in the chosen scenario for the space-time points.

The real problem is that we do not know the real DOs of GR, because we are not able to solve the super-Hamiltonian and super-momentum constraints.

In GR it was possible to derive regularized equations of motion of the particles in the non-inertial rest-frame and to study their post-Minkowskian (PM) limit in the Hamiltonian post-Minkowskian (HPM) linearization in the 3-orthogonal gauges and the emission of HPM gravitational waves (GWs) (with the energy balance under control even in the absence of self-forces). In Ref. [97] there is the HPM evaluation of the time-like and null geodesics, of the red-shift, of the geodesic deviation equation, and of the luminosity distance. Then the PN limit of these PM equations allows one to recover the known 1PN results of harmonic gauges. The more surprising result is that in the post-Newtonian (PN) expansion of the PM equations of motion there is a 0.5PN term in the forces depending upon the York time. This opens the possibility to describe dark matter as a relativistic inertial effect [36–38], implying that the effective inertial mass of particles in the 3-spaces is the bigger of the gravitational masses because it depends on the York time (i.e., on the shape of the 3-space as a 3-sub-manifold of the space-time). This leads to a violation of the Newtonian equivalence principle, because it is based on the Euclidean nature of the 3-spaces in the Galilei space-time of Newton gravity, which is incompatible with Einstein gravity. The future data of the GAIA mission on the rotation curve of the Milky Way [562] will allow testing of this prediction.

The proposed solution to the gauge problem in GR is based on the conventions of relativistic metrology for the International Celestial Reference System (ICRS) and the results on the reinterpretation of dark matter as a relativistic inertial effect arising as a consequence of a convention on the York time in an extended PM ICRS push toward the necessity of similar reinterpretation also of dark energy in cosmology [420, 421, 563–571]. As shown in Refs. [95–97], the identification of the tidal and inertial degrees of freedom of the gravitational field can be reformulated in the framework of the Lagrangian first-order ADM equations by means of the replacement of the Hamiltonian momenta with the expansion and the shear of the Eulerian observers associated with the 3+1 splitting of the space-time. Therefore, this identification can also be applied to the cosmological space-times that do not admit a Hamiltonian formulation; also in them the identification of the instantaneous 3-spaces Σ_τ, now labeled by a cosmic time, requires a conventional choice of clock synchronization, i.e., a convention on the York time ${}^3K$ defining the shape of the 3-spaces as 3-sub-manifolds of the space-time, and of 3-coordinates (the 3-orthogonal ones are acceptable also in cosmology).

In the standard ΛCDM cosmological model the class of cosmological solutions of Einstein equations is restricted to Friedmann–Robertson–Walker (FRW) space-times with nearly Euclidean 3-spaces (i.e., with a small internal 3-curvature). In them, the Killing symmetries connected with homogeneity and isotropy imply (τ is the cosmic time, $a(\tau)$ the scale factor) that the York time is no longer a gauge variable but coincides with the Hubble constant: ${}^3K(\tau) = -\frac{\dot{a}(\tau)}{a(\tau)} = -H(\tau)$. However at the first order in cosmological perturbations (see Ref. [572] for a review) one has ${}^3K = -H + {}^3K_{(1)}$ with ${}^3K_{(1)}$ being again an inertial gauge variable to be fixed with a metrological convention. Therefore, the York time has a central role also in cosmology and one needs to know the dependence on it of the main quantities, like the red-shift and the luminosity distance from supernovae, which require the introduction of the notion of dark energy to explain the 3-universe and its accelerated expansion in the framework of the standard ΛCDM cosmological model.

In inhomogeneous space-times without Killing symmetries like the Szekeres ones [573–577] the York time remains an arbitrary inertial gauge variable. Therefore the main open problem of the present approach is to see whether it is possible to find a 3-orthogonal gauge in an inhomogeneous Einstein space-time (at least in a PM approximation) in which the convention on the inertial gauge variable York time allows one to accomplish the following two tasks simultaneously: (1) to eliminate both dark matter and dark energy through the choice of a 4-coordinate system (suggested by astrophysical data) to be used in a consistent PM reformulation of ICRS; and (2) to save the main good properties of the standard ΛCDM cosmological model due to the inertial and dynamical properties of the space-time. As matter one will take the dust, whose description in the York canonical basis is given in Ref. [104].

Also in the back-reaction approach [578–583] to cosmology, according to which dark energy is a byproduct of the non-linearities of GR when one considers spatial averages of 3-scalar quantities in the 3-spaces on large scales to get a cosmological description of the universe taking into account its observed inhomogeneity, one gets that the spatial average of the product of the lapse function and of the York time (a 3-scalar gauge variable) gives the effective Hubble constant. Since this approach starts from the Hamiltonian description of an asymptotically flat space-time and since all the canonical variables in the York canonical basis, except the angles θ^i, are 3-scalars, the formalism presented in this book will allow study of the spatial average of nearly all the Hamilton equations and not only of the super-Hamiltonian constraint and of the Hamilton equation for the York time, as in the existing formulation of the approach.

The recent point of view of Ref. [584], taking into account the relevance of the voids among the clusters of galaxies, has to be reformulated in terms of the York time.

Finally, one should find the dependence upon the York time of the Landau–Lifchitz energy–momentum pseudo-tensor [585] and reexpress it as the effective energy–momentum tensor of a viscous pseudo-fluid. One will have to check whether, for certain choices of the York time, the resulting effective equation of state of the fluid has negative pressure, realizing also in this way a simulation of dark energy.

Other open problems in GR under investigation are:

1. Find the second order of the HPM expansion to see whether in PM space-times there is the emergence of hereditary terms [11, 422] like the ones present in harmonic gauges.
2. Study the HPM equations of motion of the transverse electromagnetic field to try to find Lienard–Wiechert-type solutions in GR. Study astrophysical problems where the electromagnetic field is relevant.
3. Find the explicit transformation from the harmonic gauges to the 3-orthogonal Schwinger time gauges and vice versa by integrating the partial differential

equations connecting the two families of gauges (see section 5.3 of Ref. [95]). Find the conditions for the global existence of the 3-orthogonal gauges.

4. Try to make a multitemporal quantization (see Refs. [103, 169, 170]) of the linearized HPM theory over the asymptotic Minkowski space-time, in which, after a Shanmugadhasan canonical transformation to a new York canonical basis adapted to all the constraints, only the tidal variables are quantized, not the inertial gauge ones. After this type of quantization, in which the lapse and shift functions remain c-numbers, the space-time would still be a classical 4-manifold: Only the two eigenvalues of the 3-metric describing GW are quantized and therefore only 3-metric properties like 3-distances, 3-areas, and 3-volumes become quantum properties. After having reexpressed the Ashtekar variables [306, 586] for asymptotically Minkowskian space-times (see appendix B of Ref. [587] and section 2.8 of Ref. [97]) in this final York canonical basis it will be possible to compare the outcomes of this new type of quantization with loop quantum gravity [18, 19], which has 3-space compact without boundary (so that there is no Poincaré algebra) and a frozen dynamics.

The main open problem in SR is the quantization of fields in non-inertial frames due to the no-go theorem of Refs. [588, 589], showing the existence of obstructions to the unitary evolution of a massive quantum Klein–Gordon field between two space-like surfaces of Minkowski space-time. It turns out that the Bogoljubov transformation connecting the creation and destruction operators on the two surfaces is not of the Hilbert–Schmidt type, i.e., that the Tomonaga–Schwinger approach in general is not unitary (the notion of the particle introduced in field theory by means of the Fock space in an inertial frame is not unitary equivalent to the notion in a non-inertial frame). One must reformulate the problem using the nice foliations of the admissible 3+1 splittings of Minkowski space-time and to try to identify all the 3+1 splittings allowing unitary evolution. This will be a prerequisite to any attempt to quantize canonical gravity, taking into account the equivalence principle (global inertial frames do not exist) with the further problem that in general the Fourier transform does not exist in Einstein space-times.

Appendix A

Canonical Realizations of Lie Algebras, Poincaré Group, Poincaré Orbits, and Wigner Boosts

Due to the importance of the Hamiltonian actions of Lie groups on the phase space of dynamical systems, we give a short review of this argument based on Ref. [590]. The most complete exposition of the standard properties of the canonical or Hamiltonian realizations of Lie algebras and groups as transformation groups of regular canonical transformations on the phase space M of dynamical systems is Ref. [123], especially chapter 14.

In addition, there is the description of the main properties of the Poincaré group, like the Poincaré orbits and the Wigner boosts needed for the Wigner covariance of the rest-frame instant form of dynamics.

A.1 Canonical Realizations of Lie Algebras

In a canonical realization, a phase space function $\bar{\mathcal{T}}_i$ corresponds to each abstract generator t_i of a Lie algebra and the Lie brackets $[t_i, t_j] = c_{ij}{}^k\, t_k$ are realized as the Poisson brackets $\{\bar{\mathcal{T}}_i, \bar{\mathcal{T}}_j\} = c_{ij}{}^k\, \bar{\mathcal{T}}_k + d_{ij}$, with $d_{ij} = -d_{ji}$ and $c_{ij}{}^m\, d_{mk} + c_{ki}{}^m\, d_{mj} + c_{jk}{}^m\, d_{mi} = 0$. When the pure numbers d_{ij} may be made to vanish, we speak of a true realization (as happens with the rotation, Euclidean, Lorentz, and Poincaré groups); otherwise we have a projective realization (as happens with the Galilei group). See chapters 22, 23, and 24 of [591] for the differential geometric definitions of symplectic and Hamiltonian actions of Lie algebras and groups. We shall assume we have a Hamiltonian Lie action on the phase space M of a dynamical system not only integrable to the action of a local Lie group, but also globally integrable to a symplectic Lie group action on M (all the vector fields $\bar{X}_i = \{., \bar{\mathcal{T}}_i\}$ are complete). This last requirement is the starting point for the method of reduction by means of the momentum map.

Here, we shall review the further properties of the canonical realizations of Lie groups developed in [172]. These properties are in general limited to the neighborhood of the identity; they also largely hold for simply connected Lie groups (for non-simply connected Lie groups they hold for the covering group).

Let M be a $2n$-dimensional symplectic manifold and Φ the action of the Lie group G on M: Φ: $G \times M \to M$. It can be shown that it is possible to construct a peculiar class of local charts on M by exploiting the local homeomorphism existing between a sub-manifold of M and the co-adjoint orbits on g^* (dual of the Lie algebra g of G) [590, 592, 593]. Each chart belonging to the above class, characterized by being adapted to the group structure, is called *typical form.*

According to the general theory [590–592], a canonical realization $\mathcal{K}$ of a Lie group G_r (of order r) can be characterized in terms of two basic schemes:

1. Scheme A depends entirely on the structure of the Lie algebra g_r (including its cohomology) and amounts to a pseudo-canonization of the generators in terms of k invariants $\bar{\mathcal{I}}_1, \ldots, \bar{\mathcal{I}}_k$ and $h = \frac{1}{2}(r-k)$ pairs of canonical variables (irreducible kernel of the scheme), functions of the generators.
2. Scheme B (or typical form) is an array of $2n$ canonical variables P_i, Q_i, defined by means of a canonical completion of scheme A. Locally, scheme B allows us to analyze any given canonical realization of G_r on the phase space of every dynamical system and to construct the most general canonical realization of G_r.

A generic scheme A can be usefully visualized by Eq. (A.1) ($r = 2h + k$ is the number of generators of the Lie algebra g_r):

$$\begin{array}{|c|c|} \hline P_1(\bar{\mathcal{T}}_i)\ldots P_h(\bar{\mathcal{T}}_i) & \bar{\mathcal{I}}_1(\bar{\mathcal{T}}_i)\ldots\bar{\mathcal{I}}_{k}(\bar{\mathcal{T}}_i) \\ \hline Q_1(\bar{\mathcal{T}}_i)\ldots Q_h(\bar{\mathcal{T}}_i) & \\ \hline \end{array}, \qquad (A.1)$$

where variables belonging to the same vertical pair are canonically conjugated, and variables belonging to different vertical lines commute. The quantities $\bar{\mathcal{I}}_i$ clearly commute with all the generators and are the invariants of the realization. Of course, any set of k functional independent functions $\bar{\mathcal{I}}'_1(\bar{\mathcal{I}}_1, \ldots, \bar{\mathcal{I}}_k), \ldots, \bar{\mathcal{I}}'_k(\bar{\mathcal{I}}_1, \ldots, \bar{\mathcal{I}}_k)$ of the invariants are good invariants as well.

A.2 The Lorentz and Poincaré Groups

Let M^4 be the Minkowski space-time with metric $\eta^{\mu\nu} = \epsilon(+---)$ and Cartesian coordinates x^μ. An element of the realization of the Poincaré group on M^4 is a pair (a, Λ) with a^μ a space-time translation and Λ^μ_ν a Lorentz transformation ($\Lambda^T \eta \Lambda = \eta$, $(\Lambda^T)^\mu_\nu = \Lambda_\nu{}^\mu$, $\det\Lambda = \pm 1$, $(\Lambda^o_o)^2 \geq 1$). Its action on M^4 is $x'^\mu = \Lambda^\mu_\nu x^\nu + a^\mu$. Its composition law is $(a', \Lambda') \cdot (a, \Lambda) = (a' + \Lambda' a, \Lambda' \Lambda)$ and the inverse of a Poincaré transformation is $(a, \Lambda)^{-1} = (-\Lambda^{-1} a, \Lambda^{-1})$. The Poincaré group is the semi-direct product of the Abelian translation group T^4 and the Lorentz group $\mathcal{L}$, $\mathcal{P} = T^4 \wedge \mathcal{L}$. It is a non-semisimple, non-compact Lie group.

The homogeneous Lorentz group $\mathcal{L} = O(3,1)$ is the semi-direct product of the discrete Klein group with the four elements $(1, I_s, I_t, I_{st})$ (with $I_s^2 = I_t^2 = I_{st}^2 = 1$, $I_s I_t = I_t I_s = I_{st}$, $I_s I_{st} = I_{st} I_s = I_t$, $I_t I_{st} = I_{st} I_t = I_s$) and the restricted

Lorentz group $\mathcal{L}_+^\uparrow$. Here, I_s is the space inversion ($I_s\,(x^o;\vec{x}) = (x^o;-\vec{x})$), I_t the time inversion ($I_t\,(x^o;\vec{x}) = (-x^o;\vec{x})$), and I_{st} the space time inversion ($I_{st}\,(x^o;\vec{x}) = (-x^o;-\vec{x})$). The doubly connected restricted group $\mathcal{L}_+^\uparrow$ is the proper ($\det\Lambda = 1$) orthocronous ($\Lambda^o_o \geq 1$) component connected with the identity of $\mathcal{L}$ and each element admits a unique polar decomposition $\Lambda = R\,L$ (R rotation, L boost). The improper ($\det\Lambda = -1$) orthocronous ($\Lambda^o_o \geq 1$) component is $\mathcal{L}_-^\uparrow = I_s\,\mathcal{L}_+^\uparrow$; the proper ($\det\Lambda = 1$) non-orthocronous ($\Lambda^o_o \leq -1$) component is $\mathcal{L}_+^\downarrow = I_{st}\,\mathcal{L}_+^\uparrow$; the improper ($\det\Lambda = -1$) non-orthocronous ($\Lambda^o_o \leq -1$) component is $\mathcal{L}_-^\downarrow = I_t\,\mathcal{L}_+^\uparrow$. The group $SO(3,1)$ describes $\mathcal{L}_+^\uparrow \cup \mathcal{L}_+^\downarrow$.

Since the action of the Lorentz group on M^4 leaves the quadratic form $x^\mu\,\eta_{\mu\nu}\,y^\nu$ invariant, for each $x \in M^4$ the restricted Lorentz group $\mathcal{L}_+^\uparrow$ defines the following types of orbits (or transitivity surfaces or homogeneous spaces) $\mathcal{L}_+^\uparrow\,x = \{\Lambda\,x; \Lambda \in \mathcal{L}_+^\uparrow\}$ and of little groups (or stationary or isotropy groups) $G_x = \{\Lambda \in \mathcal{L}_+^\uparrow; \Lambda\,x = x\}$:

- Ia. the future light-cone: $x^2 = 0$, $x^o > 0$ with $G_x = E(2)$; the reference vector of the future light-like orbits is $\frac{\omega}{2}\,(1;0,0,1)$;
- Ib. the past light-cone: $x^2 = 0$, $x^o < 0$ with $G_x = E(2)$; the reference vector of the past light-like orbits is $\frac{\omega}{2}\,(-1;0,0,1)$;
- IIa. the future (upper) branch of the two-sheeted hyperboloid: $x^2 = a^2$, $x^o > 0$ with $G_x = O(3)$; the reference vector of the future time-like orbits is $(|a|;\vec{0})$;
- IIb. the past (lower) branch of the two-sheeted hyperboloid: $x^2 = a^2$, $x^o < 0$ with $G_x = O(3)$; the reference vector of the future time-like orbits is $(-|a|;\vec{0})$;
- III. the one-sheeted hyperboloid: $x^2 = -a^2$ with $G_x = O(2,1)$; the reference vector of the space-like orbits is $(0;0,0,a)$;
- IV. the exceptional orbit (the vertex of the light-cone): $x^\mu = 0$ with $G_x = O(3,1)$.

The layers of $\mathcal{L}_+^\uparrow$ are the orbits $Ia \cup Ib$ and $IIa \cup IIb$ of $\mathcal{L}$: each of them contains all the points that have conjugate little groups.

The simply connected universal covering group of $\mathcal{L}_+^\uparrow$ is $SL(2,C)$. The canonical homomorphism $x^\mu \mapsto \hat{x} = \tilde{\sigma}_\mu\,x^\mu = x^o + \vec{\sigma}\cdot\vec{x}$ ($\tilde{\sigma}_\mu = (1;\vec{\sigma})$, $\sigma_\mu = (1;-\vec{\sigma})$ with $\vec{\sigma} = \{\sigma_i\}$ the Pauli matrices; $\det\hat{x} = x^2$, $x_\mu = \frac{1}{2}\,\mathrm{Tr}\,(\hat{x}\,\sigma_\mu))$ between M^4 and the set of 2×2 Hermitian complex matrices implies that every pair of 2×2 unimodular complex matrices $\pm V \in SL(2,C)$ induces the same Lorentz transformation $\Lambda(V)$: $\sigma_\mu\,\Lambda(V)^\mu_\nu\,x^\nu = [\pm V]\,\tilde{\sigma}_\mu\,x^\mu\,[\pm V^\dagger]$ with $\Lambda(V)^\mu_\nu = \frac{1}{2}\,\eta^{\mu\rho}\,\mathrm{Tr}\left(\sigma_\rho\,V\,\tilde{\sigma}_\nu\,V^\dagger\right) = \Lambda(V^\dagger)^\nu_\mu$. The polar decomposition $\Lambda(V) = R\,L$ (R rotation, L pure boost) becomes $V = U\,H$ with $U \in SU(2)$ and H Hermitian and positive definite ($H = \sqrt{V^\dagger\,V}$, $U = V\,(V^\dagger\,V)^{-1/2}$, $R = \Lambda(U)$, $L = \Lambda(H)$). By writing $H = U'\,D_H\,U'^\dagger$ with D_H diagonal, we get the Euler decomposition of $\Lambda \in \mathcal{L}_+^\uparrow$, i.e., $\Lambda = R_1\,L_{03}\,R_2$ as the product of three non-independent matrices ($R_1 = \Lambda(U\,U')$, $R_2 = \Lambda(U'^\dagger)$; $L_{03} = \Lambda(D_H)$ is a boost in direction 3).

An infinitesimal Poincaré transformation $x'^{\mu} = x^{\mu} - \epsilon^{\mu}_{\nu}\, x^{\nu} + \alpha^{\mu} = \Big[1 + i\, \alpha_{\nu}\, \check{P}^{\nu} + \frac{i}{2}\, \epsilon_{\alpha\beta}\, \check{J}^{\alpha\beta}\Big]\, x^{\mu}$ $[\Lambda^{\mu}_{\nu} = \delta^{\mu}_{\nu} - \epsilon^{\mu}_{\nu}$ with $\epsilon_{\mu\nu} = -\epsilon_{\nu\mu}]$ allows us to identify the Hermitian generators $\check{P}^{\mu} = -i\, \frac{\partial}{\partial x_{\mu}}$, $\check{J}^{\mu\nu} = -i\, \Big(x^{\mu}\, \frac{\partial}{\partial x_{\nu}} - x^{\nu}\, \frac{\partial}{\partial x_{\mu}}\Big)$ of the Lie algebra of the component $\mathcal{P}^{\uparrow}_{+} = T^4 \wedge \mathcal{L}^{\uparrow}_{+}$ of the Poincaré group in its unitary realization on Minkowski space-time. The Lie algebra of the Poincaré group is ($C^{\alpha\beta\mu\nu}_{\rho\sigma} = \eta^{\alpha\mu}\, \delta^{\beta}_{\rho}\, \delta^{\nu}_{\sigma} + \eta^{\beta\nu}\, \delta^{\alpha}_{\rho}\, \delta^{\mu}_{\sigma} - \eta^{\alpha\nu}\, \delta^{\beta}_{\rho}\, \delta^{\mu}_{\sigma} - \eta^{\beta\mu}\, \delta^{\alpha}_{\rho}\, \delta^{\nu}_{\sigma}$ are the structure constants of the Lorentz Lie algebra)

$$
\begin{aligned}
&[\,\check{P}^{\mu}, \check{P}^{\nu}] = 0,\\
&[\,\check{J}^{\alpha\beta}, \check{P}^{\mu}] = i\, (\eta^{\mu\alpha}\, \check{P}^{\beta} - \eta^{\mu\beta}\, \check{P}^{\alpha}),\\
&[\check{J}^{\alpha\beta}, \check{J}^{\mu\nu}] = i\, (\eta^{\alpha\mu}\, \check{J}^{\beta\nu} + \eta^{\beta\nu}\, \check{J}^{\alpha\mu} - \eta^{\alpha\nu}\, \check{J}^{\beta\mu} - \eta^{\beta\mu}\, \check{J}^{\alpha\nu}) = i\, C^{\alpha\beta\mu\nu}_{\rho\sigma}\, \check{J}^{\rho\sigma}.
\end{aligned}
\tag{A.2}
$$

By introducing the new generators $\check{J}^{r} = -\frac{1}{2}\, \epsilon^{ruv}\, \check{J}^{uv} = i\, (\vec{x} \times \vec{\partial})^{r}$ of space rotations and $\check{K}^{r} = \check{J}^{ro}$ of boosts, the Lie algebra takes the form

$$
\begin{aligned}
&[\check{J}^{r}, \check{J}^{s}] = i\, \epsilon^{rsu}\, \check{J}^{u},\\
&[\check{K}^{r}, \check{K}^{s}] = -i\, \check{J}^{rs} = i\, \epsilon^{rsu}\, \check{J}^{u},\\
&[\check{J}^{r}, \check{K}^{s}] = [\check{K}^{r}, \check{J}^{s}] = i\, \epsilon^{rsu}\, \check{K}^{u}.
\end{aligned}
\tag{A.3}
$$

There are two Poincaré Casimir invariants $\check{P}^2$ and $\check{W}^2$. If p^{μ} are the eigenvalues of $\check{P}^{\mu}$, the range of $\check{P}^2$ is $p^2 \in (-\infty, +\infty)$. The Casimir $\check{W}^2$ is built with the Pauli–Lubanski operator,

$$
\begin{aligned}
&\check{W}_{\mu} = \frac{1}{2}\, \epsilon_{\mu\nu\rho\sigma}\, \check{P}^{\nu}\, \check{J}^{\rho\sigma} = \Big(\vec{\check{P}} \cdot \vec{\check{J}}; \check{P}^{o}\, \vec{\check{J}} + \vec{\check{P}} \times \vec{\check{K}}\Big), \qquad \check{P}_{\mu}\, \check{W}^{\mu} = 0,\\
&[\check{J}^{\rho\nu}, \check{W}^{\mu}] = i\, (\eta^{\rho\mu}\, \check{W}^{\nu} - \eta^{\nu\mu}\, \check{W}^{\rho}),\\
&[\check{W}_{\mu}, \check{W}_{\nu}] = i\, \epsilon_{\mu\nu\rho\sigma}\, \check{W}^{\rho}\, \check{P}^{\sigma}.
\end{aligned}
\tag{A.4}
$$

In a canonical realization of the Poincaré group on the phase space (q, p) of a dynamical system, the Hamiltonian action of the Poincaré–Lie algebra is realized through functions $\bar{P}^{\mu}(q, p)$, $\bar{J}^{\mu\nu}(q, p)$ satisfying the Poisson-Lie algebra (we assume that a position variable $\bar{X}^{\mu}(q, p)$ canonically conjugated to $\bar{P}^{\mu}(q, p)$ satisfies $\{\bar{X}^{\mu}, \bar{P}^{\nu}\} = -\eta^{\mu\nu}$):

$$
\begin{aligned}
&\{\bar{P}^{\mu}, \bar{P}^{\nu}\} = 0,\\
&\{\bar{J}^{\alpha\beta}, \bar{P}^{\mu}\} = -\Big(\eta^{\mu\alpha}\, \bar{P}^{\beta} - \eta^{\mu\beta}\, \bar{P}^{\alpha}\Big),\\
&\{\bar{J}^{\alpha\beta}, \bar{J}^{\mu\nu}\} = C^{\alpha\beta\mu\nu}_{\rho\sigma}\, \bar{J}^{\rho\sigma}.
\end{aligned}
\tag{A.5}
$$

The Casimir invariants become $\bar{P}^2$ and $\bar{W}^2$ with $\bar{W}_{\mu} = \frac{1}{2}\, \epsilon_{\mu\nu\alpha\beta}\, \bar{P}^{\nu}\, \bar{J}^{\alpha\beta}$ being the Pauli–Lubanski 4-vector.

Referring to Refs. [594–596] for the theory of the unitary representations of the Poincaré group based on Mackey's induced representation method [597], let us review the notations leading to the Wigner states of the representation [598, 599].

Since the restricted Poincaré group $\mathcal{P}_+^\uparrow = T^4 \wedge SO(3,1)_+^\uparrow$ is the semi-direct product of the Abelian translation subgroup T^4 with the restricted Lorentz subgroup $\mathcal{L}_+^\uparrow = SO(3,1)_+^\uparrow$, the inducing subgroup is chosen to be $T^4 \wedge K$, where K is some closed subgroup of the restricted Lorentz subgroup. The right coset space $C_+^\uparrow = \mathcal{P}_+^\uparrow / T^4 \wedge K = SO(3,1)_+^\uparrow / K$ may be identified with the space of parameters (group manifold) of T^4.

A continuous unitary representation $U(\mathcal{P}_+^\uparrow)$ of $\mathcal{P}_+^\uparrow$ on Wigner states $\omega(c)$, $c \in C_+^\uparrow$, is induced by given continuous unitary representations $D(k)$, $k \in K$, of K and $\chi(a)$ of T^4 according to the transformation laws ($c = \mathring{p} \in C_+^\uparrow$; Λ^T is the transpose of Λ)

$$\Big(U(1,\Lambda)\,\omega\Big)(\mathring{p}) = D(R(\mathring{p},\Lambda))\,\omega(\Lambda^T\,\mathring{p}), \qquad \Big(U(a,e)\,\omega\Big)(\mathring{p}) = \chi_{\mathring{p}}(a)\,\omega(\mathring{p}), \quad \text{(A.6)}$$

with a suitable inner product. The elements $R(\mathring{p},\Lambda)$ of K are called Wigner rotations. Since the unitary irreducible representations $\chi(a)$ of T^4 are 1-dimensional, $\chi(a) = e^{i\,\mathring{p}^\mu\,a_\mu}$, they are characterized by the contro-variant four-vectors $\mathring{p}^\mu$ and are denoted $\chi_{\mathring{p}}(a)$. The inducing representations used to build the irreducible unitary Poincaré representations are chosen to be direct products $D(a,k) = \chi(a) \otimes D(k)$, with $D(k)$ a continuous unitary representation of K. This direct product representation is compatible with the semi-direct structure of $\mathcal{P}_+^\uparrow$ if $\chi(a) = \chi(k\,a)$ for $k \in K$. Usually, K is chosen to be the maximal closed subgroup of $\mathcal{L}_+^\uparrow = SO(3,1)_+^\uparrow$ leaving $\chi(a)$ invariant: in this case K is called the little group of $\chi(a)$. We get $\chi_{\mathring{p}}(a) = \chi(k\,a)$ for k in the coset $\mathring{p} \in C_+^\uparrow$ and $\chi_{\Lambda^T\,\mathring{p}}(a) = \chi(\Lambda^T\,a)$ otherwise. The representation $\chi_{\mathring{p}}(a)$ is said to form an orbit of $\chi(a)$; all the representations in the same orbit have the same little group K and the unitary representations of $\mathcal{P}_+^\uparrow$ induced by $\{T^4 \wedge K, \chi_{\mathring{p}} \otimes D(k)\}$ for fixed $D(k)$ are not only irreducible and exhaustive, but also equivalent. The orbits and the $D(k)$ completely determine the induced unitary representations of $\mathcal{P}_+^\uparrow$.

A.3 The Poincaré Orbits

The Poincaré orbits are the geometrical orbits $p^\mu = \Lambda^\mu{}_\nu\,\mathring{p}^\nu$ (we have redefined Λ^T as Λ) with $\mathring{p}^\mu = L^\mu{}_\nu(k,\mathring{p})\,\mathring{p}^\nu$,[1] namely the little group is the stability group of the standard 4-vector $\mathring{p}^\mu$ identifying the representation of T^4.

Usually the restricted Lorentz group $SO(3,1)_+^\uparrow$ is replaced by the covering group $SL(2,C)$, so that $\mathring{p}^\mu \mapsto \hat{\mathring{p}} = \mathring{p}^\mu\,\sigma_\mu$. The Poincaré orbits are (in this Appendix we put $c = 1$):

Ia. the future null- or light-like orbits (upper branch of the light-cone): $\mathring{p}^2 = 0$, $\epsilon\,\mathring{p}^o > 0$ with $K = \tilde{E}(2)$, the covering group of $E(2)$; the reference vector of the future light-like orbits is $\mathring{p}^\mu = \frac{\omega}{2}\,(1;0,0,1)$ or $\hat{\mathring{p}} = 2\omega \begin{pmatrix} 0 & 0 \\ 0 & 1 \end{pmatrix}$;

[1] $L(k,\mathring{p})$ is called the standard Wigner boost of the orbit.

Ib. the past null- or light-like orbits (lower branch of the light-cone): $\overset{\circ}{p}^{2} = 0$, $\epsilon \overset{\circ}{p}^{o} < 0$ with $K = \tilde{E}(2)$, the covering group of $E(2)$; the reference vector of the future light-like orbits is $\overset{\circ}{p}^{\mu} = \frac{\omega}{2}(-1; 0, 0, 1)$ or $\hat{\overset{\circ}{p}} = -2\,\omega \begin{pmatrix} 0 & 0 \\ 0 & 1 \end{pmatrix}$;

IIa. the future time-like orbits (upper branch of the two-sheeted mass hyperboloid): $\epsilon \overset{\circ}{p}^{2} = m^2$, $\overset{\circ}{p}^{o} > 0$ with $K = SU(2)$; the reference vector of the future time-like orbits is $\overset{\circ}{p} = m\,(1; \vec{0})$ or $\hat{\overset{\circ}{p}} = m \begin{pmatrix} 1 & 0 \\ 0 & 1 \end{pmatrix}$;

IIb. the past time-like orbits (lower branch of the two-sheeted mass hyperboloid): $\epsilon \overset{\circ}{p}^{2} = m^2$, $\overset{\circ}{p}^{o} < 0$ with $K = SU(2)$; the reference vector of the future time-like orbits is $\overset{\circ}{p} = m\,(-1; \vec{0})$ or $\hat{\overset{\circ}{p}} = -m \begin{pmatrix} 1 & 0 \\ 0 & 1 \end{pmatrix}$;

III. the space-like orbits (one-sheeted imaginary mass hyperboloid): $\epsilon \overset{\circ}{p}^{2} = -m^2$ with $K = SU(1,1)$, the covering group of $O(2,1)$; the reference vector of the space-like orbits is $\overset{\circ}{p}^{\mu} = (0; 0, 0, m)$ or $\hat{\overset{\circ}{p}} = m \begin{pmatrix} 1 & 0 \\ 0 & -1 \end{pmatrix}$;

IV. the exceptional (infrared) orbit (the vertex of the light-cone): $\overset{\circ}{p}^{\mu} = 0$ with $K = SL(2, C)$.

A.4 The Wigner Boosts

If $L(p, \overset{\circ}{p})$ denotes the standard Wigner boost of an orbit $(L^{\mu}_{\nu}(p, \overset{\circ}{p})\,\overset{\circ}{p}^{\nu} = p^{\mu})$, then the Wigner rotation is given by $R(\Lambda, \overset{\circ}{p}) = L^{-1}(p, \overset{\circ}{p})\,\Lambda^{-1}\,L(\Lambda\, p, \overset{\circ}{p})$ with $R^{\mu}_{\nu}(\Lambda, \overset{\circ}{p})\,\overset{\circ}{p}^{\nu} = \overset{\circ}{p}^{\mu}$.

The standard boost is defined modulo a transformation of the little group. Due to the importance of time- and light-like orbits for the applications, we shall discuss the choice of standard boosts in these two cases. For $\epsilon p^2 > 0$ with $K = SU(2)$, we can use either the Euler decomposition $\Lambda = R_1\, L_3\, R_2 \in SU(2) L_3 SU(2)$ and obtain the standard Wigner boost for time-like orbits (R basis), in which $\check{J}^3$ is diagonal, or to define the (Jacob–Wick) helicity bases (H basis, existing for $|\vec{p}| \neq 0$) in which the helicity is diagonal. For $p^2 = 0$ with $K = \tilde{E}(2)$, the Euler decomposition does not work (being independent of $E(2)$). The Iwasawa decomposition's consequence $\Lambda = E(2) L_3 SU(2)$ allows defining the helicity basis and then an R basis

Let us give the standard Wigner boosts for the Poincaré orbits IIa ∪ IIB and Ia ∪ Ib in the two quoted bases.

1. Time-like orbits. We have $K = SU(2)$ and $\hat{\overset{\circ}{p}} = \eta\, m \begin{pmatrix} 1 & 0 \\ 0 & 1 \end{pmatrix}$ with $\eta = \mathrm{sign}\, p^o$. A generic standard boost satisfies $\hat{L}_U(p, \overset{\circ}{p})\,\hat{\overset{\circ}{p}}\,\hat{L}^{\dagger}_U(p, \overset{\circ}{p}) = \hat{p}$ and this implies $\hat{L}_U(p, \overset{\circ}{p})\,\hat{L}^{\dagger}_U(p, \overset{\circ}{p}) = \hat{H}^2 = \frac{p^{\mu}\,\tilde{\sigma}_{\mu}}{m}$, with $\hat{H}$ a positive definite Hermitian matrix. Therefore, we get $\hat{L}_U(p, \overset{\circ}{p}) = \hat{H}\,\hat{U} = \sqrt{\frac{p^{\mu}\,\tilde{\sigma}_{\mu}}{m}}\,\hat{U}$ with $\hat{U}^{\dagger} = \hat{U}$.

1a. R basis. In this basis $\hat{U}=1$ and we have the standard Wigner boost without rotation $\hat{L}(p,\overset{\circ}{p})=\sqrt{\frac{p^{\mu}\,\tilde{\sigma}_{\mu}}{m}}$, with inverse $\hat{L}(\overset{\circ}{p},p)=L^{-1}(p,\overset{\circ}{p})=\sqrt{\frac{p^{\mu}\,\sigma_{\mu}}{m}}$. The corresponding Lorentz transformation is denoted $L^{\mu}_{\nu}(p,\overset{\circ}{p})$ and its inverse is used to define the rest-frame Pauli–Lubanski 4-vector $\overset{\circ}{\check{W}}_{\mu}=L_{\mu\nu}(\overset{\circ}{p},p)\,\check{W}^{\nu}=-m\,(0;\vec{\check{S}}_{T})$. Here, $\check{S}_{T\,i}=-\frac{1}{m}\,(\check{W}_{i}-\frac{\check{P}_{i}}{\check{P}^{o}+m}\,\check{W}_{o})$ is the rest-frame Thomas spin, satisfying $[\check{S}_{T\,i},\check{S}_{T\,j}]=i\,\epsilon_{ijk}\,\check{S}_{T\,k}$. In the R basis $\check{S}_{T\,3}$ is diagonal.

We now give the $SO(3,1)$ matrix corresponding to the Wigner boost by replacing m with $\sqrt{\epsilon\,p^{2}}$, since it is this expression that is needed when we study relativistic particles satisfying the mass shell constraint $\epsilon\,p^{2}-m^{2}\approx 0$. The rest-frame form of the time-like 4-vector p^{μ} is $\overset{\circ}{p}{}^{\mu}=\eta\,\sqrt{\epsilon\,p^{2}}\,(1;\vec{0})=\eta^{\mu o}\,\eta\,\sqrt{p^{2}}$, $\overset{\circ}{p}{}^{2}=p^{2}$, where $\eta=\mathrm{sign}\,p^{o}$.

The standard Wigner boost transforming $\overset{\circ}{p}{}^{\mu}$ into p^{μ} is [598–600]

$$
\begin{aligned}
L^{\mu}_{\nu}(p,\overset{\circ}{p}) &= \frac{1}{2}\,\eta^{\mu\rho}\,\mathrm{Tr}\left[\sigma_{\rho}\,\hat{L}(p,\overset{\circ}{p})\,\tilde{\sigma}_{\nu}\,\hat{L}^{\dagger}(p,\overset{\circ}{p})\right]=\epsilon^{\mu}_{\nu}(u(p))\\
&= \eta^{\mu}_{\nu}+2\,\frac{p^{\mu}\overset{\circ}{p}_{\nu}}{p^{2}}-\frac{(p^{\mu}+\overset{\circ}{p}{}^{\mu})\,(p_{\nu}+\overset{\circ}{p}_{\nu})}{p\cdot\overset{\circ}{p}+p^{2}}\\
&= \eta^{\mu}_{\nu}+2\,u^{\mu}(p)\,u_{\nu}(\overset{\circ}{p})-\frac{(u^{\mu}(p)+u^{\mu}(\overset{\circ}{p}))\,(u_{\nu}(p)+u_{\nu}(\overset{\circ}{p}))}{1+u^{o}(p)},\\
\nu=0 \quad & \epsilon^{\mu}_{o}(u(p))=u^{\mu}(p)=\frac{p^{\mu}}{\eta\,\sqrt{\epsilon\,p^{2}}},\\
\nu=r \quad & \epsilon^{\mu}_{r}(u(p))=\left(-u_{r}(p);\delta^{i}_{r}-\frac{u^{i}(p)\,u_{r}(p)}{1+u^{o}(p)}\right).
\end{aligned}
\tag{A.7}
$$

The inverse of $L^{\mu}_{\nu}(p,\overset{\circ}{p})$ is $L^{\mu}_{\nu}(\overset{\circ}{p},p)$, the standard boost to the rest-frame, defined by

$$
L^{\mu}_{\nu}(\overset{\circ}{p},p)=L^{\mu}_{\nu}(p,\overset{\circ}{p})=L^{\mu}_{\nu}(p,\overset{\circ}{p})|_{\vec{p}\rightarrow-\vec{p}}. \tag{A.8}
$$

Therefore, we can define the following vierbeins:[2]

$$
\begin{aligned}
&\epsilon^{\mu}_{A}(u(p))=L^{\mu}_{A}(p,\overset{\circ}{p}),\\
&\epsilon^{A}_{\mu}(u(p))=L^{A}_{\mu}(\overset{\circ}{p},p)=\eta^{AB}\,\eta_{\mu\nu}\,\epsilon^{\nu}_{B}(u(p)),\\
&\epsilon^{\bar{o}}_{\mu}(u(p))=\eta_{\mu\nu}\,\epsilon^{\nu}_{o}(u(p))=u_{\mu}(p),\\
&\epsilon^{r}_{\mu}(u(p))=-\delta^{rs}\,\eta_{\mu\nu}\,\epsilon^{\nu}_{r}(u(p))=(\delta^{rs}\,u_{s}(p);\delta^{r}_{j}-\delta^{rs}\,\delta_{jh}\,\frac{u^{h}(p)\,u_{s}(p)}{1+u^{o}(p)}),\\
&\epsilon^{A}_{o}(u(p))=u_{A}(p),
\end{aligned}
\tag{A.9}
$$

[2] The $\epsilon^{\mu}_{r}(u(p))$ are also called polarization vectors [601]; the indices r, s will be used for $A=1,2,3$ and $\bar{o}$ for $A=0$.

which satisfy

$$\epsilon^A_\mu(u(p))\,\epsilon^\nu_A(u(p)) = \eta^\mu_\nu,$$
$$\epsilon^A_\mu(u(p))\,\epsilon^\mu_B(u(p)) = \eta^A_B,$$
$$\eta^{\mu\nu} = \epsilon^\mu_A(u(p))\,\eta^{AB}\,\epsilon^\nu_B(u(p)) = u^\mu(p)\,u^\nu(p) - \sum_{r=1}^{3}\epsilon^\mu_r(u(p))\,\epsilon^\nu_r(u(p)),$$
$$\eta_{AB} = \epsilon^\mu_A(u(p))\,\eta_{\mu\nu}\,\epsilon^\nu_B(u(p)),$$
$$p_\alpha\,\frac{\partial}{\partial p_\alpha}\,\epsilon^\mu_A(u(p)) = p_\alpha\,\frac{\partial}{\partial p_\alpha}\,\epsilon^A_\mu(u(p)) = 0. \tag{A.10}$$

The Wigner rotation corresponding to the Lorentz transformation Λ is

$$R^\mu_\nu(\Lambda,p) = [L(\overset{\circ}{p},p)\,\Lambda^{-1}\,L(\Lambda p,\overset{\circ}{p})]^\mu_\nu = \begin{pmatrix} 1 & 0 \\ 0 & R^i_j(\Lambda,p)\end{pmatrix},$$

$$R^i_j(\Lambda,p) = (\Lambda^{-1})^i_j - \frac{(\Lambda^{-1})^i_o\,p_\beta\,(\Lambda^{-1})^\beta_j}{p_\rho\,(\Lambda^{-1})^\rho_o + \eta\sqrt{\epsilon\,p^2}}$$
$$-\frac{p^i}{p^o+\eta\sqrt{\epsilon\,p^2}}\left[(\Lambda^{-1})^o_j - \frac{((\Lambda^{-1})^o_o - 1)\,p_\beta\,(\Lambda^{-1})^\beta_j}{p_\rho\,(\Lambda^{-1})^\rho_o + \eta\sqrt{\epsilon\,p^2}}\right]. \tag{A.11}$$

The polarization vectors transform under the Poincaré transformations (a,Λ) in the following way

$$\epsilon^\mu_r(u(\Lambda p)) = (R^{-1})^s_r\,\Lambda^\mu_\nu\,\epsilon^\nu_s(u(p)). \tag{A.12}$$

Some further useful formulas are ($\epsilon' = \eta\sqrt{\epsilon\,p^2}$):

$$\begin{aligned}\frac{\partial}{\partial p_\mu}\quad \epsilon^B_\rho(u(p)) &= \frac{\partial}{\partial p_\mu}\,L^B_\rho(\overset{\circ}{p},p)\\ &= \frac{2}{\epsilon'}\,\eta^B_o\left(\eta^\mu_\rho - \frac{p^\mu\,p_\rho}{\epsilon'^2}\right) - \frac{1}{\epsilon'^2\,(p^o+\epsilon')}\Big[(p^B+\epsilon'\,\eta^B_o)\,(p^\mu\,\eta^o_\rho + \epsilon'\,\eta^\mu_\rho)\\ &\quad + (p^\mu\,\eta^B_o + \epsilon'\,\eta^{\mu B})\,(p_\rho + \epsilon'\,\eta^o_\rho)\Big]\\ &\quad + \frac{(p^B+\epsilon'\,\eta^B_o)(p_\rho+\epsilon'\,\eta^o_\rho)\,[(p^o+2\,\epsilon')\,p^\mu + \epsilon'^2\,\eta^\mu_o]}{\epsilon'^3\,(p^o+\epsilon')^2},\end{aligned} \tag{A.13}$$

$$\begin{aligned}\frac{1}{2}\,\epsilon^\nu_A(u(p))\,\eta^{AB}\,\frac{\partial\epsilon^\rho_B(u(p))}{\partial p_\mu}\,S_{\rho\nu} &= -\frac{1}{2}\,\eta_{AC}\,\frac{\partial\epsilon^C_\rho(u(p))}{\partial p_\mu}\,\epsilon^\rho_B(u(p))\,\bar{S}^{AB}\\ &= -\frac{1}{\epsilon'}\left[\eta^\mu_A\left(\bar{S}^{\bar{o}A} - \frac{\bar{S}^{Ar}\,p^r}{p^o+\epsilon'}\right) + \bar{S}^{\bar{o}r}\,p^r\,\frac{p^\mu + 2\,\epsilon'\,\eta^\mu_o}{\epsilon'\,(p^o+\epsilon')}\right]\\ &= -\frac{1}{\epsilon'\,(p^o+\epsilon')}\left[p_\rho\,S^{\rho\mu} + \epsilon'\left(S^{o\mu} - S^{o\rho}\,\frac{p_\rho\,p^\mu}{\epsilon'^2}\right)\right],\end{aligned} \tag{A.14}$$

$$\frac{\partial}{\partial p_\nu} R^i_k(\Lambda, p) = \frac{(\Lambda^{-1})^i_o}{(p_\rho\,(\Lambda^{-1})^\rho_o + \epsilon')^2}$$
$$\times \left[\frac{1}{\epsilon'}\,(p^\nu + \epsilon'\,(\Lambda^{-1})^\nu_o)\,p_\beta\,(\Lambda^{-1})^\beta_k - (p_\rho\,(\Lambda^{-1})^\rho_o + \epsilon')\,(\Lambda^{-1})^\nu_k\right]$$
$$- \frac{\eta^{i\nu}}{p^o + \epsilon'}\left[(\Lambda^{-1})^o_k - \frac{((\Lambda^{-1})^o_o - 1)\,p_\beta\,(\Lambda^{-1})^\beta_k}{p_\rho\,(\Lambda^{-1})^\rho_o + \epsilon'}\right]$$
$$+ \frac{p^i}{(p^o + \epsilon')^2}\left[\frac{1}{\epsilon'}\,(p^\nu + \epsilon'\,\eta^{\nu o})\,(\Lambda^{-1})^o_k + \right.$$
$$+ \frac{p^o + \epsilon'}{p_\rho\,(\Lambda^{-1})^\rho_o + \epsilon'}\,((\Lambda^{-1})^o_o - 1)$$
$$\times[(\Lambda^{-1})^\nu_k - \frac{1}{\epsilon'}\left(\frac{p^\nu + \epsilon'\,\eta^{\nu o}}{p^o + \epsilon'}\right.$$
$$\left.\left. + \frac{p^\nu + \epsilon'\,(\Lambda^{-1})^\nu_o}{p_\rho\,(\Lambda^{-1})^\rho_o + \epsilon'}\right)\,p_\beta\,(\Lambda^{-1})^\beta_k]\right]. \tag{A.15}$$

1b. H basis. It is defined by the sequence of transformations $\overset{\circ}{p}{}^\mu = \eta\, m\,(1;\vec{0}) \mapsto \bar{p}^\mu = (p^o; 0, 0, |\vec{p}|) \mapsto p^\mu = (p^o; \vec{p})$. If $\vec{p}$ is identified by the angles (θ, ϕ) and $\hat{H}_3$ is a boost in direction 3, then the helicity boost is $\hat{L}_H(p, \overset{\circ}{p}) = e^{-i\,\sigma_3\,\phi/2}\,e^{i\,\sigma_2\,\theta/2}\,e^{i\,\sigma_3\,\phi/2}\,\sqrt{\frac{\bar{p}^\mu\,\tilde{\sigma}_\mu}{m}} = \hat{U}\,\hat{H}_3 = \sqrt{\frac{p^\mu\,\tilde{\sigma}_\mu}{m}}\,e^{-i\,\sigma_3\,\phi/2}\,e^{i\,\sigma_2\,\theta/2}\,e^{i\,\sigma_3\,\phi/2} = \hat{L}(p, \overset{\circ}{p})\,U$. In this basis the helicity operator $\check{\Sigma} = \frac{\vec{p}\cdot\vec{J}}{|\vec{p}|}$ is diagonal (for $m \neq 0$ it is invariant only under rotations).

2. Light-like orbits. For $K = \tilde{E}(2)$ and $\overset{\circ}{p}{}^\mu = \eta\,2\,\omega\begin{pmatrix} 0 & 0 \\ 0 & 1\end{pmatrix}$ with $\eta = \mathrm{sign}\,p^o$, the standard boost $\hat{L}_E(p, \overset{\circ}{p})$ satisfies $\omega\,\hat{L}_E(p, \overset{\circ}{p})\,(1 + \sigma_3)\,\hat{L}^\dagger_E(p, \overset{\circ}{p}) = p^o + \vec{p}\cdot\vec{\sigma}$.

Usually we use only the H basis. In it the helicity boost is chosen to be ${}_H\hat{L}(p, \overset{\circ}{p}) = e^{-i\,\sigma_3\,\phi/2}\,e^{i\,\sigma_2\,\theta/2}\,e^{i\,\sigma_3\,\phi/2}\,e^{\frac{1}{2}\,\sigma_3\,\ln\omega}\,E$ with $E \in E(2)$. Following Ref. [175], we give the expression of the Lorentz matrix for the helicity boost.

For $p_s^2 = 0$, we have $p^\mu_s = (|\vec{p}_s| = \sqrt{\vec{p}_s^{\,2}}; \vec{p}_s)$.[3] By choosing a reference vector $\overset{\circ}{p}{}^\mu_s = \omega_s\,(1; 0, 0, 1)$, where ω_s is a dimensional parameter, the standard Wigner helicity boost ${}_H L^\mu{}_\nu(p_s, \overset{\circ}{p}_s)$, such that $p^\mu_s = {}_H L^\mu{}_\nu(p_s, \overset{\circ}{p}_s)\,\overset{\circ}{p}_s$ is

$${}_H L^\mu{}_\nu(p_s, \overset{\circ}{p}_s) = \begin{pmatrix} \frac{1}{2}\left(\frac{|\vec{p}_s|}{\omega_s} + \frac{\omega_s}{|\vec{p}_s|}\right) & 0 & \frac{1}{2}\left(\frac{|\vec{p}_s|}{\omega_s} - \frac{\omega_s}{|\vec{p}_s|}\right) \\ \frac{1}{2}\left(\frac{|\vec{p}_s|}{\omega_s} - \frac{\omega_s}{|\vec{p}_s|}\right)\frac{p^a_s}{|\vec{p}_s|} & \delta^a_b + \frac{p^a_s\,p_{sb}}{|\vec{p}_s|\,(|\vec{p}_s| + p^3_s)} & \frac{1}{2}\left(\frac{|\vec{p}_s|}{\omega_s} + \frac{\omega_s}{|\vec{p}_s|}\right)\frac{p^a_s}{|\vec{p}_s|} \\ \frac{1}{2}\left(\frac{|\vec{p}_s|}{\omega_s} - \frac{\omega_s}{|\vec{p}_s|}\right)\frac{p^3_s}{|\vec{p}_s|} & \frac{p_{sb}}{|\vec{p}_s|} & \frac{1}{2}\left(\frac{|\vec{p}_s|}{\omega_s} + \frac{\omega_s}{|\vec{p}_s|}\right)\frac{p^3_s}{|\vec{p}_s|} \end{pmatrix},$$

$${}_H\epsilon^\mu_{\tilde{A}}(\vec{p}_s) = {}_H L^\mu{}_\nu(p_s, \overset{\circ}{p}_s)\,\overset{\circ}{\epsilon}{}^\nu_{\tilde{A}} = {}_H L^\mu{}_{\tilde{A}}(p_s, \overset{\circ}{p}_s), \qquad \tilde{A} = (\tilde{\tau}, \tilde{r}),$$
$$\overset{\circ}{\epsilon}{}^\mu_{\tilde{A}}: \quad (1;0,0,0), \quad (0;1,0,0), \quad (0;0,1,0), \quad (0;0,0,1),$$
$$\eta^{\mu\nu} = {}_H\epsilon^\mu_{\tilde{A}}(\vec{p}_s)\,\eta^{\tilde{A}\tilde{B}}\,{}_H\epsilon^\nu_{\tilde{B}}(\vec{p}_s). \tag{A.16}$$

[3] We have chosen positive energy $p^o_s > 0$; more in general we should put $|\vec{p}_s| \mapsto \eta_s\,|\vec{p}_s|$ with $\eta_s = \pm 1 = \mathrm{sign}\,p^o_s$.

In this way, we get a helicity tetrad ${}_H\epsilon^\mu_{\tilde{A}}(\vec{p}_s)$ and a null basis:

$$
\begin{aligned}
p^\mu_s &= (|\vec{p}_s|;\vec{p}_s) = \omega_s\Big[{}_H\epsilon^\mu_{\tilde{\tau}}(\vec{p}_s) + {}_H\epsilon^\mu_{\tilde{3}}(\vec{p}_s)\Big], \qquad \text{when}\quad p^2_s = 0,\\
k^\mu_s(\vec{p}_s) &= \frac{1}{2\,|\vec{p}_s|^2}\,(|\vec{p}_s|;-\vec{p}_s) = \frac{1}{2\,\omega_s}\Big[{}_H\epsilon^\mu_{\tilde{\tau}}(\vec{p}_s) - {}_H\epsilon^\mu_{\tilde{3}}(\vec{p}_s)\Big],\\
{}_H\epsilon^\mu_{\tilde{\lambda}}(\vec{p}_s) &= \left(0;\,\delta^a_{\tilde{\lambda}} + \frac{p^a_s\,p_{s\tilde{\lambda}}}{|\vec{p}_s|\,(|\vec{p}_s| + p^3_s)},\,\frac{p_{s\tilde{\lambda}}}{|\vec{p}_s|}\right), \qquad \tilde{\lambda} = \tilde{1},\tilde{2},\\
p^2_s &= k^2_s(\vec{p}_s) = 0, \quad p_s\cdot k_s(\vec{p}_s) = 1,\\
p_s &\cdot\ {}_H\epsilon_{\tilde{\lambda}}(\vec{p}_s) = k_s(\vec{p}_s)\cdot{}_H\epsilon_{\tilde{\lambda}}(\vec{p}_s) = 0, \quad {}_H\epsilon_{\tilde{\lambda}}(\vec{p}_s)\cdot{}_H\epsilon_{\tilde{\lambda}'}(\vec{p}_s) = -\delta_{\tilde{\lambda}\tilde{\lambda}'},\\
\eta^{\mu\nu} &= p^\mu_s\,k^\nu_s(\vec{p}_s) + p^\nu_s\,k^\mu_s(\vec{p}_s) - \sum_{\tilde{\lambda}=\tilde{1}}^{\tilde{2}}{}_H\epsilon^\mu_{\tilde{\lambda}}(\vec{p}_s)\;{}_H\epsilon^\nu_{\tilde{\lambda}}(\vec{p}_s),\\
{}_H\epsilon^\mu_{\tilde{\tau}}(\vec{p}_s) &= \frac{p^\mu_s}{2\,\omega_s} + \omega_s\,k^\mu_s(\vec{p}_s),\\
{}_H\epsilon^\mu_{\tilde{3}}(\vec{p}_s) &= \frac{p^\mu_s}{2\,\omega_s} - \omega_s\,k^\mu_s(\vec{p}_s).
\end{aligned}
\tag{A.17}
$$

From Refs. [602–605] and from appendix A of Ref. [175] we get that the transformation properties of the polarization vectors are

$$
\begin{aligned}
&{}_H\epsilon^\mu_{\tilde{A}}((\vec{\Lambda p})_s) = \Lambda^\mu_{\ \nu}\;{}_H\epsilon^\nu_{\tilde{B}}(\vec{p}_s)\,\mathcal{R}(\vec{p}_s,\Lambda)^{\tilde{B}}_{\ \tilde{A}},\\
&\mathcal{R}(\vec{p}_s,\Lambda) = [{}_HL^{-1}(p_s,\overset{\circ}{p}_s)]\,\Lambda^{-1}\,[{}_HL(\Lambda p_s,\overset{\circ}{p}_s)] \in E_2 \subset O(3,1),
\end{aligned}
\tag{A.18}
$$

where Λ is a Lorentz transformation in $O(3,1)$. Therefore, we have to find the Wigner matrix belonging to the little group E_2 of $\overset{\circ}{p}{}^\mu_s = \omega_s\,(1;0,0,1)$. If Λ is obtained from the $SL(2,C)$ matrix $\begin{pmatrix}\alpha & \beta\\ \gamma & \delta\end{pmatrix}$ with $\alpha\,\delta - \beta\,\gamma = 1$, we obtain the following parametrization of the Wigner matrix:

$$
\begin{aligned}
&\mathcal{R}(\vec{p}_s,\Lambda) = \begin{pmatrix}
1+\frac{1}{2}\,\mathbf{u}^2 & u^1 & u^2 & -\frac{1}{2}\,\mathbf{u}^2\\
u^1\,\cos 2\theta + u^2\,\sin 2\theta & \cos 2\theta & \sin 2\theta & -u^1\,\cos 2\theta - u^2\,\sin 2\theta\\
-u^1\,\sin 2\theta + u^2\,\cos 2\theta & -\sin 2\theta & \cos 2\theta & u^1\,\sin 2\theta - u^2\,\cos 2\theta\\
\frac{1}{2}\,\mathbf{u}^2 & u^1 & u^2 & 1-\frac{1}{2}\,\mathbf{u}^2
\end{pmatrix},\\
&\quad \mathbf{u}^2 = (u^1)^2 + (u^2)^2, \qquad e^{i\,\theta} = \frac{d^*}{|d|},\\
&\quad u^1\,\cos 2\theta + u^2\,\sin 2\theta + i\left(u^1\,\sin 2\theta - u^2\,\cos 2\theta\right) = \frac{\omega_s}{|\vec{p}_s|}\,\frac{a\,c^* + b\,d^*}{|c|^2 + |d|^2},\\
&\quad a = \delta\,(|\vec{p}_s| + p^3_s) - \gamma\,(p^1_s - i\,p^2_s),\\
&\quad b = -\beta\,(|\vec{p}_s| + p^3_s) + \alpha\,(p^1_s - i\,p^2_s),\\
&\quad c = -\gamma\,(|\vec{p}_s| + p^3_s) - \delta\,(p^1_s + i\,p^2_s),\\
&\quad d = \alpha\,(|\vec{p}_s| + p^3_s) + \beta\,(p^1_s + i\,p^2_s).
\end{aligned}
\tag{A.19}
$$

Therefore, we get

$$
\begin{aligned}
(\Lambda p_s)^\mu &= \Lambda^\mu_\nu p^\nu_s,\\
{}_H\epsilon^\mu_{\tilde\lambda=\tilde 1}(\vec{\Lambda p_s}) &= \Lambda^\mu_\nu \Big[\cos 2\theta\ {}_H\epsilon^\nu_{\tilde\lambda=\tilde 1}(\vec p_s) - \sin 2\theta\ {}_H\epsilon^\nu_{\tilde\lambda=\tilde 2}(\vec p_s) + u^1\, p^\nu_s\Big],\\
{}_H\epsilon^\mu_{\tilde\lambda=\tilde 2}(\vec{\Lambda p_s}) &= \Lambda^\mu_\nu \Big[\sin 2\theta\ {}_H\epsilon^\nu_{\tilde\lambda=\tilde 1}(\vec p_s) + \cos 2\theta\ {}_H\epsilon^\nu_{\tilde\lambda=\tilde 2}(\vec p_s) + u^2\, p^\nu_s\Big],\\
k^\mu_s(\vec{\Lambda p_s}) &= \Lambda^\mu_\nu \Big[k^\nu_s(\vec p_s) + \frac{1}{2}\,\mathbf{u}^2\, p^\nu_s + \frac{u^1\,\cos 2\theta + u^2\,\sin 2\theta}{\omega_s}\ {}_H\epsilon^\nu_{\tilde\lambda=\tilde 1}(\vec p_s)\\
&\quad - \frac{u^1\,\sin 2\theta - u^2\,\cos 2\theta}{\omega_s}\ {}_H\epsilon^\nu_{\tilde\lambda=\tilde 2}(\vec p_s)\Big],\\
{}_H\epsilon^\mu_{\tilde\tau}(\vec{\Lambda p_s}) &= \frac{1}{2\,\omega_s}\,(\Lambda\, p_s)^\mu + \omega_s\, k^\mu_s(\vec{\Lambda p_s}) = \Lambda^\mu_\nu \Big[{}_H\epsilon^\nu_{\tilde\tau}(\vec p_s) + \frac{\omega_s}{2}\,\mathbf{u}^2\, p^\nu_s\\
&\quad + (u^1\,\cos 2\theta + u^2\,\sin 2\theta)\ {}_H\epsilon^\nu_{\tilde\lambda=\tilde 1}(\vec p_s)\\
&\quad - (u^1\,\sin 2\theta - u^2\,\cos 2\theta)\ {}_H\epsilon^\nu_{\tilde\lambda=\tilde 2}(\vec p_s)\Big],\\
{}_H\epsilon^\mu_{\tilde 3}(\vec{\Lambda p_s}) &= \frac{1}{2\,\omega_s}\,(\Lambda\, p_s)^\mu - \omega_s\, k^\mu_s(\vec{\Lambda p_s}) = \Lambda^\mu_\nu \Big[{}_H\epsilon^\nu_{\tilde 3}(\vec p_s) - \frac{\omega_s}{2}\,\mathbf{u}^2\, p^\nu_s\\
&\quad - (u^1\,\cos 2\theta + u^2\,\sin 2\theta)\ {}_H\epsilon^\nu_{\tilde\lambda=\tilde 1}(\vec p_s)\\
&\quad + (u^1\,\sin 2\theta - u^2\,\cos 2\theta)\ {}_H\epsilon^\nu_{\tilde\lambda=\tilde 2}(\vec p_s)\Big].
\end{aligned} \tag{A.20}
$$

A circular basis is

$$
\begin{aligned}
{}_H\epsilon^\mu_{(\pm)}(\vec p_s) &= \frac{1}{\sqrt 2}\Big[{}_H\epsilon^\mu_{\tilde\lambda=\tilde 1}(\vec p_s) \pm i\ {}_H\epsilon^\mu_{\tilde\lambda=\tilde 2}(\vec p_s)\Big],\\
{}_H\epsilon^\mu_{(\pm)}(\vec{\Lambda p_s}) &= \Lambda^\mu_\nu \Big(e^{\pm 2\, i\,\theta}\ {}_H\epsilon^\nu_{(\pm)}(\vec p_s) + \frac{u^1 \pm i\, u^2}{\sqrt 2}\, p^\nu_s\Big).
\end{aligned} \tag{A.21}
$$

Under the infinitesimal transformations $\Lambda = 1 + \zeta \approx 1 + \vec\rho\cdot\vec K + \vec\sigma\cdot\vec J = 1 + \sigma^3\, J^3 + \rho^3\, K^3 + \vec e_\perp\cdot\vec E + \vec f_\perp\cdot\vec F$[4] generated by the $SL(2,C)$ matrix $\begin{pmatrix} 1+\alpha^1+i\,\alpha^2 & \beta^1+i\,\beta^2\\ \gamma^1+i\,\gamma^2 & 1-\alpha^1-i\,\alpha^2\end{pmatrix}$,[5] we get

$$
\begin{aligned}
\mathcal{R}(\vec p_s,\Lambda) &\approx 1 + 2\,\theta\, J^3 - 2\,\vec u\cdot\vec E,\\
\theta &= -\frac{1}{2}\Big(\sigma^3 + \frac{\vec\sigma\cdot\vec p_{s\perp} - \epsilon^{ab}\, p^a_s\,\rho^b}{|\vec p_s| + p^3_s}\Big),\\
u^a(\vec\rho) &= \frac{\omega_s}{2\,|\vec p_s|^2\,(|\vec p_s| + p^3_s)}\Big[(\rho^3\, p^a_s - 2\,\rho^a\,|\vec p_s|)\,(|\vec p_s| + p^3_s) + 2\, p^a_s\,\vec\rho\cdot\vec p_{s\perp}\Big],\\
(\Lambda p_s)^\mu &\approx (\eta^\mu_\nu + \zeta^\mu_\nu)\, p^\nu_s,\\
k^\mu_s(\vec{\Lambda p_s}) &\approx (\eta^\mu_\nu + \zeta^\mu_\nu)\, k^\nu_s(\vec p_s) + \frac{1}{\omega_s}\Big(u^1\ {}_H\epsilon^\mu_{\tilde\lambda=\tilde 1}(\vec p_s) + u^2\ {}_H\epsilon^\mu_{\tilde\lambda=\tilde 2}(\vec p_s)\Big),
\end{aligned}
$$

[4] $\vec J$ and $\vec K$ are the generators of rotations and boosts, respectively, while E^a, F^a, $a = 1, 2$, are their linear combinations in the null plane basis.

[5] So that we have $\rho^1 = \gamma^1 + \beta^1$, $\rho^2 = \gamma^2 - \beta^2$, $\rho^3 = 4\,\alpha^1$, $\sigma^1 = \beta^2 + \gamma^2$, $\sigma^2 = \beta^1 - \gamma^1$, $\sigma^3 = 2\,\alpha^3$; $e^1_\perp = -(\sigma^2 + \rho^1)$, $e^2_\perp = \sigma^1 - \rho^2$, $f^1_\perp = \frac{1}{2}\,(\sigma^2 - \rho^1)$, $f^2_\perp = -\frac{1}{2}\,(\sigma^1 + \rho^2)$.

$$
{}_H\epsilon^\mu_{\tilde\lambda}(\Lambda \vec{p}_s) \approx (\eta^\mu_\nu + \zeta^\mu_\nu)\ {}_H\epsilon^\nu_{\tilde\lambda}(\vec{p}_s) + u^a\, p^\mu_s + \left(\sigma^3 - \frac{\epsilon^{ab}\, e^a_\perp\, p^b_s}{|\vec{p}_s| + p^3_s}\right) \epsilon_{\tilde\lambda\tilde\lambda'}\ {}_H\epsilon^\mu_{\tilde\lambda'}(\vec{p}_s),
$$

$$
{}_H\epsilon_{(\pm)}(\Lambda \vec{p}_s) \approx (\eta^\mu_\nu + \zeta^\mu_\nu)\ {}_H\epsilon^\nu_{pm)}(\vec{p}_s) + \frac{u^1 \pm i\, u^2}{\sqrt{2}}\, p^\mu_s \mp i \left(\sigma^3 - \frac{\epsilon^{ab}\, e^a_\perp\, p^b_s}{|\vec{p}_s| + p^3_s}\right)\ {}_H\epsilon^\mu_{(\pm)}(\vec{p}_s),
$$

$$
{}_H\epsilon^\mu_{\tilde\tau}(\Lambda \vec{p}_s) \approx (\eta^\mu_\nu + \zeta^\mu_\nu)\ {}_H\epsilon^\nu_{\tilde\tau}(\vec{p}_s) + u^1\ {}_H\epsilon^\mu_{\tilde\lambda=\tilde 1}(\vec{p}_s) + u^2\ {}_H\epsilon^\mu_{\tilde\lambda=\tilde 2}(\vec{p}_s),
$$

$$
{}_H\epsilon^\mu_{\tilde 3}(\Lambda \vec{p}_s) \approx (\eta^\mu_\nu + \zeta^\mu_\nu)\ {}_H\epsilon^\nu_{\tilde 3}(\vec{p}_s) - u^1\ {}_H\epsilon^\mu_{\tilde\lambda=\tilde 1}(\vec{p}_s) - u^2\ {}_H\epsilon^\mu_{\tilde\lambda=\tilde 2}(\vec{p}_s). \tag{A.22}
$$

The covariance properties of the indices $\tilde A = (\tau, \tilde r) = (\tau, \lambda, \tilde 3)$ is reduced to the covariance under the little group $E(2)$ of light-like orbits. Geometrically the time-like quantity $l^\mu = {}_H\epsilon^\mu_{\tilde\tau}(\vec{p}_s)$ is not a 4-vector. For massless isolated systems with $p^2_s \approx 0$, we can define non-covariant Wigner helicity space-like hyper-planes as those hyper-planes whose normal is $l^\mu = {}_H\epsilon^\mu_{\tilde\tau}(\vec{p}_s)$. This allows an instant form description, the non-covariant helicity instant form, for massless isolated systems.

The stability group $E_2 \approx T^2 \times O(2)$ of $\overset{\circ}{p}{}^\mu_s = \omega_s\,(1; 0, 0, 1)$ is generated by J^3 and E^a. It is the group of rotations (J^3) and translations (E^a) in the two-dimensional Euclidean plane. From Eq. (A.28) we get $W^2 = -4\,\omega^2[(E^1)^2 + (E^2)^2]$. For $W^2 \neq 0$, the associated irreducible representations of the Poincaré group are infinite dimensional of two types: those with any integer value of J^3 and those with any half-integer value of J^3 (these representations do not correspond to any known particle). For $W^2 = 0$ the stability group is reduced to $O(2)$ ($E^a = 0$), we have $W^\mu = \lambda\, p^\mu$ ($\lambda = \frac{W^o}{p^o} = \frac{\vec{p}\cdot\vec{J}}{p^o}$ is the helicity), and the associated irreducible representations of the Poincaré group are specified by $p^2 = 0$, $\mathrm{sign}\, p^o$ and $\lambda = 0, \pm\frac{1}{2}, \pm 1, \dots$

The null plane basis is defined in the following way for any 4-vector A^μ:

$$
A^- = \frac{1}{2}\,(A^o - A^3), \qquad A^+ = A^o + A^3, \qquad \vec{A}_\perp = (A^1, A^2),
$$

$$
A^2 = (A^o)^2 - \vec{A}^2 = 2\,A^+\,A^- - \vec{A}^2_\perp, \qquad A^o = \frac{1}{2}\,A^+ + A^-, \qquad A^3 = \frac{1}{2}\,A^+ - A^-. \tag{A.23}
$$

For the coordinates we have: $x^- = \tau$ is the null plane, x^+ is the Newtonian time, while $ds^2 = (dx^o)^2 - d\vec{x}^2 = 2\,dx^+\,dx^- - d\vec{x}^2_\perp$ is the line element.[6] The Poisson brackets $\{x^\mu, p^\nu\} = -\epsilon\,\eta^{\mu\nu}$ become $\{x^-, p^+\} = \{x^+, p^-\} = -1$, $\{x^a, p^b\} = \delta^{ab}$.

The generators of the Lorentz algebra $J^i = L^i + S^i = \frac{1}{2}\,\epsilon^{ijk}\,J^{jk}$, $K^i = K^i_L + K^i_S = J^{io}$ are replaced by the following combinations ($\epsilon^{ab} = -\epsilon^{ba}$, $\epsilon^{12} = 1$):

$$
J^3, \qquad K^3 \overset{def}{=} J^{+-},
$$

$$
E^a \overset{def}{=} J^{-a} = -\frac{1}{2}\,(\epsilon^{ab}\,J^b + N^a) = E^a_L + E^a_S,
$$

$$
F^a \overset{def}{=} J^{+a} = \epsilon^{ab}\,J^b - K^a = F^a_F + F^a_S,
$$

[6] Instead, with $u = x^o + |\vec{x}|$, $v = x^o - |\vec{x}|$ we have $ds^2 = du\,dv - \frac{1}{4}\,(v-u)^2\,(d\theta^2 + \sin^2\theta\,d\phi^2)$.

$$\{J^3, E^a\} = \epsilon^{ab} E^b, \qquad \{K^3, E^a\} = -E^a,$$
$$\{J^3, F^a\} = \epsilon^{ab} F^b, \qquad \{K^3, F^a\} = F^a, \qquad \{E^a, F^b\} = \epsilon^{ab} J^3 - \delta^{ab} K^3. \tag{A.24}$$

Their action on a 4-vector A^μ is $(a = 1, 2)$

$$\{J^3, A^+\} = 0, \qquad \{J^3, A^-\} = 0, \qquad \{J^3, A^a\} = \epsilon^{ab} A^b,$$
$$\{K^3, A^+\} = A^+, \qquad \{K^3, A^-\} = -A^-, \qquad \{K^3, A^a\} = 0,$$
$$\{E^a, A^+\} = -A^a, \qquad \{E^a, A^-\} = 0, \qquad \{E^a, A^b\} = -\delta^{ab} A^-,$$
$$\{F^a, A^+\} = 0, \qquad \{F^a, A^-\} = -A^a, \qquad \{F^a, A^b\} = -\delta^{ab} A^+. \tag{A.25}$$

The stability group of the null plane $x^- = \tau$,[7] is a seven-parameter subgroup S_+ of the Poincaré group with generators by K^3 ($\in R^+$), J^3, and E^a ($\in E_2$), and by P^a, P^- ($\in T^3$).[8] The group S_+ has a semi-direct product structure $S_+ = G_o| \times G_+$ with $E^a, P^+ \in G_+$ and $K^3, J^3, P^a \in G_o$. Both G_o, and G_+ are subgroups of the Poincaré group. The null plane helicity,

$$\mathcal{S}^3 = J^3 + \frac{1}{P^+} \epsilon^{ab} E^a P^b = \frac{W^+}{P^+} = \frac{W^o + W^3}{P^o + P^3}, \tag{A.26}$$

is a Casimir invariant of the stability group S_+, because it has zero Poisson bracket with all its generators.

The (interaction-dependent) Hamiltonians of the front form of the dynamics are only three, P^+, F^a, $(a = 1, 2)$ and form an Abelian subgroup G_- of the Poincaré group. $P^+ > 0$ gives the evolution in τ, while the F^a rotate the null plane around the light-cone. These three Hamiltonians may be replaced with a $U(2)$ dynamical algebra [605] with generators $\bar{M} = \sqrt{\epsilon P^2}$ (the invariant mass) and a dynamical spin with components $\mathcal{S}^3$ and

$$\mathcal{S}^a = \frac{1}{|\vec{P}|} \Big(W^a - \frac{P^a}{P^+} W^+ \Big), \tag{A.27}$$

with the algebra $\{\bar{M}, \mathcal{S}^i\} = 0$, $\{\mathcal{S}^i, \mathcal{S}^j\} = \epsilon^{ijk} \mathcal{S}^k$. This dynamical algebra[9] commutes with all the generators of the stability group S_+ except J^3,[10] and is not a subalgebra of the Poincaré algebra ($\bar{M}$ and $\vec{\mathcal{S}}$ are non-linear in the Poincaré

[7] Tangent to the light-cone $2\,(x^- + \tau)\, x^+ - \vec{x}_\perp^2 = 0$ along a directrix.

[8] $S_=$ acts transitively not only on the null plane, but also on the mass-shell $p^2 = m^2$, $p^o > 0$.

[9] Let us note that also in the rest-frame instant form of dynamics in the gauge $\vec{K} \approx 0$, implying $\vec{q}_+ \approx 0$, the internal Poincaré algebra is reduced to a $U(2)$ algebra with generators $\bar{M}$ and $\vec{\bar{S}}$. However, only $\bar{M}$ is a reduced (interaction-dependent) Hamiltonian. The four Hamiltonians of the instant form, namely the external energy and the external boosts, are functions of the $U(2)$ generators.

[10] The null plane helicity $\mathcal{S}^3$ is intermediate between a purely kinematical quantity like the generators of S_+ and a Hamiltonian: on the one hand it involves only generators of the stability group S_+, on the other hand it is needed to close the dynamical algebra $U(2)$.

generators). The Hamiltonians $P^+ = \frac{1}{2P^-}(\bar{M}^2 - \vec{P}_\perp^2)$ and $F^a = \frac{1}{P^-}[P^a K^3 + P^+ E^a - \epsilon^{ab}(P^b \mathcal{S}^3 + \bar{M}\mathcal{S}^b)]$ do not generate an invariant subgroup of the Poincaré group. It is clear that the real (so-called) reduced Hamiltonians replacing P^+, F^a are $\bar{M}^2$ and $\bar{M}\mathcal{S}^a$.

For the Pauli–Lubanski 4-vector we get the following presentations:

$$
\begin{aligned}
W^\mu &= \frac{1}{2}\epsilon^{\mu\nu\rho\sigma} P_\nu J_{\rho\sigma} = (\vec{P}\cdot\vec{S}; P^o\vec{S} + \vec{P}\times\vec{K}_S)\\
&= \Big(\Big(\frac{1}{2}P^+ - P^-\Big) S^3 + \epsilon^{ab} P^a \Big(E^b_S - \frac{1}{2}F^b_S\Big);\\
&\qquad \epsilon^{ab}(P^+ E^b_S - P^- F^b_S + P^a K^3_S), \Big(\frac{1}{2}P^+ + P^-\Big) S^3 - \epsilon^{ab} P^a\Big(E^b_S + \frac{1}{2}F^b_S\Big)\Big),\\
\tilde{W}^\mu &= {}_HL^{-1}(p_s, \overset{\circ}{p}_s)^\mu{}_\nu W^\nu = \left(\omega_s \frac{\vec{P}\cdot\vec{S}}{|\vec{P}|}; \mathcal{S}^a = W^a - P^a \frac{W^o + W^3}{|\vec{P}| + P^3}, \omega_s \frac{\vec{P}\cdot\vec{S}}{|\vec{P}|}\right),\\
&\qquad \text{(analogous of the Thomas spin in the time}-\text{like case)},\\
\overset{\circ}{W}{}^\mu &= \omega_s(\overset{\circ}{S}{}^{\tilde{3}}; 2\,\epsilon^{ab}\overset{\circ}{E}{}^b_S, \overset{\circ}{S}{}^{\tilde{3}}), \quad \tilde{W}^\mu \text{ in } \overset{\circ}{p}_s\text{–frame},\\
&\qquad \text{(it depends on the spin part of } E_2). \qquad\qquad (A.28)
\end{aligned}
$$

In Ref. [605] there is the definition of a null plane 3-position: (1) $Q^a_\perp = \frac{E^a}{P^-}$ conjugate to P^a; (2) some Q^-, conjugate to P^-. However, whichever definition we use for Q^-, it cannot be quantized to a self-adjoint operator [605], as happens for the photon [175].

Let us note that for an isolated system with $p^2 \approx 0$, a canonical 4-center-of-mass variable $\tilde{x}^\mu_s$ has a reduced covariance either under E_2 (continuous spin case) or under $O(2)$ (discrete spin case).

Usually the helicity is defined as $\Sigma = S_{\mu\nu}\, {}_H\epsilon^\mu_{\tilde{1}}(\vec{p}_s)\, {}_H\epsilon^\nu_{\tilde{2}}(\vec{p}_s)$[11] and we have $W^\mu \approx P^\mu \Sigma$, $\tilde{W}^\mu \approx \overset{\circ}{W}{}^\mu \approx \overset{\circ}{P}{}^\mu\, \Sigma$.

A.5 A Canonical Transformation Induced by the Wigner Boost

In all the particle models there is a canonical realization of the Poincaré group, whose ten conserved generators and Casimir invariants are ($W^\mu = \frac{1}{2}\epsilon^{\mu\nu\rho\sigma} J_{\nu\rho}\, p_\sigma$ is the Pauli–Lubanski 4-vector and $\vec{S}_T$ is the rest-frame Thomas spin for time-like Poincaré orbits)

$$
\begin{aligned}
p^\mu &= p^\mu_1 + p^\mu_2, \qquad J^{\mu\nu} = x^\mu_1 p^\nu_1 - x^\nu_1 p^\mu_1 + x^\mu_2 p^\nu_2 - x^\nu_2 p^\mu_2,\\
p^2&, \qquad W^2 = -p^2\, \vec{S}^2_T, \quad \text{when } \epsilon p^2 > 0. \qquad\qquad (A.29)
\end{aligned}
$$

[11] We have $S^{\mu\nu} = {}_H\epsilon^\mu_{\tilde{\lambda}}(\vec{p}_s)\, {}_H\epsilon^\nu_{\tilde{\lambda}'}\, S_{\tilde{\lambda}\tilde{\lambda}'}$, $S_{\tilde{\lambda}\tilde{\lambda}'} = \epsilon_{\tilde{\lambda}\tilde{\lambda}'}\,\Sigma$, , $S^i = \frac{P^i_s}{|\vec{p}_s|}\Sigma$, $K^i_S = 0$.

A first canonical transformation, defining a naive canonical, covariant center-of-mass 4-variable x^μ, is

$$
\begin{aligned}
x^\mu &= \frac{1}{2}\,(x_1^\mu + x_2^\mu), \qquad p^\mu = p_1^\mu + p_2^\mu,\\
R^\mu &= x_1^\mu - x_2^\mu, \qquad Q^\mu = \frac{1}{2}\,(p_1^\mu - p_2^\mu), \qquad Q_\perp^\mu = \left(\eta^{\mu\nu} - \frac{p^\mu\, p^\nu}{p^2}\right) Q_\nu,\\
x_1^\mu &= x^\mu + \frac{1}{2}\,R^\mu, \qquad x_2^\mu = x^\mu - \frac{1}{2}\,R^\mu,\\
p_1^\mu &= \frac{1}{2}\,p^\mu + Q^\mu p_2^\mu = \frac{1}{2}\,p^\mu - Q^\mu,\\
&\{x^\mu, p^\nu\} = \{R^\mu, Q^\nu\} = -{}^4\eta^{\mu\nu},\\
J^{\mu\nu} &= L^{\mu\nu} + S^{\mu\nu}, \qquad L^{\mu\nu} = x^\mu\, p^\nu - x^\nu\, p^\mu, \quad S^{\mu\nu} = R^\mu\, Q^\nu - R^\nu\, Q^\mu. \qquad \text{(A.30)}
\end{aligned}
$$

Let us now consider the stratum of the constraint sub-manifold of phase space defined by the mass-shell constraints $\epsilon\, p_i^2 - m_i^2 \approx 0$ containing the configurations belonging to time-like Poincaré orbits. A second canonical transformation [27] generates relative canonical variables adapted to the time-like Poincaré orbits, namely boosted at rest with the standard Wigner boost $L^\mu_\nu(\overset{\circ}{p}, p)$ sending p^μ to $\overset{\circ}{p}{}^\mu = \eta\,\sqrt{\epsilon p^2}\,(1; \vec{0})$, $\eta = \operatorname{sign} p^o$, whose columns are the vierbeins $\epsilon^\mu_A(u(p))$. As shown in Ref. [27] the exponential form of the boost (Eq. A.7) is ($\vec{\beta} = \frac{\vec{p}}{\eta\, p^o}$ is the parameter of the Lorentz transformation; we have $\det|L(\overset{\circ}{p}, p)| = +1$ for both signs of the energy p^o):

$$
\begin{aligned}
L^\mu_\nu(\overset{\circ}{p}, p) &= \left(e^{-\omega(p)\, I(p)}\right)^\mu_{\ \nu}\\
&= \left[\cosh\left(\omega(p) I(p)\right) + \sinh\left(\omega(p) I(p)\right)\right]^\mu_{.\nu}\\
&= \left[1 - I^2(p) + I^2(p)\cosh\omega(p) + I(p)\sinh\omega(p)\right]^\mu_{.\nu},\\
I^\mu_\nu(p) &= \begin{pmatrix} 0, & -\frac{p_j}{|\vec{p}|}\\ \frac{p^i}{|\vec{p}|} & 0 \end{pmatrix},\\
\cosh\omega(p) &= \frac{p^o}{\eta\,\sqrt{\epsilon\, p^2}} = \gamma = (1 - \vec{\beta}^2)^{-1/2}, \qquad \sinh\omega(p) = \frac{|\vec{p}|}{\eta\,\sqrt{\epsilon\, p^2}} = \eta\,|\vec{\beta}|\,\gamma,\\
I_{\mu\nu}(p) &= -I_{\nu\mu}(p), \qquad I^3(p) = I(p). \qquad \text{(A.31)}
\end{aligned}
$$

Then, the canonical transformation

$$
\begin{aligned}
e^{\{\bar{\psi},.\}}\,\bar{f}(x, p, R, Q) &= \bar{f} + \{\bar{\psi}, \bar{f}\} + \frac{1}{2}\,\{\bar{\psi}, \{\bar{\psi}, \bar{f}\}\} + \dots,\\
\bar{\psi} &= \frac{1}{2}\,\omega(p)\, I_{\mu\nu}(p)\, S^{\mu\nu} \qquad \text{(A.32)}
\end{aligned}
$$

is a point transformation in p^μ, linear in x^μ, R^μ, Q^μ, to the new canonical basis $\tilde{x}^\mu$, p^μ, $\tilde{R}^\mu = L^\mu_\nu(\overset{\circ}{p}, p)\, R^\nu = (T_R; \vec{\rho})$, $\tilde{Q}^\mu = L^\mu_\nu(\overset{\circ}{p}, p)\, Q^\nu = (\epsilon_R; \vec{\pi})$ (with $R^\mu = T_R\,\frac{p^\mu}{\eta\,\sqrt{p^2}} - \sum_{r=1}^3 \rho_r\,\epsilon^\mu_r(u(p))$, $Q^\mu = \epsilon_R\,\frac{p^\mu}{\eta\,\sqrt{p^2}} - \sum_{r=1}^3 \pi_r\,\epsilon^\mu_r(u(p))$):

$$\tilde{x}^\mu = x^\mu + \frac{1}{2}\,\epsilon^A_\nu(u(p))\,\eta_{AB}\,\frac{\partial\epsilon^B_\rho(u(p))}{\partial p_\mu}\,S^{\nu\rho}$$
$$= x^\mu - \frac{1}{\eta\,\sqrt{\epsilon\,p^2}\,(p^o + \eta\,\sqrt{\epsilon\,p^2})}\,\Big[p_\nu\,S^{\nu\mu} + \eta\,\sqrt{\epsilon\,p^2}\,\Big(S^{o\mu} - S^{o\nu}\,\frac{p_\nu\,p^\mu}{p^2}\Big)\Big],$$

p^μ,

$$T_R = \epsilon^o_\mu(u(p))\,R^\mu = \frac{p\cdot R}{\eta\,\sqrt{\epsilon\,p^2}},$$
$$\epsilon_R = \epsilon^o_\mu(u(p))\,Q^\mu = \frac{p\cdot Q}{\eta\,\sqrt{\epsilon\,p^2}},$$
$$\rho^r = \epsilon^r_\mu(u(p))\,R^\mu = R^r - \frac{p^r}{\eta\,\sqrt{\epsilon\,p^2}}\,\Big(R^o - \frac{\vec{p}\cdot\vec{R}}{p^o + \eta\,\sqrt{\epsilon\,p^2}}\Big),$$
$$\pi^r = \epsilon^r_\mu(u(p))\,Q^\mu = Q^r - \frac{p^r}{\eta\,\sqrt{\epsilon\,p^2}}\,\Big(Q^o - \frac{\vec{p}\cdot\vec{Q}}{p^o + \eta\,\sqrt{\epsilon\,p^2}}\Big),$$
$$\{\tilde{x}^\mu, p^\nu\} = -\eta^{\mu\nu},\quad \{T_R, \epsilon_R\} = -1,\quad \{\rho^r, \pi^s\} = \delta^{rs},$$
$$p^\mu,\qquad J^{ij} = \tilde{L}^{ij} + \tilde{S}^{ij},\qquad J^{io} = \tilde{L}^{io} + \tilde{S}^{io} = \tilde{L}^{io} + \frac{\tilde{S}^{ij}\,p^j}{\eta\,\sqrt{\epsilon\,p^2} + p^o},$$
$$\tilde{L}^{\mu\nu} = \tilde{x}^\mu\,p^\nu - \tilde{x}^\nu\,p^\mu,\qquad \tilde{S}^{ij} = \rho^i\,\pi^j - \rho^j\,\pi^i. \tag{A.33}$$

Since we have $\eta^{\mu\nu} - u^\mu(u(p))\,u^\nu(u(p)) = -\sum_{r=1}^3 \epsilon^\mu_r(u(p))\,\epsilon^\nu(u(p))$, we get $R^\mu_\perp = -\sum_{r=1}^3 \rho_r\,\epsilon^\mu_r(u(p))$ and $Q^\mu_\perp = -\sum_{r=1}^3 \pi_r\,\epsilon^\mu_r(u(p))$. The series giving $\tilde{x}^\mu$ can be re-summed due to $I^3(p) = I(p)$. We have also shown the new form of the Poincaré generators. The inverse canonical transformation is

$$x^o = \tilde{x}^o + \frac{1}{p^2}\,\Big(T_R\,\vec{p}\cdot\vec{\pi} - \epsilon_R\,\vec{p}\cdot\vec{\rho}\Big),$$
$$\vec{x} = \vec{\tilde{x}} + \frac{1}{\eta\,\sqrt{\epsilon\,p^2}}\,\Big(T_R\,\vec{\pi} - \epsilon_R\,\vec{\rho}\Big) + \frac{(\vec{p}\cdot\vec{\rho})\,\vec{\pi} - (\vec{p}\cdot\vec{\pi})\,\vec{\rho}}{\eta\,\sqrt{\epsilon\,p^2}\,(\eta\,\sqrt{\epsilon\,p^2} + p^o)} + \frac{T_R\,\vec{p}\cdot\vec{\pi} - \epsilon_R\,\vec{p}\cdot\vec{\rho}}{p^2\,(\eta\,\sqrt{\epsilon\,p^2} + p^o)}\,\vec{p},$$
$$R^o = \frac{1}{\eta\,\sqrt{\epsilon\,p^2}}\,\Big(T_R\,p^o + \vec{p}\cdot\vec{\rho}\Big),\qquad \vec{R} = \vec{\rho} + \frac{\vec{p}}{\eta\,\sqrt{\epsilon\,p^2}}\,\Big(T_R + \frac{\vec{p}\cdot\vec{\rho}}{\eta\,\sqrt{\epsilon\,p^2} + p^o}\Big),$$
$$Q^o = \frac{1}{\eta\,\sqrt{\epsilon\,p^2}}\,\Big(\epsilon_R\,p^o + \vec{p}\cdot\vec{\pi}\Big),\qquad \vec{Q} = \vec{\pi} + \frac{\vec{p}}{\eta\,\sqrt{\epsilon\,p^2}}\,\Big(\epsilon_R + \frac{\vec{p}\cdot\vec{\pi}}{\eta\,\sqrt{\epsilon\,p^2} + p^o}\Big),$$

$$\Rightarrow R^2_\perp = -\vec{\rho}^2,\qquad Q^2_\perp = -\vec{\pi}^2,\qquad R_\perp\cdot Q_\perp = -\vec{\rho}\cdot\vec{\pi}. \tag{A.34}$$

Under a Poincaré transformation with parameters (a, Λ) the variables (Eq. A.33) transform in the following way:

$$\epsilon^\mu_r(u(\Lambda p)) = (R^{-1}(\Lambda, p))_r^{\ s}\,\Lambda^\mu_\nu\,\epsilon^\nu_s(u(p)),$$
$$p'^\mu = \Lambda^\mu_\nu\,p^\nu,$$

$$\begin{aligned}\tilde{x}'^{\mu} &= \Lambda^{\mu}_{\nu}\left[\tilde{x}^{\nu} + \frac{1}{2}\,\bar{S}_{rs}\,R^{r}_{k}(\Lambda,p)\,\frac{\partial}{\partial p_{\nu}}\,R^{s}_{k}(\Lambda,p)\right] + a^{\mu}\\ &= \Lambda^{\mu}_{\nu}\left\{\tilde{x}^{\nu} + \frac{\bar{S}_{rs}}{\Lambda^{o}_{\alpha}\,p^{\alpha} + \eta\,\sqrt{\epsilon\,p^{2}}}\left[\eta^{\nu}_{r}\left(\Lambda^{o}_{s} - \frac{(\Lambda^{o}_{o}-1)\,p_{s}}{p^{o}+\eta\,\sqrt{\epsilon\,p^{2}}}\right)\right.\right.\\ &\quad\left.\left. - \frac{(p^{\nu}+\eta^{\nu}_{o}\,\eta\,\sqrt{\epsilon\,p^{2}})\,p_{r}\,\Lambda^{o}_{s}}{\eta\,\sqrt{\epsilon\,p^{2}}\,(p^{o}+\eta\,\sqrt{\epsilon\,p^{2}})}\right]\right\} + a^{\mu},\end{aligned}$$

$$\begin{aligned}&T^{'}_{R} = T_{R}, \qquad \epsilon^{'}_{R} = \epsilon_{R},\\ &\rho^{'}_{r} = \rho_{s}\,R^{s}_{r}(\Lambda,p), \qquad \pi^{'}_{r} = \pi_{s}\,R^{s}_{r}(\Lambda,p),\end{aligned} \tag{A.35}$$

where $R^{\mu}_{\nu}(\Lambda,p)$ is the Wigner rotation of Eq. (A.11).

Therefore, $\tilde{x}^{\mu}$ is not a 4-vector and $\vec{\rho}_{a}, \vec{\pi}_{a}$ are Wigner spin-1 3-vectors.

Appendix B

Grassmann Variables and Pseudo-Classical Lagrangians

Pseudo-classical mechanics allows us to give a semi-classical description of quantities like the spin, the electric charge, or the sign of the energy by using Grassmann variables. Berezin and Marinov [174, 606] show that the quantization reproduces the standard description of a quantum spin and of quantum electric charge and that moreover the introduction of a suitable distribution function allows us to make an average reproducing classical physics without the divergences associated with self-reaction. In this Appendix I give the tools needed for this formulation.

B.1 Grassmann Variables

A Grassmann algebra G_n with n generators $\xi^A \in G_n$, $A = 1, \ldots, n$, is defined by the following properties:

1. G_n is a vector space over the complex numbers C ($a\,\xi^A + b\,\xi^B \in G_n$, $a, b \in C$).
2. A product is defined over G_n, which is associative and bilinear with respect to addition and multiplication by scalars ($(a\,\xi^A + b\,\xi^B)\,\xi^C = a\,\xi^A\,\xi^C + b\,\xi^B\,\xi^C$, $(a\,\xi^A)\,(b\,\xi^B) = a\,b\,\xi^A\,\xi^B$).
3. There is a unit element $1 \in G_n$ for the product ($1\,\xi^a = \xi^A$).
4. G_n is generated by n elements ξ^A, $A = 1, \ldots, n$, obeying

$$\xi^A\,\xi^B + \xi^B\,\xi^A = 0, \qquad \Rightarrow\ (\xi^A)^2 = 0. \tag{B.1}$$

A generic element of G_n may be written in the form $g = g_o\,1 + g_A\,\xi^A + g_{AB}\,\xi^A\,\xi^B + \cdots + g_{A_1 \ldots A_n}\,\xi^{A_1} \ldots\ \xi^{A_n}$ with the complex numbers $g_{AB}, \ldots, g_{A_1 \ldots A_n}$ completely antisymmetric in their indices. An example is the Grassmann algebra of differential forms on a manifold M ($\dim M = n$): $g = g_o(x)\,1 + g_i(x)\,dx^i + g_{ij}(x)\,dx^i \wedge dx^j + \cdots + g_{i_1 \ldots i_n}(x)\,dx^{i_1} \wedge \ldots \wedge dx^{i_n}$.

We shall add the properties of Grassmann algebras useful for the applications. For a more general treatment, see Ref. [26].

A general dynamical system may have both even or bosonic (q^i) and odd or fermionic (θ^a) variables satisfying $q^i\, q^j - q^j\, q^i = 0$, $\theta^a\, q^i - q^i\, \theta^a = 0$, $\theta^a\, \theta^b + \theta^b\, \theta^a = 0$. For a function $f(q,\theta)$, we have $f(q,\theta) = f_o(q) + f_a(q)\, \theta^a + f_{ab}(q)\, \theta^a\, \theta^b + \ldots = f_E(q,\theta) + f_O(q,\theta)$, where $f_E(q,\theta) = f_o(q) + f_{ab}(q)\, \theta^a\, \theta^b + \ldots$ and $f_O(q,\theta) = f_a(q)\, \theta^a + f_{abc}(q)\, \theta^a\, \theta^b\, \theta^c + \ldots$ are the even and the odd part of the function respectively. We can introduce the Grassmann parity (a grading): (1) $\epsilon_f = 0\,(\text{mod}\,2)$ if $f = f_E$; (2) $\epsilon_f = 1\,(\text{mod}\,2)$ if $f = f_O$. If $f(q,\theta)$ has parity ϵ_f and $g(q,\theta)$ has parity ϵ_g, then $f\, g = (-)^{\epsilon_f\, \epsilon_g}\, g\, f$.

On functions $f(q,\theta)$ we can define both left derivatives $\frac{\partial}{\partial\theta^a} = \frac{\partial^L}{\partial\theta^a}$, corresponding to the variations $\delta f = \delta_L f = \delta\theta^a\, \frac{\partial f}{\partial\theta^a}$, and right derivatives $\frac{\partial^R}{\partial\theta^a}$, corresponding to the variations $\delta_R f = \frac{\partial^R f}{\partial\theta^a}\, \delta\theta^a$. If f has parity ϵ_f, both the left and right derivatives of f have parity $\epsilon_f - 1$, so that we have $\delta_R f = -(-)^{\epsilon_f}\, \delta\theta^a\, \frac{\partial^R f}{\partial\theta^a}$, i.e., $\frac{\partial f}{\partial\theta^a} = -(-)^{\epsilon_f}\, \frac{\partial^R f}{\partial\theta^a}$. Moreover, we have $\frac{\partial}{\partial\theta^a}\, \frac{\partial}{\partial\theta^b} + \frac{\partial}{\partial\theta^b}\, \frac{\partial}{\partial\theta^a} = 0$, $\frac{\partial}{\partial\theta^a}\, \frac{\partial^R}{\partial\theta^b} + \frac{\partial^R}{\partial\theta^b}\, \frac{\partial}{\partial\theta^a} = 0$. If f is odd ($\epsilon_f = 1$), we have $\frac{\partial}{\partial\theta^a}\, (f\, g) = \frac{\partial f}{\partial\theta^a}\, g - f\, \frac{\partial g}{\partial\theta^a}$. We shall always use left derivatives.

The complex conjugation (an involution) is defined as $(\theta^a\, \theta^b)^* = \theta^{a\,*}\, \theta^{b\,*}$, $(\theta^{a\,*})^* = \theta^a$, $(a\, \theta^a)^* = a^*\, \theta^{a\,*}$. The element θ^a is real if $\theta^{a\,*} = \theta^a$ and imaginary if $\theta^{a\,*} = -\theta^a$. If θ^a and θ^b are real, then $\theta^a\, \theta^b$ is imaginary.

We can also consider matrices $M(q,\theta) = M_o(q) + M_a(q)\, \theta^a + M_{ab}(q)\, \theta^a\, \theta^b + \ldots$ where $M_o(q)$, $M_a(q)$, $M_{ab}(q) = -M_{ba}(q)\ \ldots$ are ordinary matrices. As for functions we can speak of even and odd matrices. One such matrix is invertible if the matrix $M_o(q)$ is invertible. In such a case the inverse satisfies $M(q,\theta)\, M^{-1}(q,\theta) = M^{-1}(q,\theta)\, M(q,\theta) = 1$ and we have $M^{-1}(q,\theta) = M_o^{-1}(q) - \Big(M_o^{-1}(q)\, M_a(q)\, M_o^{-1}(q)\Big)\, \theta^a + \Big(\frac{1}{2}\, M_o^{-1}(q)\, (M_b(q)\, M_o^{-1}(q)\, M_a(q)\, M_o^{-1}(q) - M_a(q)\, M_o^{-1}(q)\, M_b(q)\, M_o^{-1}(q)) - M_o^{-1}(q)\, M_{ab}(q)\, M_o^{-1}(q)\Big)\, \theta^a\, \theta^b + \ldots$

A super-matrix M has the form $M = \begin{pmatrix} A & B \\ C & D \end{pmatrix}$ with A, D even matrices and B, C odd matrices. The super-trace is defined as $\text{Str}\, M = \text{Tr}\, A - \text{Tr}\, D$ and we have $\text{Str}\,(M_1 + M_2) = \text{Str}\, M_1 + \text{Str}\, M_2$, $\text{Str}\,(M_1\, M_2) = \text{Str}\,(M_2\, M_1)$, $\text{Str}\,[M_1, M_2] = 0$. The super-determinant is defined as $\text{Sdet}\, M = e^{\text{Str}\ln M}$. We have $\text{Sdet}\,(M_1\, M_2) = (\text{Sdet}\, M_1)\, (\text{Sdet}\, M_2)$, while for $M = 1 + \epsilon$ we get $\text{Sdet}\, M = 1 + \text{Str}\, \epsilon$. If A^{-1} and D^{-1} exist, so that M is invertible, we can write $M = \begin{pmatrix} A & 0 \\ C & 1 \end{pmatrix} \begin{pmatrix} 1 & A^{-1}\, B \\ 0 & D - C\, A^{-1}\, B \end{pmatrix} = \begin{pmatrix} 1 & B \\ 0 & D \end{pmatrix} \begin{pmatrix} A - B\, D^{-1}\, C & 0 \\ D^{-1}\, C & 1 \end{pmatrix}$, and we get $\text{Sdet}\, M = \det A\, \Big(\det\,(D - C\, A^{-1}\, B)\Big)^{-1} = (\det D)^{-1}\, \det\,(A - B\, D^{-1}\, C)$. The super-transpose of M is $M^T = \begin{pmatrix} A^T & C^T \\ -B^T & D^T \end{pmatrix}$, with the properties $(M_1\, M_2(^T = M_2^T\, M_1^T$, $\text{Str}\, M^T = \text{Str}\, M$, $(M^{-1})^T = (M^T)^{-1}$, $\text{Sdet}\, M^T = \text{Sdet}\, M$.

An allowed change of variables $q^{'\,i} = Q^{'\,i}(q,\theta)$, $\theta^{'\,a} = \theta^{'\,a}(q,\theta)$ must have the super-Jacobian matrix $J(q,\theta) = \frac{\partial^R (q^{'},\theta^{'})}{\partial(q,\theta)} = \begin{pmatrix} A & B \\ C & D \end{pmatrix}$ with the even matrices A and D invertible.[1] The inverse super-Jacobian matrix may be written in the following forms: $J^{-1} = \frac{\partial^R(q,\theta)}{\partial(q^{'},\theta^{'})} = \begin{pmatrix} (A - B\,D^{-1}\,C)^{-1} & 0 \\ -D^{-1}\,C\,(A - B\,D^{-1}\,C)^{-1} & 1 \end{pmatrix}$ $\begin{pmatrix} 1 & -B\,D^{-1} \\ 0 & D^{-1} \end{pmatrix} = \begin{pmatrix} 1 & -A^{-1}\,B\,(D - C\,A^{-1}\,B)^{-1} \\ 0 & (D - C\,A^{-1}\,B)^{-1} \end{pmatrix} \begin{pmatrix} A^{-1} & 0 \\ -C\,A^{-1} & 1 \end{pmatrix}$.

Berezin's rules for the integration over a Grassmann variable θ are $\int d\theta = 0$, $\int \theta\, d\theta = 1$. The integral of a function $f(\theta) = f_o + f_a\,\theta^a + \cdots + f_{n\ldots 1}\,\theta^n \ldots\, \theta^1$ is $\int f(\theta)\, d\theta = \int f(\theta)\, d\theta^1 \ldots\, d\theta^n = f_{n\ldots 1}$ and $\int \frac{\partial f}{\partial \theta^a}\, d\theta = 0$. The integration of functions $f(q,\theta)$ is well defined only if they are of compact support (contained in a region U) of the even q-space. In this case we have $\int f(q,\theta)\, dq\, d\theta = \int_U f_{n\ldots 1}(q)\, dq$. We have $\int \frac{\partial f(q,\theta)}{\partial \theta^a}\, dq\, d\theta = 0$, which implies that $\int f(q,\theta)\, dq\, d\theta$ is invariant under translations $\theta^a \mapsto \theta^a + \beta^a$. Inside Berezin integrals we have $d\theta^a\, d\theta^b + d\theta^b\, d\theta^a = 0$.

If $(q,\theta) \rightarrow (q^{'},\theta^{'})$ is an invertible change of variables and $f(q,\theta)$ is a function of compact support contained in U, then we have $\int f(q,\theta)\, dq\, d\theta = \int f(q(q^{'},\theta^{'}),\theta(q^{'},\theta^{'}))\, \mathrm{Sdet}\, \frac{\partial^R(q,\theta)}{\partial(q^{'},\theta^{'})}\, dq^{'}\, d\theta^{'}$.

The delta function for Grassmann variables, such that we have $\int f(\theta)\,\delta(\theta - \theta_o)\, d\theta = f(\theta_o)$, has the following presentations: $\delta(\theta - \theta_o) = \theta - \theta_o = \int e^{i\,(\theta-\theta_o)\,\pi}\, \frac{d\pi}{i}$ (π is an odd variable). Under an invertible change of variables we have $\prod_i \delta(q^i(q^{'},\theta^{'}) - q^i_o) \prod_a \delta(\theta^a(q^{'},\theta^{'}) - \theta^a_o) = \left(\mathrm{Sdet}\, \frac{\partial^R(q,\theta)}{\partial(q^{'},\theta^{'})}|_{q^{'}_o,\theta^{'}_o}\right)^{-1} \prod_i \delta(q^{'\,i} - q^{'\,i}_o) \prod_a \delta(\theta^{'\,a} - \theta^{'\,a}_o)$, with $q^i_o = q^i(q^{'\,k}_o, \theta^{'\,c}_o)$, $\theta^a_o = \theta^a(q^{'\,k}_o, \theta^{'\,c}_o)$.

From the Gaussian integrals $\int e^{-\frac{1}{2}\,q^i\,A_{ij}\,q^j}\, dq = (2\pi)^{\frac{n}{2}}\,(\det A)^{-\frac{1}{2}}$ (n is the number of even degrees of freedom) and $\int e^{-\theta^a_1\,D_{ab}\,\theta^b_2}\, d\theta_1\, d\theta_2 = \det D$, we get $\int e^{-\frac{1}{2}\,z^A\,M_{AB}\,z^B}\, dz = (2\pi)^{\frac{n}{2}}\,(\mathrm{Sdet}\, M)^{-\frac{1}{2}}$, where $z^A = (q^i, \theta^a)$, $dz = dq\, d\theta$ and M is a symmetric super-matrix.

B.2 Pseudo-Classical Lagrangian and Hamiltonian Theory

When we use mixed bosonic and fermionic variables for the semi-classical description of a dynamical system, the pseudo-classical even Lagrangian must contain the kinetic terms for the Grassmann variables. While in the framework of BRST theory regular Lagrangians are used [26], which imply a second-order dynamics for the fermionic degrees of freedom, for the description of particles with either spin or internal charges singular Lagrangians are employed [607, 608] and the Grassmann variables are described by a first-order formalism.

If $L(q^i, \dot{q}^i, \theta^a, \dot{\theta}^a) = L^*$ is a pseudo-classical Lagrangian depending on the bosonic configurational variables q^i and on the fermionic (real Grassmann) ones θ^a, the canonical momenta are defined as $p_i = \frac{\partial L}{\partial \dot{q}^i}$, $\pi_a = \frac{\partial^L L}{\partial \dot{\theta}^a}$, while the even Hamiltonian is $\bar{H} = \bar{H}^* = \dot{q}^i\, p_i + \dot{\theta}^a\, \pi_a - L$. We have $d\bar{H} = \dot{q}^i\, dp_i - \dot{\theta}^a$

[1] We have $A = \left(\frac{\partial q^{'\,i}}{\partial q^j}\right)$, $B = \left(\frac{\partial^R q^{'\,i}}{\partial \theta^b} = -\frac{\partial q^{'\,i}}{\partial \theta^b}\right)$, $C = \left(\frac{\partial \theta^{'\,a}}{\partial q^j}\right)$, $D = \left(\frac{\partial^R \theta^{'\,a}}{\partial \theta^b} = \frac{\partial \theta^{'\,a}}{\partial \theta^b}\right)$.

$d\pi_a - dq^i \frac{\partial L}{\partial q^i} - d\theta^a \frac{\partial^L L}{\partial \theta^a}$. This implies the Hamilton equations $\dot{q}^i \overset{\circ}{=} \frac{\partial \bar{H}}{\partial p_i}$, $\dot{\theta}^a \overset{\circ}{=} -\frac{\partial^L \bar{H}}{\partial \pi_a} = \frac{\partial^R \bar{H}}{\partial \pi_a}$, $\dot{p}_i \overset{\circ}{=} -\frac{\partial \bar{H}}{\partial q^i}$, $\dot{\pi}_a \overset{\circ}{=} -\frac{\partial^L \bar{H}}{\partial \theta^a}$.

To introduce Poisson brackets, let us consider a (even or odd) function $\bar{F}(q,\theta)$ and let us ask:

$$\frac{d\bar{F}(q,\theta)}{dt} \overset{\circ}{=} \sum_i \Big(\frac{\partial \bar{H}}{\partial p_i}\frac{\partial \bar{F}}{\partial q^i} - \frac{\partial \bar{H}}{\partial q^i}\frac{\partial \bar{F}}{\partial p_i}\Big) - \sum_a \Big(\frac{\partial^L \bar{H}}{\partial \pi_a}\frac{\partial^L \bar{f}}{\partial \theta^a} + \frac{\partial^L \bar{h}}{\partial \theta^a}\frac{\partial^L \bar{F}}{\partial \pi_a}\Big) \overset{def}{=} \{\bar{F}, \bar{H}\}. \tag{B.2}$$

As shown in Refs. [607, 608], this implies the following Poisson brackets for functions $\bar{F}$ and $\bar{G}$ with Grassmann parities ϵ_F, ϵ_g ($\epsilon_F = 0\,(\mathrm{mod}\,2)$ if $\bar{F}$ is an even function, $\epsilon_F = 1\,(\mathrm{mod}\,2)$ if $\bar{F}$ is an odd function):

$$\begin{aligned}
\{\bar{F}, \bar{G}\} &= \sum_i \left(\frac{\partial \bar{F}}{\partial q^i}\frac{\partial \bar{G}}{\partial p_i} - \frac{\partial \bar{F}}{\partial p_i}\frac{\partial \bar{G}}{\partial q^i}\right) + (-)^{\epsilon_F}\left(\frac{\partial^L \bar{F}}{\partial \theta^a}\frac{\partial^L \bar{G}}{\partial \pi_a} + \frac{\partial^L \bar{F}}{\partial \pi_a}\frac{\partial^L \bar{G}}{\partial \theta^a}\right),\\
&\epsilon_{\{\bar{F},\bar{G}\}} = \epsilon_F + \epsilon_G,\\
&\{\bar{F}, \bar{G}\} = -(-)^{\epsilon_F\,\epsilon_G}\{\bar{G}, \bar{F}\},\\
&\{\bar{F}, \bar{G}_1\,\bar{G}_2\} = \{\bar{F}, \bar{G}_1\}\,\bar{G}_2 + (-)^{\epsilon_F\,\epsilon_{G_1}}\,\bar{G}_1\,\{\bar{F}, \bar{G}_2\},\\
&\{\{\bar{F}_1, \bar{F}_2\}, \bar{F}_3\} + (-)^{\epsilon_{F_1}\,(\epsilon_{F_2}+\epsilon_{F_3})}\{\{\bar{F}_2, \bar{F}_3\}, \bar{F}_1\}\\
&\quad +(-)^{\epsilon_{F_3}\,(\epsilon_{F_1}+\epsilon_{F_2})}\{\{\bar{F}_3, \bar{F}_1\}, \bar{F}_2\} = 0.
\end{aligned} \tag{B.3}$$

In Ref. [26] there is a more complete treatment of the properties of the symplectic structures on these phase spaces and of the related notions of supersymmetry and super-manifolds.

Therefore, the fundamental Poisson brackets for pseudo-classical mechanics are

$$\{q^i, p_j\} = -\{p_j, q^i\} = \delta^i_j, \qquad \{\theta^a, \pi_b\} = \{\pi_b, \theta^a\} = -\delta^a_b. \tag{B.4}$$

The kinetic term for a set of free real Grassmann variables $\theta^a = \theta^{a\,*}$ is

$$L = \frac{i}{2}\sum_a \theta^a\,\dot{\theta}^a, \qquad S = \int dt\, L, \tag{B.5}$$

while the one for a set of complex Grassmann variables θ^a, $\theta^{a\,*}$ is

$$L = \frac{i}{2}\sum_a \Big(\theta^{a\,*}\,\dot{\theta}^a - \dot{\theta}^{a\,*}\,\theta^a\Big). \tag{B.6}$$

Let us study the Lagrangian (Eq. B.5). The canonical momenta are $\pi_a = \frac{\partial^L L}{\partial \dot{\theta}^a} = -\frac{i}{2}\,\delta_{ab}\,\theta^b$, while the Hamiltonian vanishes, $\bar{H} = 0$. We see that all the fermionic momenta can be eliminated by going to Dirac brackets due to the second-class constraints:[2]

[2] It is only with Grassmann variables and with the associated symmetric Poisson bracket for odd variables that an odd constraint may be second class by itself without having a partner, so that the reduced phase space can be odd-dimensional.

$$\bar{\chi}_a = \pi_a + \frac{i}{2}\,\delta_{ab}\,\theta^b \approx 0, \qquad \{\bar{\chi}_a, \bar{\chi}_b\} = -i\delta_{ab},$$
$$\Rightarrow \{\theta^a, \theta^b\}^* = -i\,\delta^{ab}. \tag{B.7}$$

Let us remark [26] that, since the action (Eq. B.5) gives rises to first-order equations of motion whose Cauchy problem requires only one set of initial data (either the initial configuration or the initial velocity, but not both), we have to add a suitable surface term to it to obtain them from a variational principle. The equations of motion $\dot{\theta}^a \overset{\circ}{=} 0$ are obtained from the modified action $S'[t_f, t_i] = \int_{t_i}^{t_f} dtL - \frac{i}{2}\sum_a \theta^a(t_i)\,\theta^a(t_f)$ by asking $\delta\, S' = 0$ under arbitrary variations $\delta\,\theta^a$ satisfying only the condition $\delta\,[\theta^a(t_i) + \theta^a(t_f)] = 0$ at the end-points. Indeed, this condition implies $\delta S' = -i\int_{t_i}^{t_f} dt \sum_a \dot{\theta}^a\,\delta\,\theta^a = 0$. Since we get $\theta^a(t) \overset{\circ}{=} \xi^a = $ const., this is compatible with the condition on the variations.

In the bosonic case, first-order equations of motion (Hamilton equations) are generated by the phase space action $\bar{S}[t_f, t_i] = \int_{t_i}^{t_f} dt\,(\dot{q}^i\,p_i - \bar{H})$. Usually they are obtained by requiring $\delta\,\bar{S} = 0$ under arbitrary variations $\delta\,q^i$, $\delta\,p_i$ with the only restriction $\delta\,q^i(t_f) = \delta\,q^i(t_i) = 0$. If we define the modified action $\bar{S}'[t_f, t_i] = \bar{S}[t_f, t_i] - \frac{1}{2}\,[q^i(t_f) - q^i(t_i)]\,[p_i(t_f) + p_i(t_i)] = \int_{t_i}^{t_f} dt\,[\frac{1}{2}\,(\dot{q}^i\,p_i - q^i\,\dot{p}_i) - \bar{H}] + \frac{1}{2}\,[q^i(t_i)\,p_i(t_f) - q^i(t_f)\,p_i(t_i)]$, Hamilton equations are a consequence of $\delta\,\bar{S}' = 0$ under arbitrary variations satisfying $\delta\,[q^i(t_f) + q^i(t_i)] = \delta\,[p_i(t_f) + p_i(t_i)] = 0$ at the end-points.

Let us remark that if we have a Grassmann algebra G_{2n} with $2n$ real generators ξ^a, $a = 1, \ldots, 2n$, we can describe it with n pairs of complex conjugate Grassmann variables $\theta^\alpha = \frac{1}{\sqrt{2}}\,(\xi^{2\alpha-1} + i\,\xi^{2\alpha})$, $\theta^{\alpha\,*} = \frac{1}{\sqrt{2}}\,(\xi^{2\alpha-1} - i\,\xi^{2\alpha})$, $\alpha = 1, \ldots, n$. In pseudo-classical mechanics the Dirac brackets $\{\xi^a, \xi^b\}^* = -i\,\delta^{ab}$ become $\{\theta^\alpha, \theta^{\beta\,*}\}^* = -i\,\delta^{\alpha\beta}$, $\{\theta^\alpha, \theta^\beta\}^* = \{\theta^{\alpha\,*}, \theta^{\beta\,*}\}^* = 0$.

The Dirac quantization of a Grassmann algebra G_N with such Dirac brackets yields a Clifford algebra C_N with N generators by replacing the Dirac brackets with anti-commutators. With real Grassmann variables we get ($\xi^a \mapsto \hat{\xi}^a$)

$$\{\xi^a, \xi^b\}^* = -i\,\delta^{ab} \quad \mapsto \quad [\hat{\xi}^a, \hat{\xi}^b]_+ = \hat{\xi}^a\,\hat{\xi}^b + \hat{\xi}^b\,\hat{\xi}^a = \hbar\,\delta^{ab}, \tag{B.8}$$

while for $N = 2k$ we can use complex conjugate Grassmann variables to get a realization of C_{2k} with k Fermi oscillators ($\theta^\alpha \mapsto b^\alpha$, $\theta^{\alpha\,*} \mapsto b^{\alpha\,\dagger}$):

$$\{\theta^\alpha, \theta^{\beta\,*}\}^* = -i\,\delta^{\alpha\beta}, \quad \{\theta^\alpha, \theta^\beta\}^* = \{\theta^{\alpha\,*}, \theta^{\beta\,*}\}^* = 0$$
$$\mapsto \quad [b^\alpha, b^{\beta\,\dagger}]_+ = \hbar\,\delta^{\alpha\beta}, \quad [b^\alpha, b^\beta]_+ = [b^{\alpha\,\dagger}, b^{\beta\,\dagger}]_+ = 0. \tag{B.9}$$

When $N = 2k$, the Clifford algebra C_{2k} has a unique 2^k-dimensional irreducible representation. Instead, for $N = 2k + 1$, the Clifford algebra C_{2k+1} has two non-equivalent 2^k-dimensional irreducible representations. See Refs. [26, 607–612] for the algebraic and group theoretical aspects. In Ref. [609], besides the path integral quantization over the Grassmann variables, there is also a detailed treatment of the quantum mechanics of Grassmann variables.

B.3 Pseudo-Classical Description of Charged and Spinning Particles

In Ref. [610] (see also Refs. [611, 612]) internal degrees of freedom of point-like particles like electric charge, isospin, color, etc. are described in pseudo-classical mechanics by using complex conjugate Grassmann variables θ_α, θ^*_α (Fermi oscillators at the quantum level) belonging to a representation $R = R_1 \otimes R^*_1$ of a Lie group G. At the level of Dirac brackets the generators T_a of the Lie algebra g ($[T_a, T_b] = C^c_{ab}\, T_c$) are realized in the given representation of G as $I_a = \theta^*_\alpha\, (T_a)_{\alpha\beta}\, \theta_\beta$, $\{I_a, I_b\}^* = C^c_{ab}\, I_c$. This allows us to define the coupling of a charged (scalar or spinning) particle to Yang–Mills fields.

In Ref. [610] there are the examples of the electric charge, $G = U(1)$, and of the color, $G = SU(3)$. For the pseudo-classical electric charge Q we need two conjugate complex Grassmann variables θ, θ^* to get $Q = e\, \theta^*\, \theta$ (e is the electric charge) with $Q^2 = 0$.

The same happens for the sign η of the energy of the particle (a topological number with the two values ± 1 for particles and antiparticles): We need θ_η and θ^*_η so that $\eta = \theta^*_\eta\, \theta_\eta$, $\eta^2 = 0$ (particles with negative energy have the opposite electric charge $\eta\, Q$).

The Lagrangian and the Euler–Lagrange equations of a charged particle coupled to an external electromagnetic field are

$$\begin{aligned} L &= \frac{i}{2}\, (\theta^*\, \dot\theta - \dot\theta^*\, \theta)(\tau) + \frac{i}{2}\, (\theta^*_\eta\, \dot\theta_\eta - \dot\theta^*_\eta\, \theta_\eta) - mc\, \theta^*_\eta\, \theta_\eta\, \sqrt{\dot x^2} - e\, \theta^*\, \theta\, \theta^*_\eta\, \theta_\eta\, \dot x^\mu\, A_\mu(x), \\ &\theta^*_\eta\, \theta_\eta\, mc\, \frac{d}{d\tau}\, \frac{\dot x_\mu}{\sqrt{\dot x^2}} \overset{\circ}{=} e\, \theta^*\, \theta\, F_{\mu\nu}(x)\, \dot x^\nu, \\ &\dot\theta_i\, e\, \dot x^\mu\, A_\mu(x)\, \theta \overset{\circ}{=} 0, \quad \Rightarrow \quad \frac{d\, Q}{d\tau} \overset{\circ}{=} 0. \end{aligned} \tag{B.10}$$

The canonical Hamiltonian vanishes, while the Dirac Hamiltonian is $\bar H_D = \lambda(\tau)\, \bar\chi$, where $\bar\chi = \epsilon\, (p_\mu - e\, \theta^*\, \theta\, A_\mu(x))^2 - m^2 c^2 \approx 0$ are the only first-class constraints. The second-class constraints on the Grassmann momenta imply the Dirac brackets $\{\theta, \theta^*\}^* = \{\theta_\eta, \theta^*_\eta\}^* = -i$. η and Q are constants of the motion.

The quantization of the pair $\theta(\tau)$, $\theta^*(\tau)$ gives rise to a fermionic oscillator $\theta^*\, \theta \rightarrow b^\dagger\, b + k$, where k is a c-number (due to ordering problems). For $k = 0$ we get a two-level system with electric charge e and 0. In the case of the sign of the energy, we use $\eta = 2\, \theta^*\, \theta$, which goes into $2\, (b^\dagger\, b - \frac{1}{2})$ after quantization, so that the two-level system has $\eta = \pm 1$.

The use of Grassmann variables to describe the spin of spinning particles started with Berezin and Marinov [606]. In the case of spin-1/2, one speaks of pseudo-classical descriptions of the electron [607, 608, 610, 613].

In Refs. [171, 172, 614], a Lagrangian was given depending on a bosonic position $x^\mu(\tau)$ and on five real Grassmann variables $\xi^\mu(\tau)$ and $\xi_5(\tau)$:[3]

$$L = -\frac{i}{2}\,\xi_5\,\dot{\xi}_5 - \frac{i}{2}\,\xi_\mu\,\dot{\xi}^\mu - m\,\sqrt{\left(\dot{x}_\mu - \frac{i}{m}\,\xi_\mu\,\dot{\xi}_5\right)^2}, \qquad S = \int d\tau\, L, \tag{B.11}$$

where $\dot{x}^\mu = \frac{\partial}{\partial\tau} x^\mu$ and $c = 1$; the canonical momenta are $p_\mu = -\frac{\partial L}{\partial \dot{x}^\mu} = m\,\frac{\dot{x}_\mu - \frac{i}{m}\,\xi_\mu\,\dot{\xi}_5}{\sqrt{\left(\dot{x}_\mu - \frac{i}{m}\,\xi_\mu\,\dot{\xi}_5\right)^2}}$, $\pi_\mu = \frac{\partial L}{\partial \dot{\xi}^\mu} = \frac{i}{2}\,\xi_\mu$, $\pi_5 = \frac{\partial L}{\partial \dot{\xi}_5} = \frac{i}{2}\,\xi_5 - \frac{i}{m}\,p_\mu\,\xi^\mu$. The geometrical interpretation is that one has a fibered structure: over each point of the time-like world-line of a scalar point particle in Minkowski space-time there is a Grassmann algebra G_5 with generators ξ^μ, ξ_5, as a standard fiber describing the spin structure.

In phase space, there are six primary constraints. Five of them are due to the fact that the Lagrangian is of first order in Grassmann velocities and this allows the elimination of the Grassmann momenta. One of these constraints is $\bar{\chi}_5^{'} = \bar{\chi}_5 + \frac{i}{m}\,(p_\mu\,\xi^\mu - m\,\xi_5) \approx 0$, with $\bar{\chi}_5 = \pi_5 + \frac{i}{2}\,\xi_5$. The quantity $\bar{\chi}_5$ turns out to be a constant of motion and it was discovered that only the restriction $\bar{\chi}_5 \approx 0$ allowed getting the same set of solutions from the Euler–Lagrange and Hamilton equations. The explanation of the necessity of this extra constraint is hidden in the Hessian super-matrix:

$$\begin{pmatrix} -\frac{m\,(\eta_{\mu\nu} - \frac{p_\mu\,p_\nu}{p^2})}{\sqrt{(\dot{x}^\gamma - \frac{i}{m}\,\xi^\gamma\,\dot{\xi}_5)^2}} & 0_{\mu\beta} & -\frac{i\,(\eta_{\mu\rho} - \frac{p_\mu\,p_\rho}{p^2})\,\xi^\rho}{\sqrt{(\dot{x}^\gamma - \frac{i}{m}\,\xi^\gamma\,\dot{\xi}_5)^2}} \\ 0_{\alpha\nu} & 0_{\alpha\beta} & 0_\alpha \\ -\frac{i\,(\eta_{\nu\rho} - \frac{p_\nu\,p_\rho}{p^2})\,\xi^\rho}{\sqrt{(\dot{x}^\gamma - \frac{i}{m}\,\xi^\gamma\,\dot{\xi}_5)^2}} & 0_\beta & 0 \end{pmatrix}. \tag{B.12}$$

It has three non-null eigenvectors and six null eigenvectors corresponding to the primary constraints: $\begin{pmatrix} p^\nu \\ 0^\beta \\ 0 \end{pmatrix}$ associated with $\bar{\chi} \approx 0$; $\begin{pmatrix} 0^\nu \\ w^\beta \\ 0 \end{pmatrix}$, with w^β an arbitrary odd 4-vector, associated with $\bar{\chi}_\mu \approx 0$; $\begin{pmatrix} -\frac{i}{m}\,\xi^\nu\,z \\ 0^\beta \\ z \end{pmatrix}$, with z an arbitrary odd scalar, associated with $\bar{\chi}_5^{'} \approx 0$. Since z could be either $p_\mu\,\xi^\mu$ or ξ_5, to avoid the existence of two independent null eigenvalues corresponding to the two possibilities we need $p \cdot \xi \approx \xi_5$. The requirement that the constant of motion $\bar{\chi}_5$ vanishes, transforms $\bar{\chi}_5^{'} \approx 0$ in the constraint $\bar{\chi}_D \approx 0$ of Eq. (B.14).

[3] The non-relativistic limit of this Lagrangian is $L_{NR} = \frac{i}{2}\,\vec{\xi}\cdot\dot{\vec{\xi}} + \frac{1}{2}\,m\,\dot{\vec{x}}^2$. This Lagrangian describes a non-relativistic Pauli particle.

In conclusion, we have seven constraints. Five of them are second-class constraints,

$$\bar{\chi}_\mu = \pi_\mu - \frac{i}{2}\,\xi_\mu \approx 0, \qquad \bar{\chi}_5 = \pi_5 + \frac{i}{2}\,\xi_5 \approx 0,$$
$$\{\bar{\chi}_\mu, \bar{\chi}_\nu\} = i\,\eta_{\mu\nu}, \qquad \{\bar{\chi}_5, \bar{\chi}_5\} = -i. \tag{B.13}$$

They are eliminated by introducing the Dirac brackets $\{x^\mu, p^\nu\}^* = -\epsilon\,\eta^{\mu\nu}$, $\{\xi^\mu, \xi^\nu\}^* = i\,\eta^{\mu\nu}$, $\{\xi_5, \xi_5\}^* = -i$, in place of the original Poisson brackets $\{x^\mu, p^\nu\} = \{\xi^\mu, \pi^\nu\} = -\epsilon\,\eta^{\mu\nu}$, $\{\pi_5, \xi_5\} = -1$. After quantization [607, 608], these Dirac brackets become $\left[\xi_\mu, \xi_\nu\right]_+ = -\hbar\,\epsilon\,\eta_{\mu\nu}$, $\left[\xi_\mu, \xi_5\right]_+ = 0$, $\left[\xi_5, \xi_5\right]_+ = \hbar$, so that the Grassmann variables give rise to the Dirac matrices $\sqrt{\frac{\hbar}{2}}\,\gamma_5\,\gamma^\mu$ and $\sqrt{\frac{\hbar}{2}}\,\gamma_5$ respectively.

Then there are the two first-class constraints:

$$\bar{\chi} = p^2 - m^2c^2 \approx 0, \qquad \bar{\chi}_D = p_\mu\,\xi^\mu - m\,\xi_5 \approx 0. \tag{B.14}$$

The first one is generated from τ-reparametrization invariance, while the other comes from the local super-symmetry invariance. Since the only non-vanishing Poisson bracket is $\{\bar{\chi}_D, \bar{\chi}_D\}^* = i\,\bar{\chi}$, it is said that $\bar{\chi}_D \approx 0$ is the square root of $\bar{\chi} \approx 0$.

There is a local super-symmetry invariance of the action, but no manifest world-line super-symmetry. The local transformations are $\xi_\mu \longrightarrow \xi_\mu + \epsilon_5(\tau)\,\frac{p_\mu}{m}$, $\xi_5 \longrightarrow \xi_5 + \epsilon_5(\tau)$, $x_\mu \longrightarrow x_\mu + \frac{i}{m^2}\,\epsilon_5(\tau)\,p_\mu\,\xi_5 - \frac{i}{m}\,\epsilon_5(\tau)\,[\xi_\mu - \frac{1}{m}\,p_\mu\,\xi_5]$ and their generator $-\epsilon_5(\tau)\,(\bar{G}_5 - \frac{p_\mu\,\bar{G}^\mu}{m}) = -\epsilon_5(\tau)\,(\bar{G}_5 - \frac{i}{m}\,\bar{\chi}_D - \frac{i}{m^2}\,\bar{\chi}\,\xi^5)$ becomes a constraint when we require $\bar{\chi}_5 \approx 0$. These transformations leave $\bar{\chi}_D$ invariant. Here, $\bar{G}_5 = -\bar{\chi}_5$, $\bar{G}_\mu = -\bar{\chi}_\mu - i\,(\xi_\mu - \frac{p_\mu}{m}\,\xi_5)$ $[\,p_\mu\,\bar{G}^\mu + m\,\bar{G}_5 = -m\,\bar{\chi}_5' + \frac{i}{m}\,\bar{\chi}\,\xi_5 \approx 0]$ are the generators of the super-symmetry algebra $\{\bar{G}_5, \bar{G}_5\} = -i$, $\{\bar{G}_\mu, \bar{G}_\nu\} = -i\,\eta_{\mu\nu}$, $\{\bar{G}_\mu, \bar{G}_5\} = i\,\frac{p_\mu}{m}$. While local super-symmetry invariance is the source of $\bar{\chi}_5' \approx 0$ (or of $\bar{\chi}_D \approx 0$ when $\bar{\chi}_5 \approx 0$), the Lagrangian is also (1) τ-reparametrization invariant (source of $\bar{\chi} \approx 0$); and (2) weakly quasi-invariant under the transformations generated by $\epsilon_\mu(\tau)\,\bar{\chi}^\mu$ (source of the second-class constraints $\bar{\chi}_\mu \approx 0$). Let us remark that neither $\bar{\chi}_5$ nor $\bar{\chi}_D$ are separately generators of local quasi-invariances.

The Dirac Hamiltonian is $\bar{H}_D = \lambda(\tau)\,\bar{\chi} + i\,\lambda_D(\tau)\,\bar{\chi}_D$, with $\lambda_D(\tau)$ an odd real Dirac multiplier. We have $\dot{\xi}^\mu \stackrel{\circ}{=} \{\xi^\mu, \bar{H}_D\}^* = \lambda_D(\tau)\,p^\mu$, $\dot{\xi}_5 \stackrel{\circ}{=} \lambda_D(\tau)\,m$: this shows that both ξ^μ and ξ_5 are gauge-dependent. The conserved Poincaré generators are p_μ, $J_{\mu\nu} = L_{\mu\nu} + S_{\mu\nu}$, $L_{\mu\nu} = -p_\mu\,x_\nu + p_\nu\,x_\mu$, $S_{\mu\nu} = -\pi_\mu\,\xi_\nu + \pi_\nu\,\xi_\mu \stackrel{\chi_\mu \equiv \chi_5 \equiv 0}{=} -i\,\xi_\mu\,\xi_\nu$,[4] with the spin 3-vector $S_i = \frac{1}{2}\,\epsilon_{ijk}\,S_{jk} \Rightarrow \vec{S} = -\frac{i}{2}\,\vec{\xi} \times \vec{\xi}$. Since $\dot{S}^{\mu\nu} \stackrel{\circ}{=} i\,\lambda_D(\tau)\,(\xi^\mu\,p^\nu - \xi^\nu\,p^\mu)$, the spin tensor is gauge dependent, as it should

[4] The components of the Lorentz generators satisfy $\{L^{\mu\nu}, L^{\alpha\beta}\}^* = C^{\mu\nu\alpha\beta}_{\gamma\delta}\,L^{\gamma\delta}$, $\{S^{\mu\nu}, S^{\alpha\beta}\}^* = C^{\mu\nu\alpha\beta}_{\gamma\delta}\,S^{\gamma\delta}$, where $C^{\mu\nu\alpha\beta}_{\gamma\delta}$ are the structure constants of the Lorentz algebra.

be because its boost part cannot be independent from the spin part for a point particle with a definite spin.

Concerning the 4-position x^μ, its equations of motion are $\dot{x}^\mu \stackrel{\circ}{=} \{x^\mu, \bar{H}_D\}^* = -2\,\lambda(\tau)\,p^\mu - i\,\lambda_D(\tau)\,\xi^\mu$. This shows that super-imposed to the free motion with velocity proportional to the 4-momentum there is a zitterbewegung proportional to ξ^μ. To get free motion we need the pseudo-classical Foldy–Wouthuysen mean 4-position x^μ_M [607, 608, 610].

See Ref. [173] for the description of the pseudo-classical photon.

In conclusion, for the massive Dirac spinning particle the Dirac quantization of the Grassmann algebra G_5 produces the representation of the Clifford algebra C_5 corresponding to the Dirac matrices: $\xi^\mu \mapsto \sqrt{\frac{\hbar}{2}}\,\gamma_5\,\gamma^\mu$, $\xi_5 \mapsto \sqrt{\frac{\hbar}{2}}\,\gamma_5$. In Ref. [609] the electron propagator is recovered by path integral quantization.

B.4 Berezin–Marinov Distribution Function for Pseudo-Classical Physics

The use of Grassmann variables in pseudo-classical mechanics to describe spin and/or internal charges[5] corresponds to a semi-classical description in the limit $\hbar \to 0$. For a quantum Hermitian operator with unbounded spectrum, like energy or angular momentum, the semi-classical description (for instance the WKB approximation) is reached in the limit in which $\hbar \to 0$ and the eigenvalues go to infinity. For the energy we have $E \approx n\,\hbar\omega$ and we take $n \to \infty$, $\hbar \to 0$, with $E \to$ const. $\neq 0$.

This is not possible for quantum operators like the spin, which have a finite spectrum of eigenvalues. As shown in Ref. [613], a semi-classical limit may be defined in the following sense: Consider quantities ξ which are of first order in the infinitesimal $\hbar$ ($\xi \approx \sqrt{\hbar}$ in the case of spin) and put $\xi^2 = 0$ to eliminate higher-order effects. If we have a vector space of such infinitesimals ξ_a, then $(\xi_a)^2 = 0$ and $(\sum_a A_a\,\xi_a)^2 = 0$ imply $\xi_a\,\xi_b + \xi_b\,\xi_a = 0$, namely the ξ_a belong to a Grassmann algebra.

In Ref. [606] Berezin and Marinov defined a relation between the pseudo-classical Grassmann-valued quantities and corresponding classical observable quantities by using a Grassmann-valued distribution function ρ over phase space. If A is a Grassmann-valued quantity, the mean value $< A > = \int \rho\,A\,d\mu$ ($d\mu$ is the measure of Berezin integral) is its classical counterpart. Namely, we mimic the theory of the classical density matrix ρ, solution of the Liouville equation $\frac{d\rho}{dt} \stackrel{\circ}{=} \frac{\partial\rho}{\partial t} + \{\rho, \bar{H}\} = 0$, to build phase space mean values $< \bar{f} > = \int_{T^*Q} \rho\,\bar{f}\,d\mu$

[5] For them, there is experimental evidence of a bounded discrete spectrum, but no theoretical demonstration that the quantum is proportional to $\hbar$, with the exception of the electric charge in the case that magnetic monopole exists (so that the product of electric and magnetic charges is proportional to $\hbar$). In these cases, we assume that the charge quantum becomes an infinitesimal.

($d\mu = d^n q\, d^n p$) for every function over the phase space of a Hamiltonian system.

In pseudo-classical mechanics the distribution function ρ over the Grassmann variables must satisfy

$$\begin{aligned} &\rho = \rho^*, \\ &\int \rho\, d\mu = 1, \quad \text{(normalization)} \\ &\int \rho\, \bar{f}^*\, \bar{f}\, d\mu \geq 0, \quad \text{for every } \bar{f} \text{ (positivity condition)}, \\ &\frac{\partial \rho}{\partial t} + \{\rho, \bar{H}\}^* = 0, \quad \text{(Liouville equation)}. \end{aligned} \tag{B.15}$$

In Ref. [609] there is the example of the non-relativistic charged spinning particle interacting with a constant magnetic field $\vec{B}$ ($\vec{A} = -\frac{1}{2}\vec{x} \times \vec{B}$). The Lagrangian is $L - \frac{i}{2}\vec{\xi} \cdot \dot{\vec{\xi}} + \frac{1}{2} m \dot{\vec{x}}^2 + e\, \dot{\vec{x}} \cdot \vec{A} + g\, \vec{S} \cdot \vec{B}$ (g is the magnetic moment and $\vec{S} = -\frac{i}{2}\vec{\xi} \times \vec{\xi}$ the pseudo-classical spin), while the Hamiltonian is $\bar{H} = \frac{1}{2m}(\vec{p} - e\vec{A})^2 - g\,\vec{S} \times \vec{B}$ and the Dirac brackets are $\{\xi^i, \xi^j\}^* = -i\,\delta^{ij}$. The equation of motion for the spin is $\dot{\vec{S}} \stackrel{\circ}{=} \{\vec{S}, \bar{H}\}^* = -g\,\vec{B} \times \vec{S}$; namely, it undergoes a precession with angular velocity $\omega = g\,|\vec{B}|$. The normalized distribution function is $\rho = -i\,\xi^1\,\xi^2\,\xi^3 + \vec{c}(t) \cdot \vec{\xi}$ ($\int \rho\, d\mu = 1$ with $d\mu = i\, d\xi^1\, d\xi^2\, d\xi^3$). The classical mean value of the spin is $< \vec{S} > = \int \rho\, \vec{S}\, d\mu = \vec{c}(t)$ and the Liouville equation consistently gives $\dot{\vec{c}}(t) \stackrel{\circ}{=} -g\,\vec{B} \times \vec{c}(t)$, i.e., the same precession of the pseudo-classical spin. However, the positivity condition $\int \rho\, f^*(\xi)\, f(\xi)\, d\mu \geq 0$ implies $\vec{c}(t) = 0$. Therefore, there is no classical observable effect of the spin of an isolated spinning particle. See also Ref. [614]. The situation can be different with condensed matter described by a continuous distribution of pseudo-classical spin.

The absence of classical effects seems to be the rule for pseudo-classical quantities described by an odd number of Grassmann variables.

The situation is different with the pseudo-classical description of internal charges where we use an even number of Grassmann variables to build a realization of the generators of a Lie group G and to get Fermi oscillators at the quantum level.

The normalized distribution function for charged particles is $\rho(\theta, \theta^*) = \rho^*(\theta, \theta^*) = a(\tau) + b(\tau)\,\theta + b^*(\tau)\,\theta^* + \theta^*\,\theta$ ($\int \rho\, d\theta\, d\theta^* = 1$), with a even and b, b^* odd. For arbitrary functions $f(\theta, \theta^*)$ the positivity condition would imply $a = b = b^* = 0$. But the natural class of functions to be used for the quantum theory of Fermi oscillators [26, 607, 608, 610] are the holomorphic functions of θ: $f(\theta) = \alpha + \beta\,\theta$ (α, β are complex numbers). Then, the positivity condition $\int \rho(\theta, \theta^*)\, f^*(\theta)\, f(\theta)\, d\theta\, d\theta^* = |\alpha|^2 + a\,|\beta|^2 + b^*\,\alpha^*\,\beta - b\,\alpha\,\beta^* \geq 0$ implies $a > 0$ and $b = b^* = 0$, so that we get $\rho(\theta, \theta^*) = a(\tau) + \theta^*\,\theta$. The Liouville equation $\frac{\partial \rho}{\partial \tau} + \{\rho, \bar{H}_D\}^* \stackrel{\circ}{=} 0$ implies $a(\tau) \stackrel{\circ}{=}$ const.

Therefore, with the choice $a = 1$ we get $< Q > = \int \rho(\theta, \theta^*)\, e\, \theta^*\, \theta\, d\theta\, d\theta^* = e$, namely the mean value of the pseudo-classical electric charge is the electric charge. The mean value of the particle EL equations of Eq. (B.11) reproduces the classical Maxwell–Lorentz equations.

Let us finally remark that, when we study the isolated system of a charged spinning particle plus a dynamical electromagnetic field (see Section 5.1), it turns out that the operations (1) first averaging the equations of motion with the distribution function and then solving them; and (2) first solving the equations of motion and then averaging do not commute due to the semi-classical property $Q^2 = 0$. In the second case $Q^2 = 0$ eliminates the classical Coulomb self-energy of the particle and the essential singularity on the world-line source of the causality problems of the Lorentz–Abraham–Dirac equation present in the first case. If $A_\mu(z) \stackrel{\circ}{=} A_{IN\,\mu}(z) + Q \int d\tau\, \dot{x}_\mu(\tau)\, D(z - x(\tau))$ $(\Box D(z) = \delta^4(z))$ is the pseudo-classical Lienard–Wiechert solution of the field equations in the Lorentz gauge $\partial^\mu A_\mu(z) = 0$ with the incoming radiation described by $A_{IN\,\mu}(z)$, the particle pseudo-classical equation of motion feels only the incoming radiation due to $Q^2 = 0$: $m \frac{d}{d\tau} \frac{\dot{x}_\mu}{\sqrt{\dot{x}^2}} \stackrel{\circ}{=} Q\, F_{IN\,\mu\nu}(x(\tau))\, \dot{x}^\nu(\tau)$, without self-reaction.

Appendix C

Relativistic Perfect Fluids and Covariant Thermodynamics

C.1 Relativistic Perfect Fluids

Let us consider a perfect fluid [213] in a curved space-time M^4 with unit 4-velocity vector field $U^\mu(z)$, Lagrangian coordinates $\tilde{\alpha}^i(z)$, particle number density $\hat{n}(z)$, energy density $\rho(z)$, entropy per particle $s(z)$, pressure $p(z)$, and temperature $T(z)$. Let $J^\mu(z) = \sqrt{{}^4g(z)}\,\hat{n}(z)\,U^\mu(z)$ be the densitized particle number flux vector field, so that we have $\hat{n} = \sqrt{\epsilon\,{}^4g_{\mu\nu}\,J^\mu\,J^\nu}/\sqrt{{}^4g}$. Other local thermodynamical variables are the chemical potential or specific enthalpy (the energy per particle required to inject a small amount of fluid into a fluid sample, keeping the sample volume and the entropy per particle s constant):

$$\mu = \frac{1}{\hat{n}}\,(\rho + p), \tag{C.1}$$

the physical free energy (the injection energy at a constant number density $\hat{n}$ and constant total entropy):

$$a = \frac{\rho}{\hat{n}} - T\,s, \tag{C.2}$$

and the chemical free energy (the injection energy at constant volume and constant total entropy):

$$f = \frac{1}{\hat{n}}\,(\rho + p) - T\,s = \mu - T\,s. \tag{C.3}$$

Since the local expression of the first law of thermodynamics is

$$d\rho = \mu\,d\hat{n} + \hat{n}\,T\,ds, \quad \text{or} \quad dp = \hat{n}\,d\mu - \hat{n}\,T\,ds, \quad \text{or} \quad d(\hat{n}\,a) = f\,d\hat{n} - \hat{n}\,s\,dT, \tag{C.4}$$

an equation of state for a perfect fluid may be given in one of the following forms:

$$\rho = \rho(\hat{n}, s), \quad \text{or} \quad p = p(\mu, s), \quad \text{or} \quad a = a(\hat{n}, T). \tag{C.5}$$

By definition, the stress–energy–momentum tensor for a perfect fluid is

$$T^{\mu\nu} = -\epsilon\,\rho\,U^\mu\,U^\nu + p\,({}^4g^{\mu\nu} - \epsilon\,U^\mu\,U^\nu) = -\epsilon\,(\rho+p)\,U^\mu\,U^\nu + p\,{}^4g^{\mu\nu}, \qquad \text{(C.6)}$$

and its equations of motion are

$$T^{\mu\nu}_{;\nu} = 0, \qquad (\hat{n}\,U^\mu)_{;\mu} = \frac{1}{\sqrt{{}^4g}}\,\partial_\mu\,J^\mu = 0. \qquad \text{(C.7)}$$

As shown in Ref. [213], an action functional for a perfect fluid depending upon $J^\mu(z)$, ${}^4g_{\mu\nu}(z)$, $s(z)$, and $\tilde{\alpha}^i(z)$ requires the introduction of the following Lagrange multipliers to implement all the required properties:

1. $\theta(z)$: This is a scalar field named thermasy; it is interpreted as a potential for the fluid temperature $T = \frac{1}{\hat{n}}\frac{\partial\rho}{\partial s}|_{\hat{n}}$. In the Lagrangian it is interpreted as a Lagrange multiplier for implementing the entropy exchange constraint, $(s\,J^\mu)_{,\mu} = 0$.
2. $\varphi(z)$: This is a scalar field; it is interpreted as a potential for the chemical free energy f. In the Lagrangian it is interpreted as a Lagrange multiplier for the particle number conservation constraint, $J^\mu_{,\mu} = 0$.
3. $\beta_i(z)$: These are three scalar fields; in the Lagrangian they are interpreted as Lagrange multipliers for the constraint $\tilde{\alpha}^i_{,\mu}\,J^\mu = 0$ that restricts the fluid 4-velocity vector to be directed along the flow lines $\tilde{\alpha}^i = \text{const}$.

Given an arbitrary equation of state of the type $\rho = \rho(\hat{n}, s)$, the action functional is

$$S[{}^4g_{\mu\nu}, J^\mu, s, \tilde{\alpha}, \varphi, \theta, \beta_i] = \int d^4z\,\Big(-\sqrt{{}^4g}\rho\,(\frac{|J|}{\sqrt{{}^4g}}, s) + J^\mu\,[\partial_\mu\,\varphi + s\,\partial_\mu\,\theta + \beta_i\,\partial_\mu\,\tilde{\alpha}^i]\Big). \qquad \text{(C.8)}$$

By varying the 4-metric, we get the standard stress–energy–momentum tensor,

$$T^{\mu\nu} = \frac{2}{\sqrt{{}^4g}}\,\frac{\delta S}{\delta\,{}^4g_{\mu\nu}} = -\epsilon\,\rho\,U^\mu\,U^\nu + p\,({}^4g^{\mu\nu} - \epsilon\,U^\mu\,U^\nu) = -\epsilon\,(\rho+p)\,U^\mu\,U^\nu + p\,{}^4g^{\mu\nu}, \qquad \text{(C.9)}$$

where the pressure is given by

$$p = \hat{n}\,\frac{\partial\rho}{\partial\hat{n}}|_s - \rho. \qquad \text{(C.10)}$$

The Euler–Lagrange (EL) equations for the fluid motion are

$$\frac{\delta S}{\delta J^\mu} = \mu\,U_\mu\, + \partial_\mu\,\varphi + s\,\partial_\mu\,\theta + \beta_i\,\partial_\mu\,\tilde{\alpha}^i = 0,$$
$$\frac{\delta S}{\delta\varphi} = -\partial_\mu\,J^\mu = 0,$$
$$\frac{\delta S}{\delta\theta} = -\partial_\mu\,(s\,J^\mu) = 0,$$

$$
\begin{aligned}
\frac{\delta S}{\delta s} &= -\sqrt{{}^4g}\,\frac{\partial \rho}{\partial s} + J^\mu\,\partial_\mu\,\theta = 0,\\
\frac{\delta S}{\delta \tilde{\alpha}^i} &= -\partial_\mu\,(\beta_i\,J^\mu) = 0,\\
\frac{\delta S}{\delta \beta_i} &= J^\mu\,\partial_\mu\,\tilde{\alpha}^i = 0.
\end{aligned} \tag{C.11}
$$

The second equation is the particle number conservation, the third one the entropy exchange constraint, and the last one restricts the fluid 4-velocity vector to being directed along the flow lines $\tilde{\alpha}^i = \text{const}$. The first equation gives the Clebsch or velocity-potential representation of the 4-velocity U_μ (the scalar fields in this representation are called Clebsch or velocity potentials). The fifth equation implies the constancy of the β_i along the fluid flow lines, so that these Lagrange multipliers can be expressed as a function of the Lagrangian coordinates. The fourth equation, after a comparison with the first law of thermodynamics, leads to the identification $T = U^\mu\,\partial_\mu\,\theta = \frac{1}{\hat{n}}\frac{\partial \rho}{\partial s}|_{\hat{n}}$ for the fluid temperature.

Moreover, we can show that the EL equations imply the conservation of the stress–energy–momentum tensor $T^{\mu\nu}_{;\nu} = 0$. This equations can be split in the projection along the fluid flow lines and in the one orthogonal to them:

1. The projection along the fluid flow lines plus the particle number conservation give $U_\mu\,T^{\mu\nu}_{;\nu} = -\frac{\partial \rho}{\partial s}\,U^\mu\,\partial_\mu\,s = 0$, which is verified due to the entropy exchange constraint. Therefore, the fluid flow is locally adiabatic – that is, the entropy per particle along the fluid flow lines is conserved.
2. The projection orthogonal to the fluid flow lines gives the Euler equations, relating the fluid acceleration to the gradient of pressure:

$$
({}^4g_{\mu\nu} - \epsilon\,U_\mu\,U_\nu)\,T^{\nu\alpha}_{;\alpha} = -\epsilon\,(\rho + p)\,U_{\mu;\nu}\,U^\nu - (\delta^\nu_\mu - \epsilon\,U_\mu\,U^\nu)\,\partial_\nu\,p. \tag{C.12}
$$

By using $p = \hat{n}\,\frac{\partial \rho}{\partial \hat{n}}|_s - \rho$, it is shown in Ref. [213] that these equations can be rewritten as

$$
2\,(\mu\,U_{[\mu})_{;\nu]}\,U^\nu = -\epsilon\,(\delta^\nu_\mu - U_\mu\,U^\nu)\,\frac{1}{\hat{n}}\frac{\partial \rho}{\partial s}|_{\hat{n}}\,\partial_\nu\,s. \tag{C.13}
$$

The use of the entropy exchange constraint allows rewriting the equations in the form

$$
2\,V_{[\mu;\nu]}\,U^\nu = T\,\partial_\mu\,s, \tag{C.14}
$$

where $V_\mu = \mu\,U_\mu$ is the Taub current (important for the description of circulation and vorticity), which can be identified with the 4-momentum per particle of a small amount of fluid to be injected in a larger sample of fluid without changing the total fluid volume or the entropy per particle. Now, from the EL equations, we get

$$
2\,V_{[\mu;\nu]}\,U^\nu = -2\,(\partial_{[\mu}\,\varphi + s\,\partial_{[\mu}\,\theta + \beta_i\,\partial_{[\mu}\,\tilde{\alpha}^i)_{;\nu]}\,U^\nu = (s\,\partial_{[\mu}\,\theta)_{;\nu]}\,U^\nu = T\,\partial_\mu\,s, \tag{C.15}
$$

and this result implies the validity of the Euler equations.

In the non-relativistic limit $(n\,U^\mu)_{;\mu} = 0$, $T^{\mu\nu}_{;\nu} = 0$ become the particle number (or mass) conservation law, the entropy conservation law, and the Euler–Newton equations. See Refs. [615, 616] for the post-Newtonian approximation.

We refer to Ref. [213] for the complete discussion. The previous action has the advantage over other actions that the canonical momenta conjugate to φ and θ are the particle number density and entropy density seen by Eulerian observers at rest in space. The action evaluated on the solutions of the equations of motion is $\int d^4z\,\sqrt{{}^4g(z)}\,p(z)$.

In Ref. [213] there is a study of a special class of global Noether symmetries of this action associated with arbitrary functions $F(\tilde{\alpha}, \beta_i, s)$. It is shown that for each F there is a conservation equation $\partial_\mu\,(F\,J^\mu) = 0$ and a Noether charge $Q[F] = \int_\Sigma d^3\sigma\,\sqrt{\gamma}\,\hat{n}\,(\epsilon\,l_\mu\,U^\mu)\,F(\tilde{\alpha}, \beta_i, s)$ (Σ is a space-like hyper-surface with future pointing unit normal l_μ and with a 3-metric with determinant $\sqrt{\gamma}$). For $F = 1$ inside a volume V in Σ we get the conservation of particle number within a flow tube defined by the bundle of flow lines contained in the volume V. The factor $\epsilon\,l_\mu\,U^\mu$ is the relativistic gamma factor characterizing a boost from the Lagrangian observers with 4-velocity U^μ to the Eulerian observers with 4-velocity l^μ; thus, $\hat{n}\,(\epsilon\,l_\mu\,U^\mu)$ is the particle number density as seen from the Eulerian observers. These symmetries describe the changes of Lagrangian coordinates $\tilde{\alpha}^i$ and the fact that both the Lagrange multipliers φ and θ are constant along each flow line (so that it is possible to transform any solution to the fluid equations of motion into a solution with $\varphi = \theta = 0$ on any given space-like hyper-surface).

However, the Hamiltonian formulation associated with this action is not trivial, because the many redundant variables present in it give rise to many first- and second-class constraints. In particular, we get:

1. second-class constraints:

   a. $\pi_{J^\tau} \approx 0$, $J^\tau - \pi_\varphi \approx 0$;
   b. $\pi_s \approx 0$, $s\,J^\tau - \pi_\theta \approx 0$;
   c. $\pi_{\beta_s} \approx 0$, $\beta_s\,J^\tau - \pi_{\alpha^s} \approx 0$.

2. first-class constraints: $\pi_{J^r} \approx 0$, so that the J^r are gauge variables.

Therefore, the physical variables are the five pairs: φ, π_φ; θ, π_θ; $\tilde{\alpha}^i$, $\pi_{\alpha^r}, \pi_{\alpha_i}$ $(i = 1, 2, 3)$ and we could study the associated canonical reduction.

In Ref. [213] (see its rich bibliography for the references) there is a systematic study of the action principles associated with the three types of equations of state present in the literature, first by using the Clebsch potentials and the associated Lagrange multipliers, then only in terms of the Lagrangian coordinates by inserting the solution of some of the EL equations in the original action and eventually by adding surface terms.

1. Equation of state $\rho = \rho(\hat{n}, s)$. One has the action

$$S[\hat{n}, U^\mu, \varphi, \theta, s, \tilde{\alpha}^r, \beta_r; {}^4g_{\mu\nu}] = -\int d^4x \sqrt{{}^4g}\Big[\rho(\hat{n}, s) - \hat{n}\, U^\mu \left(\partial_\mu \varphi - \theta\, \partial_\mu s + \beta_r\, \partial_\mu \tilde{\alpha}^r\right)\Big]. \quad (C.16)$$

If we know $s = s(\tilde{\alpha}^r)$ and $J^\mu = J^\mu(\tilde{\alpha}^r) = -\sqrt{{}^4g}\, \epsilon^{\mu\nu\rho\sigma}\, \partial_\nu \tilde{\alpha}^1\, \partial_\rho \tilde{\alpha}^2 \partial_\sigma \tilde{\alpha}^3$ $\eta_{123}(\tilde{\alpha}^r)$, we can define $\tilde{S} = S - \int d^4x\, \partial_\mu [(\varphi + s\,\theta)\, J^\mu]$, and we can show that it has the form

$$\tilde{S} = \tilde{S}[\tilde{\alpha}^r] = -\int d^4x \sqrt{{}^4g}\, \rho\left(\frac{|J|}{\sqrt{{}^4g}}, s\right). \quad (C.17)$$

2. Equation of state: $p = p(\mu, s)$ ($V^\mu = \mu U^\mu$ Taub vector):

$$\begin{aligned} S_{(p)} &= S_{(p)}[V^\mu, \varphi, \theta, s, \tilde{\alpha}^r, \beta_r; {}^4g_{\mu\nu}] \\ &= \int d^4x \sqrt{{}^4g}\left[p(\mu, s) - \frac{\partial p}{\partial \mu}\left(|V| - \frac{V^\mu}{|V|}\left(\partial_\mu \varphi + s\, \partial_\mu \theta + \beta_r\, \partial_\mu \tilde{\alpha}^r\right)\right)\right], \end{aligned} \quad (C.18)$$

or by using one of its EL equations $V_\mu \stackrel{\circ}{=} -(\partial_\mu \varphi + s\, \partial_\mu \theta + \beta_r\, \partial_\mu \tilde{\alpha}^r)$ to eliminate V^μ, we get Schutz's action (μ determined by $\mu^2 = -V^\mu V_\mu$):

$$\tilde{S}_{(p)}[\varphi, \theta, s, \tilde{\alpha}^r, \beta_r; {}^4g_{\mu\nu}] = \int d^4x \sqrt{{}^4g}\, p(\mu, s). \quad (C.19)$$

3. Equation of state $a = a(\hat{n}, T)$. The action is

$$S_{(a)}[J^\mu, \varphi, \theta, \tilde{\alpha}^r, \beta_r; {}^4g_{\mu\nu}] = \int d^4x \left[|J|\, a\left(\frac{|J|}{\sqrt{{}^4g}}, \partial_\mu \theta\, J^\mu\right) - J^\mu \left(\partial_\mu \varphi + \beta_r\, \partial_\mu \tilde{\alpha}^r\right)\right], \quad (C.20)$$

or $\tilde{S}_{(a)}[\varphi, \theta, s, \tilde{\alpha}^r, \beta_r] = S_{(a)} - \int d^4x \left[|J|\, a\left(\frac{|J|}{\sqrt{{}^4g}}, \frac{J^\mu}{|J|}\, \partial_\mu \theta\right)\right]$.

At the end of Ref. [213] there is the action for isentropic fluids and for their particular case of a dust (used in Ref. [617] as a reference fluid in canonical gravity).

The isentropic fluids have equation of state $a(\hat{n}, T) = \frac{\rho(\hat{n})}{\hat{n}} - sT$ with $s =$ const. (constant value of the entropy per particle). By introducing $\varphi' = \varphi + s\,\theta$, the action can be written in the form

$$S_{(isentropic)}[J^\mu, \varphi', \tilde{\alpha}^r, \beta_r, {}^4g^{\mu\nu}] = \int d^4x \left[-\sqrt{{}^4g}\, \rho\left(\frac{|J|}{\sqrt{{}^4g}}\right) + J^\mu \left(\partial_\mu \varphi' + \beta_r\, \partial_\mu \tilde{\alpha}^r\right)\right], \quad (C.21)$$

or

$$\tilde{S}_{(isentropic)}[\tilde{\alpha}^r; {}^4g_{\mu\nu}] = -\int d^4x \sqrt{{}^4g}\, \rho\left(\frac{|J|}{\sqrt{{}^4g}}\right). \quad (C.22)$$

The dust has equation of state $\rho(\hat{n}) = \mu\,\hat{n}$, namely $a(\hat{n}, T) = \mu - s\,T$ so that we get zero pressure $p = \hat{n}\,\frac{\partial\rho}{\partial\hat{n}} - \rho = 0$. Again with $\varphi^{'} = \varphi + s\,\theta$, the action becomes

$$S_{(dust)}[J^\mu, \varphi^{'}, \tilde{\alpha}^r, \beta_r, {}^4g^{\mu\nu}] = \int d^4x \left[-\mu\,|J| + J^\mu\,(\partial_\mu\,\varphi^{'} + \beta_r\,\partial_\mu\,\tilde{\alpha}^r)\right], \quad \text{(C.23)}$$

or with $U_\mu = -\frac{1}{\mu}\,(\partial_\mu\,\varphi^{'} + \beta_r\,\partial_\mu\,\tilde{\alpha}^r)$. [In Ref. [617]: $M = \mu\,\hat{n}$ rest mass (energy) density and $T = \varphi^{'}/\mu$, $W_r = -\beta_r$, $Z^r = \tilde{\alpha}^r$; $U_\mu = -\partial_\mu\,T + W_r\,\partial_\mu\,Z^r$]:

$$S^{'}_{(dust)}[T, Z^r, M, W_r; {}^4g_{\mu\nu}] = -\frac{1}{2}\int d^4x\,\sqrt{{}^4g}\,(\mu\,\hat{n})\,\Big(U_\mu\,{}^4g^{\mu\nu}\,U_\nu - \epsilon\Big), \quad \text{(C.24)}$$

or

$$\tilde{S}_{(dust)}[\tilde{\alpha}^r; {}^4g_{\mu\nu}] = -\int d^4x\,\mu\,|J|. \quad \text{(C.25)}$$

In Ref. [617] there is a study of the action (Eq. C.24) since the dust is used as a reference fluid in general relativity (GR). At the Hamiltonian level we get:

1. three pairs of second-class constraints ($\pi^r_{\vec{W}}(\tau, \vec{\sigma}) \approx 0$, $\pi_{\vec{Z}\,r}(\tau, \vec{\sigma}) - W_r(\tau, \vec{\sigma})\,\pi_T(\tau, \vec{\sigma}) \approx 0$), which allow the elimination of $W_r(\tau, \vec{\sigma})$ and $\pi^r_{\vec{W}}(\tau, \vec{\sigma})$;
2. a pair of second-class constraints ($\pi_M(\tau, \vec{\sigma}) \approx 0$ plus the secondary $M(\tau, \vec{\sigma}) - \frac{\pi_T^2}{\sqrt{\gamma}\,\sqrt{\pi_T^2 + {}^3g^{rs}\,(\pi_T\,\partial_r\,T + \pi_{\vec{Z}\,u}\,\partial_r\,Z^u)\,(\pi_T\,\partial_s\,T + \pi_{\vec{Z}\,v}\,\partial_s\,Z^v)}}(\tau, \vec{\sigma}) \approx 0$), which allow the elimination of M, π_M.

C.2 Covariant Relativistic Thermo-dynamics of Equilibrium and Non-Equilibrium

In this section we shall collect some results on relativistic fluids, which are well known but scattered in the specialized literature. We shall use essentially Ref. [219], which has to be consulted for the relevant bibliography. See also Ref. [618].

First, we recall some notions of covariant thermodynamics of equilibrium.

Let us remember that given the stress–energy–momentum tensor of a continuous medium $T^{\mu\nu}$, the densities of energy and momentum are T^{oo} and $c^{-1}\,T^{ro}$ respectively (so that $dP^\mu = c^{-1}\,\eta\,T^{\mu\nu}\,d\Sigma_\nu$ is the 4-momentum that crosses the 3-area element $d\Sigma_\nu$ in the sense of its normal [$\eta = -1$ if the normal is space-like, $\eta = +1$ if it is time-like]); instead, $c\,T^{or}$ is the energy flux in the positive r direction, while T^{rs} is the r component of the stress in the plane perpendicular to the s direction (a pressure, if it is positive). A local observer with time-like 4-velocity u^μ ($u^2 = \epsilon\,c^2$) will measure energy density $c^{-2}\,T^{\mu\nu}\,u_\mu\,u_\nu$ and energy flux $\epsilon\,T^{\mu\nu}\,u_\mu\,n_\nu$ along the direction of a unit vector n^μ in his rest-frame.

For a fluid at thermal equilibrium with $T^{\mu\nu} = \rho\,U^\mu\,U^\nu - \epsilon\,\frac{p}{c^2}\,({}^4g^{\mu\nu} - \epsilon\,U^\mu\,U^\nu)$ (U^μ is the hydrodynamical 4-velocity of the fluid) with particle number density $\hat{n}$, specific volume $V = \frac{1}{\hat{n}}$, and entropy per particle $s = \frac{k_B S}{\hat{n}}$ (k_B is Boltzmann's constant) in its rest-frame, the energy density is

$$\rho c^2 = \hat{n}\,(m c^2 + e), \tag{C.26}$$

where e is the mean internal (thermal plus chemical) energy per particle and m is the particle's rest mass.

From a non-relativistic point of view, by writing the equation of state in the form $s = s(e, V)$, the temperature and the pressure emerge as partial derivatives from the first law of thermodynamics in the form (Gibbs equation)

$$ds(e, V) = \frac{1}{T}\,(de + p\,dV). \tag{C.27}$$

If $\mu_{clas} = e + p\,V - T\,s$ is the non-relativistic chemical potential per particle, its relativistic version is

$$\mu^{'} = m\,c^2 + \mu_{class} = \mu - T\,s, \tag{C.28}$$

($\mu = \frac{\rho\,c^2 + p}{\hat{n}}$ is the specific enthalpy, also called chemical potential) and we get

$$\begin{aligned} \mu^{'}\,\hat{n} &= \rho\,c^2 + p - \hat{n}\,T\,s = \rho\,c^2 + p - k_B\,T\,S, \\ k_B\,T\,dS &= d(\rho\,c^2) - \mu^{'}\,d\hat{n} = d(\rho\,c^2) - (\mu - T\,s)\,d\hat{n}, \qquad \text{or} \\ d(\rho\,c^2) &= \mu^{'}\,d\hat{n} + T\,d(\hat{n}s) = \mu\,d\hat{n} + \hat{n}\,T\,ds. \end{aligned} \tag{C.29}$$

By introducing the thermal potential $\alpha = \frac{\mu^{'}}{k_B\,T} = \frac{\mu - T\,s}{k_B\,T}$ and the inverse temperature $\beta = \frac{c^2}{k_B\,T}$, these two equations take the form

$$\begin{aligned} S &= \frac{\hat{n}\,s}{k_B} = \beta\,\left(\rho + \frac{p}{c^2}\right) - \alpha\,\hat{n}, \\ dS &= \beta\,d\rho - \alpha\,d\hat{n}. \end{aligned} \tag{C.30}$$

Let us remark that in Refs. [615, 619, 620] different notations are used, some of which are given in the following equation (in Ref. [615] ρ is denoted e and $\rho^{'}$ is denoted r):

$$\rho + \frac{p}{c^2} = \hat{n}\,(m\,c^2 + e) + \frac{p}{c^2} = \rho^{'}\,h = \rho^{'}\,\left(c^2 + e^{'} + \frac{p}{c^2\,\rho^{'}}\right) = \rho^{'}\,(c^2 + h^{'}), \tag{C.31}$$

where $\rho^{'} = \hat{n}\,m$ is the rest-mass density ($r_* = \sqrt{^4g}\,\rho^{'}$ is called the coordinate rest-mass density) and $e^{'} = e/mc$ is the specific internal energy (so that $\rho^{'}h$ is the effective inertial mass of the fluid; in the post-Newtonian (PN) approximation of Ref. [615] it is shown that $\sigma = c^{-2}\,(T^{oo} + \sum_s T^{ss}) + O(c^{-4}) = c^{-2}\,\sqrt{^4g}\,(-T^o_o + T^s_s) + O(c^{-4})$ has the interpretation of equality of the passive and the active gravitational mass). For the specific enthalpy or chemical potential we get ($\mu/m = h = c^2 + h^{'}$ is called enthalpy)

$$\mu = \frac{1}{\hat{n}}\,\left(\rho + \frac{p}{c^2}\right) = \frac{mc^2}{\rho^{'}}\,\left(\rho + \frac{p}{c^2}\right) = m\,h = m\,(c^2 + h^{'}). \tag{C.32}$$

See Ref. [619] for a richer table of conversion of notations.

Relativistically, we must consider, besides the stress–energy–momentum tensor $T^{\mu\nu}$ and the associated 4-momentum $P^\mu = \int_V d^3\Sigma_\nu\, T^{\mu\nu}$, a particle flux density $\hat{n}^\mu$ (one $\hat{n}^\mu_a$ for each constituent a of the system) and the entropy flux density s^μ. At thermal equilibrium, all these a priori unrelated 4-vectors must be parallel to the hydrodynamical 4-velocity,

$$\hat{n}^\mu = \hat{n}\, U^\mu, \qquad s^\mu = s\, U^\mu, \qquad P^\mu = P\, U^\mu, \tag{C.33}$$

Analogously, we have $V^\mu = V\, U^\mu$ ($V = 1/\hat{n}$ is the specific volume), $\beta^\mu = \beta\, U^\mu = \frac{c^2}{k_B\, T}\, U^\mu$ (a related 4-vector is the equilibrium parameter 4-vector $i^\mu = \mu'\, \beta^\mu$).

Since $\epsilon\, U_\mu\, T^{\mu\nu} = \rho\, U^\nu$, we get the final manifestly covariant form of the previous two equations (now the hydrodynamical 4-velocity is considered as an extra thermodynamical variable):

$$\begin{aligned} S^\mu &= S\, U^\mu = \frac{\hat{n}\, s}{k_B}\, U^\mu = \frac{p}{c^2}\, \beta^\mu - \alpha\, \hat{n}^\mu - \epsilon\, \beta_\nu\, T^{\nu\mu}, \\ dS^\mu &= -\alpha\, d\hat{n}^\mu - \epsilon\, \beta_\nu\, dT^{\nu\mu}. \end{aligned} \tag{C.34}$$

Global thermal equilibrium imposes

$$\partial_\mu\, \alpha = \partial_\mu\, \beta_\nu + \partial_\nu\, \beta_\mu = 0. \tag{C.35}$$

As a consequence, we get

$$d\left(\frac{p}{c^2}\, \beta^\mu\right) = \hat{n}^\mu\, d\alpha + \epsilon\, T^{\nu\mu}\, d\beta_\nu, \tag{C.36}$$

namely the basic variables $\hat{n}^\mu$, $T^{\nu\mu}$, and S^μ can all be generated from partial derivatives of the fugacity 4-vector (or thermodynamical potential):

$$\begin{aligned} &\phi^\mu(\alpha, \beta_\lambda) = \frac{p}{c^2}\, \beta^\mu, \\ &\hat{n}^\mu = \frac{\partial \phi^\mu}{\partial \alpha}, \qquad T^{\nu\mu}_{(mat)} = \frac{\partial \phi^\mu}{\partial \beta_\nu}, \qquad S^\mu = \phi^\mu - \alpha\, \hat{n}^\mu - \beta_\nu\, T^{\nu\mu}_{(mat)}, \end{aligned} \tag{C.37}$$

once the equation of state is known. Here, $T^{\nu\mu}_{(mat)}$ is the canonical or material (in general non-symmetric) stress tensor, ensuring that reversible flows of field energy are not accompanied by an entropy flux.

This final form remains valid (at least to first order in deviations) for states that deviate from equilibrium, when the 4-vectors S^μ, $\hat{n}^\mu$, ... are no more parallel; the extra information in this equation is precisely the standard linear relation between entropy flux and heat flux. The second law of thermodynamics for relativistic systems is $\partial_\mu\, S^\mu \geq 0$, which becomes a strict equality in equilibrium.

The fugacity 4-vector ϕ^μ is evaluated by using the covariant relativistic statistical theory for thermal equilibrium [219] starting from a grand canonical ensemble with density matrix $\hat{\rho}$ by maximizing the entropy $S = -\mathrm{Tr}(\hat{\rho} \ln \hat{\rho})$ subject to the constraints $Tr\hat{\rho} = 1$, $Tr(\hat{\rho}\,\hat{n}) = \hat{n}$, $Tr(\hat{\rho}\,\hat{P}^{\lambda}) = P^{\lambda}$. This gives (in the large volume limit)

$$\hat{\rho} = Z^{-1}\, e^{\alpha\,\hat{n} + \beta_\mu\,\hat{P}^\mu},$$

with

$$\ln Z = \int_{\triangle\Sigma} \epsilon\,\phi^\mu\, d\Sigma_\mu, \qquad \hat{n} = \int_{\triangle\Sigma} \epsilon\,\hat{n}^\mu\, d\Sigma_\mu, \qquad P^\mu = \int_{\triangle\Sigma} \epsilon\, T^{\mu\nu}_{(mat)}\, d\Sigma_\nu \tag{C.38}$$

(it is assumed that the members of the ensemble are small [macroscopic] subregions of one extended body in thermal equilibrium, whose world-tubes intersect an arbitrary space-like hyper-surface in small 3-areas $\triangle\Sigma$). Therefore, we have to find the grand canonical partition function,

$$Z(V_\mu, \beta_\mu, i_\mu) = \sum_n e^{i_\mu\,\hat{n}^\mu}\, Q_n(V_\mu, \beta_\mu), \quad \text{where} \quad Q_n(V_\mu, \beta_\mu) = \int_{V_\mu} d\sigma_n(q,p)\, e^{-\beta_\mu\, P^\mu}. \tag{C.39}$$

It is the canonical partition function for fixed volume V_μ and $d\sigma_n$ is the invariant micro-canonical density of states. For an ideal Boltzmann gas of N free particles of mass m it is

$$d\sigma_n(p,m) = \frac{1}{N!} \int \delta^4\left(P - \sum_{i=1}^{N} p_i\right) \prod_{i=1}^{N} 2\, V_\mu\, p_i^\mu\, \theta(p_i^o)\, \delta(p_i^2 - \epsilon\, m^2)\, d^4p_i. \tag{C.40}$$

Following Ref. [621], Q_n can be evaluated in the rest-frame instant form on the Wigner hyper-plane (this method can be extended to a gas of molecules, which are N-body bound states):

$$Q_n = \frac{1}{N!} \left[\frac{V\, m^2}{2\,\pi^2\,\beta}\, K_2(m\,\beta)\right]^N. \tag{C.41}$$

The same results may be obtained by starting from the covariant relativistic kinetic theory of gas (see Refs. [622–624]; see Ref. [219] for a short review) whose particles interact only by collisions by using Synge's invariant distribution function $N(q,p)$ [625] (the number of particle world-lines with momenta in the range $(p_\mu, d\omega)$ that cross a target 3-area $d\Sigma_\mu$ in M^4 in the direction of its normal is given by $dN = N(q,p)\, d\omega\, \eta\, v^\mu\, d\Sigma_\mu$ [$= N\, d^3q d^3p$ for the 3-space q^o = const.]; $d\omega = d^3p/v^o\, \sqrt{{}^4g}$ is the invariant element of 3-area on the mass-shell). One arrives at a transport equation for $N, \frac{dN}{d\tau} = \nabla_\mu\,(N\, v^\mu)$ (v^μ is the particle velocity obtained from the Hamilton equation implied by the one-particle Hamiltonian $\bar{H} = \sqrt{\epsilon\, {}^4g^{\mu\nu}(q)\, p_\mu\, p_\nu} = m$ [it is the energy after the gauge fixing $q^o \approx \tau$ to the first-class contsraint ${}^4g^{\mu\nu}(q)\, p_\mu\, p_\nu - \epsilon\, m^2c^2 \approx 0$]; ∇_μ is the covariant gradient holding the 4-vector p_μ [not its components] fixed) with a collision term $C[N]$ describing the collisions; for a dilute simple gas dominated by binary collisions we arrive at the Boltzmann equation (for $C[N] = 0$, one solution is the relativistic version of the Maxwell–Boltzmann distribution function, i.e., the classical Jüttner–Synge one $N = \text{const.}\, e^{-\beta_\mu\, P^\mu} / 4\,\pi\, m^2\, K_2(m\,\beta)$ for the Boltzmann gas [625]). The H-theorem ($\nabla_\mu\, S^\mu \geq 0$, where $S^\mu(q) = -\int [N \ln(N\, h^3) - N]\, v^\mu\, d\omega$ is the entropy flux) and the results at thermal equilibrium emerge (from the balance law $\nabla_\mu\, (\int\, N\, f\, v^\mu\, d\omega) = \int\, f\, C[N]\, d\omega$ [f is an arbitrary tensorial function] we can deduce the conservation laws $\nabla_\mu\, \hat{n}^\mu = \nabla_\mu\, T^{\nu\mu} = 0$, where $\hat{n}^\mu = \int\, N\, v^\mu\, d\omega$, $T_\nu{}^\mu = \int\, N\, p_\nu\, v^\mu\, d\omega$; the vanishing of entropy production at local thermal equilibrium gives $N_{eq}(q,p) = h^{-3}\, e^{\alpha(q) + \beta_\nu(q)\, P^\nu}$ in the case of the Boltzmann statistic and we get ($U^\mu = \beta^\mu / \beta$) $n^\mu_{eq} = \int\, N_{eq} v^\mu\, d\omega = n\, U^\mu$, $T^{\nu\mu}_{eq} = \rho\, U^\nu\, U^\mu - \epsilon\, p\, ({}^4g^{\nu\mu} - \epsilon\, U^\nu\, U^\mu)$, $S^\mu_{eq} = p\,\beta^\mu - \alpha\, n^\mu_{eq} - \beta_\nu\, T^{\mu\nu}_{eq}$; we obtain the equations for the Boltzmann ideal gas).

One can study the small deviations from thermal equilibrium ($N = N_{eq}(1+f)$, where N_{eq} is an arbitrary local equilibrium distribution) with the linearized Boltzmann equation and then by using either the Chapman–Enskog ansatz of quasi-stationarity of small deviations (this ignores the gradients of f and gives the standard Landau–Lifshitz and Eckart phenomenological laws; we get the Fourier equation for heat conduction and the Navier–Stokes equation for the bulk and shear stresses; however, we have parabolic and not hyperbolic equations, implying non-causal propagation) or with the Grad method in the 14-moment approximation. This method retains the gradients of f (there are five extra thermodynamical variables, which can be explicitly determined from 14 moments among the infinite set of moments $\int\, N\, p^\mu\, p^\nu\, p^\rho\, \ldots\, d^4p$ of kinetic theory; no extra auxiliary state variables are introduced to specify a non-equilibrium state besides $T^{\mu\nu}$, $\hat{n}^\mu$, S^μ) and gives phenomenological laws which are the kinetic equivalent of Müller extended thermodynamics and its various developments. Now the equations are hyperbolic there is no causality problem, but there are problems with shock waves. See Ref. [219] for the bibliography and for a review of the non-equilibrium phenomenological laws (see also Ref. [626]) of Eckart and Landau–Lifshitz, of the various formulations of extended thermodynamics, of non-local thermodynamics.

While in Ref. [627] it is said that the difference between causal hyperbolic theories and a-causal parabolic ones is unobservable, in Ref. [628] (see also Ref. [629]) there is a discussion of the cases in which hyperbolic theories are relevant. See also the numerical codes of Refs. [630–632].

In phenomenological theories, the starting point are the equations $\partial_\mu\, T^{\mu\nu} = \partial_\mu\, \hat{n}^\mu = 0$, $\partial_\mu\, S^\mu \geq 0$. There is the problem of how to define a 4-velocity and a rest-frame for a given non-equilibrium state. Another problem is how to specify a non-equilibrium state completely at the macroscopic level: a priori we could need an infinite number of auxiliary quantities (vanishing at equilibrium) and an equation of state depending on them. The basic postulate of extended thermodynamics is the absence of such variables.

Regarding the rest-frame problem, there are two main solutions in the literature connected with the relativistic description of heat flow:

1. Eckart theory. One considers a local observer in a simple fluid who is at rest with respect to the average motion of the particles: its 4-velocity $U^{\mu}_{(eck)}$ is parallel by definition to the particle flux $\hat{n}^{\mu}$, namely

$$\hat{n}^{\mu} = \hat{n}_{(eck)}\, U^{\mu}_{(eck)}. \tag{C.42}$$

This local observer sees heat flow as a flux of energy in his rest-frame:

$$\begin{aligned}
\epsilon\, U_{(eck)\mu}\, T^{\mu\nu} &= \rho_{(eck)}\, U^{\mu}_{(eck)} + q^{\mu}_{(eck)},\\
&\text{so that we get}\\
T^{\mu\nu} &= \rho_{(eck)}\, U^{\mu}_{(eck)}\, U^{\nu}_{(eck)} + q^{\mu}_{(eck)}\, U^{\nu}_{(eck)} + U^{\mu}_{(eck)}\, q^{\nu}_{(eck)} + P^{\mu\nu}_{(eck)},\\
P^{\mu\nu}_{(eck)} &= P^{\nu\mu}_{(eck)} = \epsilon\,(p + \pi_{(eck)})\,({}^4g^{\mu\nu} - \epsilon\, U^{\mu}_{(eck)}\, U^{\nu}_{(eck)} + \pi^{\mu\nu}_{(eck)},\\
&P^{\mu\nu}_{(eck)}\, U_{(eck)\nu} = q_{(eck)\mu}\, U^{\mu}_{(eck)} = 0,\\
&\pi_{(eck)\mu\nu}\,({}^4g^{\mu\nu} - \epsilon\, U^{\mu}_{(eck)}\, U^{\nu}_{(eck)}) = 0,
\end{aligned} \tag{C.43}$$

where p is the thermodynamic pressure, $\pi_{(eck)}$ the bulk viscosity, and $\pi^{\mu\nu}_{(eck)}$ the shear stress.

This description has the particle conservation law $\partial_{\mu}\, \hat{n}^{\mu} = 0$.

2. Landau–Lifshitz theory. One considers a different observer (drifting slowly in the direction of heat flow with a 3-velocity $\vec{v}_D = \vec{q}/\hat{n}\, m\, c^2$) whose 4-velocity $U^{\mu}_{(ll)}$ is by definition such to give a vanishing heat flow, i.e., there is no net energy flux in his rest-frame: $U^{\mu}_{(ll)}\, T^{\nu}_{\mu}\, \hat{n}_{\nu} = 0$ for all vectors $\hat{n}_{\mu}$ orthogonal to $U^{\mu}_{(ll)}$. This implies that $U^{\mu}_{(ll)}$ is the time-like eigenvector of $T^{\mu\nu}$, $T^{\mu\nu}\, U_{(ll)\nu} = \epsilon\, \rho_{(ll)}\, U^{\mu}_{(ll)}$, which is unique if $T^{\mu\nu}$ satisfies a positive energy condition. Now we get

$$\begin{aligned}
T^{\mu\nu} &= \rho_{(ll)}\, U^{\mu}_{(ll)}\, U^{\nu}_{(ll)} + P^{\mu\nu}_{(ll)},\\
P^{\mu\nu}_{(ll)} &= P^{\nu\mu}_{(ll)} = \epsilon\,(p + \pi_{(ll)})\,({}^4g^{\mu\nu} - \epsilon\, U^{\mu}_{(ll)}\, U^{\nu}_{(ll)} + \pi^{\mu\nu}_{(ll)},\\
&P^{\mu\nu}_{(ll)}\, U_{(ll)\nu} = 0, \qquad \pi_{(ll)\mu\nu}\,({}^4g^{\mu\nu} - \epsilon\, U^{\mu}_{(ll)}\, U^{\nu}_{(ll)}) = 0,\\
\hat{n}^{\mu} &= \hat{n}_{(ll)}\, U^{\mu}_{(ll)} + j^{\mu}_{(ll)}, \qquad j_{(ll)\mu}\, U^{\mu}_{(ll)} = 0 \qquad (\vec{j} = -\hat{n}\, \vec{v}_D = -\vec{q}/m\, c^2).
\end{aligned} \tag{C.44}$$

This observer in his rest-frame does not see a heat flow but a particle drift. This description has the simplest form of the energy–momentum tensor.

One has $\hat{n}_{(eck)} = \hat{n}_{(ll)}\, ch\, \varphi$, $\rho_{(eck)} = \rho_{(ll)}\, ch^2\, \varphi + p_{(ll)}\, sh^2\, \varphi = \pi^{\mu\nu}\, j_{\mu}\, j_{\nu}/\hat{n}^2_{(eck)}$, with $ch\, \varphi = U^{\mu}_{(ll)}\, U_{(eck)\mu}$ (the difference is a Lorentz factor $\sqrt{1 - \vec{v}^2_D/c^2}$, so that there are insignificant differences for many practical purposes if deviations from equilibrium are small). The angle $\varphi \approx j/\hat{n} \approx v_D/c \approx q/\hat{n}\, m\, c^2$ is a dimensionless measure of the deviation from equilibrium ($\hat{n}_{(ll)} - \hat{n}_{(eck)}$ and $\rho_{(ll)} - \rho_{(eck)}$ are of order φ^2).

One can decompose $T^{\mu\nu}$, $\hat{n}^{\mu}$ in terms of any 4-velocity U^{μ} that falls within a cone of angle $\approx \varphi$ containing $U^{\mu}_{(eck)}$ and $U^{\mu}_{(ll)}$; each choice U^{μ} gives a particle density $\hat{n}(U) = \epsilon\, u_{\mu}\, \hat{n}^{\mu}$ and energy density $\rho(U) = U_{\mu}\, U_{\nu}\, T^{\mu\nu}$, which are independent of U^{μ} if we neglect terms of order φ^2. Therefore, we have:

1. if $S_{eq}(\rho(U), n(U))$ is the equilibrium entropy density, then $S(U) = \epsilon\, U_\mu\, S^\mu = S_{eq} + O(\varphi^2)$;
2. if $p(U) = -\frac{\partial \rho/\hat{n}}{\partial 1/\hat{n}}|_{S/\hat{n}}$ is the (reversible) thermodynamical pressure defined as work done in an isentropic expansion (off-equilibrium this definition allows us to separate it from the bulk stress $\pi(U)$ in the stress–energy–momentum tensor), and p_{eq} is the pressure at equilibrium, then $p(U) = P_{eq}(\rho(U), n(U)) + O(\varphi^2)$.

By postulating that the covariant Gibbs relation remains valid for arbitrary infinitesimal displacements $(\delta\,\hat{n}^\mu, \delta\, T^{\mu\nu}, \ldots)$ from an equilibrium state, we get a covariant off-equilibrium thermodynamics based on the equation

$$\begin{aligned} S^\mu &= p(\alpha, \beta)\, \beta^\mu - \alpha\, \hat{n}^\mu - \beta_\nu\, T^{\mu\nu} - Q^\mu\, (\delta\, \hat{n}^\nu, \delta\, T^{\nu\rho}, \ldots), \\ \nabla_\mu\, S^\mu &= -\delta\, \hat{n}^\mu\, \partial_\mu\, \alpha - \delta\, T^{\mu\nu}\, \nabla_\nu\, \beta_\mu - \nabla_\mu\, Q^\mu \geq 0, \end{aligned} \tag{C.45}$$

with Q^μ of second order in the displacements and α, β_μ arbitrary. At equilibrium we recover $S^\mu_{eq} = p\, \beta^\mu - \alpha\, \hat{n}^\mu_{eq} - \beta_\nu\, T^{\mu\nu}_{eq}$, $\nabla_\mu\, S^\mu_{eq} = 0$ (with $U^\mu = \beta^\mu/\beta$ and [for viscous heat-conducting fluids, but not for super-fluids] $\partial_\mu\, \alpha = \nabla_\mu\, \beta_\nu + \nabla_\nu\, \beta_\mu = 0$).

If we choose $\beta^\mu = U^\mu/k_B\, T$ parallel to $\hat{n}^\mu$ of the given off-equilibrium state, we are in the Eckart frame, $U^\mu = U^\mu_{(eck)}$, and we get

$$\begin{aligned} S &= \epsilon\, U_{(eck)\mu}\, S^\mu = S_{eq} + \epsilon\, U_{(eck)\mu}\, Q^\mu, \\ \sigma^\mu_{(eck)} &= ({}^4g^{\mu\nu} - \epsilon\, U^\mu_{(eck)}\, U^\nu_{(eck)})\, S_\nu = \beta\, q^\mu_{(eck)} - ({}^4g^{\mu\nu} - \epsilon\, U^\mu_{(eck)}\, U^\nu_{(eck)})\, Q_\nu, \\ q^\mu_{(eck)} &= -({}^4g^{\mu\nu} - \epsilon\, U^\mu_{(eck)}\, U^\nu_{(eck)})\, T^\rho_\nu\, U_{(eck)\rho}, \end{aligned} \tag{C.46}$$

so that to linear order we get the standard relation between entropy flux $\vec{\sigma}_{(eck)}$ and heat flux $\vec{q}_{(eck)}$

$$\vec{\sigma}_{(eck)} = \frac{\vec{q}_{(eck)}}{k_B\, T} + (\text{possible second} - \text{order term}). \tag{C.47}$$

If we choose $U^\mu = U^\mu_{(ll)}$, the time-like eigenvector of $T^{\mu\nu}$, so that $U_{(ll)\mu}\, T^\mu_\nu\, ({}^4g^\nu_\rho - \epsilon\, U^\nu_{(ll)}\, U_{(ll)\rho}) = 0$, we are in the Landau–Lifshitz frame and we get

$$\begin{aligned} \sigma^\mu_{(ll)} &= ({}^4g^{\mu\nu} - \epsilon\, U^\mu_{(ll)}\, U^\nu_{(ll)})\, S_\nu = -\alpha\, j^\mu_{(ll)} - ({}^4g^{\mu\nu} - \epsilon\, U^\mu_{(ll)}\, U^\nu_{(ll)})\, Q_\nu, \\ j^\mu_{(ll)} &= ({}^4g^{\mu\nu} - \epsilon\, U^\mu_{(ll)}\, U^\nu_{(ll)})\, \hat{n}_\nu, \end{aligned} \tag{C.48}$$

so that at linear order we get the standard relation between entropy flux $\vec{\sigma}_{(ll)}$ and diffusive flux $\vec{j}_{(ll)}$

$$\vec{\sigma}_{(ll)} = -\frac{\mu}{k_B\, T}\, \vec{j}_{(ll)} + (\text{possible second} - \text{order term}). \tag{C.49}$$

In the Landau–Lifshitz frame heat flow and diffusion are contained in the diffusive flux $\vec{j}_{(ll)}$ relative to the mean mass–energy flow.

The entropy inequality becomes (each term is of second order in the deviations from local equilibrium)

$$0 \leq \nabla_\mu S^\mu = -\delta\, \hat{n}^\mu\, \partial_\mu\, \alpha - \delta\, T^{\mu\nu}\, \nabla_\nu\, \beta_\mu - \nabla_\mu\, Q^\mu, \tag{C.50}$$

with the fitting conditions $\delta\, \hat{n}^\mu\, U_\mu = \delta\, T^{\mu\nu}\, U_\mu\, U_\nu = 0$, which contain all information about the viscous stresses, heat flow, and diffusion in the off-equilibrium state (they are dependent on the arbitrary choice of the 4-velocity U^μ).

Once a detailed form of Q^μ is specified, linear relations between irreversible fluxes $\delta\, T^{\mu\nu}$, $\delta\, \hat{n}^\mu$ and gradients $\nabla_{(\mu}\, \beta_{\nu)}$, $\partial_\mu\, \alpha$ follow.

C.2.1 $Q^\mu = 0$ *(as in the Non-relativistic Case)*

The spatial entropy flux $\vec{\sigma}$ is only a strictly linear function of heat flux $\vec{q}$ and diffusion flux $\vec{j}$. In this case, the off-equilibrium entropy density $S = \epsilon\, U_\mu\, S^\mu$ is given by the equilibrium equation of state $S = S_{eq}(\rho, \hat{n})$. We have $0 \leq \nabla_\mu\, S^\mu = -\delta\, \hat{n}^\mu\, \partial_\mu\, \alpha - \delta\, T^{\mu\nu}\, \nabla_\nu\, \beta_\mu$, with fitting conditions $\delta\, \hat{n}^\mu\, U_\mu = \delta\, T^{\mu\nu}\, U_\mu\, U_\nu = 0$ and with U^μ still arbitrary at first order.

1. Landau–Lifshitz frame and theory. $U^\mu = U^\mu_{(ll)}$ is the time-like eigenvector of $T^{\mu\nu}$. This and the fitting conditions imply $\delta\, T^{\mu\nu}\, U_{(ll)\nu} = 0$. The shear and bulk stresses $\pi^{\mu\nu}_{(ll)}$, $\pi_{(ll)}$ are identified by the decomposition

$$\delta T^{\mu\nu} = \pi^{\mu\nu}_{(ll)} + \pi_{(ll)}\, ({}^4g^{\mu\nu} - \epsilon\, U^\mu_{(ll)}\, U^\nu_{(ll)}), \qquad \pi^{\mu\nu}_{(ll)}\, U_{(ll)\nu} = \pi^\mu_{(ll)\mu} = 0. \tag{C.51}$$

The inequality $\nabla_\mu\, S^\mu \geq 0$ becomes

$$-j^\mu_{(ll)}\, \partial_\mu\, \alpha - \beta\, \pi^{\mu\nu}_{(ll)} < \nabla_\nu\, \beta_\mu > -\beta\, \pi_{(ll)}\, \nabla_\mu\, U^\mu_{(ll)} \geq 0, \qquad j^\mu_{(ll)} = \delta\, \hat{n}^\mu,$$

$$\begin{aligned} < X_{\mu\nu} > = {} & [({}^4g^\alpha_\mu - \epsilon\, U^\alpha_{(ll)}\, U_{(ll)\mu})\, ({}^4g^\beta_\nu - \epsilon\, U^\beta_{(ll)}\, U_{(ll)\nu}) \\ & - \frac{1}{3}\, ({}^4g_{\mu\nu} - \epsilon\, U_{(ll)\mu}\, U_{(ll)\nu})\, ({}^4g^{\alpha\beta} - \epsilon\, U^\alpha_{(ll)}\, U^\beta_{(ll)})]\, X_{\alpha\beta}, \end{aligned} \tag{C.52}$$

(the $< \dots >$ operation extracts the purely spatial, trace-free part of any tensor).

If the equilibrium state is isotropic (Curie's principle) and if we assume that $(j^\mu_{(ll)}, \pi^{\mu\nu}_{(ll)}, \pi_{(ll)})$ are linear and purely local functions of the gradients, $\nabla_\mu\, S^\mu \geq 0$ implies

$$j^\mu_{(ll)} = -\kappa\, ({}^4g^{\mu\nu} - \epsilon\, U^\mu_{(ll)}\, U^\nu_{(ll)})\, \partial_\mu\, \alpha, \quad \kappa > 0, \tag{C.53}$$

(it is a mixture of Fourier's law of heat conduction and of Fick's law of diffusion, stemming from the relativistic mass–energy equivalence), and the standard Navier–Stokes equations (ζ_S, ζ_V are shear and bulk viscosities)

$$\pi_{(ll)\mu\nu} = -2\, \zeta_S < \nabla_\nu\, \beta_\mu >, \qquad \pi_{(ll)} = \frac{1}{3}\, \zeta_V\, \nabla_\mu\, U^\mu_{(ll)}. \tag{C.54}$$

2. Eckart frame and theory. $U^{\mu}_{(eck)}$ parallel to $\hat{n}^{\mu}$. Now we have the fitting condition $\delta\,\hat{n}^{\mu}=0$. The heat flux appears in the decomposition of $\delta\,T^{\mu\nu}$ ($a_{(eck)\mu}=U^{\nu}_{(eck)}\,\nabla_{\nu}\,U_{(eck)\mu}$ is the 4-acceleration):

$$\delta\,T^{\mu\nu}=q^{\mu}_{(eck)}\,U^{\nu}_{(eck)}+U^{\mu}_{(eck)}\,q^{\nu}_{(eck)}+\pi^{\mu\nu}_{(eck)}+\pi_{(eck)}\,({}^{4}g^{\mu\nu}-\epsilon\,U^{\mu}_{(eck)}\,U^{\nu}_{(eck)}). \tag{C.55}$$

The inequality $\nabla_{\mu}\,S^{\mu}\geq 0$ becomes

$$q^{\mu}_{(eck)}\,(\partial_{\mu}\,\alpha-\beta\,a_{(eck)\mu})-\beta\,(\pi^{\mu\nu}_{(eck)}\,\nabla_{\nu}\,U_{(eck)\mu}+\pi_{(eck)}\,\nabla_{\mu}\,U^{\mu}_{(eck)})\geq 0. \tag{C.56}$$

With the simplest assumption of linearity and locality, we obtain Fourier's law of heat conduction (it is not strictly equivalent to the Landau–Lifshitz one, because they differ by spatial gradients of the viscous stresses and the time-derivative of the heat flux)

$$q^{\mu}_{(eck)}=-\kappa\,({}^{4}g^{\mu\nu}-\epsilon\,U^{\mu}_{(eck)}\,U^{\nu}_{(eck)})\,(\partial_{\nu}\,T+T\,\partial_{\tau}\,U_{(eck)\nu}), \tag{C.57}$$

(the term depending on the acceleration is sometimes referred to as an effect of the inertia of heat), and the same form of the Navier–Stokes equations for $\pi^{\mu\nu}_{(eck)}$, $\pi_{(eck)}$ (they are not strictly equivalent to the Landau–Lifshitz ones because they differ by gradients of the drift $\vec{v}_D=\vec{q}/\hat{n}\,m\,c^2$).

For a simple fluid, Fourier's law and Navier–Stokes equations (nine equations) and the conservation laws $\nabla_{\mu}\,T^{\mu\nu}=\nabla_{\mu}\,\hat{n}^{\mu}=0$ (five equations) determine the 14 variables $T^{\mu\nu}$, $\hat{n}^{\mu}$ from suitable initial data. However, these equations are of mixed parabolic–hyperbolic– elliptic type and, as said, we get a-causality and instability.

Kinetic theory gives

$$Q^{\mu}=-\frac{1}{2}\int N_{eq}f^2\,p^{\mu}\,d\omega\neq 0, \tag{C.58}$$

for a gas up to second order in the deviation $(N-N_{eq})=N_{eq}f$ ($Q^{\mu}=0$ requires small gradients and quasi-stationary processes). Two alternative classes of phenomenological theories are as follows.

C.2.2 Linear Non-local Thermodynamics (NLT)

This theory gives a rheomorphic rather than causal description of the phenomenological laws: transport coefficients at an event x are taken to depend not on the entire causal past of x, but only on the past history of a comoving local fluid element. It is a linear theory restricted to small deviations from equilibrium, which can be derived from the linearized Boltzmann equation by projector–operator techniques (and probably inherits its causality properties). Instead of writing $(\delta\,\hat{n}_{\nu}(x),\delta\,T_{\mu\nu}(x))=\sigma(U,T)\,(-\partial_{\mu}\,\alpha(x),-\nabla_{(\mu}\,\beta_{\nu)}(x))$, this local phenomenological law is generalized to $(\delta\,\hat{n}_{\mu}(\vec{x},x^o),\delta\,T_{\mu\nu}(\vec{x},x^o))=\int_{-\infty}^{\infty}dx^{o'}\,\sigma\,(x^o-x^{o'})\,(-\partial_{\mu}\,\alpha\,(\vec{x}\,x^{o'}),-\nabla_{(\mu}\,\beta_{\nu)}\,(\vec{x}\,x^{o'}))$.

C.2.3 Local Non-linear Extended Thermodynamics (ET)

This is more relevant for relativistic astrophysics, where correlation and memory effects are not of primary interest and, instead, we need a tractable and consistent transport theory co-extensive at the macroscopic level with the Boltzmann equation. It is assumed that the second-order term $Q^{\mu}\,(\delta\,\hat{n}^{\nu}, \delta T^{\nu\rho}, \ldots)$ does not depend on auxiliary variables vanishing at equilibrium: this ansatz is the phenomenological equivalent of Grad's 14-moment approximation in kinetic theory. These theories are called second-order theories and many of them are analyzed in Ref. [633]; when the dissipative fluxes are subject to a conservation equation, these theories are said to be of causal divergence type, like the ones of Refs. [634–636]. Another type of theory (extended irreversible thermodynamics; in general these theories are not of divergence type) was developed in Refs. [637–640]: here, there are transport equations for the dissipative fluxes rather than conservation laws.

For small deviations we retain only the quadratic terms in the Taylor expansion of Q^{μ} (leading to linear phenomenological laws): This implies five new undetermined coefficients,

$$Q^{\mu} = \frac{1}{2}\, U^{\mu}\,[\beta_o\,\pi^2 + \beta_1\, q^{\mu}\, q_{\mu} + \beta_2\, \pi^{\mu\nu}\, \pi_{\mu\nu}] - \alpha_o\,\pi\, q^{\mu} - \alpha_1\, \pi^{\mu\nu}\, q_{\nu}, \tag{C.59}$$

with $\beta_i > 0$ from $\epsilon\, U_{\mu}\, Q^{\mu} > 0$ (the β_i are "relaxation times"). A first-order change of rest-frame produces a second-order change in Q^{μ} (going from the Landau–Lifshitz frame to the Eckart one, we get $\alpha_{(eck)i} - \alpha_{(ll)i} = \beta_{(ll)1} - \beta_{(eck)1} = [(\rho + p)\,T]^{-1}$, $\beta_{(ll)0} = \beta_{(eck)o}$, $\beta_{(ll)2} = \beta_{(eck)2}$, and the phenomenological laws are now invariant to first order).

In the Eckart frame the phenomenological laws take the form

$$\begin{aligned} q^{\mu}_{(eck)} &= -\kappa\, T\,({}^4g^{\mu\nu} - \epsilon\, U^{\mu}_{(eck)}\, U^{\nu}_{(eck)})\,[T^{-1}\,\partial_{\nu}\, T + a_{(eck)\nu} + \beta_{(eck)1}\,\partial_{\tau}\, q_{(eck)\nu} \\ &\quad - \alpha_{(eck)o}\,\partial_{\nu}\,\pi_{(eck)} - \alpha_{(eck)1}\,\nabla_{\rho}\,\pi^{\rho}_{(eck)\nu}], \end{aligned}$$

$$\pi_{(eck)\mu\nu} = -2\,\zeta_S\,[< \nabla_{\nu}\, U_{(eck)\mu} > + \beta_{(eck)2}\,\partial_{\tau}\,\pi_{(eck)\mu\nu} - \alpha_{(eck)1} < \nabla_{\nu}\, q_{(eck)\mu} >],$$

$$\pi_{(eck)} = -\frac{1}{3}\,\zeta_V\,[\nabla_{\mu}\, U^{\mu}_{(eck)} + \beta_{(eck)o}\,\partial_{\tau}\,\pi_{(eck)} - \alpha_{(eck)o}\,\nabla_{\mu}\, q^{\mu}_{(eck)}], \tag{C.60}$$

which reduce to the equation of the standard Eckart theory if the five relaxation (β_i) and coupling (α_i) coefficients are equal to zero. See, for instance, Ref. [641] for a complete treatment and also Ref. [642]. For appropriate values of these coefficients, these equations are hyperbolic and, therefore, causal and stable. The transport equations can be understood [628] as evolution equations for the dissipative variables as they describe how these fluxes evolve from an initial arbitrary state to a final steady one (the time parameter τ is usually interpreted as the relaxation time of the dissipative processes). In the case of a gas the new coefficients can be found explicitly in Ref. [637] (see also Ref. [643–645] for a recent approach to relativistic interacting gases starting from the Boltzmann equation), and they are purely thermodynamical functions. Wave-front speeds are finite and comparable with the speed of sound. A problem with these theories is that they do not admit a regular shock structure (like the Navier–Stokes equations) once the speed of the shock front exceeds the highest characteristic velocity (a subshock will form within a shock layer for speeds exceeding the wave-front velocities of thermoviscous effects). The situation slowly ameliorates if more moments are taken into account [641].

In the approach reviewed in Ref. [641] the extra indeterminacy associated with the new five coefficients is eliminated (at the price of high non-linearity) by annexing to the usual conservation and entropy laws a new phenomenological assumption (in this way we obtain a causal divergence type theory):

$$\nabla_\rho A^{\mu\nu\rho} = I^{\mu\nu}, \tag{C.61}$$

in which $A^{\rho\mu\nu}$ and $I^{\mu\nu}$ are symmetric tensors with the following traces:

$$A^{\mu\nu}_{\nu} = -n^\mu, \qquad I^\mu_\mu = 0. \tag{C.62}$$

These conditions are modeled on kinetic theory, in which $A^{\rho\mu\nu}$ represents the third moment of the distribution function in momentum space, and $I^{\mu\nu}$ the second moment of the collision term in the Boltzmann equation. The previous equations are central in the determination of the distribution function in Grad's 14-moment approximation. The phenomenological theory is completed by the postulate that the state variables S^μ, $A^{\rho\mu\nu}$, $I^{\mu\nu}$ are invariant functions of $T^{\mu\nu}$, $\hat{n}^\mu$ only. The theory is an almost exact phenomenological counterpart of the Grad approximation. See Ref. [620] for the beginning (only non-viscous heat conducting materials are treated) of a derivation of extended thermodynamics from a variational principle.

Everything may be rephrased in terms of the Lagrangian coordinates of the fluid. What is lacking in the non-dissipative case of heat conduction is the functional form of the off-equilibrium equation of state reducing to $\rho = \rho(\hat{n}, s)$ at thermal equilibrium. In the dissipative case the system is open and $T^{\mu\nu}$, P^μ, $\hat{n}^\mu$ are not conserved.

See Ref. [646] for attempts to define a classical theory of dissipation in the Hamiltonian framework and Ref. [647] about Hamiltonian molecular dynamics for the addition of an extra degree of freedom to an N-body system to transform it into an open system (with the choice of a suitable potential for the extra variable, the equilibrium distribution function of the N-body subsystem is exactly the canonical ensemble).

However, the most constructive procedure is to get (starting from an action principle) the Hamiltonian form of the energy–momentum of a closed system, as has been done in Ref. [149] for a system of N charged scalar particles, in which the mutual action-at-a-distance interaction is the complete Darwin potential extracted from the Lienard–Wiechert solution in the radiation gauge (the interactions are momentum- and, therefore, velocity-dependent). In this case we can define an open (in general dissipative) subsystem by considering a cluster of $n < N$ particles and assigning to it a non-conserved energy–momentum tensor built with all the terms of the original energy–momentum tensor, which depend on the canonical variables of the n particles (the other $N - n$ particles are considered as external fields).

References

[1] M.H. Soffel, *Relativity in Astrometry, Celestial Mechanics and Geodesy* (Springer, Berlin, 1989).

[2] J. Kovalevski, I.I. Mueller, and B. Kolaczek, *Reference Frames in Astronomy and Geophysics* (Kluwer, Dordrecht, 1989).

[3] O.J. Sovers and J.L. Fanselow, Astrometry and Geodesy with Radio Interferometry: Experiments, Models, Results, *Rev. Mod. Phys.* **70**, 1393 (1998).

[4] R.M. Wald, *General Relativity* (University of Chicago Press, Chicago, 1984).

[5] A. Ashtekar, *New Perspectives in Canonical Gravity* (Bibliopolis, Naples, 1988).

[6] A. Ashtekar, *Lectures on Non-Perturbative Canonical Gravity* (World Scientific, Singapore, 1991).

[7] R. D'Inverno, *Introducing Einstein Relativity* (Oxford University Press, Oxford, 1992).

[8] D. Christodoulou and S. Klainerman, *The Global Nonlinear Stability of the Minkowski Space*, (Princeton University Press, Princeton, 1993).

[9] J. Stewart *Advanced General Relativity* (Cambridge University Press, Cambridge, 1993).

[10] I. Ciufolini and J.A. Wheeler, *Gravitation and Inertia* (Princeton University Press, Princeton, 1995).

[11] M. Maggiore, *Gravitational Waves* (Oxford University Press, Oxford, 2008).

[12] E. Gourgoulhon, *Special Relativity in General Frames* (Springer, Berlin, 2013).

[13] E. Gourgoulhon, 3+1 *Formalism in General Relativity* (Springer, Berlin, 2012).

[14] N. Straumann, *General Relativity with Applications to Astrophysics* (Springer, Berlin, 2004).

[15] C.M. Will, *Theory and Experiment in Gravitational Physics* (Cambridge University Press, New York, 1993).

[16] C. M. Will, The Confrontation between General Relativity and Experiment, *Living Rev Rel.* **9**, 1 (2006); **17**, 1 (2014) (arXiv 1403.7377)

[17] C. M. Will, *Was Einstein Right? A Centenary Assessment* (2014) (arXiv 1409.7871).

[18] C. Rovelli, *Quantum Gravity* (Cambridge University Press, Cambridge, 2004).

[19] T. Thiemann, *Modern Canonical Quantum General Relativity* (Cambridge University Press, Cambridge, 2007).

[20] P.A.M. Dirac, *Lectures on Quantum Mechanics* (Yeshiva University, New York, 1964);

[21] P. A. M. Dirac, Generalized Hamiltonian Dynamics, *Can. J. Math.* **2**, 129 (1950).

[22] J.L. Anderson and P. G. Bergmann, Constraints in Covariant Field Theories, *Phys. Rev.* **83**, 1018 (1951).

[23] P.G. Bergmann and J. Goldberg, Dirac Bracket Transformations in Phase Space, *Phys. Rev.* **98**, 531 (1955).

[24] P.G. Bergmann, Observables in General Relativity, *Rev. Mod. Phys.* **33**, 510 (1961).

[25] A.J. Hanson, T. Regge, and C. Teitelboim, Constrained Hamiltonian Systems, in *Contributi del Centro Linceo Interdisciplinare di Scienze Matematiche*, et al. (Accademia Nazionale dei Lincei, Roma, 1975).

[26] M. Henneaux and C. Teitelboim, *Quantization of Gauge Systems* (Princeton University Press, Princeton, 1992).
[27] K. Sundermeyer, *Constraint Dynamics with Applications to Yang-Mills Theory, General Relativity, Classical Spin, Dual String Model* (Springer, Berlin, 1982).
[28] D.M. Gitman and I.V. Tyutin, *Quantization of Fields with Constraints* (Springer, Berlin, 1990).
[29] J. Govaerts, *Hamiltonian Quantization and Constrained Dynamics* (Leuwen University Press, Leuwen, 1991).
[30] L. Lusanna, Dirac–Bergmann Constraints in Physics: Singular Lagrangians, Hamiltonian Constraints and the Second Noether Theorem, *Int. J. Geom. Meth. Mod. Phys.* **15**, 1830004 (2018) (arXiv 1702.07598).
[31] B. Guinot, Applications of General Relativity to Metrology, *Metrologia* **42**, 261 (2005).
[32] M. Soffel, S.A. Klioner, G. Petit, et al., The IAU 2000 Resolutions for Astrometry, Celestial Mechanics and Metrology in the Relativistic Framework: Explanatory Supplement, Astron. J. **126**, 2687, (2003) (arXiv astro-ph/0303376).
[33] D.D. McCarthy and G. Petit, *IERS Conventions (2003)* (Verlag des BKG, Frankfurt am Main, 2004)
[34] G.H. Kaplan, The IAU Resolutions on Astronomical Reference Systems, Time Scales and Earth Rotation Models, U.S. Naval Observatory Circular No. 179 (2005) (arXiv astro-ph/0602086).
[35] T.D. Moyer, *Formulation for Observed and Computed Values of Deep Space Network Data Types for Navigation* (John Wiley, New York, 2003).
[36] L. Lusanna, Relativistic Metrology: From Earth to Astrophysics, in *Modern Metrology Concerns*, ed. L. Cocco (InTech, London, 2012)
[37] L. Lusanna and R. Stanga, Relativistic Celestial Metrology: Dark Matter as an Inertial Gauge Effect, in *Trends in Modern Cosmology*, ed. A.J. Capistrano de Souza (Intech, Rijeka, 2017) (arXiv 1502.06801).
[38] M. Villani, Constraints on ADM Tetrad Gravity Parameter Space from S2 Star in the Center of the Galaxy and from the Solar System, 2015 (arXiv:1502.06801).
[39] M. Pauri and G. M. Prosperi, Canonical Realizations of the Rotation Group, *J. Math. Phys.* **8**, 2256 (1967)
[40] M. Pauri and G. M. Prosperi, Canonical Realizations of the Galilei Group, *J. Math. Phys.* **9**, 1146 (1968)
[41] M. Pauri and G. M. Prosperi, Canonical Realizations of the Poincare Group: 1. General Theory, *J. Math. Phys.* **16**, 1503 (1975)
[42] M. Pauri and G. M. Prosperi, Canonical Realization of the Poincare Group I. Space-Time Description of Two Particles Interacting at a Distance, Newtonian-Like Equations of Motion and Approximately Relativistic Lagrangian Formulation, *J. Math. Phys.* **17**, 1468 (1976).
[43] M. Pauri, Canonical Realizations of the Poincare Group with Increasing Mass-Spin Trajectories, in *Group Theoretical Methods in Physics*, ed. K. B. Wolf (Springer, Berlin, 1980).
[44] R. G. Littlejohn and M. Reinsch, Gauge Fields in the Separation of Rotations and Internal Motions in the N-Body Problem, *Rev. Mod. Phys.* **69**, 213 (1997).
[45] D. Alba, L. Lusanna, and M. Pauri, Dynamical Body Frames, Orientation–Shape Variables and Canonical Spin Bases for the Nonrelativistic N-Body Problem, *J. Math. Phys.* **43**, 373 (2002).
[46] R. DePietri, L. Lusanna, and M. Pauri, Gauging Kinematical and Internal Symmetry Groups for Extended Systems: The Galilean One-Time and Two-Times Harmonic Oscillators, *Class. Quantum Grav.* **11**, 1417 (1996).
[47] D. Alba, L. Lusanna, and M. Pauri, Multipolar Expansions for Closed and Open Systems of Relativistic Particles, J. Math. Phys. **46** (2005).
[48] G. Longhi, L. Lusanna, and J.M. Pons, On the Many-Time Formulation of Classical Particle Mechanics, *J. Math. Phys.*, 1893 **30** (1989).

[49] R. De Pietri, L. Lusanna, and M. Pauri, Standard and Generalized Newtonian Gravities as 'Gauge' Theories of the Extended Galilei Group: I. The Standard Theory. *Class. Quantum Grav.* **12**, 219 (1995).

[50] R. De Pietri, L. Lusanna, and M. Pauri, Standard and Generalized Newtonian Gravities as 'Gauge' Theories of the Extended Galilei Group: II. Dynamical 3-Spaces Theories, *Class. Quantum Grav.* **12**, 255 (1995).

[51] D. Alba, Quantum Mechanics in Noninertial Frames with a Multitemporal Quantization Scheme: II. Nonrelativistic Particles, *Int. J. Mod. Phys.* **A21**, 3917 (2006) (arXiv hep-th/0504060).

[52] V. Perlick, Characterization of Standard Clocks by Means of Light Rays and Freely Falling Particles, *Gen. Rel. Grav.* **19**, 1059 (1987)

[53] V. Perlick, Characterization of Standard Clocks in General Relativity, in *Semantic Aspects of Space-Time Theories*, eds. U. Majer and H.J. Schimdt (Bl-Wissenschaftsverlag, Mannheim, 1994).

[54] J. Ehlers, F.A.E. Pirani, and A. Schild, The Geometry of Free-Fall and Light Propagation in General Relativity, in *Papers in Honor of J.L. Synge*, ed. L. O'Raifeartaigh (Oxford University Press, London, 1972).

[55] B. Mashhoon, The Hypothesis of Locality and Its Limitations, in *Relativity in Rotating Frames*, eds. G. Rizzi and M. L. Ruggiero (Kluwer, Dordrecht, 2003) (arXiv gr-qc/0303029).

[56] B. Mashhoon, Limitations of Spacetime Measurements, *Phys. Lett.* **A143**, 176 (1990).

[57] B. Mashhoon, The Hypothesis of Locality in Relativistic Physics, *Phys. Lett.* **A145**, 147 (1990).

[58] B. Mashhoon, Measurement Theory and General Relativity, in *Black Holes: Theory and Observation*, eds. F. W. Hehl, C. Kiefer, and R. J. K. Metzler (Springer, Heidelberg, 1998).

[59] B. Mashhoon, Acceleration-Induced Nonlocality, in *Advances in General Relativity and Cosmology*, ed. G. Ferrarese (Pitagora, Bologna, 2003) (arXiv gr-qc/0301065).

[60] B. Mashhoon and U. Muench, Length Measurement in Accelerated Systems, *Ann. Phys. (Leipzig)* **11**, 532 (2002).

[61] D. Alba and L. Lusanna, Simultaneity, Radar 4-Coordinates and the 3+1 Point of View about Accelerated Observers in Special Relativity (2003) (arXiv gr-qc/0311058)

[62] D. Alba and L. Lusanna, Generalized Radar 4-Coordinates and Equal-Time Cauchy Surfaces for Arbitrary Accelerated Observers, *Int. J. Mod. Phys.* **D16**, 1149 (2007) (arXiv gr-qc/0501090).

[63] S. Boersma and T. Dray, Slicing, Threading and Parametric Manifolds, *Gen. Rel. Grav.* **27**, 319 (1995) (arXiv gr-qc/9407020).

[64] R. J. Jantzen, P. Carini, and D. Bini, The Many Faces of Gravito-Magnetism, *Ann. Phys. (N.Y.)* **215**, 1 (1992) (arXiv gr-qc/0106043).

[65] R.J. Jantzen, P. Carini, and D. Bini, Relative Observer Kinematics in General Relativity, *Class. Quantum Grav.* **12**, 2549 (1995).

[66] R.J. Jantzen, P. Carini, and D. Bini, The Intrinsic Derivatives and Centrifugal Forces in General Relativity: 1. Theoretical Foundations, *Int. J. Mod. Phys.* **D6**, 1 (1997) (arXiv gr-qc/0106013).

[67] R.J. Jantzen, P. Carini, and D. Bini, The Intrinsic Derivatives and Centrifugal Forces in General Relativity: 2. Applications to Circular Orbits in Some Familiar Stationary Axisymmetric Space-Times, *Int. J. Mod. Phys.* **D6**, 143 (1997) (arXiv gr-qc/0106014).

[68] R.J. Jantzen, P. Carini, and D. Bini, The Inertial Forces: Test Particle Motion Game, 1998 (arXiv gr-qc/9710051).

[69] D. Bini, A. Merloni, and R. T. Jantzen, Adapted Frames for Space-Time Splittings with an Additional Observer Family, *Nuovo Cimento* **113B**, 611 (1998).
[70] D. Bini and R. T. Jantzen, Circular Holonomy, Clock Effects and Gravito-Magnetism: Still Going Around in Circles after All These Years, in *Proceedings of the 9th ICRA Workshop on Fermi and Astrophysics, 2001*, eds. R. Ruffini and C. Sigismondi (World Scientific, Singapore, 2002) (arXiv gr-qc/0202085).
[71] R.K. Sachs and H. Wu, *General Relativity for Mathematicians* (Springer, Berlin, 1977).
[72] E. Fermi, Sopra i Fenomeni che Avvengono in Vicinanza di una Linea Oraria, *Atti Acad. Naz. Lincei Rend. Cl. Sci. Fiz. Mat. Nat.* **31**, 184, 306 (1922).
[73] F.K. Manasse and C.W. Misner, Fermi Normal Coordinates and Some Basic Concepts in Differential Geometry, *J. Math. Phys.* **4**, 735 (1963).
[74] W.T. Ni and M. Zimmermann, Inertial and Gravitational Effects in the Proper Reference Frame of an Accelerated, Rotating Observer, *Phys. Rev.* **D17**, 1473 (1978).
[75] K.P. Marzlin, On the Physical Meaning of Fermi Coordinates, *Gen. Rel. Grav.* **26**, 619 (1994).
[76] K.P. Marzlin, What is the Reference Frame of an Accelerated Observer?, *Phys. Lett.* **A215**, 1 (1996).
[77] R.F. Marzke and J.A. Wheeler, Gravitation as Geometry: I. The Geometry of the Space-Time and the Geometrodynamical Standard Meter, in *Gravitation and Relativity*, eds. H.Y. Chiu and W.F. Hoffman (Benjamin, New York, 1964).
[78] C.W. Misner, K.S. Thorne and J.A. Wheeler, *Gravitation* (Freeman, New York, 1973).
[79] M. Pauri and M. Vallisneri, Marzke–Wheeler Coordinates for Accelerated Observers in Special Relativity, *Found. Phys. Lett.* **13**, 401 (2000) (arXiv gr-qc/0006095).
[80] G. Rizzi and M.L. Ruggiero, eds., *Relativity in Rotating Frames: Relativistic Physics in Rotating Reference Frames* (Kluwer, Dordrecht, 2003).
[81] G. Rizzi and A. Tartaglia, Speed of Light on Rotating Platforms, *Found. Phys.* **28**, 1663 (1998) (arXiv gr-qc/9805089).
[82] G. Rizzi and A. Tartaglia, On Local and Global Measurements of the Speed of Light on Rotating Platforms, *Found. Phys. Lett.* **12**, 179 (1999).
[83] G.E. Stedman, Ring-Laser Tests of Fundamental Physics and Geophysics, *Rep. Prog. Phys.* **60**, 615 (1997).
[84] E.J. Post, Sagnac Effect, *Rev. Mod. Phys.* **39**, 475 (1967).
[85] G.H. Derrick, Classical Mechanics with Respect to an Observer's Past Light Cone, *J. Math. Phys.* **38**, 64 (1987).
[86] P. Havas, Simultaneity, Conventionalism, General Covariance and the Special Theory of Relativity, *Gen. Rel. Grav.* **19**, 435 (1987).
[87] C.M. Møller, *The Theory of Relativity* (Oxford University Press, Oxford, 1957)
[88] C.M. Møller, Sur la Dinamique des Syste'mes ayant un Moment Angulaire Interne, *Ann. Inst. H. Poincaré* **11**, 251 (1949).
[89] P.A.M. Dirac, Forms of Relativistic Dynamics, *Rev. Mod. Phys.* **21**, 392 (1949).
[90] H.W. Crater and L. Lusanna, Non-Inertial Frames in Minkowski Space-Time, Accelerated either Mathematical or Dynamical Observers and Comments on Non-Inertial Relativistic Quantum Mechanics, *Int. J. Geom. Met. Mod. Phys.* **11**, 1450086 (2014) (arXiv 1405.3257).
[91] H. Bondi, *Assumption and Myth in Physical Theory* (Cambridge University Press, Cambridge, 1967).
[92] D. Bini, L. Lusanna and B. Mashhon, Limitations of Radar Coordinates, *Int.J.Mod.Phys.* **D14**, 1 (2005) (arXiv gr-qc/0409052).
[93] L. Lusanna, Classical Yang -Mills Theory with Fermions: I. General Properties of a System with Constraints, *Int. J. Mod. Phys.* **A10**, 3531 (1995).

[94] L. Lusanna, Classical Yang -Mills Theory with Fermions: II. Dirac's Observables, *Int. J. Mod. Phys.* **A10**, 3675 (1995).

[95] D. Alba and L. Lusanna, The Einstein -Maxwell -Particle System in the York Canonical Basis of ADM Tetrad Gravity: I. The Equations of Motion in Arbitrary Schwinger Time *Gauges, Canad. J. Phys.* **90**, 1017 (2012) (arXiv 4087.0907).

[96] D. Alba and L. Lusanna, The Einstein -Maxwell -Particle System in the York Canonical Basis of ADM Tetrad Gravity: II. The Weak Field Approximation in the 3-Orthogonal Gauges and Hamiltonian Post-Minkowskian Gravity - the N-Body Problem and Gravitational Waves with Asymptotic Background, *Canad.J. Phys.* **90**, 1077 (2012) (arXiv 1003.5143).

[97] D. Alba and L. Lusanna, The Einstein -Maxwell -Particle System in the York Canonical Basis of ADM Tetrad Gravity: III. The Post-Minkowskian N-Body Problem, its Post-Newtonian Limit in Non-Harmonic 3-Orthogonal Gauges and Dark Matter as an Inertial Effect, *Canad. J. Phys.* **90**, 1131 (2012) (arXiv 1009.1794).

[98] D. Alba and L. Lusanna, Charged Particles and the Electro-Magnetic Field in Non-Inertial Frames: I. Admissible 3+1 Splittings of Minkowski Spacetime and the Non-Inertial Rest Frames, *Int. J. Geom. Methods Mod. Phys.* **7**, 33 (2010) (arXiv 0908.0213).

[99] D. Alba and L. Lusanna, Charged Particles and the Electro-Magnetic Field in Non-Inertial Frames: II. Applications: Rotating Frames, Sagnac Effect, Faraday Rotation, Wrap-up Effect, *Int. J. Geom. Methods Mod. Phys.* **7**, 185 (2010) (arXiv 0908.0215).

[100] D. Alba, H.W. Crater, and L. Lusanna, Towards Relativistic Atom Physics: I. The Rest-Frame Instant Form of Dynamics and a Canonical Transformation for a System of Charged Particles plus the Electro-Magnetic Field, *Canad. J. Phys.* **88**, 379 (2010) (arXiv 0806.2383).

[101] D. Alba, H.W. Crater, and L. Lusanna, Towards Relativistic Atom Physics: II. Collective and Relative Relativistic Variables for a System of Charged Particles plus the Electro-Magnetic Field, *Canad. J. Phys.* **88**, 425 (2010) (arXiv 0811.0715).

[102] N. Stergioulas, Rotating Stars in Relativity, *Living Rev. Relativ.* **6**, 3 (2003).

[103] D. Alba and L. Lusanna, Quantum Mechanics in Noninertial Frames with a Multitemporal Quantization Scheme: I. Relativistic Particles, *Int. J. Mod. Phys.* **A21**, 2781 (2006) (arXiv hep-th/0502194).

[104] D. Alba and L. Lusanna, Dust in the York Canonical Basis of ADM Tetrad Gravity: The Problem of Vorticity, *Int. J. Geom. Meth. Mod. Phys.* **12**, 1550076 (2015) (arXiv 1106.0403).

[105] L. Lusanna, The N- and 1-Time Classical Descriptions of N-Body Relativistic Kinematics and the Electromagnetic Interaction, *Int. J. Mod. Phys.* **A12**, 645 (1997).

[106] E. Schmutzer and J. Plebanski, Quantum Mechanics in Noninertial Frames of Reference, *Fortschr. Phys.* **25**, 37 (1978).

[107] L. Lusanna and M. Materassi, A Canonical Decomposition in Collective and Relative Variables of a Klein–Gordon Field in the Rest-Frame Wigner-Covariant Instant Form, *Int. J. Mod. Phys.* **A15**, 2821 (2000) (arXiv hep-th/9904202).

[108] F. Bigazzi and L. Lusanna, Dirac Fields on Spacelike Hypersurfaces, Their Rest-Frame Description and Dirac Observables, *Int. J. Mod. Phys.* **A14**, 1877 (1999) (arXiv hep-th/9807054).

[109] S. Weinberg, *The Theory of Fields*, 3 volumes (Cambridge University Press, Cambridge, 1995, 1996, and 2000).

[110] L.H. Thomas, The Relativistic Dynamics of a System of Particles Interacting at a Distance, *Phys. Rev.* **85**, 868 (1952).

[111] B. Bakamjian and L.H. Thomas, Relativistic Particle Dynamics II, *Phys. Rev.* **92**, 1300 (1953).

[112] L.L. Foldy, Relativistic Particle Systems with Interactions, *Phys. Rev.* **122**, 275 (1961).

[113] L.L. Foldy, Relativistic Particle Systems, *Phys. Rev.* **D15**, 3044 (1977).

[114] L. Bel, Dynamique des Systmes de N Particules Ponctuelles en Relativit Restreinte, *Ann. Inst. Henri Poincaré* **A12**, 307 (1970).

[115] L. Bel, Predictive Relativistic Mechanics, Ann. Inst. Henri Poincaré A14, 189 (1971). Hamiltonian Poincaré Invariant Systems, *Ann. Inst. Henri Poincaré* **18**, 57 (1973).

[116] L. Bel and J. Martin, Formes Hamiltoniennes et Systemes Conservatifs, *Ann. Inst. Henri Poincaré* **22**, 173 (1975).

[117] L. Bel and X. Fustero, Mcanique Relativiste Predictive des Systmes de N Particules, *Ann. Inst. Henri Poincaré* **25**, 411 (1976).

[118] D.G. Currie, Poincaré-Invariant Equations of Motion for Classical Particles, *Phys. Rev.* **142**, 817 (1966).

[119] R.N. Hill, Canonical Formulation of Relativistic Mechanics, *J. Math. Phys.* **8**, 201 (1967).

[120] D.G. Currie, T.F. Jordan, and E.C.G. Sudarshan, Relativistic Invariance and Hamiltonian Theories of Interacting Particles, *Rev. Mod. Phys.* **35**, 350 (1965).

[121] H. Leutwyler, A No-Interaction Theorem in Classical Relativistic Hamiltonian Particle Mechanics, *Nuovo Cimento* **37**, 556 (1965).

[122] S. Chelkowski, J. Nietendel, and R. Suchanek, The No-Interaction Theorem in Relativistic Particle Mechanics, *Acta Phys. Pol.* **B11** (1980) 809.

[123] E.C.G. Sudarshan and N. Mukunda, Classical Mechanics: A Modern Perspective (Wiley, New York 1974).

[124] Ph. Droz Vincent, Hamiltonian Systems of Relativistic Particles, *Rept. Math. Phys.* **8**, 79 (1975).

[125] Ph. Droz Vincent, Two-Body Relativistic Systems, *Ann. Inst. H. Poincare Phys. Theor.* **27**, 407 (1977).

[126] Ph. Droz Vincent, N Body Relativistic Systems, *Ann. Inst.H. Poincare Phys. Theor.* **32**, 377 (1980).

[127] Ph. Droz Vincent, Two-Body Relativistic Scattering of Directly Interacting Particles, *Phys. Rev.* **D29**, 687 (1984).

[128] Ph. Droz Vincent, Hamiltonian Two-Body System in Special Relativity, *Int. J. Theor. Phys.* **50**, 3481 (2011).

[129] I.T. Todorov, On the Quantification of a Mechanical System with Second Class Constraints, *Ann. Inst. Henri Poincaré* **28A**, 207 (1978).

[130] I.T. Todorov, Constraint Hamiltonian Approach to Relativistic Point Particle Dynamics, SISSA report, 1980.

[131] I.T. Todorov, Constraint Hamiltonian Dynamics of Directly Interacting Relativistic Point Particle, in *Quantum Theory, Groups, Fields and Particles*, ed. A.O. Barut (Reidel, Dordrecht, 1983).

[132] V.A. Rizov, H. Sazdjian, and I.T. Todorov, On the Relativistic Quantum Mechanics of Two Interacting Spinless Particles, *Ann. Phys. (N.Y.)* **165**, 59 (1985).

[133] A. Komar, Constraint Formalism of Classical Mechanics, *Phys. Rev.* **D18**, 1881 (1978).

[134] A. Komar, Interacting Relativistic Particles, *Phys. Rev.* **D18**, 1887 (1978).

[135] A. Komar, Space-Time Orbits for Interacting Relativistic Particles: Syntactic Versus Semantic Observables, *Phys. Rev.* **D18**, 3617 (1978).

[136] A. Komar, Constraints, Hermiticity and Correspondence, *Phys. Rev.* **D19**, 2908 (1979).

[137] L. Lusanna, A Model for N Classical Relativistic Particles, s*Nuovo Cimento* **65B**, 135 (1981).

[138] L. Lusanna, From Relativistic Mechanics towards Green's Functions: Multi-Temporal Dynamics, in *Proc. of the VII Seminar on Problems of High Energy Physics and Quantum Field Theory*, Protvino USSR 1984, vol. I.

[139] G. Longhi and L. Lusanna, Bound-State Solutions, Invariant Scalar Products, and Conserved Currents for a Class of Two-Body Relativistic Systems, *Phys. Rev.* **D34**, 3707 (1986).

[140] D. Alba, H.W. Crater, and L. Lusanna, Relativistic Quantum Mechanics and Relativistic Entanglement in the Rest-Frame Instant Form of Dynamics, *J. Math. Phys.* **52**, 062301 (2011) (arXiv 0907.1816).

[141] D. Alba, H.W. Crater, and L. Lusanna, Hamiltonian Relativistic Two-Body Problem: Center of Mass and Orbit Reconstruction, *J. Phys.* **A40**, 9585 (2007) (arXiv gr-qc/0610200).

[142] D. Alba, L. Lusanna, and M. Pauri, Centers of Mass and Rotational Kinematics for the Relativistic N-Body Problem in the Rest-Frame Instant Form, *J. Math. Phys.* **43**, 1677 (2002) (arXiv hep-th/0102087).

[143] L. Lusanna, The Chrono-Geometrical Structure of Special and General Relativity: A Re-Visitation of Canonical Geometrodynamics, Lectures at 42nd Karpacz Winter School of Theoretical Physics: Current Mathematical Topics in Gravitation and Cosmology, Ladek, Poland, 6 -11 February 2006, *Int. J. Geom. Meth. Mod. Phys.* **4**, 79 (2007) (arXiv gr-qc/0604120).

[144] L. Lusanna, The Chronogeometrical Structure of Special and General Relativity: Towards a Background-Independent Description of the Gravitational Field and Elementary Particles, in *General Relativity Research Trends*, ed. A. Reiner (Nova Science, New York, 2005) (arXiv gr-qc/0404122).

[145] L. Lusanna, From Clock Synchronization to Dark Matter as a Relativistic Inertial Effect, Lecture at the Black Objects in Supergravity School BOSS2011, Frascati, 9 -13 May 2011, *Springer Proc. Phys.* **144**, 267 (2013) (arXiv 1205.2481).

[146] L. Lusanna, Non-Inertial Frames in Special and General Relativity, in *Gravity: Where do We Stand?*, eds. R. Peron, M. Colpi, V. Gorini, and U. Moschella (Springer, Berlin, 2016) (arXiv 1310.4465).

[147] G. Veneziano, Quantum Strings and the Constants of Nature, in *The Challenging Questions*, ed. A. Zichichi (Plenum Press, New York, 1990).

[148] H. Epstein, V. Glaser, and A. Jaffe, Nonpositivity of the Energy Density in Quantized Field Theories, *Nuovo Cimento* **36**, 1016 (1965).

[149] H.W. Crater and L. Lusanna, The Rest-Frame Darwin Potential from the Lienard-Wiechert Solution in the Radiation Gauge, *Ann. Phys. (N.Y.)* **289**, 87 (2001).

[150] D. Alba, H.W. Crater, and L. Lusanna, The Semiclassical Relativistic Darwin Potential for Spinning Particles in the Rest Frame Instant Form: Two-Body Bound States with Spin 1/2 Constituents, *Int. J. Mod. Phys.* **A16**, 3365 (2001) (arXiv hep-th/0103109).

[151] D. Alba and L. Lusanna, The Lienard–Wiechert Potential of Charged Scalar Particles and their Relation to Scalar Electrodynamics in the Rest-Frame Instant Form, *Int. J. Mod. Phys.* **A13**, 2791 (1998) (arXiv hep-th/0708156).

[152] W.G. Dixon, Extended Objects in General Relativity: Their Description and Motion, in *Isolated Gravitating Systems in General Relativity, Proceedings of the International School of Physics, Enrico Fermi LXVII*, ed. J. Ehlers (North-Holland, Amsterdam, 1979).

[153] W.G. Dixon, Mathisson's New Mechanics: Its Aims and Realisation, *Acta Physica Polonica B Proc. Suppl.* **1**, 27 (2008)

[154] W.G. Dixon, Description of Extended Bodies by Multipole Moments in Special Relativity, *J. Math. Phys.* **8**, 1591 (1967).

[155] D. Alba, L. Lusanna, and M. Pauri, New Directions in Non-Relativistic and Relativistic Rotational and Multipole Kinematics for N-Body and Continuous

Systems, in *Atomic and Molecular Clusters: New Research*, ed. Y.L. Ping (Nova Science, New York, 2006) (arXiv hep-th/0505005).
[156] G.C. Hegerfeldt, Remark on Causality and Particle Localization, *Phys. Rev.* **D10**, 3320 (1974).
[157] G.C. Hegerfeldt, Violation of Causality in Relativistic Quantum Theory?, *Phys. Rev. Lett.* **54**, 2395 (1985).
[158] D.G.N. Fleming, Covariant Position Operators, Spin and Locality, *Phys. Rev.* **137B**, 188 (1965).
[159] D.G.N. Fleming, Non-Local Properties of Stable Particles, *Phys. Rev.* **139B**, 963 (1965).
[160] D.G.N. Fleming, A Manifestly Covariant Description of Arbitrary Dynamical Variables in Relativistic Quantum Theory, *J. Math. Phys.* **7**, 1959 (1966).
[161] G.N. Fleming and H. Bennet, Hyperplane Dependence in Relativistic Quantum Mechanics, *Found. Phys.* **19**, 231 (1989).
[162] J. Butterfield and G. Fleming, Strange Positions, in *From Physics to Philosophy*, eds. J. Butterfield and C. Pagonis (Cambridge University Press, Cambridge, 1999).
[163] W.H. Zurek, Decoherence, Einselection and the Quantum Origins of the Classical, *Rev. Mod. Phys.* **75**, 715 (2003).
[164] A. Einstein, B. Podolski, and N. Rosen, Can Quantum-Mechanical Description of Physical Reality be Considered Complete?, *Phys. Rev.* **47**, 777 (1935).
[165] H.R. Brown, *Physical Relativity: Space-Time Structure from a Dynamical Perspective* (Oxford University Press, Oxford, 2005).
[166] L. Lusanna and M. Pauri, On the Transition from the Quantum to the Classical Regime for Massive Scalar Particles: A Spaziotemporal Approach, *Eur. Phys. J. Plus* **129**, 178 (2014) (arXiv 1207.1248).
[167] H.W. Crater and L. Lusanna, On Relativistic Entanglement and Localization of Particles and on their Comparison with the Nonrelativistic Theory, *Int. J. Mod. Phys.* **A29**, 1450163 (2014) (arXiv 1306.6524).
[168] H.W. Crater and L. Lusanna, Non-Inertial Frames in Minkowski Space-Time, Accelerated Either Mathematical or Dynamical Observers and Comments on Non-Inertial Relativistic Quantum Mechanics, *Int. J. Geom. Met. Mod. Phys.* **11**, 1450086 (2014) (arXiv 1405.3257).
[169] L. Lusanna, Classical Observables of Gauge Theories from the Multitemporal Approach, *Contemp. Math.* **132**, 531 (1992)
[170] L. Lusanna, From Relativistic Mechanics towards Green's Functions: Multitemporal Dynamics, *Proceedings of the VII Seminar on Problems of High Energy Physics and Quantum Field Theory, Protvino 1984* (Protvino University Press, Protvino, 1984).
[171] F. Bigazzi and L. Lusanna, Spinning Particles on Spacelike Hypersurfaces and their Rest-Frame Description, *Int. J. Mod. Phys.* **A14**, 1429 (1999).
[172] A. Lucenti, L. Lusanna and M. Pauri, Dirac Observables and Spin Bases for n Relativistic Spining Particles, *J. Phys.* **A31**, 1633 (1998).
[173] D. Bini and L. Lusanna, Spin-Rotation Couplings: Spinning Test Particles and Dirac Fields, *Gen. Rel. Grav.* **40**, 1145 (2008) (arXiv 0710.0791).
[174] F.A. Berezin and M.S. Marinov, Classical Spin and Grassmann Algebra, *JEPT Lett.* **21**, 375 (1975).
[175] A. Barducci and L. Lusanna, The Photon in Pseudoclassical Mechanics, *Nuovo Cimento* **77A**, 39 (1983).
[176] D. Alba, H.W. Crater, and L. Lusanna, Massless Particles Plus Matter in the Rest-Frame Instant Form of Dynamics, *J. Phys.* **A43**, 405203 (2010) (arXiv 1005.5521).
[177] D. Alba, H.W. Crater, and L. Lusanna, A Relativistic Version of the Two-Level Atom in the Rest-Frame Instant Form of Dynamics, J. Phys. **A46**, 195303 (2013) (arXiv 1107.1669).

[178] D. Alba, H.W. Crater, and L. Lusanna, The Rest-Frame Instant Form and Dirac Observables for the Open Nambu String, *Eur. Phys. J. Plus* **126**, 26 (2011) (arXiv 1005.3659).
[179] F. Colomo, G. Longhi and L. Lusanna, Classical Solutions of the Many-Time Functional Equations of Motion of the Nambu String, *Int. J. Mod. Phys.* **A5**, 3347 (1990).
[180] F. Colomo, G. Longhi, and L. Lusanna, Classical Canonical Observables of the Nambu String, *Mod. Phys. Lett.* **A5**, 17 (1990).
[181] F. Colomo and L. Lusanna, Wigner Theory of the Nambu String: I. The Open String, *Int. J. Mod. Phys.* **A7**, 1705 (1992).
[182] F. Colomo and L. Lusanna, Wigner Theory of the Nambu String: II. The Closed String, *Int. J. Mod. Phys.* **A7**, 4107 (1992).
[183] L. Lusanna and P. Valtancoli, Dirac's Observables for the Higgs Model: I. The Abelian Case, *Int. J. Mod. Phys.* **A12**, 4769 (1997).
[184] L. Lusanna and P. Valtancoli, Dirac's Observables for the Higgs Model: II. The Non-Abelian SU(2) Case, *Int. J. Mod. Phys.* **A12**, 4797 (1997).
[185] D. Alba and L. Lusanna, The Classical Relativistic Quark Model in the Rest-Frame Wigner-Covariant Coulomb Gauge, *Int. J. Mod. Phys.* **A13**, 3275 (1998) (arXiv hep-th/9705156).
[186] L. Lusanna and P. Valtancoli, Dirac's Observables for the $SU(3) \otimes SU(2) \otimes U(1)$ Standard Model, *Int. J. Mod. Phys.* **A13**, 4605 (1998).
[187] G. Longhi and M. Materassi, A Canonical Decomposition in Collective and Relative Variables of a KleinGordon Field in the Rest-Frame Wigner-Covariant Instant Form, *Int. J. Mod. Phys.* **A14**, 3387 (1999) (arXiv hep-th/9809024).
[188] G. Longhi and M. Materassi, A Canonical Realization of the BMS Algebra, *J. Math. Phys.* **40**, 480 (1999) (arXiv hep-th/9803128).
[189] P.A.M. Dirac, Gauge Invariant Formulation of Quantum Electrodynamics, *Can. J. Phys.* **33**, 650 (1955).
[190] C. Cohen-Tannoudji, J. Dupont-Roc and G. Grynberg, *Atom–Photon Interactions: Basic Processes and Applications* (Wiley, New York, 1992).
[191] C. Cohen-Tannoudji, J. Dupont-Roc, and G. Grynberg, *Photons and Atoms: Introduction to Quantum Electrodynamics* (Wiley, New York, 1989).
[192] F. Rohrlich, *Classical Charged Particles* (Addison-Wesley, Reading, MA, 1965).
[193] S. Parrot, *Relativistic Electrodynamics and Differential Geometry* (Springer, New York, 1987).
[194] E. Poisson, An Introduction to the Lorentz-Dirac Equation (arXiv gr-qc/9912045).
[195] R.P. Feynman and J.A. Wheeler, Interaction with the Absorber as the Mechanism of Radiation, *Rev. Mod. Phys.* **17**, 157 (1945).
[196] R.P. Feynman and J.A. Wheeler, Classical Electrodynamics in Terms of Direct Interparticle Action, *Rev. Mod. Phys.* **21**, 425 (1949).
[197] F.A. Berezin, *The Method of Second Quantization* (Academic Press, New York, 1966).
[198] C. Itzykson and J.B. Zuber, *Quantum Field Theory* (McGraw-Hill, Singapore, 1987).
[199] L. O'Raifeartaigh, *Group Structure of Gauge Theories* (Cambridge University Press, Cambridge, 1986).
[200] R.M. Wilcox, Exponential Operators and Parameter Differentiation in Quantum Physics, *J. Math. Phys.* **8**, 962 (1967).
[201] A. Chakrabarti, Relativistic Position Operator for Free Particles, *J. Math. Phys.* **4**, 1223 (1963).
[202] J.M. Arms, J.E. Marsden, and V. Moncrief, Symmetry and Bifurcations of Momentum Mappings, *Commun. Math. Phys.* **78**, 455 (1981).
[203] J.M. Arms, Symmetry and Solution Set Singularities in Hamiltonian Field Theories, *Acta Phys. Pol.* **B17**, 499 (1986).

[204] V.N. Gribov, Quantization of Nonabelian Gauge Theories, *Nucl. Phys.* **B139**, 1 (1978).
[205] V. Moncrief, Gribov Degeneracies: Coulomb Gauge Conditions and Initial Value Constraints, *J. Math. Phys.* **20**, 579 (1979).
[206] F. Canfora, F. deMicheli, P. Salgado-Robelledo, and J. Zanelli, Gribov Ambiguity and Degenerate Systems, *Phys. Rev.* **D90**, 044065 (2014) (arXiv 1405.0394).
[207] N. Vandersickel and D. Zwanziger, The Gribov Problem and QCD Dynamics, *Phys. Report* **520**, 175 (2012) (arXiv 1202.1491).
[208] S. Kobayashi and K. Nomizu, *Foundations of Differential Geometry* (Interscience, New York, 1963).
[209] A. Trautman, *Differential Geometry for Physicists* (Bibliopolis, Napoli, 1984).
[210] R. Schmidt, *Infinite Dimensional Hamiltonian Systems* (Bibliopolis, Napoli, 1987).
[211] R. Abraham, J.E. Marsden, and T. Ratiu, *Manifolds, Tensor Analysis and Applications* (Springer, Berlin, 1988).
[212] E. Binz, J. Śniatycki, and H. Fischer, *Geometry of Classical Fields* (North-Holland, Amsterdam, 1988).
[213] L. Lusanna and D. Nowak-Szczepaniak, The Rest-Frame Instant Form of Relativistic Perfect Fluids with Equation of State $\rho = \rho(\eta, s)$ and of Nondissipative Elastic Materials. *Int. J. Mod. Phys.* **15**, 4943 (2000).
[214] D. Alba and L. Lusanna, Generalized Eulerian Coordinates for Relativistic Fluids: Hamiltonian Rest-Frame Instant Form, Relative Variables, Rotational Kinematics. *Int. J. Mod. Phys.* **19**, 3025 (2004) (arXiv hep-th/0209032).
[215] D. Alba, H.W. Crater, and L. Lusanna, On the Relativistic Micro-Canonical Ensemble and Relativistic Kinetic Theory for N Relativistic Particles in Inertial and Non-Inertial Rest Frames. *Int. J. Geom. Methods Mod. Phys.* **12**, 1550049 (2015) (arXiv 1202.4667 version 2).
[216] D. Alba, H.W. Crater, and L. Lusanna, On the Relativistic Micro-Canonical Ensemble and Relativistic Kinetic Theory for N Relativistic Particles in Inertial and Non-Inertial Rest Frames. *Int. J. Geom. Methods Mod. Phys.* **12**, 1550049 (2015) (arXiv 1202.4667 version 1)
[217] L. Lusanna, From Relativistic Mechanics towards Relativistic Statistical Mechanics, *Entropy* **19**, 436 (2017).
[218] J.D. Brown, Action Functionals for Relativistic Perfect Fluids, *Class. Quantum Grav.* **10**, 1579 (1993).
[219] W. Israel, Covariant Fluid Mechanics and Thermodynamics: An Introduction, in *Relativistic Fluid Dynamics*, eds. A. Anile and Y. Choquet-Bruhat (Springer, Berlin, 1989).
[220] C.C. Lin, Hydro-dynamics of Helium II, in *Liquid Helium*, ed. G. Careri (Academic Press, New York, 1963).
[221] J. Serrin, Mathematical Principles of Classical Fluid Mechanics, in *Handbuch der Physik* vol. 8, eds. S. Flügge and C. Truesdell (Springer, Berlin, 1959).
[222] J. Kijowski and W.M. Tulczyjew, *A Symplectic Framework for Field Theories*, (Springer, Berlin, 1979).
[223] J. Kijowski and W.M. Tulczyjew, Relativistic Hydrodynamics of Isentropic Flows, *Mem. Acad. Sci. Torino V* **6**, 3 (1982).
[224] H.P. Künzle and J.M. Nester, Hamiltonian Formulation of Gravitating Perfect Fluids and the Newtonian Limit, *J. Math. Phys.* **25**, 1009 (1984).
[225] J. Kijowski and G. Magli, Unconstrained Hamiltonian Formulation of General Relativity with Thermo-Elastic Sources, *Class. Quantum Grav.* **15**, 3891 (1998).
[226] S. Adler and T. Buchert, Lagrangian Theory of Structure Formation in Pressure-Supported Cosmological Fluids, *Astron. Astrophys.* **343**, 317 (1999) (arXiv astro-ph/9806320).
[227] D. Bao, J. Marsden, and R. Walton, The Hamiltonian Structure of General Relativistic Perfect Fluids, *Commun. Math. Phys.* **99**, 319 (1985).

[228] D.D. Holm, Hamiltonian Techniques for Relativistic Fluid Dynamics and Stability Theory, in *Relativistic Fluid Dynamics*, eds. A. Anile and Y. Choquet-Bruhat (Springer, Berlin, 1989).

[229] K. Huang, *Statistical Mechanics* (Wiley, New York, 1987).

[230] S.R. De Groot, W.A. van Leeuwen, C.G. van Weert, *Relativistic Kinetic Theory: Principles and Applications* (North-Holland, Amsterdam, 1980).

[231] J.M. Steward, *Non-Equilibrium Relativistic Kinetic Theory* (Springer, Berlin/Heidelberg, 1971).

[232] R. Balescu, Relativistic Statistical Thermodynamics, *Physica* **40**, 309 (1968).

[233] R.L. Liboff, Kinetic Theory: Classical, Quantum and Relativistic Descriptions, 2nd ed. (Wiley, New York, 1998).

[234] R. Hakim, *Introduction to Relativistic Statistical Mechanics: Classical and Quantum* (World Scientific, Singapore, 2011)

[235] R. Hakim, Remarks on Relativistic Statistical Mechanics I, *J. Math. Phys.* **8**, 1315 (1967).

[236] R. Hakim and A. Mangeney, Relativistic Kinetic Equations Includin Radiation Effects, I: Vlasov Approximation, *J. Math. Phys.* **9**, 116 (1968).

[237] E. Lehmann, Covariant Equilibrium Statistical Mechanics, *J. Math. Phys.* **47**, 023303 (2006).

[238] L.P. Horwitz, W.C. Schieve, C. Piron, Gibbs Ensembles in Relativistic Classical and Quantum Mechanics, *Ann. Phys.* **137**, 306 (1981).

[239] L.P. Horwitz, S. Shashoua, W.C. Schieve, A Manifestly Covariant Relativistic Boltzmann Equation for the Evolution of a System of Events. *Physica* **161**, 300 (1989).

[240] G. Ares de Parga and B. López-Carrera, Relativistic Statistical Mechanics vs. Relativistic Thermodynamics, *Entropy* **13**, 1664 (2011).

[241] G. Rickayzen, and J. G. Powles, Temperature in the Classical Microcanonical Ensemble, *J. Chem. Phys.* **114**, 4333 (2001).

[242] H.H. Rugh, Dynamical Approach to Temperature, *Phys. Rev. Lett.* **78**, 772 (1997).

[243] H.H. Rugh, A Geometric, Dynamical Approach to Thermodynamics, *J. Phys.* **31**, 7761 (1998) (arXiv chao-dyn/9703013).

[244] J.U. Andersen, E. Bonderup, K. Hansen, On the Concept of Temperature for a Small Isolated System, *J. Chem. Phys.* **114**, 6518 (2001).

[245] E.D. Belega, E.A. Cheremukhin, P.V. Elyutin, and D.N. Trubnikov, On the Definition of the Microcanonical Temperature of Small Weakly Bound Molecular Clusters, *Chem. Phys. Lett.* **496**, 167 (2010).

[246] C. Giardiná and R. Livi, Ergodic Properties of Microcanonical Observables, *J. Stat. Phys.* **91**, 1027 (1998).

[247] E.V. Votyakov, H.T. Hidmi, A. DeMartino and D.H.E. Gross, Microcanonical Mean-Field Thermodynamics of Self-Gravitating and Rotating Systems, *Phys. Rev. Lett.* **89**, 031101 (2002).

[248] F. Becattini and L. Ferroni, The Microcanonical Ensemble of the Ideal Relativistic Quantum Gas with Angular Momentum Conservation, *Eur. Phys. J.* **52**, 597 (2007).

[249] A. Campa, T. Dauxois, and S. Ruffo, Statistical Mechanics and Dynamics of Solvable Models with Long-Range Interactions, *Phys. Rep.* **480**, 57 (2009) (arXiv 0907.0323).

[250] A. Campa, T. Dauxois, D. Fanelli, and S. Ruffo, *Physics of Long-Range Interacting Systems* (Oxford University Press, London, 2014).

[251] T.M. Rocha Filho, M.A. Amato, and A. Figueiredo, A Novel Approach to the Determination of Equilibrium Properties of Classical Hamiltonian Systems with Long-Range Interactions, *J. Phys.* **42**, 165001 (2009).

[252] H. Touchette, The Large Deviation Approach to Statistical Mechanics, *Phys. Rep.* **478**, 1 (2009).

[253] N.G. Van Kampen, Relativistic Thermodynamics of Moving Systems, *Phys. Rev.* **173**, 295 (1968).
[254] T.S. Bíró and P.Ván, About the Temperature of Moving Bodies, *Europhys. Lett.* **89**, 30001 (2010).
[255] D. Mi, H.Y. Zhong, D.M. Tong, The Existing Different Proposals for Relativistic Temperature Transformations: The Whys and Wherefores, *Mod. Phys. Lett.* **24**, 73 (2009).
[256] Z.C. Wu, Inverse Temperature 4-Vector in Special Relativity, *Europhys. Lett.* **88**, 20005 (2009).
[257] G.I. Sewell, Note on the Relativistic Thermodynamics of Moving Bodies, *J. Phys.* **43**, 485001 (2010) (arXiv 1010.2045).
[258] C.K. Yuen, Lorentz Transformations of Thermodynamic Quantities, *Am. J. Phys.* **38**, 246 (1970).
[259] T.K. Nakamura, Three Views of a Secret in Relativistic Thermodynamics, *Prog. Theor. Phys.* **128**, 463 (2012) (arXiv 0812.3725).
[260] G.A. De Parga and B. López-Carrera, Redefined Relativistic Thermodynamics based on the Nakamura Formalism, *Physica* **388**, 4345 (2009).
[261] S. Hilbert, P. Hänggi, J. Dunkel, Thermodynamical Laws in Isolated Systems, *Phys. Rev.* **90**, 062116 (2014).
[262] E. Poisson, The Motion of Point Particles in Curved Spacetime, *Living Rev. Relativ.* **7**, 6 (2004) (arXiv gr-qc/0306052).
[263] M. Tuckerman, The Ideal Gas: Microcanonical Ensemble Treatment, Lectures in Statistical Mechanics (2003). Available online: www.nyu.edu/classes/tuckerman/stat.mech/lectures/lecture_-6/node2.htlm.
[264] P. Bergmann, Generalized Statistical Mechanics, *Phys. Rev.* **84**, 1026 (1951).
[265] T. Matolcsi, J. Kristóf, M. Székely, On the Momentum Distribution of Molecules of an Ideal Gas, *Publ. Appl. Anal.* **7**, 1 (1996).
[266] B. Boisseau and W.A. Van Leeuwen, Relativistic Boltzmann Theory in D+1 Spacetime Dimensions, *Ann. Phys.* **195**, 376 (1989).
[267] C. Cercignani and G.M. Kreimer, The Relativistic Boltzmann Equation: Theory and Applications (Birkhäuser, Basel, 2002).
[268] R.M. Strain, Global Newtonian Limit for the Relativistic Boltzmann Equation near Vacuum, *SIAM J. Math. Anal.* **42**, 1568 (2010).
[269] R.M. Strain, Coordinates in the Relativistic Boltzmann Theory, *Kinet. Relat. Model.* **4**, 345 (2011) (arXiv 1011.5093).
[270] J. Norton, General Covariance and the Foundations of General Relativity: Eight Decades of Dispute, *Rep. Prog. Phys.* **56**, 791 (1993).
[271] L. Lusanna and M. Pauri, Explaining Leibniz Equivalence as Difference of Non-inertial Appearances: Dis-solution of the Hole Argument and Physical Individuation of Point-Events, *Hist. Philos. Mod. Phys.* **37**, 692 (2006) (arXiv gr-qc/0604087).
[272] L. Lusanna and M. Pauri, The Physical Role of Gravitational and Gauge Degrees of Freedom in General Relativity. I: Dynamical Synchronization and Generalized Inertial Effects, *Gen. Rel. Grav.* **38**, 187 (2006) (arXiv gr-qc/0403081).
[273] L. Lusanna and M. Pauri, The Physical Role of Gravitational and Gauge Degrees of Freedom in General Relativity. II: Dirac versus Bergmann Observables and the Objectivity of Space-Time, *Gen. Rel. Grav.* **38**, 229 (2006) (arXiv gr-qc/0407007).
[274] L. Lusanna and M. Pauri, Dynamical Emergence of Instantaneous 3-Spaces in a Class of Models of General Relativity, in *Relativity and the Dimensionality of the World*, ed. A. van der Merwe (Springer, Berlin, 2007) (arXiv gr-qc/0611045).
[275] L. Cacciapuoti and C. Salomon, ACES: Mission Concept and Scientific Objective, ESA document (Estec, March 28, 2007).
[276] L. Blanchet, C. Salomon, P. Teyssandier, and P. Wolf, Relativistic Theory for Time and Frequency Transfer to Order $1/c^3$, *Astron. Astrophys.* **370**, 320 (2000).

[277] L. Lusanna, Dynamical Emergence of 3-Space in General Relativity: Implications for the ACES Mission, in *Proceedings of the 42th Rencontres de Moriond Gravitational Waves and Experimental Gravity*, La Thuile (Italy), March 11–18, 2007.

[278] R. Arnowitt, S. Deser and C.W. Misner, Canonical Variables for General Relativity, *Phys. Rev.* **117**, 1595 (1960)

[279] R. Arnowitt, S. Deser, and C.W. Misner, The Dynamics of General Relativity, in *Gravitation: An Introduction to Current Research*, ed. L.Witten (Wiley, New York, 1962) (arXiv gr-qc/0405109).

[280] L. Lusanna, The Rest-Frame Instant Form of Metric Gravity. *Gen. Rel. Grav.* **33**, 1579, (2001) (arXiv gr-qc/0101048).

[281] L. Lusanna and S. Russo, A New Parametrization for Tetrad Gravity, *Gen. Rel. Grav.* **34**, 189 (2002) (arXiv gr-qc/0102074).

[282] R. De Pietri, L. Lusanna, L. Martucci and S. Russo, Dirac's Observables for the Rest-Frame Instant Form of Tetrad Gravity in a Completely Fixed 3-Orthogonal Gauge, *Gen. Rel. Grav.* **34**, 877 (2002) (arXiv gr-qc/0105084).

[283] J. Agresti, R. De Pietri, L. Lusanna, and L. Martucci, Hamiltonian Linearization of the Rest-Frame Instant Form of Tetrad Gravity in a Completely Fixed 3-Orthogonal Gauge: A Radiation Gauge for Background-Independent Gravitational Waves in a Post-Minkowskian Einstein Spacetime, *Gen. Rel. Grav.* **36**, 1055 (2004) (arXiv gr-qc/0302084).

[284] D. Alba and L. Lusanna, The York Map as a Shanmugadhasan Canonical Transformation in Tetrad Gravity and the Role of Non-Inertial Frames in the Geometrical View of the Gravitational Field, *Gen. Rel. Grav.* **39**, 2149 (2007) (arXiv gr-qc/0604086 v2).

[285] L. Lusanna and M. Villani, Hamiltonian Expression of Curvature Tensors in the York Canonical Basis: I) Riemann Tensor and Ricci Scalars, *Int. J. Geom. Meth. Mod. Phys.* **11**, 1450052 (2014) (arXiv 1401.1370).

[286] L. Lusanna and M. Villani, Hamiltonian Expression of Curvature Tensors in the York Canonical Basis: II) Weyl Tensor, Weyl Scalars, Weyl Eigenvalues and the Problem of the Observables of the Gravitational Field, *Int. J. Geom. Meth. Mod. Phys.* **11**, 1450053 (2014) (arXiv 1401.1375).

[287] L.Lusanna, Killing Symmetries as Hamiltonian Constraints, *Int. J. Geom. Meth. Mod. Phys.* **13**, 1650044 (2016) (arXiv 1108.3224).

[288] L. Lusanna, Canonical ADM Tetrad Gravity: From Metrological Inertial Gauge Variables to Dynamical Tidal Dirac Observables, *Int. J. Geom. Meth. Mod. Phys.* **12**, 1530001 (2015) (arXiv 1401.1375).

[289] C.J. Isham and K. Kuchar, Representations of Spacetime Diffeomorphisms: I. Canonical Parametrized Field Theories, *Ann. Phys. (N.Y.)* **164**, 288 (1984).

[290] C.J. Isham and K. Kuchar, Representations of Spacetime Diffeomorphisms: II. Canonical Geometrodynamics, *Ann. Phys. (N.Y.)* **164**, 316 (1984).

[291] K. Kuchar, Canonical Geometrodynamics and General Covariance, *Found. Phys.* **16**, 193 (1986).

[292] M. Nakahara, *Geometry, Topology and Physics* (IOP, Bristol, 1990).

[293] B. O'Neil, *Semi-Riemannian Geometry* (Academic Press, New York, 1983).

[294] T. Regge and C. Teitelboim, Role of Surface Integrals in the Hamiltonian Formulation of General Relativity, *Ann. Phys. (N.Y.)* **88**, 286 (1974).

[295] R. Beig and Ó. Murchadha, The Poincaré Group as the Symmetry Group of Canonical General Relativity, *Ann. Phys. (N.Y.)* **174**, 463 (1987).

[296] L. Andersson, Momenta and Reduction for General Relativity, *J. Geom. Phys.* **4**, 289 (1987).

[297] T. Thiemann, Generalized Boundary Conditions for General Relativity for the Asymptotically Flat Case in Terms of Ashtekar's Variables, *Class. Quantum Grav.* **12**, 181 (1995).

[298] A. Ashtekar, Asymptotic Structure of the Gravitational Field at Spatial Infinity, in *General Relativity and Gravitation*, vol. 2, ed. A. Held (Plenum, New York, 1980).

[299] A. Ashtekar and R.O. Hansen, A Unified Treatment of Null and Spatial Infinity in General Relativity: I. Universal Structure, Asymptotic Symmetries, and Conserved Quantities at Spatial Infinity, *J. Math. Phys.* **19**, 1542 (1978).

[300] A. Ashtekar and A. Magnon, From i^o to the 3+1 Description of Spatial Infinity, *J. Math. Phys.* **25**, 2682 (1984).

[301] A. Ashtekar and J.D. Romano, Spatial Infinity as a Boundary of Spacetime, *Class. Quantum Grav.* **9**, 1069 (1992).

[302] R. Geroch, Spinor Structure of Space-Times in General Relativity I, *J. Math. Phys.* **9**, 1739 (1968).

[303] D. Bleecker, *Gauge Theory and Variational Principles* (Addison-Wesley, London, 1981).

[304] A. Sen, Quantum Theory of Spin-3/2 Field in Einstein Spaces, *Int. J. Theor. Phys.* **21**, 1 (1982).

[305] P. Sommers, Space Spinors, *J. Math. Phys.* **21**, 2567 (1980).

[306] A. Ashtekar, *New Perspectives in Canonical Gravity* (Bibliopolis, Napoli, 1988).

[307] Y. Choquet-Bruhat and J.W. York, Jr., The Cauchy Problem, in *General Relativity and Gravitation*, vol. 1, ed. A. Held (Plenum, New York, 1980).

[308] P.G. Bergmann, Observables in General Relativity, *Rev. Mod. Phys.* **33**, 510 (1961).

[309] P.G. Bergmann, The General Theory of Relativity, in *Handbuch der Physik*, vol. IV, 4 ed. S. Flugge (Springer, Berlin, 1962).

[310] P.G. Bergmann and A. Komar, Poisson Brackets between Locally Defined Observables in General Relativity, *Phys. Rev. Lett.* **4**, 432 (1960)

[311] P.G. Bergmann and A. Komar, The Coordinate Group Symmetries of General Relativity, *Int. J. Theor. Phys.* **5**, 15 (1972).

[312] J.A. Wheeler, Geometrodynamics and the Issue of the Final State, in *Relativity, Groups and Topology*, eds. B.S. De Witt and C. De Witt (Gordon and Breach, London, 1964);

[313] J.A. Wheeler, Superspace and the Nature of Quantum Geometrodynamics, in *Battelle Rencontres 1967*, eds. C. De Witt and J.A. Wheeler (Springer-Verlag, New York, 1968).

[314] A.E. Fischer, The Theory of Superspace, in *Relativity*, eds. M. Carmeli, L. Fickler and L. Witten (Plenum, New York, 1970).

[315] A.E. Fischer, The Unfolding of Singularities in Superspace, *Gen. Rel. Grav.* **15**, 1191 (1983).

[316] A.E. Fischer, Resolving the Singularities in the Space of Riemannian Geometries, *J. Math. Phys.* **27**, 718 (1986).

[317] M. Rainer, The Moduli Space of Local Homogeneous 3-Geometries, presentation at the Pacific Conference on Gravitation and Cosmology, Seoul 1996.

[318] S. Timothy Swift, Natural Bundles: I. A Minimal Resolution of Superspace, *J. Math. Phys.* **33**, 3723 (1992).

[319] S. Timothy Swift, Natural Bundles: II. Spin and the diffeomorphism group, *J. Math. Phys.* **34**, 3825 (1993).

[320] S. Timothy Swift, Natural Bundles: III. Resolving the Singularities in the Space of Immersed Submanifolds, *J. Math. Phys.* **34**, 3841 (1993).

[321] J.M. Arms, J.E. Marsden and V. Moncrief, Symmetry and Bifurcations of Momentum Mappings, *Commun. Math. Phys.* **78**, 455 (1981).

[322] D. Giulini, On the Configuration Space Topology in General Relativity, *Helv. Phys. Acta* **68**, 86 (1995).

[323] J. Lee and R.M. Wald, Local Symmetries and Constraints, *J. Math. Phys.* **31**, 725 (1990).

[324] R. Beig, The Classical Theory of Canonical General Relativity, in *Canonical Gravity: From Classical to Quantum*, eds. J. Ehlers and H. Friedrich (Springer, Berlin, 1994).

[325] A. Lichnerowicz, L'Integration des Equations de la Gravitation Relativiste et le Probleme des N Corps, *J. Math. Pure Appl.* **23**, 37 (1944).

[326] Y. Choquet-Bruhat, Sur l'Intgration des Equations d'Einstein, *C. R. Acad. Sci. Paris* **226**, 1071 (1948).

[327] Y. Choquet-Bruhat, *The Cauchy Problem in Gravitation: An Introduction to Current Research*, ed. L. Witten (Wiley, New York, 1962).

[328] J.W. York, Jr., Gravitational Degrees of Freedom and the Initial-Value Problem, *Phys. Rev. Lett.* **26**, 1656 (1971).

[329] J.W. York, Jr., Role of Conformal Three-Geometry in the Dynamics of Gravitation, *Phys. Rev. Lett.* **28**, 1082 (1972).

[330] J.W. York, Jr., Mapping onto Solutions of the Gravitational Initial Value Problem, *J. Math. Phys.* **13**, 125 (1972)

[331] J.W. York, Jr., Conformally Invariant Orthogonal Decomposition of Symmetric Tensors on Riemannian Manifolds and the Initial-Value Problem of General Relativity, *J. Math. Phys.* **14**, 456 (1972)

[332] J.W. York, Jr., Covariant Decompositions of Symmetric Tensors in the Theory of Gravitation, *Ann. Ins. H. Poincare* **21**, 318 (1974).

[333] N. O'Murchadha and J.W. York, Jr., Existence and Uniqueness of Solutions of the Hamiltonian Constraint of General Relativity on Compact Manifolds, *J. Math. Phys.* **14**, 1551 (1972)

[334] N. O'Murchadha and J.W. York, Jr., The Initial-Value Problem of General Relativity, *Phys. Rev.* **D10**, 428 (1974).

[335] J.W. York Jr., Kinematics and Dynamics of General Relativity, in *Sources of Gravitational Radiation*, ed. L.L. Smarr (Cambridge University Press, Cambridge, 1979).

[336] A. Qadir and J.A. Wheeler, York's Cosmic Time Versus Proper Time, in *From SU(3) to Gravity, Y. Ne'eman's Festschrift*, eds. E. Gotsma and G. Tauber (Cambridge University Press, Cambridge, 1985).

[337] P.J. McCarthy, Structure of the Bondi -Metzner -Sachs Group, *J. Math. Phys.* **13**, 1837 (1972).

[338] P.J. McCarthy, Representations of the Bondi -Metzner -Sachs Group I, *Proc. Roy. Soc. London* **A330**, 517 (1972).

[339] P.J. McCarthy, Representations of the Bondi -Metzner -Sachs Group II, *Proc. Roy. Soc. London* **A333**, 317 (1973).

[340] P.J. McCarthy, Asymptotically Flat Spacetimes and Elementary Particles, *Phys. Rev. Lett.* **29**, 817 (1972).

[341] P.J. McCarthy and M. Crampin, Representations of the Bondi -Metzner -Sachs Group III: Poincare Spin Multiplicities and Irreducibility, *Proc. Roy. Soc. London* **A335**, 301 (1973).

[342] J. Winicour, Angular Momentum in General Relativity, in *General Relativity and Gravitation*, vol. 2, ed. A. Held (Plenum, New York, 1980).

[343] J.L. Jaramillo and E. Gourgoulhon, Mass and Angular Momentum in General Relativity, in *Mass and Motion in General Relativity*, eds. L. Blanchet, A. Spallicci, and B. Whiting, (Springer, Berlin, 2011).

[344] P.A.M. Dirac, The Hamiltonian Form of Field Dynamics, *Canad. J. Math.* **3**, 1 (1951).

[345] Y. Choquet-Bruhat, A. Fischer, and J.E. Marsden, Maximal Hypersurfaces and Positivity of A New Proof of the Positive Energy Theorem Mass, in *LXVII E.Fermi Summer School of Physics Isolated Gravitating Systems in General Relativity*, ed. J. Ehlers (North-Holland, Amsterdam, 1979).

[346] J. Frauendiener, Triads and the Witten Equation, *Class. Quantum Grav.* **8**, 1881 (1991).

[347] A.Sen, On the Existence of Neutrino Zero-Modes in Vacuum Spacetimes, *J. Math. Phys.* **22**, 1781 (1981).
[348] A. Sen, Gravity as a Spin System, *Phys. Lett.* 119B, 89 (1982).
[349] E. Witten, A New Proof of the Positive Energy Theorem, *Commun. Math. Phys.* **80**, 381 (1981).
[350] S. Deser and C.J. Isham, Canonical Vierbein Form of General Relativity, *Phys. Rev.* **D14**, 2505 (1976).
[351] C.J. Isham, Canonical Quantum Gravity and the Problem of Time, in *Integrable Systems, Quantum Groups and Quantum Field Theories*, eds. L.A. Ibort and M.A. Rodriguez (Kluwer, London, 1993).
[352] C.J. Isham, Conceptual and Geometrical Problems in Quantum Gravity, in *Recent Aspects of Quantum Fields*, eds. H. Mitter and H. Gausterer (Springer, Berlin, 1991).
[353] C.J. Isham, Prima Facie Questions in Quantum Gravity and Canonical Quantum Gravity and the Question of Time, in *Canonical Gravity: From Classical to Quantum*, eds. J. Ehlers and H. Friedrich (Springer, Berlin, 1994).
[354] C.W. Misner, K.S. Thorne, and J.A. Wheeler, *Gravitation* (Freeman, New York, 1973).
[355] S. Weinberg, Gravitation and Cosmology (J. Wiley, New York, 1972).
[356] J. Isenberg and J. Nester, Canonical Gravity, in *General Relativity and Gravitation*, vol. 1, ed. A. Held (Plenum, New York, 1980).
[357] A.D. Rendall, Local and Global Existence Theorems for the Einstein Equations, *Living Rev. Rel.*, **1**, 4 (1998) and **3**, 1 (2000) (arXiv gr-qc/0001008).
[358] H. Friedrich and A.D. Rendall, The Cauchy Problem for Einstein Equations, in *Einstein's Field Equations and their Physical Interpretation*, ed. B.G. Schmidt (Springer, Berlin, 2000).
[359] C. Teitelboim, The Hamiltonian Structure of Space-Time, in *General Relativity and Gravitation*, vol 1, ed. A. Held, (Plenum, New York, 1980).
[360] K. Kuchar, I: Geometry of Hyperspace, *J. Math. Phys.* **17**, 777, (1976).
[361] K. Kuchar, II: Kinematics of Tensor Fields in Hyperspace, *J. Math. Phys.* **17**, 792 (1976).
[362] K. Kuchar, III: Dynamics of Tensor Fields in Hyperspace, *J. Math. Phys.* **17**, 801 (1976).
[363] K. Kuchar, IV: Geometrodynamics with Tensor Sources, *J. Math. Phys.* **18**, 1589 (1977).
[364] J.M. Pons and L. Shepley, Evolutionary Laws, Initial Conditions and Gauge Fixing in Constrained Systems, *Class. Quantum Grav.* **12**, 1771 (1995) (arXiv gr-qc/9508052).
[365] J.M. Pons, D.C. Salisbury, and L.C. Shepley, Gauge Transformations in the Lagrangian and Hamiltonian Formalisms of Generally Covariant Theories, *Phys. Rev.* **D55**, 658 (1997) (arXiv gr-qc/9612037).
[366] R. Sugano, Y. Kagraoka, and T. Kimura, Gauge Transformations and Gauge-Fixing: Condition in Constraint System, *Int. J. Mod. Phys.* **A7**, 61 (1992).
[367] B. Dittrich, Partial and Complete Observables for Hamiltonian Constrained Systems, *Gen. Rel. Grav.* **39**, 1891 (2007) (arXiv gr-qc/0411013).
[368] B. Dittrich, Partial and Complete Observables for Canonical General Relativity, *Gen. Rel. Grav.* **23**, 6155 (2006) (arXiv gr-qc/0507106).
[369] B. Dittrich, P.A. Hohn, T.A. Koslowski, and M.I. Nelson, Chaos, Dirac Observables and Constraint Quantization, preprint (2015) (arXiv 1508.01947).
[370] T. Thiemann Reduced Phase Space Quantization and Dirac Observables, *Class.Quantum Grav.* **23**, 1163 (2006) (arXiv 0411031).
[371] J.M. Pons, D.C. Salisbury, and K.A. Sundermeyer, Revisiting Observables in Generally Co-variant Theories in the Light of Gauge Fixing Methods, *Phys. Rev. D80*, 084015 (2009) (arXiv 0905.4564).

[372] J.M. Pons, D.C. Salisbury, and K.A. Sundermeyer, Observables in Classical Canonical Gravity: Folklore Demystified (2010) (arXiv 1001.272).
[373] C. Rovelli, Relational Quantum Mechanics, *Int. J. Theor. Phys.* **35**, 1637 (1996) (arXiv quant-ph/9609002).
[374] C.Rovelli and M.Smerlak, Relational EPR, *Found. Phys.* **37**, 427 (2007) (arXiv quant-ph/0604064).
[375] J. Schwinger, Quantized Gravitational Field, *Phys. Rev.* **130**, 1253 (1963).
[376] J.L. Synge, *Relativity: The General Theory* (North-Holland, Amsterdam, 1964).
[377] H. Stephani, *General Relativity*, 2nd edition (Cambridge University Press, Cambridge, 1996).
[378] B.F. Schutz, *A First Course in General Relativity* (Cambridge University Press, Cambridge, 1985).
[379] K. Thorne, Gravitational Radiation, in *Three Hundred Years of Gravitation*, ed. S. Hawking and W. Israel (Cambridge University Press, Cambridge, 1987).
[380] K. Thorne, The Theory of Gravitational Radiation: An Introductory Review, in *Gravitational Radiation*, ed. N. Deruelle and T. Piran (North-Holland, Amsterdam, 1983).
[381] K. Thorne, Multipole Expansions of Gravitational Radiation, *Rev. Mod. Phys.* **52**, 299 (1980).
[382] M.E. Pati and C.M. Will, Post-Newtonian Gravitational Radiation and Equations of Motion via Direct Integration of the Relaxed Einstein Equations: Foundations, *Phys. Rev.* **D62**, 124015 (2000)
[383] M.E. Pati and C.M. Will, II: Two-Body Equations of Motion to Second Post-Newtonian Order and Radiation Reaction to 3.5 Post-Newtonian Order, *Phys. Rev.* **D65**, 104008 (2002).
[384] C. Will, III: Radiation Reaction for Binary Systems with Spinning Bodies, *Phys. Rev.* **D71**, 084027 (2005).
[385] H. Wang and C. Will, IV: Radiation Reaction for Binary Systems with Spin -Spin Coupling, *Phys. Rev.* **D75**, 064017 (2007).
[386] T. Mitchell and C. Will, V: Evidence for the Strong Equivalence Principle in Second Post-Newtonian Order, *Phys. Rev.* **D75**, 124025 (2007).
[387] C.M. Will, On the Unreasonable Effectiveness of the Post-Newtonian Approximation in Gravitational Physics, *Proc. Nat. Acad. Sci. USA* **108**, 5938 (2011) (arXiv 1102.5192).
[388] T. Damour and A. Nagar, The Effective One Body Description of the Two-Body Problem (2009).
[389] T. Damour, Introductory Lectures on the Effective One Body Formalism (2008) (arXiv 0802.4047).
[390] G. Schaefer, Post-Newtonian Methods: Analytic Results on the Binary Problem, in *Mass and Motion in General Relativity, Proceedings of the 2008 CNRS School in Orleans/France*, eds. L. Blanchet, A. Spallicci, and B. Whiting (Springer, Berlin, 2011) (arXiv 0910.2857).
[391] G. Schaefer, The Gravitational Quadrupole Radiation-Reaction Force and the Canonical Formalism of ADM, *Ann. Phys. (N.Y.)* **161**, 81 (1985).
[392] G. Schaefer, The ADM Hamiltonian and the Postlinear Approximation, *Gen. Rel. Grav.* **18**, 255 (1985).
[393] T. Ledvinka, G. Schaefer, and J. Bicak, Relativistic Closed-Form Hamiltonian for Many-Body Gravitating Systems in the Post-Minkowskian Approximation, *Phys. Rev. Lett.* **100**, 251101 (2008) (arXiv 0807.0214).
[394] J. Ehlers and E. Rundolph, Dynamics of Extended Bodies in General Relativity: Center-of-Mass Description and Quasi-Rigidity, *Gen. Rel. Grav.* **8**, 197 (1977).
[395] W. Beiglboeck, The Center of Mass in Einstein's Theory of Gravitation, *Commun. Math. Phys.* **5**, 106 (1967).
[396] R. Schattner, The Center of Mass in General Relativity, *Gen. Rel. Grav.* **10**, 377 (1978).

[397] R. Schattner, The Uniqueness of the Center of Mass in General Relativity, *Gen. Rel. Grav.* **10**, 395 (1979).
[398] J. Ehlers and R. Geroch, Equation of Motion of Small Bodies in Relativity, *Ann. Phys.* **309**, 232 (2004).
[399] S. Kopeikin and I. Vlasov, Parametrized Post-Newtonian Theory of Reference Frames, Multipolar Expansions and Equations of Motion in the N-body Problem, *Phys. Rep.* **400**, 209 (2004) (arXiv gr-qc/0403068).
[400] J. Steinhoff and D. Puetzfeld, Multipolar Equations of Motion for Extended Test Bodies in General Relativity (2009) (arXiv 0909.3756).
[401] E.E. Flanagan and S.A. Hughes, The Basics of Gravitational Wave Theory, *New J. Phys.* **7**, 204 (2005) (arXiv gr-qc/0501041).
[402] Y. Mino, M. Sasaki, and T. Tanaka, Gravitational Radiation Reaction to a Particle Motion, *Phys. Rev.* **D55**, 3457 (1997) (arXiv gr-qc/9606018).
[403] T.C. Quinn and R.M. Wald, *Phys. Rev.* **D56**, 3381 (1997) (arXiv gr-qc/9610053).
[404] S. Detweiler and B.F. Whiting, Self-Force via a Green's Function Decomposition, *Phys. Rev.* **D67**, 024025 (2003) (arXiv gr-qc/0202086).
[405] R.M. Wald, Introduction to Gravitational Self-Force, in *Mass and its Motion, Proceedings of the 2008 CNRS School in Orleans/France*, eds. L. Blanchet, A. Spallicci, and B. Whiting (Springer, Berlin, 2011) (arXiv 0907.0412).
[406] S.E. Gralla and R.M. Wald, Derivation of Gravitational Self-Force, in *Mass and its Motion, Proceedings of the 2008 CNRS School in Orleans/France*, eds. L. Blanchet, A. Spallicci, and B. Whiting (Springer, Berlin, 2011) (arXiv 0907.0414).
[407] A. Pound, A New Derivation of the Gravitational Self-Force (2009) (arXiv 0907.5197).
[408] I. Ciufolini and E.C. Pavlis, A Confirmation of the General Relativistic Prediction of the Lense-Thirring Effect, *Nature* **431**, 958 (2004).
[409] C.W.F. Everitt and B.W. Parkinson, Gravity Probe B Science Results: NASA Final Report (2008). Available online: http://einstein:stanford:edu/content/finalreport/GPBFinalNASAReport-020509-web.pdf.
[410] S.M. Kopeikin and E.B. Fomalont, Gravitomagnetism, Causality and Aberration of gravity in the Gravitational Light-Ray Detection Experiments, *Gen. Rel. Grav.* **39**, 1583 (2007) (arXiv gr-qc/0510077).
[411] S.M. Kopeikin and E.B. Fomalont, Aberration and the Fundamental Speed of Gravity in the Jovian Detection Experiment, *Found. Phys.* **36**, 1244 (2006) (arXiv astro-ph/0311063).
[412] S.M. Kopeikin and V.V. Makarov, Gravitational Bending of Light by Planetary Multipoles and its Measurement with Microarcsecond Astronomical Interferometers, *Phys. Rev.* **D75**, 062002 (2007) (arXiv astro-ph/0611358).
[413] S. Carlip, Model-Dependence of Shapiro Time Delay and the "Speed of Gravity/Speed of Light" Controversy, *Class. Quantum Grav.* **21**, 3803 (2004) (arXiv gr-qc/0403060).
[414] C. Will, Propagation Speed of Gravity and the Relativistic Time Delay, *Astrophys. J.* **590**, 683 (2003) (arXiv astro-ph/0301145).
[415] L. Lindegren and D. Dravins, The Fundamental Definition of 'Radial Velocity', *Astron. Astrophys.* **401**, 1185 (2003) (arXiv astro-ph/0302522).
[416] P. Teyssandier, C. Le Poncin-Lafitte, and B. Linet, A Universal Tool for Determining the Time Delay and the Frequency Shift of Light: Synge's World Function, *Astrophys. Space Sci. Libr.* **349**, 153 (2007) (arXiv 0711.0034).
[417] P. Teyssandier and C. Le Poncin-Lafitte, General Post-Minkowskian Expansion of Time Transfer Functions, *Class. Quantum Grav.* **25**, 145020 (2008) (arXiv 0803.0277).
[418] M.H. Bruegmann, Light Deflection in the Postlinear Gravitational Field of Bounded Pointlike Masses, *Phys. Rev.* **D72**, 024012 (2005) (arXiv gr-qc/0501095).

[419] E. Harrison, The Redshift–Distance and Velocity–Distance Laws, *Astrophys. J.* **403**, 28 (1993).
[420] M. Bartelmann, The Dark Universe, *Rev. Mod. Phys.* **82**, 331 (2010) (arXiv 0906.5036).
[421] R. Bean, TASI Lectures on Cosmic Acceleration (2009) (arXiv 1003.4468).
[422] M. Favata, Post-Newtonian Corrections to the Gravitational-Wave Memory for Quasi-Circular, Inspiralling Compact Binaries (2009) (arXiv 0812.0069).
[423] J.L. Jaramillo, J.A.V. Kroon, and E. Gourgoulhon, From Geometry to Numerics: Interdisciplinary Aspects in Mathematical and Numerical Relativity, *Class. Quantum Grav.* **25**, 093001 (2008).
[424] I. Hinder, The Current Status of Binary Black Hole Simulations in Numerical Relativity, (2010) (arXiv 1001.5161).
[425] L. Blanchet, Gravitational Radiation from Post-Newtonian Sources and Inspiralling Compact Binaries, *Living Rev. Rel.* **9**, 4 (2006)
[426] L. Blanchet, Post-Newtonian Theory and the Two-Body Problem (2009) (arXiv 0907.3596).
[427] T. Damour, Gravitational Radiation and the Motion of Compact Bodies, in *Gravitational Radiation*, eds. N. Deruelle and T. Piran (North-Holland, Amsterdam, 1983).
[428] T. Damour, The Problem of Motion in Newtonian and Einsteinian Gravity, in *Three Hundred Years of Gravitation*, eds. S. Hawking and W. Israel (Cambridge University Press, Cambridge, 1987).
[429] E. Poisson, The Motion of Point Particles in Curved Space-Time, *Living Rev. Rel.* **7**, 6 (2004) (arXiv gr-qc/0306052).
[430] E. Poisson, Constructing the Self-Force, in *Mass and its Motion, Proceedings of the 2008 CNRS School in Orleans/France*, eds. L. Blanchet, A. Spallicci, and B. Whiting (Springer, Berlin, 2011).
[431] L. Infeld and J. Plebanski, *Motion ande Relativity*, (Pergamon, Oxford, 1960).
[432] R. Geroch and J. Traschen, Strings and Other Distributional Sources in General Relativity, *Phys. Rev.* **D36**, 1017 (1987).
[433] R. Steinbauer and J.A. Vickers, The Use of Generalized Functions and Distributions in General Relativity, *Class. Quantum Grav.* **23**, R91 (2006) (arXiv gr-qc/0603078).
[434] A. Barducci, R. Casalbuoni, and L. Lusanna, Energy–Momentum Tensor of Extended Relativistic Systems, *Nuovo Cim.* **54A**, 340 (1979).
[435] C. Moni Bidin, G. Carraro, G.A. Me'ndez and W.F. van Altena, No Evidence for a Dark Matter Disk within 4 kpc from the Galactic Plane (2010) (arXiv 1011.1289).
[436] T. Damour T. and N. Deruelle, General Relativistic Celestial Mechanics of Binary Systems: I. The Post-Newtonian Motion, *Ann. Inst.H. Poincaré* **43**, 107 (1985).
[437] T. Damour T. and N. Deruelle, General Relativistic Celestial Mechanics of Binary Systems. II. The Post-Newtonian Timing Formula, *Ann. Inst.H. Poincaré* **44**, 263 (1986).
[438] M.S. Longair, *Galaxy Formation* (Springer, Berlin, 2008).
[439] M. Ross, Dark Matter: The Evidence from Astronomy, Astrophysics and Cosmology (2010) (arXiv 1001.0316).
[440] K.Garret and G. Duda, Dark Matter: A Primer, *Adv. Astron.* **2011**, 968283 (2011) (arXiv 1006.2483).
[441] P. Schneider, J. Ehlers, and E.E. Falco, *Gravitational Lenses* (Springer, Berlin, 1992).
[442] M. Bartelmann and P. Schneider, Weak Gravitational Lensing, Phys.Rev. **D340**, 291 (2001) (arXiv astro-ph/9912508).
[443] E. Battaner and E. Florido, The Rotation Curve of Spiral Galaxies and its Cosmological Implications, *Fund. Cosmic Phys.* **21**, 1 (2000) (arXiv astro-ph/0010475).

[444] D.G. Banhatti, Disk Galaxy Rotation Curves and Dark Matter Distribution, *Curr. Sci.* **94**, 986 (2008) (arXiv astro-ph/0703430).
[445] W.J.G. de Blok and A. Bosma, High Resolution Rotation Curves of Low Surface Brightness Galaxies, *Astron. Astrophys.* **385**, 816 (2002) (arXiv astro-ph/0201276).
[446] M. Milgrom, New Physics at Low Accelerations (MOND): An Alternative to Dark Matter, (2009) (arXiv 0912.2678).
[447] S. Capozziello, V.F. Cardone, and A. Troisi, Low Surface Brightness Galaxies Rotation Curves in the Low Energy Limit of R^n Gravity: No Need for Dark Matter?, *Mon. Not. R. Astron. Soc.* **375**, 1423 (2007) (arXiv astro-ph/0603522).
[448] J.L. Feng, Dark Matter Candidates from Particle Physics and Methods of Detection, (2010) (arXiv 1003.0904).
[449] F.I. Cooperstock and S. Tieu, General Relativistic Velocity: The Alternative to Dark Matter, *Mod. Phys. Lett.* **A23**, 1745 (2008) (arXiv 0712.0019).
[450] J. Kovalevski, I.I. Mueller, and B. Kolaczek, *Reference Frames in Astronomy and Geophysics* (Kluwer, Dordrecht, 1989).
[451] O.J. Sovers and J.L. Fanselow, Astrometry and Geodesy with Radio Interferometry: Experiments, Models, Results, *Rev. Mod. Phs.* **70**, 1393 (1998).
[452] C. Ma, E.F. Arias, T.M. Eubanks, et al., The International Celestial Reference Frame as Realized by Very Long Baseline Interferometry, *Astron. J.* **116**, 516 (1998).
[453] K.J. Johnstone and Chr. de Vegt, Reference Frames in Astronomy, *Ann. Rev. Astron. Astrophys.* **37**, 97 (1999).
[454] A. Fey, G. Gordon, and C. Jacobs (eds.), The Second Realization of the International Celestial Reference Frame by Very Long Baseline Interferometry, IERS Technical Note 35 (2009).
[455] UCLA, The ABC's of Distances (2007). Available online: http://www.astro.ucla.edu/wright/distance:htm.
[456] B.W. Carroll and D.A. Ostlie, *An Introduction to Modern Astrophysics*, 2nd edition (Pearson, Upper Saddle River, NJ, 2007).
[457] J.L. Anderson and P.G. Bergmann, Constraints in Covariant Field Theories, *Phys. Rev.* **83**, 1018 (1951).
[458] P.G. Bergmann and J. Goldberg, Dirac Bracket Transformations in Phase Space, *Phys. Rev.* **98**, 531 (1955).
[459] M.J. Gotay, J.M. Nester, and G. Hinds, Presymplectic Manifolds and the Dirac -Bergmann Theory of Constraints, *J. Math. Phys.* **19**, 2388 (1978).
[460] M.J. Gotay and J.M. Nester, Presymplectic Lagrangian Systems. I: The Constraint Algorithm and the Equivalence Theorem, *Ann. Inst. Henri Poincaré* **A30**, 129 (1979).
[461] M.J. Gotay and J.M. Nester, Presymplectic Lagrangian Systems. II: The Second-Order Equation Problem, *Ann. Inst. Henri Poincaré* **A32**, 1 (1980).
[462] C. Battle, J. Gomis, J.M. Pons, and N. Roma'n-Roy, Equivalence between the Lagrangian and Hamiltonian Formalism for Constrained Systems, *J. Math. Phys.* **27**, 2953 (1986).
[463] A. Echeverri'a-Enri'quez, M.C. Muñoz-Lecanda, and N. Roma'n-Roy, Reduction of Presymplectic Manifolds with Symmetry, *Rev. Math. Phys.* **11**, 1209 (1999)
[464] A. Echeverri'a-Enri'quez, M.C. Muñoz-Lecanda, and N. Roma'n-Roy, A Geometrical Analysis of the Field Equations in Field Theory, *Int. J. Math. Meth. Sc.* **29**, 687 (2002) (arXiv math-ph/0105018).
[465] A. Echeverri'a-Enri'quez, M.C. Muñoz-Lecanda, and N. Roma'n-Roy, Geometry of Multisymplectic Hamiltonian First-Order Field Theories, *J. Math. Phys.* **41**, 7402 (2000) (arXiv math-ph/0004005).
[466] M. de Le'on, J. Mari'n-Solano, J.C. Marrero, M.C. Muñoz-Lecanda, and N.Roma'n-Roy, Singular Lagrangian Systems on Jet Bundles, *Fortsch. Phys.* **50**, 105 (2002).

[467] H. Cendra, M. Etchechoury, and S.J. Ferraro, An Extension of the Dirac and Gotay–Nester Theories of Constraints for Dirac Dynamical Systems, *J. Geom. Mech.* **6**, 167 (2014) (arXiv 1106.3354).
[468] R. Schmidt, *Infinite Dimensional Hamiltonian Systems* (Bibliopolis, Napoli, 1987).
[469] F. Gantmacher, *Lectures in Analytical Mechanics* (Mir, Moscow, 1970).
[470] V.I. Arnold, *Mathematical Methods of Classical Mechanics* (Springer, New York, 1978).
[471] P. Liebermann and C.M. Marle, Symplectic Geometry and Analytical Mechanics (Reidel, Dordrecht, 1987).
[472] M. Nakahara, *Geometry, Topology and Physics* (Institute of Physics Publishing, Bristol, 1990)
[473] L. Lusanna, An Enlarged Phase Space for Finite-Dimensional Constrained Systems, Unifying Their Lagrangian, Phase- and Velocity-Space Descriptions, *Phys. Report* **185**, 1 (1990).
[474] B.M. Barbashov and V.V. Nesterenko, Continuous Symmetries in Field Theory, *Fortschr. Phys.* **31**, 535 (1983).
[475] W. Sarlet and F.Cantrijn, Symmetries and Conservation Laws for Generalized Hamiltonian Systems, *SIAM Rev.* **23**, 467 (1981).
[476] J.D. Logan, *Invariant Variational Principles* (Academic Press, New York, 1977).
[477] N.P. Konopleva and V.N. Popov, *Gauge Fields* (Harwood, New York, 1981).
[478] J.F. Cariñena, J.A. Lazaro-Cami and E. Martinez, On Second Noether's Theorem and Gauge Symmetries in Mechanics, *Int. J. Geom. Meth. Mod. Phys.* **3**, 471 (2006) (arXiv math/0511180).
[479] E. Candotti, C. Palmieri, and B. Vitale, On the Inversion of Noether's Theorem in the Lagrangian Formalism, *Nuovo Cimento* **70**, 233 (1970).
[480] E. Candotti, C. Palmieri, and B. Vitale, Universal Noether's Nature of Infinitesimal Transformations in Lorentz Covariant Field Theories, *Nuovo Cimento* **A7**, 271 (1972).
[481] R.L. Anderson and N.H. Ibragimov, *Lie-Bäcklund Transformations in Applications* (SIAM, Philadelphia, PA, 1979).
[482] J.E. Marsden and A. Weinstein, Reduction of Symplectic Manifolds with Symmetry, *Rep. Math. Phys.* **5**, 121 (1974).
[483] R. Sjamaar and E. Lerman, Stratified Symplectic Spaces and Reduction, *Ann. Math.* **134**, 375 (1991).
[484] L. Lusanna, The Second Noether Theorem as the Basis of the Theory of Singular Lagrangians and Hamiltonian Constraints, *Riv. Nuovo Cimento* **14**, 1 (1991).
[485] L. Lusanna, The Shanmugadhasan Canonical Transformation, Function Groups and the Second Noether Theorem, *Int. J. Mod. Phys.* **A8**, 4193 (1993).
[486] L. Lusanna, The Relevance of Canonical Transformations in Gauge Theories and General Relativity, *Lecture Notes of Seminario Interdisciplinare di Matematica (Basilicata Univ.)* **5**, 125 (2006).
[487] L. Lusanna, Classical Observables of Gauge Theories from the Multitemporal Approach, *Contemp. Math.* **132**, 531 (1992).
[488] F. Loran, Non-Abelianizable First Class Constraints, *Commun. Math. Phys.* **254**, 167 (2005) (arXiv hep-th/0303014).
[489] A. Cabo and D.L. Martinez, On Dirac's Conjecture for Hamiltonian Systems with First- and Second-Class Constraints, *Phys. Rev.* **42**, 2726 (1990).
[490] M. Chaichian, D. Louis Martinez, and L. Lusanna, Dirac's Constrained Systems: The Classification of Second-Class Constraints, *Ann. Phys. (N.Y.)* **232**, 40 (1994).
[491] R. Beig, Asymptotic Structure of Isolated Systems, in *Highlights in Gravitation and Cosmology*, eds. B.R. Iyer, A. Kembhavi, J.V. Narlikar, and C.V. Vishveshwara (Cambridge University Press, Cambridge, 1988).

[492] R. Penrose, Asymptotic Properties of Fields and Space-Times, *Phys. Rev. Lett.* **10**, 66 (1963).
[493] R. Penrose, Zero Rest-Mass Fields Including Gravitation: Asymptotic Behavior, *Proc. Roy. Soc. London* **A284**, 159 (1965).
[494] R. Geroch and G.T. Horowitz, Asymptotically Simple does not Imply Asymptotically Minkowskian, *Phys. Rev. Lett.* **40**, 203 (1978).
[495] R. Geroch and B.C. Xanthopoulous, Asymptotic Simplicity is Stable, *J. Math. Phys.* **19**, 714 (1978).
[496] R. Geroch, Structure of the Gravitational Field at Spatial Infinity, *J. Math. Phys.* **13**, 956 (1972).
[497] P. Sommers, The Geometry of the Gravitational Field at Spacelike Infinity, *J. Math. Phys.* **19**, 549 (1978).
[498] P. Cruściel, On the Structure of Spatial Infinity: The Geroch structure. *J. Math. Phys.* **30**, 2090 (1990).
[499] P. Cruściel, Geodesically Regular Ashtekar -Hansen Structures, *J. Math. Phys.* **30**, 2094 (1990).
[500] P.G. Bergmann, "Gauge-Invariant" Variables in General Relativity, *Phys. Rev.* **124**, 274 (1961).
[501] H. Friedrich, Gravitational Fields Near Space-like and Null Infinity, *J. Geom. Phys.* **24**, 83 (1998).
[502] H. Friedrich, On the Conformal Structure of Gravitational Fields in the Large, in *Highlights in Gravitation and Cosmology*, eds. B.R. Iyer, A. Kembhavi, J.V. Narlikar, and C.V. Vishveshwara (Cambridge University Press, Cambridge, 1988).
[503] H. Friedrich, Asymptotic Structure of Space-Time, in *Recent Advances in General Relativity*, eds. A.I. Janis and J.R. Porter (Birkhauser, Basel, 1992).
[504] J. Frauendiener, Conformal Infinity, *Liv. Rev. Rel.* **3**, 4 (2000).
[505] J. Bicák, Radiative Spacetimes: Exact Approaches, in *Relativistic Gravitation and Gravitational Radiation*, eds. J.A. Marck and J.P. Lasota (Cambridge University Press, Cambridge, 1997).
[506] J. Winicour, Radiative Space-Times: Physical Properties and Parameters, in *Highlights in Gravitation and Cosmology*, eds. B.R. Iyer, A.Kembhavi, J.V. Narlikar, and C.V. Vishveshwara (Cambridge University Press, Cambridge, 1988).
[507] R. Beig and B.G. Schmidt, Einstein's Equations Near Spatial Infinity, *Commun. Math. Phys.* **87**, 65 (1982).
[508] R. Beig, Integration of Einstein's Equations Near Spatial Infinity, *Proc. Roy. Soc. London* **A391**, 295 (1984).
[509] H. Bondi, Gravitational Waves in General Relativity, *Nature* **186**, 535 (1960).
[510] H. Bondi, M.G. van der Burg, and A.W.K. Metzner, Gravitational Waves in General Relativity VII: Waves from Isolated Axisymmetric Systems, *Proc. Roy. Soc. London* **A269**, 21 (1962).
[511] R.K. Sachs, Gravitational Waves in General Relativity VI: The Outgoing Radiation Condition, *Proc. Roy. Soc. London* **A264**, 309 (1962).
[512] R.K. Sachs, Gravitational Waves in General Relativity VIII: Waves in Asymptotically Flat Spacetimes, **A270**, 103 (1962).
[513] R.K. Sachs, Asymptotic Symmetries in Gravitational Theory, *Phys. Rev.* **128**, 2851 (1962).
[514] P.J. McCarthy, Structure of the Bondi-Metzner-Sachs Group, *J. Math. Phys.* **13**, 1837 (1972); Representations of the Bondi-Metzner-Sachs Group. I.Determination of the Representations and II. Properties and Classification of Representations, *Proc. Roy. Soc.* London **A330**, 517 (1972) and **A333**, 317 (1973); Asymptotically Flat Space-times and Elementary Particles, *Phys. Rev. Lett.* **29**, 817 (1972).
[515] P.J. McCarthy and M. Crampin, Representations of the Bondi-MetznerSach Group, III, *Proc. Roy. Soc.* London **A335**, 301 (1973).

[516] J. Isenberg and J.E. Marsden, The York Map is a Canonical Transformation, *J. Geom. Phys.* **1**, 85 (1984).

[517] N. O'Murchadha, Total Energy Momentum in General Relativity, *J. Math. Phys.* **27**, 2111 (1986).

[518] D. Christodoulou and N. O'Murchadha, The Boost Problem in General Relativity, *Commun. Math. Phys.* **80**, 271 (1981).

[519] D. Giulini, C. Kiefer, and H.D. Zeh, Symmetries, Superselection Rules and Decoherence, *Phys. Lett.* **A199**, 291 (1995) (arXiv gr-qc/9410029).

[520] J. Hartle, R. Laflamme, and D. Marolf, Conservation Laws in the Quantum Mechanics of Closed Systems, *Phys. Rev.* **D51**, 7007 (1995).

[521] V. Moncrief, Gribov Degeneracies: Coulomb Gauge Conditions and Initial Value Constraints, *J. Math. Phys.* **20**, 579 (1979).

[522] M. Cantor, Elliptic Operators and the Decomposition of Tensor Fields, *Bull. Am. Math. Soc.* **5**, 235 (1981).

[523] S.W. Hawking and G.T. Horowitz, The Gravitational Hamiltonian, Action, Entropy and Surface Terma, *Class. Quantum Grav.* **13**, 1487 (1996).

[524] D. Marolf, Mass Superselection, Canonical Gauge Transformations and Asymptotically Flat Variational Principles, *Class. Quantum Grav.* **13**, 1871 (1996).

[525] J. Barbour, General Relativity as a Perfectly Machian Theory, in *Mach's Principle: From Newton's Bucket to Quantum Gravity*, eds. J.B. Barbour and H. Pfister, (BirkHäuser, Boston, MA, 1995).

[526] B.S. De Witt, Quantum Theory of Gravity, *Phys. Rev.* **160**, 1113 (1967).

[527] B.S. De Witt, Quantum theory of Gravity: II. The Manifestly Covariant Theory, *Phys. Rev.* **162**, 1195 (1967).

[528] B.S. De Witt, The Dynamical Theory of Groups and Fields (Gordon and Breach, New York, 1967).

[529] R. Schoen and S.T. Yau, Positivity of the Total Mass of a General Space-Time, *Phys. Rev. Lett.* **43**, 1457 (1979).

[530] R. Schoen and S.T. Yau, On the Proof of the Positive Mass Conjecture in General Relativity, *Commun. Math. Phys.* **65**, 45 (1979).

[531] R. Schoen and S.T. Yau, Proof of the Positive Mass Theorem. II, *Commun. Math. Phys.* **79**, 231 (1980).

[532] E. Witten, A New Proof of the Positive Energy Theorem, *Commun. Math. Phys.* **80**, 381 (1981).

[533] Y. Choquet-Bruhat, Positive Energy Theorems, in *Relativity, Groups and Topology II*, eds. B.S. De Witt and R. Stora (North-Holland, Amsterdam, 1984).

[534] G.T. Horowitz, The Positive Energy Theorem and its Extensions, in *Asymptotic Behaviour of Mass and Spacetime Geometry*, ed. F.J. Flaherty (Springer, Berlin, 1984).

[535] L. Landau and E. Lifschitz, *The Classical Theory of Fields* (Addison-Wesley, Cambridge, MA, 1951).

[536] Y.L. Wang, Z.X. Wu, H.Z. Pan, W.T. Lu, H. Jiang, and L. Chen, The Limit of Noether Conserved Charges is the Number of Primary First-Class Constraints in a Constrained System, *Commun. Theor. Phys.* **58**, 539 (2012).

[537] Y.L. Wang, C.T. Xu, H. Jiang, W.T. Lu, and H.Z. Pan, The Dirac Conjecture and the Non-Uniqueness of Lagrangian, preprint (2013) (arXiv 1306.3580).

[538] R. Sugano, Y. Kagraoka, and T. Kimura, Gauge Transformations and Gauge-Fixing: Condition in Constraint System, Int. J. Mod. Phys. **7**, 61 (1992).

[539] L.D. Faddeev and S.L. Shatashvili, Realization of the Schwinger Term in the Gauss Law and the Possibility of Correct Quantization of a Theory with Anomalies, *Phys. Lett.* **B167**, 225 (1986).

[540] I.A. Batalin and E.S. Fradkin, Operator Quantization of Dynamical Systems with Irreducible First- and Second-Class Constraints, *Phys. Lett.* **B180**, 157 (1986).

[541] I.A. Batalin and E.S. Fradkin, Operational Quantization of Dynamical Systems Subject to Second Class Constraints, *Nucl. Phys.* **B279**, 514 (1987).

[542] I.A. Batalin and I.V. Tyutin, Existence Theorem for the Effective Gauge Algebra in the Generalized Canonical Formalism with Abelian Conversion of Second-Class Constraints, *Int. J. Mod. Phys.* **A6**, 3255 (1991).
[543] S. Shanmugadhasan, Canonical Formalism for Degenerate Lagrangians, *J. Math. Phys.* **14**, 677 (1973).
[544] R.O. Fulp and J.A. Marlin, Function Groups Associated with Constraint Submanifolds, *Rep. Math. Phys.* **18**, 295 (1980).
[545] L. Castellani, D. Dominici, and G. Longhi, Canonical Transformations and Quantization of Singular Lagrangian Systems, Nuovo Cimento **48A**, 91 (1978).
[546] D. Dominici and J. Gomis, Poincaré–Cartan Integral Invariant and Canonical Transformations for Singular Lagrangians, *J. Math. Phys.* **21**, 2124 (1980).
[547] L. Lusanna, An Extension of the Second Noether Theorem, *Nuovo Cimento* **B52**, 141 (1979).
[548] P.G. Bergmann, Non-Linear Field Theories, *Phys. Rev.* **75**, 680 (1949).
[549] A. Komar, Field Theoretic Constraint Formalism, *Found. Phys.* **15**, 473 (1985).
[550] V. Moncrief, Gribov Degeneracies: Coulomb Gauge Conditions and Initial Value Constraints, *J. Math. Phys.* **20**, 579 (1979).
[551] L. Lusanna, On the BRS's, *J. Math. Phys.* **31**, 428 (1990).
[552] P.J. Steinhardt, Problems of Quantization in the Infinite Momentum Frame, *Ann. Phys. (N.Y.)* **128**, 425 (1980).
[553] R. Benguria, P. Cordero, and C. Teitelboim, Aspects of the Hamiltonian Dynamics of Interacting Gravitational Gauge and Higgs Fields with Applications to Spherical Symmetry, *Nucl. Phys.* **B122**, 61 (1977).
[554] W.M. Seiler and R.W. Tucker, Involution and Constrained Dynamics I: The Dirac Approach, *J. Phys.* **A28**, 4431 (1995).
[555] R. Utiyama, Invariant Theoretical Interpretation of Interaction, *Phys. Rev.* **101**, 1597 (1956)
[556] R. Utiyama, Theory of Invariant Variation and the Generalized Canonical Dynamics, *Prog. Theor. Phys. Suppl.* **9**, 19 (1959).
[557] A. Trautman, Conservation Laws in General Relativity, in *Gravitation*, ed. L. Witten (Wiley, New York, 1962).
[558] A. Trautman, Foundations and Current Problems of General Relativity, in *Lectures on General Relativity, Brandeis Summer Institute in Theoretical Physics* (Prentice Hall, Englewood Cliffs, 1964).
[559] D.L. Karatas and K.L.Kowalski, Noether's Theorem for Local Gauge Transformations, *Am. J. Phys.* **58**, 123 (1990).
[560] H.A. Al-Kuwari and M.O. Taha, Noether Theorem and Local Gauge Invariance, *Am. J. Phys.* **59**, 363 (1991).
[561] K. Brading and H.R. Brown, Noether's Theorems and Gauge Symmetries, preprint (2000), (arXiv hep-th/0009058).
[562] M. Crosta, M. Gianmaria, M.G. Lattanzi, and E. Poggio, Shedding Light on the Milky Way Rotation Curve with Gaia DR2 (2018) (arXiv 1810.04445)
[563] G.F.R.Ellis and H.van Elst, Cosmological Models, Cargese Lectures 1998, *NATO Adv. Stud. Inst. Ser. C. Math. Phys. Sci.* **541**, 1 (1999) (arXiv gr-qc/9812046).
[564] G.F.R. Ellis, Inhomogeneity Effects in Cosmology (2011) (arXiv 1103.2335).
[565] C. Clarkson, G.F.R. Ellis, A. Faltenbacher, R. Maartens, O. Umeh, and J.P. Uzan, (Mis-)Interpreting Supernovae Observations in a Lumpy Universe (2011) (arXiv 1109.2484).
[566] T. Buchert and G.F.R. Ellis, The Universe Seen at Different Scales, *Phys. Lett.* **A347**, 38 (2005) (arXiv gr-qc/0506106).
[567] R.Maartens, Is the Universe Homogeneous? (2011) (arXiv 1104.1300).
[568] C.G. Tsagas, A. Challinor and R. Maartens, Relativistic Cosmology and Large-Scale Structure, *Phys. Rep.* **465**, 61 (2008) (arXiv 0705.4397).
[569] C. Clarkson and R. Maartens, Inhomogeneity and the Foundations of Concordance Cosmology, *Class. Quantum Grav.* **27**, 124008 (2010) (arXiv 1005.2165).
[570] R. Durrer, What Do We Really Know about Dark Energy? (2011) (arXiv 1103.5331).

[571] C. Bonvin and R. Durrer, What Galaxy Surveys Really Measure (2011) (arXiv 1105.5280).
[572] K. Nakamura, Second-Order Gauge-Invariant Cosmological Perturbation Theory: Current Status, *Ad. Astron.* **2010**, 576273 (2010) (arXiv 1001.2621).
[573] W.B. Bonnor, A.H. Sulaiman, and N. Tanimura, Szekeres's Space-Times have no Killing Vectors, *Gen. Rel. Grav.* **8**, 549 (1977).
[574] B.K. Berger, D.M. Eardley, and D.W. Olson, Notes on the Spacetimes of Szekeres, *Phys. Rev.* **D16**, 3086 (1977).
[575] A. Nwankwo, M. Ishak, and J. Thomson, Luminosity Distance and Redshift in the Szekeres Inhomogeneous Cosmological Models (2010) (arXiv 1005.2989).
[576] J. Plebański and A. Krasiński, *An Introduction to General Relativity and Cosmology* (Cambridge University Press, Cambridge, 2006).
[577] M.N. Celerier, Effects of Inhomogeneities on the Expansion of the Universe: A Challenge to Dark Energy? (2012) (arXiv 1203.2814).
[578] T. Buchert, Dark Energy from Structure: A Status Report, *Gen. Rel. Grav.* **40**, 467 (2008) (arXiv 0707.2153).
[579] A. Wiegand and T. Buchert, Multiscale Cosmology and Structure-Emerging Dark Energy: A Plausibility Analysis, *Phys. Rev.* **D82**, 023523 (2010) (arXiv 1002.3912).
[580] J. Larena, Spatially Averaged Cosmology in an Arbitrary Coordinate System, *Phys. Rev.* **D79**, 084006 (2009) (arXiv 0902.3159).
[581] R.J. van den Hoogen, Averaging Spacetime: Where Do We Go from Here? (2010) (arXiv 1003.4020).
[582] S. Räsänen, Backreaction: Directions of Progress, *Class. Quantum Grav.* **28**, 164008 (2011) (arXiv 1102.0408).
[583] T. Buchert and S. Räsänen, Backreaction in Late-Time Cosmology, *Ann. Rev. Nucl. Particle Sci.* **62**, 57 (2012) (arXiv 1112.5335).
[584] D.L. Wiltshire, What is Dust? Physical Foundations of the Averaging Problem in Cosmology (2011) (arXiv 1106.1693).
[585] L. Fatibene, M. Ferraris, M. Francaviglia, and L. Lusanna, ADM Pseudotensors, Conserved Quantities and Covariant Conservation Laws in General Relativity, *Ann. Phys.* **327**, 1593 (2012) (arXiv 1007.4071).
[586] M. Henneaux, J.E. Nelson, and C. Schomblond, Derivation of Ashtekar Variables from Tetrad Gravity, *Phys. Rev.* **D39**, 434 (1989).
[587] L. Lusanna, Canonical Gravity and Relativistic Metrology: From Clock Synchronization to Dark Matter as a Relativistic Inertial Effect (2012) (arXiv 1108.3224, version 1).
[588] C.G. Torre and M. Varadarajan, Functional Evolution of Free Quantum Fields, *Class. Quantum Grav.* **16**, 2651 (1999)
[589] C.G. Torre and M. Varadarajan, Quantum Fields at Any Time, *Phys. Rev.* **D58**, 064007 (1998).
[590] M. Pauri and G.M. Prosperi, Canonical Realizations of Lie Symmetry Groups, *J. Math. Phys.* **7**, 366 (1966).
[591] G. Marmo, E.J. Saletan, A. Simoni, and B. Vitale, Dynamical Systems: A Differential Geometric Approach to Symmetry and Reduction (Wiley, Chicester, 1985).
[592] M. Abraham and J.M. Marsden, *Foundations of Mechanics*, 2nd edition (Benjamin, Reading, MA, 1979).
[593] J.M.Souriau, *Structure des syste'mes dynamiques* (Dunod, Paris, 1970).
[594] U.H. Niederer and L. O'Raifeartaigh, Realizations of the Unitary Representations of the Inhomogeneous Space-Time Groups I: General Structure, Fortschr. Phys. **22**, 111 (1974).
[595] U.H. Niederer and L. O'Raifeartaigh, Realizations of the Unitary Representations of the Inhomogeneous Space-Time Groups II: Covariant Realizations of the Poincaré Group, *Fortschr. Phys.* **22**, 131 (1974).
[596] U.H. Niederer, Massless Fields as Unitary Representations of the Poincaré Group, *Fortschr. Phys.* **27**, 191 (1979).

[597] G. Mackey, Induced Representations of Groups and Quantum Mechanics (Benjamin, New York, 1968).
[598] E. Wigner, On Unitary Representations of the Inhomogeneous Lorentz Group, *Ann. Math.* **40**, 149 (1939)
[599] E. Wigner, *Group Theory* (Academic Press, New York, 1959).
[600] D. Dominici, J. Gomis, and G. Longhi, A Lagrangian for Two Interacting Relativistic Particles: Canonical Formulation, *Nuovo Cimento* **48A**, 257 (1978).
[601] S. Weinberg, Feynman Rules for Any Spin, *Phys. Rev.* **133B**, 1318 (1964).
[602] G.C. D'Emma, On Quantization of the Electromagnetic Field, *Helv. Phys. Acta* **53**, 535 (1980).
[603] D. Han, Y.S. Kim, and D. Son, Photon Spin as a Rotation in Gauge Space, *Phys. Rev.* **D25**, 461 (1982).
[604] S. Weinberg, Photons and Gravitons in S-Matrix Theory: Derivation of Charge Conservation and Equality of Gravitational and Inertial Mass, *Phys. Rev.* **135B**, 1049 (1964).
[605] Y. Ohnuki, *Unitary Representations of the Poincare Group and Relativistic Wave Equations* (World Scientific, Singapore, 1988).
[606] F.A. Berezin and M.S. Marinov, Particle Spin Dynamics as the Grassmann Variant of Classical Mechanics, *Ann. Phys. (N.Y.)* **104**, 336 (1977).
[607] R. Casalbuoni, On the Quantization of Systems with Anticommuting Variables, *Nuovo Cimento* **33A**, 115 (1976).
[608] R. Casalbuoni, The Classical Mechanics for Bose -Fermi Systems, *Nuovo Cimento* **33A**, 389 (1976).
[609] A. Barducci, F. Bordi, and R. Casalbuoni, Path Integral Quantization of Spinning Particles Interacting with Crossed External Electromagnetic Fields, *Nuovo Cimento* **64B**, 287 (1986).
[610] A. Barducci, R. Casalbuoni, and L. Lusanna, Anticommuting Variables, Internal Degrees of Freedom, and the Wilson Loop, *Nucl. Phys.* **B180**, 141 (1981).
[611] A. Barducci, F. Buccella, R. Casalbuoni, L. Lusanna, and E. Sorace, Quantized Grassmann Variables and Unified Theories, *Phys. Lett.* **67B**, 344 (1977)
[612] A. Barducci, F. Buccella, R. Casalbuoni, L. Lusanna, and E. Sorace, Fermi Oscillators for Vectorlike Unified Theories with Unitary Groups, *Lett. Nuovo Cimento* **19**, 335 (1977).
[613] A. Barducci, R. Casalbuoni, and L. Lusanna, A Possible Interpretation of Theories Involving Grassmann Variables, *Lett. Nuovo Cimento* **19**, 581 (1977).
[614] J. Gomis and M. Novell, Pseudoclassical Description for a Nonrelativistic Spinning Particle: II. Classical Content, *Phys. Rev.* **D33**, 2220 (1986).
[615] L. Blanchet, T. Damour, and G. Schäfer, Post-Newtonian Hydrodynamics and Post-Newtonian Gravitational Wave Generation for Numerical Relativity, *Mon. Not. R. Astr. Soc.* **242**, 289 (1990).
[616] H. Asada, M. Shibata, and T. Futamase, Post-Newtonian Hydrodynamic Equations Using the (3+1) Formalism in General Relativity, Osaka University preprint (1996) (arXiv gr-qc/9606041).
[617] J.D. Brown and K. Kuchar, Dust as a Standard of Space and Time in Canonical Quantum Gravity, *Phys. Rev.* **D51**, 5600 (1995).
[618] R. Maartens, Causal Thermodynamics in Relativity, Lectures at the H.Rund Workshop on Relativity and Thermodynamics (1996) (arXiv astreo-ph/9609119).
[619] L. Smarr, C. Taubes, and J.R. Wilson, General Relativistic Hydrodynamics: The Comoving, Eulerian and Velocity Potential Formalisms, in *Essays in General Relativity: A Festschrift for Abraham Taub*, ed. F.J. Tipler (Academic Press, New York, 1980).
[620] B. Carter, Covariant Theory of Conductivity in Ideal Fluid or Solid Media, in *Relativistic Fluid Dynamics*, eds. A. Anile and Y. Choquet-Bruhat (Springer, Berlin, 1989).

[621] F. Karsch and D.E. Miller, Covariant Structure of Relativistic Gases in Equilibrium, *Phys. Rev.* **D24**, 2564 (1981).
[622] J.M. Stewart, *Non-Equilibrium Relativistic Kinetic Theory*, (Springer, Berlin, 1971).
[623] R. Hakim, Remarks on Relativistic Statistical Mechanics: I, *J. Math. Phys.* **8**, 1315, (1967).
[624] R. Hakim, Relativistic Kinetic Equations Including Radiation Effects: I. Vlasov Approximation, *J. Math. Phys.* **8**, 1379 (1967).
[625] J.L. Synge, *The Relativistic Gas* (North-Holland, Amsterdam, 1957).
[626] Ch. G.van Weert, Some Problems in Relativistic Hydrodynamics, in *Relativistic Fluid Dynamics*, eds. A. Anile and Y. Choquet-Bruhat (Springer, Berlin, 1989).
[627] L. Lindblom, The Relaxation Effect in Dissipative Relativistic Fluid Theories, *Ann. Phys. (N.Y.)* **247**, 1 (1996).
[628] A.M. Anile, D. Pavón, and V. Romano, The Case for Hyperbolic Theories of Dissipation in Relativistic Fluids (1998) (arXiv gr-qc/9810014).
[629] A.M. Anile, *Relativistic Fluids and Magnetofluids* (Cambridge University Press, Cambridge, 1989).
[630] J.M. Martí and E. Müller, Numerical Hydrodynamics in Special Relativity (1990) (arXiv astro-ph/9906333).
[631] J.A. Font, N. Stergioulas, and K.D. Kokkotas, Nonlinear Hydrodynamical Evolution of Rotating Relativistic Stars: Numerical Methods and Code Tests (1999) (arXiv gr-qc/9908010).
[632] T.W. Baumgarte, S.A. Hughes, L. Rezzolla, S.L. Shapiro, and M. Shibata, Implementing Fully Relativistic Hydrodynamics in Three Dimensions (1999) (arXiv gr-qc/9907098).
[633] R. Geroch and L. Lindblom, Causal Theories of Dissipative Relativistic Fluids, *Ann. Phys. (N.Y.)* **207**, 394 (1991).
[634] I. Liu, I. Müller, and T.Ruggeri, Relativistic Thermodynamics of Gases, *Ann. Phys. (N.Y.)* **169**, 191 (1986).
[635] R. Geroch and L. Lindblom, Dissipative Relativistic Fluid Theories of Divergence Type, *Phys. Rev.* **D41**, 1855 (1990).
[636] E. Calzetta, Relativistic Fluctuating Hydrodynamics, *Class. Quantum Grav.* **15**, 653 (1998) (arXiv gr-qc/9708048).
[637] I. Müller, Zum Paradoxon der Wrmeleitungstheorie, *Z. Phys.* **198**, 329 (1967).
[638] W. Israel, Nonstationary Irreversible Thermodynamics: A Causal Relativistic Theory, *Ann. Phys. (N.Y.)* **100**, 310 (1976).
[639] W.Israel and J.M.Stewart, Transient Relativistic Thermodynamics and Kinetic Theory, *Ann. Phys. (N.Y.)* **118**, 341 (1979).
[640] D. Jou, G. Leblon, and J. Casas Vasquez, *Extended Thermodynamics*, 2nd edition (Springer, Heidelberg, 1996).
[641] I. Müller and T. Ruggeri, *Rational Extended Thermodynamics*, 2nd edition (Springer, Berlin, 1998).
[642] J. Peitz and S. Appl, 3+1 Formulation of Non-Ideal Hydrodynamics, *Class. Quantum Grav.* **16**, 979 (1999) (arXiv gr-qc/9710107).
[643] J.P. Uzan, Dynamics of Relativistic Interacting Gases: From a Kinetic to a Fluid Description, Class. *Quantum Grav.* **15**, 1063 (1998).
[644] J.P. Uzan, Comment on Dynamics of Relativistic Interacting Gases, *Class. Quantum Grav.* **15**, 3737 (1998).
[645] J.M. Stewart, Comment: The Dynamics of Relativistic Interacting Gases, *Class. Quantum Grav.* **15**, 3731 (1998).
[646] P.J. Morrison, A Paradigm for Joined Hamiltonian and Dissipative Systems, *Physica* **18D**, 410 (1986).
[647] S. Nosé, A Unified Formulation of the Constant Temperature Molecular Dynamics Methods, *J. Chem. Phys.* **81**, 511 (1984).

Index

action-at-a-distance, 42
ADM formalism, 131, 141
ADM Poincaré generators, 144, 165
Anderson, 211
approach
  global 3+1, 12
  local 1+3, 7
approximation in general relativity
  Hamiltonian post-Minkowskian, 167
  Hamiltonian post-Newtonian, 178
Ashtekar–Hanson, 206, 210
Ashtekar–Romano, 206
asymptotically flat, 124, 206
atomic relativistic physics, 71

back-reaction, 246
Beig–O'Murchadha, 215
Beig–Schmidt, 207
Bel, 28
Berezin, 79, 267
Berezin–Marinov, 265, 270, 273
  distribution function, 273
Bergmann, 127, 207, 211
Bianchi identity, 131, 152
binary stars, 180
Boltzmann equation, 116
Bondi, 207
boundary conditions, 59, 65, 80, 99, 204

Candotti–Palmieri–Vitale, 193
canonical coordinates, 29
Cauchy problem, 9
center of mass in relativity
  external, 35
  internal, 36
  gravity, 177
  problem, 36
Christodoulou–Klainerman space-times, 126, 204, 208, 212
Chrusciel, 213
clock synchronization, 6, 12, 19
coordinate singularities in Minkowski space-time, 8
constraint Hamiltonian theory of Dirac–Bergmann
  diagonalization of chains, 200
  Dirac field, 85
  Electromagnetism, 73
  field theory, 236
  first- and second-class, 199
  fluids, 111
  Klein–Gordon field, 55
  metric gravity, 131, 133
  tetrad gravity, 143, 152
  third- and fourth-class, 234
cosmology, 245
Coulomb cloud, 74
Currie–Hill, 29
Currie–Jordan–Sudarshan, 29

dark matter, 181
Darwin potential, 76
diffeomorphism
  active, 119
  passive, 119, 125
Dirac
  algebra, 134
  algorithm, 197
  brackets, 24, 203
  conjecture, 225
  field, 79
  field observable, 79, 94
  forms of dynamics, 10
  Hamiltonian, 22, 52, 55, 67, 85, 110, 133, 143, 148, 149, 196
  multipliers, 196
  observable, 194, 223, 228
Droz Vincent, 29
dust, 108, 145

Eckart, 285
Einstein relativity principle
  Minkowski, 6
  gravity, 119
Einstein space-times, 123
  Einstein equations, 132
electromagnetic field, 65, 72, 155
  Dirac observables, 70, 94
embedding functions of the 3+1 approach, 12

energy–momentum tensor, 6
  electromagnetic field, 67, 75
  particles, 22
  tetrad gravity, 142
  Yang–Mills field, 100
entanglement, 46
Epstein–Glaser–Jaffe, 39
equation of state, 108
equivalence principle
  gravity, 119
  Newton, 4
Euler–Lagrange equations, 107, 132
expansion, 17, 153
extrinsic curvature, 130, 143, 147

Fermi coordinates, 8
Field equations
  Dirac, 95
  Electromagnetic, 71, 74, 96
  Klein–Gordon, 57
fluids
  Euler–Lagrange equations, 107
  Lagrangian coordinates, 106
  micro–canonical ensemble, 112
  perfect, 105, 276
  relativistic, 104
Fokker–Pryce center of inertia, 36
frames
  inertial Galilean, 4
  inertial Minkowski, 5
  non-inertial Galilean, 5
  non-inertial Gravity, 127, 144
  non-inertial Minkowski, 12, 15, 23
Frauendiener, 127
Friedrich, 207

Galilei
  group, 4
  space-time, 3
gauge copies, 98
gauge transformations in field theory and general relativity, 204
  improper, 98, 205
gauge transformations in parametrized Minkowski theories, 20
gauges in tetrad gravity
  3-orthogonal, 149
  harmonic, 150, 171
  Schwinger, 148
  traceless transverse, 171
Gauss's law, 67, 84, 101
Geroch, 206
Giulini, 214
globally hyperbolic space-time, 123
Grad, 290, 291

Grassmann variables, 265
  Dirac field, 79
  electric charge, 72
  energy sign, 21
Gravitational waves, 151, 167, 169, 171
Gravito-magnetic, 14, 148, 151, 169
gravity
  metric, 123
  tetrad, 138, 155
Gribov ambiguity, 98, 125

Haag theorem, 78
Hadamard theorem, 125
Hamilton–Dirac equations, 57, 71, 95, 134
Havas, 10
Hegerfeld, 44
Hessian matrix, 190, 194, 231
Hopf–Rinow theorem, 124
hyper-plane, space-like or 3-space
  gravity, 123
  Minkowski, 6

inertial forces
  Minkowskian, 14
  Newtonian, 5
  tetrad gravity, 148
inertial gauge variables
  parametrized Minkowski theories, 14
  tetrad gravity, 148
Infeld–Plebanski, 180
isolated system in general relativity, 205

Jacobi variables, 34

Killing symmetries, 155
kinetic relativistic theory, 113, 116
Klein–Gordon field, 54
  center of mass, 57
  relative variables, 64
Komar, 30

Lagrangian
  regular, 189
  singular, 194
Landau–Lifshitz, 286
lapse function, 13, 129, 149
length definition, 1
Lichnerowitz–ChoquetBruhat–York, 125
Lichnerowitz equation, 147
Lie algebras, 248
Lie–Bäcklund transformations, 193
Lienard–Wiechert potential, 74
light 1-way and 2-way velocities, 6

Longhi–Materassi, 58
Lorentz group, 249

Marolf, 213
metric tensor
  metric gravity, 128
  Minkowski, 6, 12
  tetrad gravity, 141
metrology, relativistic, 1, 6
  barycentric, 9
  celestial, 9, 245
  geocentric, 9
micro-canonical relativistic ensemble, 112
Minkowski space-time, 1, 6
Misner internal intrinsic time, 147
Mőller center of energy, 36
Mőller center of energy, 37
Mőller world-tube and radius, 38

Newton–Wigner center of mass, 35
Noether
  first theorem, 192
  second theorem, 235
no-interaction theorem, 29

observer
  Eulerian congruence, 17, 153
  inertial, 6, 12
  mathematical or dynamical, 47
  non-inertial, 8
  skew congruence, 18
O'Murchadha, 210, 212

parametrized Galilei theories, 25
parametrized Minkowski theories, 21
particles, relativistic, 14, 18, 20, 28, 39, 72, 145, 155, 174, 178
Penrose, 206
Poincare' group generators, 249
  ADM gravity, 144
  Minkowski, 7, 22
Poincare' orbits, 252
Predictive coordinates, 28
pseudo-classical mechanics, 270
  electric charge, 270
  energy sign, 270
  spinning particles, 270

quantum relativistic mechanics, 44

radar 4-coordinates, 12
Raychaudhuri equation, 152
Regge–Teitelboim, 210, 214, 219
relative times, 27, 39
relative variables of particles
  HPM tetrad gravity, 178
  special relativity, 40
rest-frame
  inertial, 25
  non-inertial, 25, 47, 127, 144
rotating disk, 8

Sachs, 217
Sagnac effect, 9
Schutz, 280
Sen–Witten, 127
Shanmugadhasan canonical transformation, 70, 145, 229
shear, 17, 153
shift functions, 13, 129, 148
spatial infinity, 14, 124, 206–208
spinor structure, 124
statistical relativistic mechanics, 116
super-translations, 123, 209
Szekeres space-time, 246

temperature, micro-canonical, 113
tetrads or vierbeins
  gravity, 138
  Minkowski, 15, 31, 254
thermodynamics, relativistic, 281
Thomas spin tensor, 90
tidal variables
  equations of motion, 161
  tetrad gravity, 147
time convention, 1
Todorov, 29
Trautman conservation law, 242
triads, 13, 140
two-body problem, 43

Utiyama, 241

vorticity, 17

Wheeler–DeWitt super-metric, 132
Wigner
  boost, 31, 253, 261
  covariance, 43
  rotation, 31, 35, 139, 253
Wigner-covariant 3-spaces, 31

Yang–Mills field, 96
  Dirac observables, 104
York, 210
York canonical basis, 146
York map, 145, 210, 211
York time, 126, 148, 173, 184

zeroth postulate of quantum mechanics, 49

www.ingramcontent.com/pod-product-compliance
Lightning Source LLC
LaVergne TN
LVHW062207200926
848484LV00006B/1337
* 9 7 8 1 1 0 8 4 8 0 8 2 6 *